APPLIED STUDIES IN CLIMATE ADAPTATION

Applied Studies in Climate Adaptation

Edited by

Jean P. Palutikof
National Climate Change Adaptation Research Facility,
Griffith University, Australia

Sarah L. Boulter
National Climate Change Adaptation Research Facility,
Griffith University, Australia

Jon Barnett
School of Geography,
University of Melbourne, Australia

David Rissik
National Climate Change Adaptation Research Facility,
Griffith University, Australia

WILEY Blackwell

Library of Congress Cataloging-in-Publication Data

Applied studies in climate adaptation / edited by Jean P. Palutikof, Sarah L. Boulter, Jon Barnett and David Rissik.
     page   cm
  Includes bibliographical references and index.
  ISBN 978-1-118-84501-1 (cloth)
1. Climatic changes–Australia.   2. Climatic changes–Environmental aspects–Australia.
3. Biodiversity–Climatic factors–Australia.   4. Climate change mitigation–Australia.
5. Human beings–Effect of climate on–Australia.   6. Human ecology–Australia.
7. Climatic changes–Economic aspects–Australia.   8. Environmental policy–Australia.
9. Environmental management–Australia.   I. Palutikof, J. P.
  QC903.2.A8A66 2014
  304.2′50994–dc23
                                               2014018374

A catalogue record for this book is available from the British Library.

Set in 9/11.5pt Trump Mediaeval by SPi Publisher Services, Pondicherry, India
Printed and bound in Singapore by Markono Print Media Pte Ltd

1   2015

Contents

List of contributors

LUCY ANDERTON *Department of Agriculture and Food WA, Australia*

VERONICA ARBON *Wilto Yerlo, University of Adelaide, Australia*

JOHN P.Y. ARNOULD *School of Life and Environmental Sciences, Deakin University, Australia*

EMMA K. AUSTIN *School of Environmental and Life Sciences, University of Newcastle, Australia*

JON BARNETT *School of Geography, University of Melbourne, Australia*

GUY BARNETT *CSIRO Land and Water Flagship, Australia*

R. MATTHEW BEATY *CSIRO Land and Water Flagship, Australia*

ANDREW BEER *Centre for Housing, Urban and Regional Planning, University of Adelaide, Australia*

BRENDAN BERGHOUT *Hunter Water Corporation, Australia*

FRANS BERKHOUT *Department of Geography, King's College, UK*

PENG BI *Discipline of Public Health, University of Adelaide, Australia*

GILAD BINO *Australian Wetlands, Rivers and Landscapes Centre, University of New South Wales, Australia*

DEANNE K. BIRD *Risk Frontiers, Macquarie University, Australia*

JOHN BOLAND *School of Information Technology and Mathematical Sciences, University of South Australia, Australia*

HELEN J. BOON *College of Arts, Society and Education, James Cook University, Australia*

SARAH L. BOULTER *National Climate Change Adaptation Research Facility, Griffith University, Australia*

PAMELA BOX *Department of Environment and Geography, Macquarie University, Australia*

GRAHAM L. BRADLEY *Griffith Climate Change Response Program, Griffith University, Australia*

IAN BURTON *Adaptation and Impacts Research Group, Meteorological Services Canada, Canada*

RODRIGO H. BUSTAMANTE *Climate Adaptation Flagship, CSIRO Marine and Atmospheric Research, Australia*

DAVE CALLAGHAN *School of Civil Engineering, University of Queensland, Australia*

TIM CAPON *CSIRO Land and Water Flagship and School of Economics, University of Sydney, Australia*

SAMANTHA J. CAPON *Australian Rivers Institute, Griffith University, Australia*

DEAN B. CARSON *School of Medicine, Flinders University and the Northern Institute, Charles Darwin University, Australia*

LYNDA E. CHAMBERS *Centre for Australian Weather and Climate Research, Australia*

DONG CHEN *CSIRO Land and Water Flagship, Australia*

EDDO COIACETTO *Griffith School of Environment, Griffith University, Australia*

AUGUSTINE CONTEH *School of Public Health and Social Work, Queensland University of Technology, Australia*

BETHANY COOPER *Centre for Water Policy and Management, La Trobe University, Australia*

IGNACIO CORREA-VELEZ *School of Public Health and Social Work, Queensland University of Technology, Australia*

PETER COWELL *School of Geosciences, University of Sydney, Australia*

LIN CRASE *Centre for Water Policy and Management, La Trobe University, Australia*

JASON CREAN *NSW Department of Primary Industries, Australia*

RYAN CROMPTON *Risk Frontiers, Macquarie University, Australia*

LIJIE CUI *School of Engineering, University of Newcastle, Australia*

JAGO DODSON *Global, Urban and Social Studies, RMIT University, Australia*

HALLIE EAKIN *School of Sustainability, Arizona State University, USA*

JANE EDWARDS *CRMA School of Commerce, University of South Australia, Australia*

MICHELLE C. ELLUL *Griffith Climate Change Response Program, Griffith University, Australia*

ANNA FARMERY *Institute for Marine and Antarctic Studies, University of Tasmania, Australia*

DAVID FELDMAN *Department of Agriculture and Food WA, Australia*

DEANE FERGIE *Anthropology of Native Title Societies, University of Adelaide, Australia*

AYSHA FLEMING *Climate Adaptation Flagship, CSIRO Marine and Atmospheric Research, Australia*

CAMERON S. FLETCHER *CSIRO Ecosystem Sciences, Australia*

ANITA FOERSTER *The University of Melbourne, Australia*

DONALD C. FRANKLIN *Research Institute for the Environment and Livelihoods, Charles Darwin University, Australia*

STEWART FRUSHER *Institute for Marine and Antarctic Studies, University of Tasmania, Australia*

K. RENUKA GANEGODAGE *The School of Economics, University of Queensland, Australia*

STEPHEN T. GARNETT *Research Institute for the Environment and Livelihoods, Charles Darwin University, Australia*

ROLF GERRITSEN *Central Australian Research Group (CARG), Charles Darwin University, Australia*

SANDRA M. GIFFORD *Swinburne Institute for Social Research, Swinburne University of Technology, Australia*

MICHELLE GRAYMORE *Water in Drylands Collaborative Research Program, University of Ballarat, Australia*

BRIDGET S. GREEN *Institute for Marine and Antarctic Studies, University of Tasmania, Australia*

LUCY HACKWORTH *Anthropology of Native Title Societies, University of Adelaide, Australia*

WADE L. HADWEN *Australian Rivers Institute, Griffith University, Australia*

EMILY HAMILTON *Australian Council of Social Services, Australia*

ALANA HANSEN *Discipline of Public Health, University of Adelaide, Australia*

BEN P. HARMAN *CSIRO Ecosystem Sciences, Australia*

NICK HARVEY *Geography, Environment and Population, University of Adelaide, Australia*

PETER HAYMAN *South Australian Research and Development Institute, Australia*

KATHARINE HAYNES *Risk Frontiers, Macquarie University, Australia*

BENJAMIN J. HENLEY *School of Engineering, University of Newcastle, Australia and School of Earth Sciences, University of Melbourne, Australia*

GREG HERTZLER *Faculty of Agriculture and Environment, University of Sydney, Australia;*

and Faculty of Natural and Agricultural Sciences, University of Pretoria, South Africa

SONJA HEYENGA *CSIRO Ecosystem Sciences, Australia*

ALISTAIR J. HOBDAY *Climate Adaptation Flagship, CSIRO Marine and Atmospheric Research, Australia*

RALPH HORNE *College of Design and Social Context, RMIT University, Australia*

MICHAEL HOWES *Griffith School of Environment, Griffith University, Australia*

GRAEME HUGO *Australian Population and Migration Research Centre, University of Adelaide, Australia*

JAMES INNES *Climate Adaptation Flagship, CSIRO Marine and Atmospheric Research, Australia*

LESLEY INSTONE *Centre for Urban and Regional Studies, University of Newcastle, Australia*

NAZRUL ISLAM *Department of Agriculture and Food WA, Australia*

KIM JENKINS *Australian Wetlands, Rivers and Landscapes Centre, University of New South Wales, Australia*

SARAH JENNINGS *School of Economics and Finance, University of Tasmania, Australia*

ANTHONY S. KIEM *School of Environmental and Life Sciences, University of Newcastle, Australia*

DAVID KING *Centre for Disaster Studies, James Cook University, Australia*

RICHARD T. KINGSFORD *Australian Wetlands, Rivers and Landscapes Centre, University of New South Wales, Australia*

ROSS KINGWELL *Australian Export Grains Innovation Centre and University of Western Australia, Australia*

MICHAEL KROEHN *Centre for Housing, Urban and Regional Planning, University of Adelaide, Australia*

GEORGE KUCZERA *School of Engineering, University of Newcastle, Australia*

NATASHA KURUPPU *Institute for Sustainable Futures, University of Technology Sydney, Australia*

JULIA LAW *Centre for Housing, Urban and Regional Planning, University of Adelaide, Australia*

LILLY LIM-CAMACHO *Climate Adaptation Flagship, CSIRO Earth System Sciences and Resource Engineering, Australia*

FELIX LIPKIN *CSIRO Ecosystem Sciences, Australia*

ADAM LOCH *CRMA School of Commerce, University of South Australia, Australia*

ANDREW MACINTOSH *ANU College of Law, Australian National University, Australia*

KARL MALLON *Climate Risk Pty Ltd, Australia*

ANDREW MARTEL *Faculty of Architecture, Building and Planning, University of Melbourne, Australia*

JOHN MARTIN *LaTrobe University, Australia*

RYAN R.J. MCALLISTER *CSIRO Ecosystem Sciences, Australia*

DELPHINE MCANENEY *Risk Frontiers, Macquarie University, Australia*

JOHN MCANENEY *Risk Frontiers, Macquarie University, Australia*

ANTHEA MCCLINTOCK *NSW Department of Primary Industries, Australia*

JAN MCDONALD *Faculty of Law, University of Tasmania, Australia*

DARRYN MCEVOY *RMIT University, Australia*

STEPHEN MCFALLAN *CSIRO Land and Water Flagship, Australia*

CELIA MCMICHAEL *School of Social Sciences, La Trobe University, Australia*

KATHLEEN J. MEE *Centre for Urban and Regional Studies, University of Newcastle, Australia*

JACQUI MEYERS *CSIRO Land and Water Flagship, Australia*

MELINDA L. MOIR *School of Plant Biology, University of Western Australia, Australia*

MOHAMMAD MORTAZAVI-NAEINI *School of Engineering, University of Newcastle, Australia*

JOHN MORTON *Natural Resources Institute, University of Greenwich, UK*

SUSANNE C. MOSER *Susanne Moser Research and Consulting, USA*

PIERRE MUKHEIBIR *Institute for Sustainable Futures, University of Technology Sydney, Australia*

JANE MULLETT *RMIT University, Australia*

JANINA MURTA *Institute for Sustainable Futures, University of Technology Sydney, Australia*

RADE MUSULIN *Aon Benfield Analytics Asia Pacific, Australia*

MONIKA NITSCHKE *Department for Health and Ageing South Australia, Australia*

ANA NORMAN-LÓPEZ *Climate Adaptation Flagship, CSIRO Marine and Atmospheric Research, Australia*

MELISSA NURSEY-BRAY *Geography, Environment and Population, University of Adelaide, Australia*

TETSUYA OKADA *Risk Frontiers, Macquarie University, Australia*

JONATHAN T. OVERPECK *Institute of the Environment, University of Arizona, USA*

JANE PALMER *Institute for Sustainable Futures, University of Technology Sydney, Australia*

ROB PALMER *AuConsulting, Plympton, South Australia, Australia*

JEAN P. PALUTIKOF *National Climate Change Adaptation Research Facility, Griffith University, Australia*

MEG PARSONS *School of Environment, University of Auckland, New Zealand*

SEAN PASCOE *Climate Adaptation Flagship, CSIRO Marine and Atmospheric Research, Australia*

NICHOLAS PAWSEY *Centre for Water Policy and Management, La Trobe University, Australia*

GRETTA T. PECL *Institute for Marine and Antarctic Studies, University of Tasmania, Australia*

ROGER PIELKE JR. *Center for Science and Technology Policy Research, University of Colorado, USA*

ÉVA E. PLAGÁNYI-LLOYD *Climate Adaptation Flagship, CSIRO Marine and Atmospheric Research, Australia*

STEPHEN PULLEN *School of Natural and Built Environments, University of South Australia, Australia*

ALICIA N. RAMBALDI *The School of Economics, University of Queensland, Australia*

ROSHANKA RANASINGHE *UNESCO-IHE, The Netherlands; and Research School of Earth Sciences, The Australian National University, Australia*

ALAN RANDALL *Faculty of Agriculture and Environment, University of Sydney, Australia*

JOSEPH P. RESER *Griffith Climate Change Response Program, Griffith University, Australia*

APRIL E. RESIDE *Centre for Tropical Biodiversity and Climate Change, James Cook University, Australia*

LESTER-IRABINNA RIGNEY *Division of the Deputy Vice Chancellor Vice President, University of Adelaide, Australia*

DAVID RISSIK *National Climate Change Adaptation Research Facility, Griffith University, Australia*

KERRYLEE ROGERS *School of Earth and Environmental Sciences, University of Wollongong, Australia*

WASIM SAMAN *School of Engineering, University of South Australia, Australia*

TODD SANDERSON *Faculty of Agriculture and Environment, University of Sydney, Australia*

PEGGY SCHROBBACK *School of Economics and Finance, Queensland University of Technology, Australia*

ARUSYAK SEVOYAN *Australian Population and Migration Research Centre, University of Adelaide, Australia*

HEATHER SHEARER *Urban Research Program, Griffith University, Australia*

WILLIAM D. SOLECKI *Institute for Sustainable Cities, CUNY Hunter College, USA*

JANE SPEIJERS *Department of Agriculture and Food WA, Australia*

CLAIRE STACEY *Australian Institute of Aboriginal and Torres Strait Islander Studies, Australia*

LISA M. STRELEIN *Australian Institute of Aboriginal and Torres Strait Islander Studies, Australia*

AARON STUART *Arabana Association, Centacare Catholic Family Services Country SA, Australia*

PAZIT TAYGFELD *Urban Research Program, Griffith University, Australia*

BRUCE M. TAYLOR *CSIRO Ecosystem Sciences, Australia*

MIKE TAYLOR *University of South Australia, Australia*

OLIVER THEBAUD *Climate Adaptation Flagship, CSIRO Marine and Atmospheric Research, Australia*

LINDA THOMAS *Climate Adaptation Flagship, CSIRO Marine and Atmospheric Research, Australia*

JOHN TIBBY *Geography, Environment and Population, University of Adelaide, Australia*

TRAN TRAN *Australian Institute of Aboriginal and Torres Strait Islander Studies, Australia*

SELINA TUALLY *Centre for Housing, Urban and Regional Planning, University of Adelaide, Australia*

EMMA TURNER *Hunter Water Corporation, Australia*

E. INGRID VAN PUTTEN *Climate Adaptation Flagship, CSIRO Marine and Atmospheric Research, Australia*

NICOLA VAUGHAN *Policy and Participation, Housing and Community Services, ACT Government, Australia*

DANIELLE C. VERDON-KIDD *School of Environmental and Life Sciences, University of Newcastle, Australia*

DAVID WAINWRIGHT *School of Civil Engineering, University of Queensland, Australia; and Whitehead and Associates, Australia*

GEORGE WALKER *Aon Benfield Analytics Asia Pacific, Australia*

ANGELA WARDELL-JOHNSON *University of the Sunshine Coast, Australia*

ELISSA WATERS *School of Geography, University of Melbourne, Australia*

JESSICA K. WEIR *Institute for Culture and Society, University of Western Sydney, Australia*

JASON WEST *Department of Accounting Finance and Economics, Griffith University, Australia*

SARAH ANN WHEELER *CRMA School of Commerce, University of South Australia, Australia*

MIRIAM WILLIAMS *Centre for Urban and Regional Studies, University of Newcastle, Australia*

COLIN WOODROFFE *School of Earth and Environmental Sciences, University of Wollongong, Australia*

VILAPHONH XAYAVONG *Department of Agriculture and Food WA, Australia*

GARY YOHE *Department of Economics, Wesleyan University, USA*

Acknowledgements

This work was carried out with the financial support of the Australian Government (Department of Climate Change and Energy Efficiency) and the National Climate Change Adaptation Research Facility. The role of NCCARF is to lead the research community in a national interdisciplinary effort to generate the information needed by decision-makers in government, business and in vulnerable sectors and communities to manage the risk of climate change impacts.

The views expressed herein are not necessarily the views of the Commonwealth or NCCARF, and neither the Commonwealth nor NCCARF accept responsibility for information or advice contained herein.

1 Introduction to the book

JEAN P. PALUTIKOF[1], SARAH L. BOULTER[1],
JON BARNETT[2] AND DAVID RISSIK[1]

[1]*National Climate Change Adaptation Research Facility, Griffith University, Australia*
[2]*School of Geography, University of Melbourne, Australia*

1.1 Why this book exists

1.1.1 NCCARF and its research programs

In 2008, in response to a clearly felt need to build national resilience and capacity to adapt to climate change, the Australian Government funded the National Climate Change Adaptation Research Facility (NCCARF) for a period of 5 years and for $50 million. The mission of NCCARF was to generate and deliver to Australian decision-makers the knowledge needed to effectively adapt to climate change. Of the $50 million, a sum of $30 million was used to fund research projects into adaptation, making this one of the largest such research programs anywhere in the world at that time. It resulted in some 120 projects, exploring every aspect of adaptation relevant to Australia.

1.1.2 How was the research program constructed?

The research program was constructed around groups of sectors, namely:
- marine biodiversity and resources;
- terrestrial and freshwater biodiversity;
- primary industries;
- settlements and infrastructure;
- human health;
- emergency management
- adaptation and Indigenous communities; and
- social, economic and institutional dimensions of adaptation.

As a first step, and for each topic, a team of researchers and practitioners in adaptation was brought together to write a research plan that identified knowledge gaps and research priorities. Each research plan went through an intensive period of consultation and review, to ensure it properly reflected the views of the adaptation community. The identified research priorities then became the basis for an open call for proposals.

In addition to these thematic topics, a cross-discipline synthesising and integrative topic was constructed. This consisted of short projects (rarely longer than 12 months) based mainly around literature reviews and stakeholder interviews, identified as being of particular interest to practitioners, sometimes in response to real-world events. Thus NCCARF funded a number of flood-related adaptation research projects in response to questions asked by stakeholders following the widespread devastating floods that affected large parts of Australia in 2010 and again in 2011. A number of chapters in this volume are based around these synthesis and integrative research projects.

A notable feature of the research program was the involvement of end-users. As emphasised by

Applied Studies in Climate Adaptation, First Edition. Edited by Jean P. Palutikof, Sarah L. Boulter, Jon Barnett and David Rissik.
© 2015 John Wiley & Sons, Ltd. Published 2015 by John Wiley & Sons, Ltd.

the title of this book, adaptation is essentially an applied field of study (Chapter 2 by Palutikof et al.). NCCARF's adaptation research aimed to be useful to the end-user community and to be delivered in accessible formats. To address this requirement, project investigators were required to write and fulfil an end-user communication plan and/or to set up a project steering committee that included strong participation from end-users.

1.1.3   Why this book?

All the final reports from the NCCARF research programs in its first phase are available on the website www.nccarf.edu.au. These are often weighty volumes, containing the detail of the relevant literature, the research methodology and analyses of extensive datasets. NCCARF has already devoted considerable effort to producing user-friendly summaries of the research findings, and continues to do so. However, these summaries are primarily directed towards Australian adaptation practitioners; they assume considerable contextual knowledge, and focus on the concerns and problems raised by adaptation in Australia.

As outlined in Chapter 2 of this volume, Australia has much experience in adapting to climate variability and change. It has one of the most (naturally) variable climates in the developed world, and already the impacts of anthropogenic climate change are being experienced. Australian decision-makers face a multiplicity of issues as they seek timely and cost-effective responses to climate change which balance the needs and concerns of different stakeholder groups, while ensuring that planning, regulatory and legislative requirements are met. These issues, and the emerging solutions to address them, are not particular to Australia. To name just a few in the area of primary industry, this book deals with how farmers are adapting to changing conditions (Chapter 14 by Hertzler et al. and Chapter 15 by Kingwell et al.), the likely outcomes of the interaction of irrigation water tariffs and markets with climate change (Chapter 17 by Cooper et al. and Chapter 18 by Wheeler et al.), and how value chains in the fishing and aqua-

culture industries can evolve profitably under climate change (Chapter 16 by Hobday et al.).

The editors of this book (and the authors of this chapter) see much value in sharing the results from NCCARF's research programs with the wider global community of adaptation researchers and end-users. In preparing their chapters, all authors were asked to consider the international context for their work and the international implications of their results.

1.2   Structure and content of the book

After this introductory chapter, the book continues with a chapter written by the editors which highlights some of their emerging thoughts and conclusions around adaptation, and how these have been influenced during the editorial process. These are presented as four loosely linked arguments. First, we argue that the Australian experience with adaptation policy and research is of global relevance. Second, we can no longer say that adaptation to climate change is not happening, even if we might disagree about its effectiveness. Third, there is limited utility, and probably only confusion, in the increasing classification of types of adaptation, for example as being 'autonomous' or 'evolutionary' in nature. Fourth, we argue that adaptation is not a 'science' as sometimes claimed, but rather a complex interdisciplinary 'field of inquiry' whose ultimate rationale must include delivering policy-relevant information to decision-makers.

The chapters that follow are divided into nine sections. To increase the international relevance of the book, we approached nine (non-Australian) experts in adaptation and asked them to contribute short 'think pieces' to open each of these nine sections. We would like to express our gratitude to these nine experts and thank them for entering into the spirit of the exercise. We gave them little in the way of guidance, simply emphasising that these should be their musings on the subject of their section, based on their experiences over the years.

The first section covers frameworks for enabling adaptation and looks at the context within which adaptation takes place and the instruments and circumstances that can act to facilitate or impede the adaptation process. Gary Yohe from the United States contributes the think-piece, providing some personal reflections. Chapter 4 (McAneney) looks at the role of insurance, concluding that the short time horizons of the industry impede its capacity to support adaptation to climate change. Macintosh and colleagues (Chapter 5) look at planning instruments for adaptation, and how these can be designed to facilitate adaptation. This is followed by a chapter on public perceptions of climate change by Reser et al. Chapter 7 (Verdon-Kidd et al.) then explores the science–policy interface and the challenges faced by those who seek to cross that interface.

This introductory framing section is followed by eight thematic sections. These are arranged in a semi-logical order, building from the natural environment through to settlements, disaster management and business. Inevitably, there is overlap between sections. For example, a number of chapters deal with heat stress and adaptation; these can be found in Section 5 on building resilience among vulnerable groups (Chapter 27 by Hansen et al.), in Section 6 on the Indigenous experience of climate change (Chapter 32 by Horne and Martel) and in Section 7 on settlements and housing (Chapter 39 by Saman et al. and Chapter 40 by Barnett et al.).

The thematic section on ecosystem management (Section 2) is opened by a think-piece by Jonathan Overpeck with the opening question: at what level does climate change pose an 'unacceptable' risk to biodiversity? There follow four chapters on adaptation management options for birds (Chapter 9 by Franklin et al.), seabirds and marine mammals (Chapter 10 by Hobday et al.), Ramsar wetlands (Chapter 11 by Bino et al.) and finally invertebrates (Chapter 12 by Moir).

Section 3 on farming has a think-piece from John Morton making the point that, although Australia is a highly developed economy, it is also (and unusually so for a developed country) largely tropical and subtropical; several features of its agriculture therefore have resonance for some poorer countries of the tropics. The five

main chapters in this section explore climate change impacts and adaptation interactions for broadacre (dryland) farming (Chapter 14 by Hertzler et al.; Chapter 15 by Kingwell et al.), irrigation water governance (Chapter 17 by Cooper et al.; Chapter 18 by Wheeler et al.) and value chains in the fisheries and aquaculture industries (Chapter 16 by Hobday et al.).

Section 4 on coasts opens with a think-piece by Susi Moser. This is a personal reflection on the stresses, both human and physical, which affect the coastal zone and a call to arms to all those researchers and decision-makers with responsibilities in the coastal zone to ensure that the coastal zone remains a place of beauty and sustenance. Four chapters follow. Woodroffe and colleagues (Chapter 20) show how climate change can be integrated into risk assessment approaches and combined with engineering models to create more comprehensive coastal risk assessments. Hadwen and Capon (Chapter 21) also present a risk-based approach to adaptation, focusing on habitats and ecosystems and emphasising the need to consider non-climatic pressures when exploring adaptation options. McEvoy and Mullett (Chapter 22) have a specific interest in climate change risks to seaports and their operation. Fletcher and colleagues (Chapter 23) explore the economics, equity and institutional arrangements for adaptation. Chapter 24 by Waters and Barnett concludes this section, exploring the views of stakeholders on where the responsibilities lie for adaptation in the coastal zone.

Section 5 covers the building of resilience among vulnerable groups. The think piece is an essay by Hallie Eakin on the 'turn' that has taken place from a focus on vulnerability to climate change, to a focus on capacity. In Chapter 26, Barnett and Palutikof explore the limits to adaptation using five example locations in Australia where the impacts of future climate change may exceed the capacity to adapt. There follow two chapters on extreme events, looking at how the impacts on vulnerable groups are experienced and can be managed. Chapters 27 (Hansen et al.) and 28 (Correa-Velez et al.) look, respectively, at the effects of heat waves on culturally and linguistically diverse groups and at the experience of the 2011 Queensland floods by refugee groups. Sevoyan and Hugo (Chapter 29) explore the relationships between social exclusion and vulnerability to climate change, and how the barriers of social exclusion can be overcome to ensure that effective adaptation can take place. Finally, in Chapter 30 Hamilton and Mallon look at the vulnerability of community organisations to extreme events, and the implications on their capacity to continue to provide community services under climate change.

Section 6 is on the Indigenous experience of climate change. The think-piece by Meg Parsons places the challenge of climate change to Australia's Indigenous communities firmly within a global context. It proposes two requirements for Indigenous communities to prosper under climate change: the first is to move beyond current conceptualisation of Indigenous as 'traditional' and consider the diversity of Indigenous communities; the second relates to future planning and the need to consider how adaptation relates to social justice and Indigenous rights. There follow four diverse chapters considering adaptation and Indigenous communities from four very different perspectives. Chapter 32 (Horne and Martel) looks at Indigenous housing in the desert climate of Alice Springs, and at what modifications may be required to ensure a level of comfort under an even hotter climate. Haynes and colleagues explore the Indigenous experience of Cyclone Tracy, which razed Darwin to the ground in December 1974, in Chapter 33. The authors seek to understand whether there were attributes of the Indigenous experience which might point to particular capacities and vulnerabilities. In Chapter 34 Tran and colleagues look at the role of native title in the governance of adaptation, using two small Indigenous settlements as case studies. Nursey-Bray and colleagues (Chapter 35) look at adaptation in one Indigenous group (the Arabana people) and, in particular, how their attachment to and engagement with their land influences their capacity to adapt.

Section 7 on settlements and housing opens with a think-piece by Bill Solecki on global urban development and climate change. There follow five chapters in this section; Chapters 37 and 38 are on settlements and Chapters 39–41 are on housing. Beer and colleagues (Chapter 37) develop an index of vulnerability for country towns, which suggests that it is the most remote communities that are the most vulnerable. Mortazavi-Naeini and colleagues test the robust multi-objective optimisation approach as a tool to assist urban water supply managers to adapt to uncertain future climate change in Chapter 38. Saman and colleagues (Chapter 39) look at how design can improve the comfort levels of houses during heat waves. Barnett and colleagues (Chapter 40) also explore thermal performance in heat waves, but with a particular emphasis on low-income housing. Finally, Instone and colleagues (Chapter 41) look at strategies for facilitating adaptation in the rental housing market, which again in Australia is primarily a low-income market.

Section 8 is on the topic of adaptation and disaster management. The think-piece is provided by Ian Burton, reflecting on the shortcoming of 'practical adaptation' and the need to ensure that responses to disasters – responses he terms 'palliative adaptation' – do not leave us more vulnerable to long-term climate change. Three chapters follow. In Chapter 43, Boon seeks to understand the factors that promote community resilience in the face of disasters by studying four towns and four different events: bushfire, flood, drought and cyclone. She finds that strong individual resilience and social connectedness promote strong community resilience. Bird and colleagues describe a similar study in Chapter 44 but in this case focused on flood disasters, looking at the respective roles of individuals, communities and government in responding to disasters and the factors that promote and inhibit resilience. Howes (Chapter 45) completes this section by investigating the potential for governments to conserve scarce resources by integrating disaster risk management and climate change adaptation.

The final section is on the subject of business. It opens with a think-piece by Frans Berkhout exploring organisational adaptation from the perspectives of utility-maximising, behavioural and institutional approaches. Shearer and colleagues (Chapter 47) look at major barriers and drivers of adaptive capacity for the urban property development industry, noting that moves towards greater sustainability are primarily driven by client demand which, in Australia at least, is generally lacking. Kuruppu and colleagues investigate adaptation among small-to-medium enterprises (SMEs) in Chapter 48. These authors conclude that SMEs, where they adopt adaptation strategies, address climate extremes rather than climate change, and that it is their past experience of extremes that is their primary motivator for adaptation. Finally, in Chapter 49 West proposes a framework to enable businesses to better consider the strengths and weaknesses of various approaches for assessing adaptation options, and includes a case study of wind damage to coastal assets.

Acknowledgements

We would like to acknowledge financial support from the Australian Government through the Department of Climate Change and Energy Efficiency, without which it would not have been possible to produce this book. Most chapters are based on research funded through this initiative. We would also like to thank all those who have worked in the Research Management Team of NCCARF over the years: Richard McKellar (Research Manager), Florence Crick, Ida Fellegara, Daniela Guitart, Ann Penny, Frank Stadler, David George and Daniel Stock. Their diligence in ensuring quality in the NCCARF research projects is carried through into this book.

2 Adaptation as a field of research and practice: notes from the frontiers of adaptation

JEAN P. PALUTIKOF[1], JON BARNETT[2],
SARAH L. BOULTER[1] AND DAVID RISSIK[1]

[1] *National Climate Change Adaptation Research Facility, Griffith University, Australia*
[2] *School of Geography, University of Melbourne, Australia*

2.1 Introduction

This chapter seeks to draw together the overarching themes from the contributions to this book, and does so through the development of four arguments.

First, we argue that the Australian experience with adaptation policy and research is of global relevance. Australia is a developed country with a highly variable climate and has invested significantly in adaptation research; as such, it has something to teach the world about adaptation. There are lessons from Australia, including about: what works and what doesn't; ways of applying the institutions and knowledge to respond to climate variability; the greater challenge of adapting to climate change; and ways of producing knowledge about adaptation. This argument is of course a key rationale for this book, which consolidates key insights from adaptation research and practice from diverse sectors and places across Australia.

Second, we can no longer say that adaptation to climate change is not happening, even if we might disagree about its effectiveness and whether it is happening in the right locations and institutions.

Climate variability and climate change are seamless, as the literature on detection and attribution of climate extremes clearly demonstrates (Peterson et al. 2013), and adaptation to one must inevitably deliver adaptation to the other. This is also an important justification for this book, which contains many present-day examples of actions in response to events and policies that cannot be said to be independent of the material and symbolic reality of climate change. Even those who would claim to be climate change sceptics act in the knowledge that climate change exists, at least as a topic for debate and as a policy issue and, like it or not, their actions must be influenced by that knowledge.

Our remaining arguments focus on adaptation as a field of research. Third, there is limited utility and probably only confusion offered by the increasing classification of types of adaptation, for example, as being 'autonomous' or 'evolutionary' in nature. We test this by exploring the explanatory power of the concepts of 'incremental' and 'transformational' adaptation when applied to two Australian examples.

Fourth, we argue that adaptation cannot be said to be a 'science'; rather, it is a complex interdisciplinary field of inquiry whose ultimate rationale must include delivering policy-relevant information to decision-makers. On the basis of our experience with adaptation research in Australia, we consider key determinants of 'good' adaptation research.

This chapter uses many examples from Australia to illustrate the points made. These examples are generally taken from the last 2–3 decades, when adaptation to climate change as well as to climate variability has become a significant policy issue across all levels of government. Nevertheless, there is a much deeper and richer history of adaptation to Australia's climate, most notably by Indigenous people who, over several tens of thousands of years, created cultured landscapes that affected the reproduction of many key ecosystems (Langton 1998). There have also been adaptations made in settler society long before formal recognition of the problem of climate change including for example: crop breeding by William Farrer in the second half of the nineteenth century to produce wheat varieties which, through drought and disease resistance, would deliver high yields and high-quality grain under Australian climates (Wrigley 1981); the introduction of large-scale formal irrigation from California in the 1880s through the efforts of Alfred Deakin (Norris 1981); evolution in governing freshwater resources (Quiggin 2001; Connell 2007; Leblanc et al. 2012); and the emergence of community-based approaches to disaster risk management (Gabriel 2003; Elsworth et al. 2009; McLennan and Handmer 2014).

2.2 Argument 1: the Australian experience

Australia is a wealthy liberal-democratic society that is exposed to high levels of climate variability (Min et al. 2013). By all measures, Australia has high adaptive capacity. The United Nations Human Development Index is a useful proxy for adaptive capacity as it reflects levels of wealth, health and education. By this measure, in 2012 Australia was ranked second in the world; this is a wealthy, healthy and well-educated society. There

is high awareness of climate change. In a 2011 nationwide survey, 87% of respondents accepted that there was some degree of human influence on climate change, 74% considered that the world's climate is already changing and half considered that Australia is already affected (Reser et al. 2012). Further, there has been a large and purposeful investment in climate change science and adaptation research (e.g. in the period 2007–2013 the Australian Government invested $129 million in adaptation research through the National Climate Change Adaptation Program). Perhaps more than anywhere else in the world, Australia has key vulnerabilities and sensitivities, a lot of evidence and a lot of capacity for adaptation. The Australian adaptation experience is therefore one that the world can learn from.

2.2.1 The Australian climate of extremes

Australia has a climate that is highly variable from year to year; statistical analysis of observations demonstrates it is more variable than equivalent climates elsewhere in the world (Nicholls et al. 1997; Love 2005). This variability leads to a climate of extremes: floods and droughts, cyclones, heat waves and bushfires. Australia is the driest inhabited continent (Head et al. 2104), and severe drought affects some part of the nation about once every 18 years. Looking into the recent past, the so-called Millennium Drought affected much of eastern Australia from 1995 to 2010 and was, in terms of intensity if not extent, one of the worst droughts recorded in Australia (Verdon-Kidd and Kiem 2009; Kiem et al. 2010). The drought terminated abruptly in 2010 with the onset of a strong La Niña event, and in December 2010–January 2011 there was widespread and severe flooding in southeast Queensland followed by a major flood event in Victoria.

The Australian climate is typified by these swings between drought and flood. The tropical North experiences cyclones, with the northwest coastline being most frequently impacted. About two cyclones cross this coastline each year on average, one of which is severe. The southern cities of Melbourne and Adelaide experience severe heat waves, and temperature extremes associated with high wind speeds and humidity levels in the single figures can lead to devastating bushfires. For example, the Black Saturday fires in Victoria in 2009 led to 173 deaths. Karoly and Boulter (2013) have explored the effects of the multiple climate-related disasters in 2010/11 on Australian society and the economy, and make the point that such multiple events are likely to become more common under climate change.

2.2.2 Disaster response and adaptation

It has been suggested (e.g. Adger et al. 2007) that adaptation follows an extreme event. The Australian experience somewhat supports this. For example, Cyclone Tracy which hit Darwin on Christmas Day 1974 and destroyed 60% of the housing beyond repair (Mason and Haynes 2010), led to a wholesale rewriting of the building codes for cyclone-prone areas of Australia. Nevertheless, what is becoming clear from the Australian experience is that adaptations following a single extreme event do not necessarily 'keep up'; subsequent events can still cause extensive damage and loss of life, even when adaptation has taken place. The potential causes are many, and likely to be some combination of these: that the adaptation actions are inadequate, that socio-economic factors such as increased housing density are affecting vulnerability, that our understanding of the return periods of extremes is flawed, or that climate change is changing the nature of extremes.

Bushfire is a good example of this failure of adaptation to protect against the severity of subsequent events, and exploration of recent bushfires in Australia provides some insights into the management of extreme events that have relevance to climate change adaptation. The recent (October 2013) severe bushfires in New South Wales, which destroyed more than 200 homes, revealed two important insights. First, it wasn't simply a matter of increased severity of the bushfire, but also of changing seasonality that contributed to the impact of this event. Because the fires

occurred very early in the season, the fire-fighting aircraft that are brought from the United States every year to help fight fires in Australia were not ready to be deployed. As the fire seasons in the two nations become extended as a result of climate change, the economies that can be achieved by sharing equipment will begin to be lost.

The second insight from the recent NSW bushfires is that houses in bushfire-prone areas are still inadequately protected, and there are two reasons for this. The first relates to governance structures in Australia: building code adoption is a matter for the states and territories rather than national government. Although Victoria adopted changes to its building codes following the Black Saturday fires, there was no requirement for the other states and territories to do so. The second relates to cost: bushfire protection requirements such as fire-resistant shutters on all windows and doors carry a considerable cost. Although this may deter people from building in at-risk areas, which is a positive outcome, it can also deter people from making less-expensive modifications to existing houses that would provide protection. Interventions by governments to promote adaptation must be carefully designed to avoid negative outcomes.

2.2.3 'Build back better' and individual preparedness

Historical experience with extreme events means that Australian emergency managers are well-prepared and that responses are generally timely and commensurate in scale with the level of risk. However, there is ongoing debate about the need for betterment or to 'build back better' during the recovery phase, that is, to create more resilient communities which are less reliant on the emergency services during disasters. We now explore each of these a little further.

'Build back better' is an approach often promoted in the recovery from disasters in developing countries by development agencies and relief organisations. It is founded on the sense that, following a disaster, development and disaster response agencies can come together to build a brighter, better future; it tends to focus on material things rather than people. But this notion is contested: responses need not necessarily be immediate, and the emphasis should arguably be as much on rebuilding people's lives as on construction projects (Schilderman and Lyons 2011).

In the developed-world context of Australia, 'build back better' means something subtly different. It is a retreat from the traditional model of rebuilding (particularly transport infrastructure) in the same way and to the same standard as before, on the assumption that this was an extreme event with a return period of 100 years or more (i.e. of similar scale to the lifetime of the structure). It is an acceptance that, for whatever reason (climate change, improper understanding of natural variability, inadequate construction or poor choice of location), improvements need to be made. Although more commonly focused on infrastructure and buildings, it can also encompass small business recovery and community projects (Mannakkara and Wilkinson 2012).

Betterment is a desirable goal in theory, but it is easier said than done. Insurance generally covers the replacement value, in which case it cannot be a source of funding for betterment. In Australia, funds for betterment have been available from the federal government under the Natural Disaster Relief and Recovery Arrangements (NDRRAs) since 2007, but have not been widely accessed. The reasons appear to be related to lack of awareness and the way that the NDRRAs are applied (Regional Australia Institute 2013).

Extending 'build back better' to encompass businesses and communities to build resilience has become a theme in Australian Government thinking around disasters in recent years. The National Disaster Resilience Framework, for example, outlined the premise that 'Individuals and communities should be more self-reliant and be prepared to take responsibility for the risks they live with' (Australian Government 2011, p. 2). This was closely followed by the National Strategy for Disaster Resilience statement that 'disaster resilience is based on individuals taking their share of responsibility for preventing, preparing for, responding to and recovering from disasters' (National Strategy for Disaster Resilience; COAG 2011, p. v). Essentially, the Framework and the Strategy emerged from recognition by government that the emergency services cannot reach all of the people under threat during a disaster, especially given the long distances and low population densities of rural Australia, and was an attempt to bring awareness of this reality to vulnerable communities. The Strategy took into account the role of climate change, saying 'climate change is making weather patterns less predictable and more extreme' (COAG 2011, p. iv).

Overall, there is a general acceptance across Australian society that climate change is a reality and that it is happening now (Leviston and Walker 2012) with the pragmatic corollary that, whether or not it is of human origin, it is probably necessary to do something about it. Howes and colleagues quote a Victorian government official talking about the emergency services: 'you do not find many climate change sceptics on the end of hoses anymore ... they are dealing with increasing numbers of fires, increasing rainfall events, increasing storm events' (Howes et al. 2013, p. 17). If there is an acceptance that climate is changing, it would be odd if nothing was being done about it. Indeed, as the following sections show, much is being done.

2.3   Argument 2: farewell to the no-adaptation world

Among the research and practice communities there is a sense that we are not yet taking action to adapt to climate change. For example, Ford and colleagues (2011, p. 327) state that they 'find limited evidence of adaptation action' and Mimura and colleagues (2014) say that adaptation is only now 'transitioning from a phase of awareness and promotion to the construction and implementation of plans, strategies, legislation and projects'. This interpretation of adaptation is, we suggest, based on a narrow view of adaptation as a finite and technical-rational process, with a formally declared beginning and clearly identified milestones and endpoints. There are three reasons to suggest that this idea of a no-adaptation world cannot be sustained.

The first reason why we can say that the no-adaptation world has vanished is because, by almost all accounts, central components of adaptation include investigating the problem and raising awareness of the issue. As explained earlier, awareness of climate change risks is high across the Australian population, and the country has made significant investments in adaptation research (the outcomes of which are covered by Argument 4; see Section 2.5). Indeed, this book is one of many products of this investment, signifying that adaptation in Australia is underway. Further, we are able to discern a significant amount of policy activity directed towards adaptation, with Australia's recent National Communication to the United Nations Framework Convention on Climate Change providing something of an inventory of this activity (DIICCSRTE 2013). While much is concerned with assessments and assessment and decision support tools, adaptation plans and agreements are being developed across all levels of governments, in most states and territories, across the seven priority sectors identified by the Council of Australian Governments Select Council on Climate Change, namely: water resources, coasts, infrastructure, natural ecosystems, agriculture, emergency management and vulnerable communities. In this respect the debate about the necessity of and evidence for adaptation is long over; adaptation is happening.

The second reason why we say that adaptation is happening follows from this research and awareness. Climate change is a now social fact that affects reason and action in multifarious ways. There is now ample evidence that some observed changes in biophysical systems in Australia and elsewhere *have* a significant climate signal. Indeed, this was well demonstrated in the 'Summary for policymakers' of the Fourth Assessment Report of the IPCC, where Figure 1 identified over 29 000 incidents of significant environmental change that were consistent with those expected from climate change (IPCC 2007). More recent research on sea-level fingerprints further confirms that there are significant changes in biophysical systems that are highly likely to be caused by climate change (Mitrovica et al. 2009). Given this evidence of impact, in many sectors – including agriculture, construction, infrastructure development, insurance, emergency management, forestry, natural resource management, property development and town planning – decisions are now being made in light of the risks of climate change. Almost all public and much private decision-making now takes place in the context of knowledge of climate change. This knowledge cannot be without effect, even if the outcome of such decisions cannot be measured and may be nil or indeed counterproductive.

Finally, we would argue that adaptation is essentially a pragmatic response to a perceived present or future imbalance between climate and the societies and environments that it affects. For many decision-makers, the reasons for this are irrelevant – they may be due to a change in climate or they may be due to social change – but the need for action to address the imbalance remains the same. Indeed, research and academic writing on adaptation (although not always by that name) predates by many decades the current emphasis on purposeful actions to address climate change. For example, we may cite the works of: Griffith Taylor (1880–1963, Professor of Geography at the University of Sydney) on the environmental limits of settlement in Australia; Bruce Davidson who, in 1965, wrote *The Northern Myth*, setting out the limits to agricultural development in tropical Australia; and George Goyder who in 1865 drew a line across the map of South Australia separating areas where cereal farming was likely to be uneconomic because of low rainfall from areas where it would be viable (Nidumolu et al. 2012).

In the following section we present two case studies of present-day adaptation in Australia.

The first, on water resource management in the Greater Perth region, is a clear example of purposeful adaptation taking place in the knowledge of scientific evidence for a diminishing water resource driven by climate change. The second looks at how the government has, over the years, tailored its drought support for farmers to the climatic realities of Australia. Although the underlying motivation is not climate change, it has nevertheless delivered a more robust and resilient farming system which is better able to withstand the rigours of climate change; it is an example of adaptation. Our final reason for maintaining that adaptation is already taking place, therefore, is that it doesn't necessarily require a climate change motivation. Adaptation is essentially pragmatic, and many actions to build resilience against a variable climate can be found that deliver risk reduction in the face of climate change.

Adaptation is now widely understood as being a matter of multiple ongoing processes of responding to changes, the 'end' points of which are far distant and will probably be reached with some new equilibrium level in the global climate system (Nelson et al. 2007; Stafford-Smith et al. 2011). If the precise beginnings of these processes are hard to identify, it is now quite clear, at least in Australia, that adaptation has begun. In this context this book presents findings from research purposefully intended to advance knowledge about adaptation across the continent, across sectors and from diverse disciplines.

2.4   Argument 3: the obfuscations of adaptation classifications

Perhaps because it is ungainly and hard to define succinctly, there is no shortage of attempts to classify adaptation as being of various types according to sectors, intentions, timing, effectiveness and so on. Looking through the glossaries of the past three IPCC Working Group II reports, for example, we find that adaptation can be: anticipatory, autonomous, community-based, ecosystem-based, evolutionary, incremental, mal(adaptive), physiological, planned, private, public, reactive and transformational. The definitions of these and the distinctions between them are all ambiguous, and so good grist for the publication mill. This cottage industry of sub-classifying adaptation creates confusion, which acts as a deterrent to new entrants to adaptation research, and does little to help decision-makers.

Most recently, distinguishing between 'incremental' and 'transformational' adaptation has been in vogue. The Glossary to the Fifth Assessment of the IPCC Working Group II defines: adaptation as 'the process of adjustment to actual or expected climate and its effects'; incremental adaptation as 'actions where the central aim is to maintain the essence and integrity of a system or process at a given scale'; and transformational adaptation as 'adaptation that changes the fundamental attributes of a system in response to climate and its effects' (IPCC, 2014). The distinction between these three is unclear. The focus of the definition of incremental adaptation is *maintaining the essence and integrity of a system*, and the focus of the definition of transformational adaptation is *changing the fundamental attributes of a system*. There is therefore no standard point of comparison; these are not exclusive, since the essence of a system can be maintained even while its fundamental attributes may change. Indeed, this is often the case in the realm of economic planning. The essential goals of all governments are to maintain employment and growth, but the attributes of economic systems can radically change in order to meet these goals as various processes of restructuring indicate.

As a means to test the salience of the concepts of 'incremental' and 'transformational' adaptation, we examine the extent to which they can help explain two Australian examples of adaptation practice. We first briefly describe these cases.

2.4.1   Case Study 1: water resource management in SW Australia

A decline in winter rainfall since the 1960s in SW Western Australia is clear from observations.

A number of initiatives were set in train in response to this decline, with dams rebuilt (e.g. the Victoria dam, rebuilt during 1990–1991) and extremes recalibrated. Partly in response to the decline, the Indian Ocean Climate Initiative (IOCI) was set up in 1998. Since 1998, IOCI has worked to understand the causes of the observed drying trend and whether the trend is likely to continue in the future under scenarios of increasing greenhouse gas concentrations. Researchers have shown that the downward trend has intensified in recent years and that the area affected has expanded. The trend can be linked to changes in atmospheric circulation that are 'consistent with what would be expected in an atmosphere influenced by increasing greenhouse gas concentrations' (IOCI 2012, p. 10). It is therefore likely that the trend will continue at least for the present century (IOCI 2012).

The Water Corporation of Western Australia is responsible for the provision of piped water supplies into the greater Perth region (population close to 2 million). Faced with the reality of a declining resource, as exemplified by the inflows to Perth reservoirs shown in Figure 2.1 and by scientific evidence linking the decline to global warming, the Corporation set out to deliver a climate-independent water supply to greater Perth. To do this, the Water Corporation had considerable resources at its disposal. With a population of only 2.5 million people, Western Australia has a Gross Domestic Product among the top 50 economies of the world, based around exports of primary products such as iron ore, gold and wheat.

In 2005–2006, half the public water supply of the greater Perth region was from surface reservoirs and half from groundwater. By 2010–2011, the effective supply from surface reservoirs had

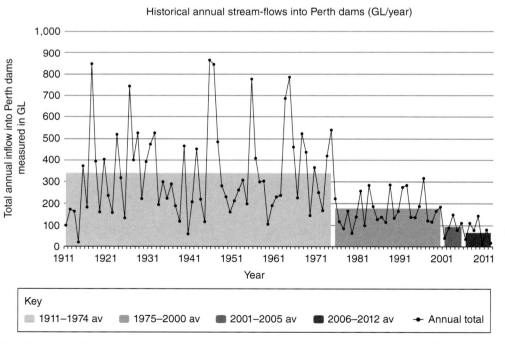

Figure 2.1 Historic inflows to Perth reservoirs. Source: www.watercorporation.com.au. Reproduced with permission of the Water Corporation of Western Australia. For colour details please see Plate 1.

dropped to 22%; desalination supplied 20% and groundwater supplied the balance. At the beginning of 2013, approximately 50% of the piped water supply was derived from desalination.

There are of course a number of well-known drawbacks of desalination that call into question its long-term sustainability as a source of water. The first is cost; desalinated water in SW Western Australia is delivered at approximately $2.20 per kL, compared to 20c per kL for water from surface reservoirs. The second is the energy requirement which is generally not less than 3 kWh m^{-3}, much higher than local freshwater supplies at 0.2 kWh m^{-3} or less. The Water Corporation has made a strong commitment to renewable energy use, taking the entire output of a 10 MW solar farm and a 55 MW wind farm.

Whatever our view of the sustainability of desalination, the fact remains that the Water Corporation has delivered on its commitment to a climate-independent supply; it has therefore adapted effectively to a lower rainfall regime that

has been linked to climate change. Looking forward, the Corporation is likely to turn its attention to wastewater recycling technologies and to demand reduction measures as it seeks to deliver a secure water supply to small inland towns.

2.4.2 Case Study 2: government support for drought-affected farmers

As we have already noted, drought is a persistent and regular feature of the Australian climate. Figure 2.2 shows a time series of drought occurrence in Australia. Nevertheless, European settlers to Australia brought with them a farming paradigm that sprung from much more benevolent climate conditions, and the process of adjustment was long and sometimes painful. Until 1989, drought was treated in policy terms as a natural disaster: a rare crisis that had to be managed so that impacted people and communities could return to 'normal' as quickly as possible. Through the 1970s and 1980s there was

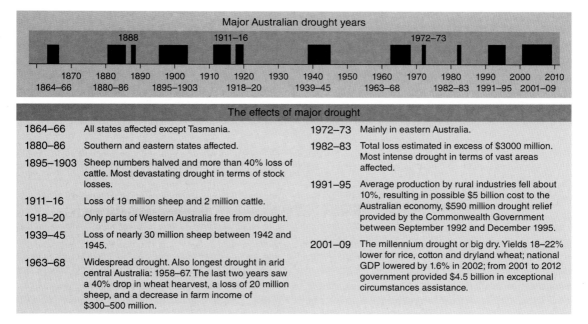

Figure 2.2 Major Australian drought episodes since the mid-nineteenth century. Source: Reproduced with permission of the Bureau of Meteorology/Commonwealth of Australia.

however a dawning realisation that this was far from the truth, and that drought would be better managed as a recurring risk to farming in broadly the same way as risks such as price fluctuations and pest and disease outbreaks (Botterill 2013).

Drought was removed from the Natural Disaster Relief Arrangements (NDRA) in 1989 and, since then, Australia has taken a risk management approach to drought. Support for farmers to achieve a sustainable industry include exit grants to encourage non-profitable farm operators to leave the industry, income smoothing strategies providing support to drought-proof farms, and educational programs. A review by the Productivity Commission (a cross-party government-funded think tank) in 2008 reinforced the view that a risk management approach to drought is appropriate. Among other things, this review recommended abolishing certain policies that encouraged moral hazards including the exceptional circumstances payments. Such payments were seen to be based on implausible delimitations of the spatial extent and recurrence of droughts, and to lead to inequities in provision (Botterill and Hayes 2012).

Thus, Australia has taken a risk management approach to drought since 1989. The changes that were introduced following the removal of drought from the NDRA were truly transformational in nature. This was transformational adaptation – to climate variability in this case, rather than climate change – but perfectly epitomising the transformational changes that will have to take place in response to climate change.

2.4.3 Incremental versus transformational adaptation?

It appears to us that the actions of the water managers in Western Australia and of the Australian Government in managing drought were both transformational in character. They required a fundamental rethinking of the challenge being faced, and therefore the appropriateness of the available strategies to manage these risks. There was an acceptance that, in Western Australia, the climate had changed and was likely to continue changing. There was an acceptance that, in Western Australia, the climate had changed and was likely to continue changing to the extent that what had been done in the past would no longer work in the future and that, in Australian farming, prior policy was not delivering a robust and resilient sector because it was poorly adapted to the Australian climate. In both cases, it was clear to decision-makers that there needed to be change to management and policy; things had to be done differently. Their thinking on the risk underwent a transformation and, in many ways, so too did their responses to adapting to climate change. In both cases however, the goal was (to use the IPCC language) to maintain 'the essence and integrity of a system'. In the case of Western Australia the goal was robust and secure water supply, and in the case of national drought policy the goal was a robust and competitive farming sector.

In looking at these cases, the distinction between transformational change and incremental change does not explain much. If anything, the cases show that adaptation is a continuum. Decision-makers at all levels of government, in the private sector and individuals do and will undertake those actions that they consider to be appropriate and feasible given the perceived level of present and future risk. Those actions will escalate as the risk is seen to increase both absolutely and in relation to other risks, since financial considerations must take account of where the money is best spent.

Indeed, the distinction between incremental and transformational may be more in the eye of the beholder. Transformational change is often equated with rapid change; in fact, it may take decades to achieve, leaving the impression with the observer that it is incremental in nature. An example is rural water policy in Australia, which is inarguably being transformed but over three decades. The process began with the water reform agreement reached by the Council of Australian Governments (COAG) in 1994. The next key point was the National Water Initiative of 2004 (see Chapter 17). More recently, the government has initiated the Sustainable Rural Water Use and Infrastructure program, with a focus on water reform in the Murray–Darling Basin. Under the Murray–Darling Basin Plan, this process will take at least a further seven years. While each step in the process may be seen as incremental, overall there is no doubt that the changes they will institute in Australia's rural water management are transformational.

Given these practical characteristics of adaptation decisions, *a priori* or *posteriori* judgements about the degree to which adaptation should be 'incremental' or 'transformational' seem somewhat arcane outside of the academic community.

2.5   Argument 4: the nature of adaptation research

There has been an explosion of research in climate change adaptation in Australia since 2007, which coincides with the aforementioned investments in research made under the National Climate Change Adaptation Program. Figure 2.3 shows the results of a search of the Scopus database for articles that use the terms 'adapt*' and 'clim*' in the title, abstract or keywords, and where the country of affiliation is Australia. The search was conducted in October 2013, covered articles published in English in the years 2004–2013 and included all subject areas and document types.

There are two important characteristics of this research. First, it spans all major subject areas, with the physical and life sciences dominating as shown in Figure 2.4. The major journals reflect this as shown in Table 2.1, where most of the major sources of publications are physical and life science journals.

The second important characteristic of this research on adaptation from Australian institutions is the diversity of centres from which it is produced, as shown in Table 2.2. No single institution clearly dominates research in this field, but the vast majority clearly come from the university sector.

This analysis, and the contents of the book, suggests that adaptation is perhaps not so easily described as a 'science' as Moss and colleagues (2013) for example suggest. Much depends on what one means by 'science' of course, and we do not seek to be definitive in this argument. However, the sheer diversity of entities at risk for which adaptation is required – ranging from single species to human cultures and from specific sites to continental scales – and the enormous diversity of disciplines and methods, journals and institutions suggests that research on adaptation is far too plural in nature to be meaningfully described as a 'science'. There may be many 'sciences' of adaptation, and these would ideally have the characteristics of post-normal science (cf. Funtowicz and Ravetz 2003) as implied by Moss and colleagues (2013). The possibility that

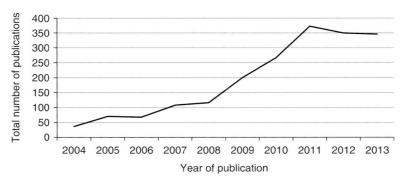

Figure 2.3   Number of articles published on clim* AND adapt* from Australian research institutions as returned from a search of the Scopus database.

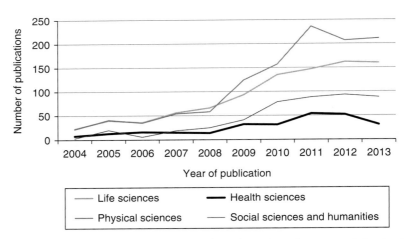

Figure 2.4 Number of articles published on clim* AND adapt* from Australian research institutions, by subject area, as returned from a search of the Scopus database. For colour details please see Plate 2.

Table 2.1 The top sources of articles published between 2004 and 2013 on clim* AND adapt* from Australian research institutions, as returned from a search of the Scopus database.

Rank	Journal	Articles
1	*Global Change Biology*	36
2	*Climatic Change*	33
3	*Plos One*	30
4	*Global Environmental Change*	26
5	*Modsim 2011 19th International Congress on Modelling and Simulation Sustaining Our Future Understanding and Living with Uncertainty*	26
6	*18th World Imacs Congress and Modsim09 International Congress on Modelling and Simulation Interfacing Modelling and Simulation with Mathematical and Computational Sciences Proceedings*	23
7	*Crop and Pasture Science*	22
8	*Mitigation and Adaptation Strategies for Global Change*	22
9	*Science*	21
10	*New Phytologist*	21

adaptation is a 'science' seems questionable however given the absence of key theories or models, dominant paradigms and established methods. Further, as Patt (2013) argues, narrowing research on adaptation into a single 'science' may come at a cost to the powerful insights that presently come from researchers drawing deeply on their experiences from a wide range of disciplines.

Noting research on other complex problems that are multi-sited, multi-scalar and cross sectoral – such as environmental or development studies – adaptation research is perhaps more accurately characterised as a field of research or an area of study (adaptation studies). This better acknowledges the diversity of disciplines, reasons, methods, theories and locations that are possible

Table 2.2 The major Australian institutions publishing research between 2004 and 2013 as returned from a search ('clim* AND adapt*') of the Scopus database.

Rank	Institution	Articles
1	University of Melbourne	191
2	University of Queensland	183
3	Australian National University	172
4	University of Western Australia	143
5	James Cook University	132
6	Monash University	118
7	Commonwealth Scientific and Industrial Research Organisation	113
8	University of Sydney	112
9	University of New South Wales	101
10	University of Tasmania	91

in research on adaptation, as well as the driving imperative to be practical, applied and useful.

This does not mean research in adaptation studies can be judged by anything less than the hallmarks of best practices that apply to any other field of research. Indeed, there is always a danger in characterising a field of research as being somehow unique (e.g. as being uniquely challenging, novel, applied or interdisciplinary) such that lesser standards of research are somehow excused. The characteristics of good adaptation research must be the same as for all research: it must be informed by theories, use rigorous methods to collect and analyse evidence, apply results to inform theories that can explain adaptation and be peer reviewed. Importantly, this means adaptation research should seek evidence and avoid excessive classifications, conceptualisations and normative prescriptions.

2.6 Conclusions

Although this book is exclusively devoted to adaptation studies from Australia, it has universal relevance for all those working to adapt our environment and societies to climate change. Because of the characteristics of the Australian environment and culture, and the links between extreme events, disaster management and adaptation, the experiences, decisions and actions of Australia today have great relevance for the global adaptation community as it seeks solutions to the challenge of climate change.

We do not subscribe to the philosophy that adaptation must be purposeful and only occurs once climate change impacts are experienced. As many chapters of this book demonstrate, adaptation abounds, reinforcing our argument that many decisions are now made in the context of knowledge of climate change whether or not the decision-maker is convinced of its reality. In this respect, abstract and ever-expanding classification of adaptation into categories such as anticipatory, autonomous, community-based, ecosystem-based, evolutionary, incremental and so on does not explain much about adaptation practices nor, we suggest, does it help decision-makers. It is perhaps more helpful, and indeed closer to reality, to consider adaptation as a continuum rather than as a series of procedurally or analytically distinct phases or types.

Given the growth of adaptation research as reflected in the literature, and the great diversity of topics and approaches under the general heading of 'adaptation', it seems inaccurate to categorise adaptation as a science; rather, it is a diverse field of study, drawing together knowledge and expertise from many disciplines in order to formulate and deliver solutions to climate change impacts. The concept of 'adaptation studies' seems more appropriate and, to this end, we hope that this book is a useful contribution to this exciting science–policy challenge.

Acknowledgements

We thank Steve Dovers (Australian National University) for his insightful comments, which have resulted in a much improved chapter.

References

Adger, W.N., Agrawala, S., Mirza, M.M.Q. et al. (2007) Assessment of adaptation practices, options, constraints and capacity. In: Parry, M.L., Canziani, O.F.,

Palutikof, J.P. et al. (eds) *Climate Change 2007: Impacts, Adaptation and Vulnerability. Contribution of Working Group II to the Fourth Assessment of the Intergovernmental Panel on Climate Change*, 1st edition. Cambridge University Press, Cambridge, pp. 717–743.

Australian Government (2011) *Natural Disaster Resilience Framework*. Commonwealth of Australia, Canberra. Available at http://www.em.gov.au/Publications/Program%20publications/Pages/NationalDisasterResilienceFramework.aspx (accessed 11 February 2014).

Botterill, L.C. (2013) Are policy entrepreneurs really decisive in achieving policy change? Drought policy in the USA and Australia. *Australian Journal of Politics and History* 59, 97–112.

Botterill, L.C. and Hayes, M.J. (2012) Drought triggers and declarations: science and policy considerations for drought risk management. *Natural Hazards* 64, 139–151.

Connell, D. (2007) *Water Politics in the Murray–Darling Basin*. The Federation Press, Sydney.

COAG (2011) *National Strategy for Disaster Resilience: Building the Resilience of our Nation to Disasters*. Council of Australian Governments, Commonwealth of Australia, Canberra.

DIICCSRTE (2013) *Australia's Sixth National Communication on Climate Change: A Report Under the United Nations Framework Convention on Climate Change*. Department of Industry, Innovation, Climate Change, Science, Research and Tertiary Education. Commonwealth of Australia, Canberra.

Elsworth, G., Gilbert, J., Rhodes, A. and Goodman, H. (2009) Community safety programs for bushfire: what do they achieve, and how? *Australian Journal of Emergency Management* 24, 17–25.

Ford, J.D., Berrang-Ford, L. and Paterson, J. (2011) A systematic review of observed climate change adaptation in developed nations. *Climatic Change* 106, 327–336.

Funtowicz, S. and Ravetz, J.R. (1993) Science for the post-normal age. *Futures* 25, 735–755.

Gabriel, P. (2003) The development of municipal emergency management planning in Victoria, Australia. *Australian Journal of Emergency Management* 18, 74–80.

Head, L. Adams, M., McGregor, H.V. and Toole, S. (2014) Climate change and Australia. *Wiley Interdisciplinary Reviews: Climate Change* 5, 175–197.

Howes, M., Grant-Smith, D., Reis, K. et al. (2013) *Rethinking Disaster Risk Management and Climate Change Adaptation*. National Climate Change Adaptation Research Facility, Gold Coast.

Indian Ocean Climate Initiative (2012) *Western Australia's Weather and Climate: A Synthesis of Indian Ocean Climate Initiative Stage 3 Research*. CSIRO and Bureau of Meteorology, Commonwealth of Australia, Canberra.

IPCC (2007) Summary for policymakers. In: Parry, M.L., Canziani, O.F., Palutikof, J.P. et al. (eds) *Climate Change 2007: Impacts, Adaptation and Vulnerability. Contribution of Working Group II to the Fourth Assessment of the Intergovernmental Panel on Climate Change*, 1st edition. Cambridge University Press, Cambridge, pp. 7–22.

IPCC (2014) Glossary. In: Field, C.B., Barros, V., Mastrandrea, M. et al. (eds) *Climate Change 2014: Impacts, Adaptation and Vulnerability. Contribution of Working Group II to the Fifth Assessment of the Intergovernmental Panel on Climate Change*, 1st edition. Cambridge University Press, Cambridge, forthcoming.

Karoly, D.J. and Boulter, S. (2013) Afterword: floods, storms, fires and pestilence – disaster risk in Australia during 2010–2011. In: Boulter, S., Palutikof, J., Karoly, D.J. and Guitart, D. (eds) *Natural Disasters and Adaptation to Climate Change*. Cambridge University Press, Cambridge, pp. 252–261.

Kiem, A.S., Askew, L.E., Sherval, M. et al. (2010) *Drought and the Future of Rural Communities: Drought Impacts and Adaptation in Regional Victoria, Australia*. National Climate Change Adaptation Research Facility, Gold Coast.

Langton, M. (1998) *Burning Questions: Emerging Environmental Issues for Indigenous Peoples in Northern Australia*. Centre for Indigenous Natural and Cultural Resources Management, Northern Territory University, Darwin.

Leblanc, M., Tweed, S., Van Dijk, A. and Timbal, B. (2012) A review of historic and future hydrological changes in the Murray-Darling Basin. *Global and Planetary Change* 80–81, 226–246.

Leviston, Z. and Walker, I. (2012) Beliefs and denials about climate change: An Australian perspective. *Ecopsychology* 4, 277–285.

Love, G. (2005) Impacts of climate variability on regional Australia. In: Nelson, R. and Love, G. (eds) *Outlook 2005 conference proceedings, climate session papers*. Australian Bureau of Agricultural and Resource Economics, Canberra. Available at http://data.daff.gov.au/brs/data/warehouse/pe_abarebrs99001170/pc13021.pdf (accessed 23 May 2014).

Mannakkara, S. and Wilkinson, S. (2012) Build back better principles for economic recovery: case study of the Victorian bushfires. *Journal of Business Continuity and Emergency Planning* 6, 164–173.

Mason, M. and Haynes, K. (2010) *Adaptation Lessons from Cyclone Tracy*. National Climate Change Adaptation Research Facility, Gold Coast.

McLennan, B. and Handmer, J. (2014) *Sharing Responsibility in Australian Disaster Management. Final Report for the Sharing Responsibility Project*. RMIT University, Melbourne.

Mimura, N., Pulwarty, R.S., Duc, D.M. et al. (2014) Adaptation Planning and Implementation. In: Field, C.B., Barros, V., Mastrandrea, M. et al. (eds) *Climate Change 2014: Impacts, Adaptation and Vulnerability. Contribution of Working Group II to the Fifth Assessment of the Intergovernmental Panel on Climate Change*, 1st edition. Cambridge University Press, Cambridge, forthcoming.

Min, S.-K., Cai, W. and Whetton, P. (2013) Influence of climate variability on seasonal extremes over Australia. *Journal of Geophysical Research: Atmospheres* 118, 643–654.

Mitrovica, J.X., Gomez, N. and Clark, P.U. (2009) The sea-level fingerprint of West Antarctic collapse. *Science* 323, 753.

Moss, R.H., Meehl, G.A., Lemos, M.C. et al. (2013) Hell and high water: Practice-relevant adaptation science. *Science* 342, 696–698.

Nelson, D., Adger, N. and Brown, K. (2007) Adaptation to environmental change: Contributions of a resilience framework. *Annual Review of Environment and Resources* 32, 395–419.

Nicholls, N., Drosdowsky, W. and Lavery, B. (1997) Australian rainfall variability and change. *Weather* 52, 66–72.

Nidumolu, U.B., Hayman, P.T., Howden, S.M. and Alexander, B.M. (2012) Re-evaluating the margin of the South Australian grain belt in a changing climate. *Climate Research* 51, 249–260.

Norris, R. (1981) *Deakin, Alfred (1856–1919)*. Australian Dictionary of Biography, National Centre of Biography, Australian National University, Canberra.

Available at http://adb.anu.edu.au/biography/deakin-alfred-5927/text10099 (accessed 23 May 2014).

Patt, A. (2013) Should adaptation be a distinct field of science? *Climate and Development* 5, 187–188.

Peterson, T.C., Hoerling, M.P., Stott P.A. and Herring, S. (eds.) (2013) Explaining extreme events of 2012 from a climate perspective. *Special Supplement to the Bulletin of the American Meteorological Society* 94 (9), S1–S74.

Quiggin, J. (2001) Environmental economics and the Murray-Darling river system. *Australian Journal of Agricultural and Resource Economics* 45, 67–94.

Regional Australia Institute (2013). *From Recovery to Renewal: Case Study Reports*. Regional Australia Institute, Canberra. Available at http://www.regionalaustralia.org.au/wp-content/uploads/2013/06/RAI-Natural-Disasters-Report-Case-Studies.pdf (accessed 23 May 2014).

Reser, J.P., Bradley, G.L., Glendon, A.I., Ellul, M.C. and Callaghan, R. (2012) *Public Risk Perceptions, Understandings, and Responses to Climate Change and Natural Disasters in Australia, 2010 and 2011*. National Climate Change Adaptation Research Facility, Gold Coast.

Schilderman, T. and Lyons, M. (2011) Resilient dwellings or resilient people? Towards people-centred reconstruction. *Environmental Hazards: Human and Policy Dimensions* 10, 218–231.

Stafford Smith, M., Horrocks, L., Harvey, A. and Hamilton, C. (2011) Rethinking adaptation for a 4°C world. *Philosophical Transactions of the Royal Society A* 369, 196–216.

Verdon-Kidd, D. C. and Kiem, A. S. (2009) Nature and causes of protracted droughts in southeast Australia: Comparison between the Federation, WWII, and Big Dry droughts. *Geophysical Research Letters* 36, L22707.

Wrigley, C.W. (1981) *Farrer, William James (1845–1906)*. Australian Dictionary of Biography, National Centre of Biography, Australian National University, Canberra. Available at http://adb.anu.edu.au/biography/farrer-william-james-6145/text10549 (accessed 23 May 2014).

Section 1
Frameworks for enabling adaptation

3   Thoughts on the context of adaptation to climate change

GARY YOHE[*]

Department of Economics, Wesleyan University, USA

Working Group II of the Intergovernmental Panel on Climate Change (IPCC 2007a) focused their attention on adaptation and vulnerability in their contribution to the IPCC's Fourth Assessment Report. It made the case that contemplating adaptation to climate change should no longer be dismissed as evidence that society is giving up on trying to ameliorate the problem at its source (by reducing emissions of heat-trapping gases of all sorts). Rather, the Working Group II report argued that adaptation must be included as an essential part of society's portfolio of responses to growing risks arising from climate change. Reports on adaptation to climate change released by the National Research Council of the United States (NRC 2010a) under the rubric of 'America's Climate Choices' adopted and reinforced this conclusion by, for example, recognising the evolving adaptation strategies of governments at all levels around the world. This is also true for the New York (City) Panel on Climate Change (NPCC 2010a, b) and the US National Climate Assessment. The latter contributed directly to President Obama's Climate Action Plan (White House 2013) by speaking of the necessity, if not the means, of increasing 'preparedness'.

[*]This essay relies heavily on Yohe (2010) and Yohe (2013), in thought as well as exposition.

Indeed, in language that was *unanimously* approved by all of the nations who have signed the United Nations Framework on Climate Change (word by word), the nations of the world closed their 'Summary for policymakers' for the Fourth Assessment Report Synthesis document by emphasising the necessity that decision-makers across the globe consider the concept of risk as their primary perspective in their international and national deliberations on responses to climate change. To be specific, they agreed that: 'Responding to climate change involves an *iterative risk management process that includes both adaptation and mitigation* and takes into account climate change damages, co-benefits, sustainability, equity and attitudes to risk' (IPCC 2007b, p. 22, emphasis added).

To be clear, national governments throughout the world have, by accepting this language, clearly stated their fundamental understanding of the urgency of responding with adaptation as well as mitigation *and* that managing risks associated with climate change must be the central theme in present and future planning and policy decisions concerning both. Moreover, they have identified critical criteria upon which they will weigh their options and they have recognised that 'mid-course corrections' must be anticipated as part of the process.

Societies notice many of the impacts of climate change by detecting increasingly intense and/or more frequent extreme weather events and attributing the observed change in weather to climate change. Long a part of the Reasons for Concern (beginning with IPCC 2001) under the title 'Risk of extreme weather events', modern analysis has carefully begun to assess relative confidence in statements of detection and attribution across extreme events by assessing evidence and agreement in the published literature (see Mastrandrea et al. 2010). These assessments are of course the foundation for using observed changes to support projections of further change over the next decades and centuries. Of particular importance here are extreme events such as heavier precipitation events (snow in the winter and rain in the summer), more intense coastal storms (at least with respect to their manifestation when they come ashore, impacts that are driven by observed and projected sea-level rise for all types of storms), and severe droughts, floods, wildfires and heat waves (with appropriate recognition of confounding factors, but also exposure of human and natural systems).

In these events, direct attribution to anthropogenic sources of climate change is difficult. The preponderance of evidence continues to lead IPCC and other assessments to focus on changes that can however, to some degree, be attributed to human activity. The magnitude of these changes will very likely be exacerbated over the near and more distant future as natural climate variability (through extreme events) is distributed around the increasingly worrisome central tendencies of climate change--especially since observed temperature increases driven by higher greenhouse-gas concentrations reflect only 50% of the corresponding equilibrium warming (Solomon et al. 2009). It follows that near-term decisions to mitigate climate change modestly (or not at all) may actually commit the planet to sudden, irreversible changes by the end of the century (Solomon et al. 2009; NRC 2010b).

Urgency in that regard is amplified by the emerging understanding that long-run equilibrium temperature is determined by the *maximum* of atmospheric concentrations (of greenhouse gases such as carbon dioxide calibrated in terms of carbon-dioxide equivalents; Solomon et al. 2009). Does this mean that converging to a lower concentration limit buys us very little? Probably, but to be clear the question raised here is 'Why should the planet waste resources to lower concentrations from an observed maximum if equilibrium temperature and therefore damages cannot be lowered significantly for thousands of years?' The answer is that even with a low discount rate, doing so would be a bad investment because temperature and associated damages will have been determined by higher concentrations. Investments designed to converge to a lower concentration target from above would produce only a few benefits that would likely be dwarfed by the mitigation costs of doing so.

Given this evidence, it is safe to say that climate *is* changing (the old normal is broken even if the new normal has not yet been established). In the absence of significant reductions in emissions of greenhouse gases designed to stabilise concentrations at some as-yet-undetermined (but higher than current) level, the climate will continue to change at an accelerating pace over the short run and into the longer run with growing, if uncertain, consequences. The manifestations of this change will therefore demand that more attention be paid to adaptation as part of plans to promote sustainable development, but without giving up on mitigation.

In interpreting this last point, it is essential to emphasise the fundamental linkages between adaptation (specifically with respect to climate change) and sustainable development more broadly defined (which includes responding to many other sources of societal stress). This point was made explicitly in chapter 20 of IPCC (2007a) where authors noted that then-recent work had confirmed the chapter 18 IPCC (2001) conclusion that any system's vulnerability to climate change, climate variability and/or any other external stress is the product of exposure and sensitivity to that stress (or to multiple sources of multiple stress, for that matter). Nothing has really changed since. It is still the case that exposure and sensitivity can be influenced positively or negatively by individual or societal responses to climate change and/or other stresses. Reducing exposure and sensitivity as well as building capacity to adapt, in combination with reductions in greenhouse gases, remain among the essential elements of responses to manage climate risks.

References

IPCC (2001) *Climate Change 2001: Impacts, Adaptation and Vulnerability. Contribution of Working Group II to the Third Assessment Report of the Intergovernmental Panel on Climate Change.* McCarthy, J.J., Canziani, O.F., Leary, N.A., Dokken, D.J. and White, K.S. (eds) Cambridge University Press, Cambridge.

IPCC (2007a) *Climate Change 2007: Impacts, Adaptation and Vulnerability. Contribution of Working Group II to the Fourth Assessment Report of the Intergovernmental Panel on Climate Change.* Parry, M.L., Canziani, O.F., Palutikof, J.P., van der Linden, P.J. and Hanson, C.E. (eds) Cambridge University Press, Cambridge.

IPCC (2007b) *Climate Change 2007: Synthesis Report. Contribution of Working Groups I, II and III to the Fourth Assessment Report of the Intergovernmental Panel on Climate Change.* Pachauri, R.K. and Reisinger, A. (eds) Cambridge University Press, Cambridge.

Mastrandrea, M., Field, C., Stocker, T. et al. (2010) *Guidance Notes for Lead Authors of the IPCC Fifth Assessment Report on Consistent Treatment of Uncertainties.* Intergovernmental Panel on Climate Change, Geneva.

National Research Council (NRC) of the United States (2010a) *Adapting to the Impacts of Climate Change.* National Academies Press, Washington, DC.

National Research Council (NRC) of the United States (2010b) *Climate Stabilization Targets: Emissions, Concentrations, and Impacts of Decades to Millennia. Prepublication.* National Academies Press, Washington, DC.

New York Panel on Climate Change (NPCC) (2010a) Climate change adaptation in New York City: Building a risk-management response. *Annals of the New York Academy of Sciences* 1196.

New York Panel on Climate Change (NPCC) (2010b) *Adaptation Assessment Guidebook. NPCC Workbook.* New York Academy of Sciences, New York.

Solomon, S., Plattner, G-K., Knutti, R. and Friedlingstein, P. (2009) Irreversible climate change due to carbon dioxide emissions. *Proceedings of the National Academies of Science* 106(6), 1704–1709.

White House (2013) *Climate Action Plan.* Executive Office of the President, Washington.

Yohe, G. (2010) Risk assessment and risk management for infrastructure planning and investment. *The Bridge* 40(3), 14–21.

Yohe, G. (2013) Climate change adaptation: a risk-management approach. In: Atkinson, G., Dietz, S., Neumayer, E. and M. Agarwala (eds) *Handbook of Sustainable Development*, 2nd edition. Edward Elgar Publishing, Cheltenham, UK.

4 Reflections on disaster loss trends, global climate change and insurance

JOHN MCANENEY[1], RYAN CROMPTON[1],
RADE MUSULIN[2], GEORGE WALKER[2],
DELPHINE MCANENEY[1] AND ROGER PIELKE JR[3]

[1]*Risk Frontiers, Macquarie University, Australia*
[2]*Aon Benfield Analytics Asia Pacific, Australia*
[3]*Center for Science & Technology Policy Research, University of Colorado, USA*

4.1 Introduction

This paper summarises salient findings from the NCCARF-funded report entitled 'Market-based mechanisms for climate change adaptation' by McAneney et al. (2013). It first examines the mechanisms responsible for the increasing cost of natural disasters, the implications of this for climate change adaptation and briefly explores the capacity of insurance markets to incentivise such adaptation. For commentary on catastrophe bonds and capital market funding of disaster losses (Froot 2001), the reader is referred to the original report (McAneney et al. 2013).

The increase in natural disaster losses has led to concerns that anthropogenic climate change is contributing to this trend. Perils of concern are those likely to cause damage to property assets, in particular tropical cyclones, storms including hailstorms, floods and bush (wild) fires. This paper summarises recent Australian scholarship on this topic as well as efforts to estimate the timescale at which an anthropogenic climate change signal might be detectable in US hurricane loss data. US hurricanes warrant special attention because: (1) of their impact on the global insurance market through the supply of and demand for reinsurance; (2) the availability of a long-term normalised economic loss history from land-falling hurricanes; and (3) high-quality modelling of the impact of anthropogenic warming on basin-wide hurricane activity.

This paper also addresses the potential for insurance to be a positive actor in helping to reduce the risk to property by extreme weather that may be influenced by future climate change. This is not a responsibility that the private sector can shoulder on its own however, so we review government involvement in the provision of natural catastrophe insurance. We conclude with some brief observations about the challenge posed by the rising toll of natural disasters and whether the insurance industry can play a role in incentivising risk reduction.

Applied Studies in Climate Adaptation, First Edition. Edited by Jean P. Palutikof, Sarah L. Boulter, Jon Barnett and David Rissik.
© 2015 John Wiley & Sons, Ltd. Published 2015 by John Wiley & Sons, Ltd.

4.2 Property losses and natural disasters due to extreme weather

Before apples-with-apples comparisons can be made between the impacts of past and more recent natural-hazard loss events, changes in various societal factors known to influence the magnitude of losses must first be accounted for. Climate-related influences stem from changes in the frequency and/or intensity of natural perils, whereas socio-economic factors comprise changes in the vulnerability and exposure to the natural hazard. Adjustment for the latter (non-climate-related) factors has become known as loss normalisation (Pielke and Landsea 1998; Pielke et al. 2008). Loss normalisation attempts to answer the question: what would be the losses if historic events were to recur under current societal conditions? In what follows we describe recent Australian responses to this question.

Crompton and McAneney (2008) normalised Australian weather-related insured losses over the period 1967–2006 to 2006 values. Loss data obtained from the Insurance Council of Australia (ICA; http://www.insurancecouncil.com.au/) were adjusted for changes in dwelling numbers and nominal dwelling values (excluding land value) since the time of the original event. In a marked departure from previous normalisation studies, an additional adjustment was applied to tropical cyclone losses to account for improvements in construction standards mandated for new buildings in tropical cyclone-prone parts of the country (McAneney et al. 2007; Mason et al. 2013). The success of improved building standards in reducing losses has been demonstrated repeatedly in more recent events including tropical cyclones Larry (in 2006) and Yasi (in 2011).

Figure 4.1a and b shows the annual aggregate losses and the annual aggregate normalised losses (2011/12 values) for weather-related events in the ICA's Disaster List. These figures are updated from those of Crompton and McAneney (2008) using a more refined methodology described in Crompton (2011). Importantly, no trend is evident in the normalised losses (Fig. 4.1b), implying that socio-economic factors alone are sufficient to explain the increase in the cost of insurance sector losses (Fig. 4.1a). In other words, it is not possible to detect the role of anthropogenic climate change once losses are normalised. We note that despite record high 2012/13 summer air temperatures across Australia, industry losses for the financial year (1 July 2012 to 30 June 2013) were very close to the long-term average normalised loss of AU\$ 1.1 billion.

Readers might imagine bushfire losses to be more sensitive to increasing air temperatures than some other perils. Following the large loss of life and building damage experienced in the 2009 bushfire in Victoria, Crompton et al. (2010) examined the history of fatalities and building damage since 1925. Figure 4.2a and b shows the actual and normalised building damage expressed in numbers of residential homes destroyed. Once building damage is adjusted for increases in dwelling numbers, no residual trend was found that might be attributed to anthropogenic climate change or other factors, for that matter. A similar result was found for fatalities.

Bouwer (2011) provides a comprehensive review of loss normalisation studies which concur that, despite widespread assertions to the contrary, it is not yet possible to detect or attribute an anthropogenic influence on disaster losses. McAneney et al. (2013) updated the Bouwer (2011) review and noted that the Special Report of the Intergovernmental Panel on Climate Change came to the same conclusion (IPCC 2012) on the lack of attribution.

After also examining these issues, Barthel and Neumayer (2012, p. 229) concluded that: 'Climate change neither is nor should be the main concern for the insurance industry. Accumulation of wealth in disaster prone areas is and will always remain by far the most important driver of future economic disaster damage.'

We agree that climate change is not likely to be a material threat for the insurance industry, but in Section 4.5 we turn the question around and ask whether or not insurance has a role in encouraging climate change adaptation.

(a)

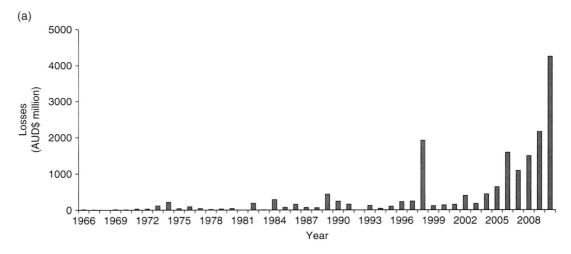

(b)

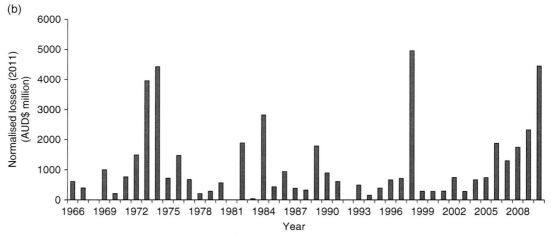

Figure 4.1 (a) Annual aggregate insured losses (AU$ million) for weather-related events in the ICA Disaster List for years beginning 1 July; (b) as in (a) but with losses normalised to 2011/12 values. Source: Crompton 2011. Reproduced with permission of Risk Frontiers.

4.3   Timescale at which an anthropogenic climate change signal might be observed in US tropical cyclone losses

Crompton and colleagues posed the following question in respect to US hurricanes: 'If changes in storm characteristics happen as projected in a warming climate, then on what time frame – the *emergence timescale* – might we expect to detect

the effects of those changes in economic loss data with some scientific certainty, say with 95% confidence?' (Crompton et al. 2011, p. 2).

The point of departure for Crompton et al. (2011) was projections of future Atlantic basin hurricane activity from the NOAA Geophysical Fluid Dynamics Laboratory (Bender et al. 2010). Combining these with the Pielke et al. (2008) normalised loss history, a bootstrapping analysis

(a)

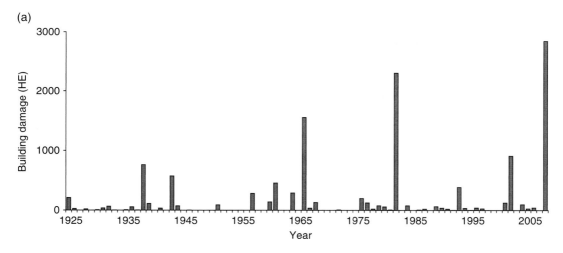

(b)

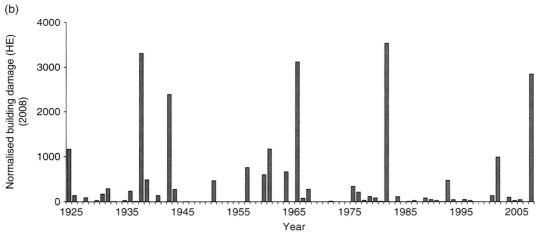

Figure 4.2 (a) Annual aggregate building damage expressed as the number of House Equivalents (HE) destroyed due to bushfire events in Australia for year beginning 1 July. (b) As in (a) but with HE normalised to 2008/09 values. The time series begins in 1925, the first year in the 20th century to experience large numbers of building destruction due to bushfires. HE refers to all building damage but in this case can be interpreted as numbers of residential dwellings. Source: Crompton et al. 2010. © American Meteorological Society. Used with permission.

showed anthropogenic signals to emerge at timescales of between 120 and 550 years. This wide range derives from the different Global Climate Models (GCMs) that set the boundary conditions for the downscaling. It took 260 years for an 18-model ensemble-based signal to emerge!

Emanuel (2011) implemented an alternative methodology to Crompton et al. (2011), also finding the time to detection to be long: in excess of a century for three of four models analysed, with one indistinguishable from background noise even after 200 years. These long time scales reflect the challenging signal-to-noise problem facing climate change attribution in disaster loss time series. It also argues strongly against using abnormally large losses from individual Atlantic hurricanes or seasons as evidence of anthropogenic climate change.

4.4   Government provision of catastrophe insurance

Government involvement in the provision of catastrophe insurance has usually arisen in the face of perceived market failures of the private market, often following a significant natural disaster. Such residual market mechanisms or pools have assumed the legacy of inappropriate land use, unrealistic risk assessment and lack of consideration to risk reduction. The issue has high currency in Australia after large economic losses caused by flooding in Queensland and Victoria in 2011 and widespread criticism of insurers whose policy covers at the time often excluded damage due to riverine flood (van den Honert and McAneney 2011).

In their examination of various government pools, McAneney et al. (2013) identify some key distinctions between private and public insurance schemes as follows.

1. *Government* insurance systems can raise funds post-event by issuing government bonds, levies on policies or new taxes, for example.

2. *Private* insurance systems must prefund all losses; it is not acceptable to have a loss and then try to collect funds to pay for it after an event (American Academy of Actuaries 2012).

3. *Private* insurance systems usually attract taxes on profits, which can mean that earnings on funds needed to pay claims from infrequent events are taxed away because they show up as income in years without extreme events. *Government* insurance systems are not bound by this constraint.

4. *Private* insurance systems operating in a competitive market increasingly set prices related to risk.

5. *Governments* can use their sovereign power to compel one group of consumers to pay too much in order to provide a subsidy to another.

6. Both *private* and *government* insurance systems can in principal encourage mitigation through premium discounts and risk-informed underwriting. However, *government* insurance systems often dilute the incentives for mitigation by subsidising high risks from low risks or by raising revenue for losses from an unrelated source, like a tax levy.

7. With financial back-up or guarantees from the state, government pools can fall back on resources not available to the private sector. This was the case for the Earthquake Commission in New Zealand after the 2010 and 2011 Christchurch earthquakes and more generally in situations where losses become so large that the private sector cannot efficiently handle them without extreme disruption (such as would be the case for a possible $100 billion US hurricane in Florida). In this case the 'actuarially sound' premium and the affordable premium diverge to the point where in effect everyone goes uninsured, which may be an undesirable outcome.

Just like their private sector counterparts, public sector insurance schemes differ around the world. In Spain, the provision of catastrophe insurance is socialised with a standard levy applied to all insurance products. In New Zealand, the Earthquake Commission covers the first NZ$ 100 000 of loss to residential property insured with the private sector and with a single premium. In the UK a new arrangement to deal with flood risk is being negotiated in which a not-for-profit organisation, Flood Re, is expected to assume responsibility for homes at risk to events more frequent than a 1-in-200 year flood. Insurance companies in France are required to cover most hazards as part of standard policies, with the government offering subsidised reinsurance as an alternative to the commercial reinsurance market. In none of the schemes examined were premiums risk-rated. In the Netherlands, which faces flooding as an existential threat, no government insurance exists to cover the threat for residential homeowners; the focus has been on national flood defences.

Government pools usually contain an inherent contradiction in trying to provide reduced (or subsidised) cost insurance to high-risk properties, and so the funding of deficits to which they are inevitably prone is important (McAneney et al. 2013). As stated above, pools can in principle minimise deficits over time by encouraging risk mitigation but, as for Kunreuther (1996, 2006), we found little evidence of this. The National Flood Insurance Program (NFIP) and Texas Windstorm Insurance Association (TWIA) were exceptions to the rule. A positive outcome of the NFIP is the high percentage of local authorities imposing floodplain management schemes based on the 100-year return period flood height; however, Burby (2001) questions the extent to which this has inhibited construction activity in flood-hazard areas or had much impact on federal disaster relief costs. Because the NFIP is voluntary, it also suffers from adverse selection with only those at high risk likely to buy cover. The TWIA has had a big effect on building standards, particularly for houses and other low-rise buildings.

An ongoing contentious issue in the US has been the degree of political influence exerted to keep premiums low and to have policyholders in low- and high-risk areas charged similar rates. In the absence of adequate regulation, this lack of financial incentives for mitigation encourages development in high-risk areas (Jaffe and Russell 2013). Calls for reform in the US may bring about some positive changes in the residual market mechanisms, but there is debate as to the place of a state-run entity in the insurance market (Jaffe and Russell 2006).

4.5　Can insurers promote climate change adaptation?

The primary goal of the insurance sector is the assessment and transfer of risk, *as it is currently understood*. Insurance policies generally have a duration of a single year, a period at odds with the lifespan of a building (generally 50–100 years) and the timescale at which climate change amplification of extreme weather might become measurable in loss data. This mismatch makes it difficult for insurers to materially influence

adaptation to *future* climate change except through the rigorous pricing of the *existing* risk. A move to risk-rated premiums could have socially desirable outcomes if it were to encourage changes by government and other actors to invest in mitigation infrastructure, better building codes and risk informed land-use planning practices, and so over time act to quell the increase in disaster losses.

When we look to ways to address the increasing trend in losses it is impossible to overlook the decisive role that poor land-use planning has played in recent disasters. We note just two examples of loss of life and buildings from Australia: the 2009 Victorian bushfires and the 2010/11 Queensland floods. In the first case, work undertaken for the 2009 Royal Commission (Chen and McAneney 2010; Crompton et al. 2010) showed that a large proportion of buildings destroyed either lay within bushland or at a very small distance from it. In fact, 25% of destroyed homes were situated within a metre of the bush and so were effectively part of the fuel load.

Similar observations pertain to the 2010/11 Queensland floods. With economic losses of some AU$ 6 billion, the flooding led to the introduction of a temporary reconstruction tax. Lost in the political debate was how similar the flooding footprint in Brisbane was to that of the 1974 floods (Fig. 4.3), and no doubt those of much bigger floods recorded in the 1800s. The area is now much more heavily developed than was the case in 1974, with the Brisbane City Council approving 1811 additional development applications since just 2005 (K. Doss, City Planning and Economic Development, pers. comm. 2011). Both of these examples point to clear failures of land-use planning practices.

4.6　So what can we do?

While it is difficult to influence the likelihood of extreme weather events, that is, the hazard, we have control over where and how we build. The success of regulated improvements to residential construction in tropical cyclone-prone parts of

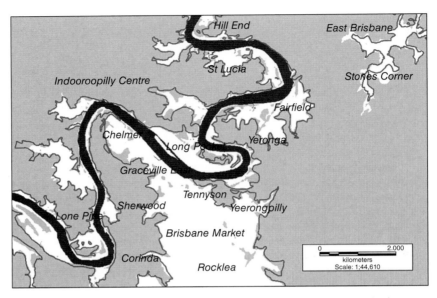

Figure 4.3 The red outline shows the extent of flooding in the 1974 Brisbane event while the area in white is our best representation of the footprint of the 2011 flood. Source: van den Honert and McAneney 2011. For colour details please see Plate 3.

Australia shows what can be achieved with political will (McAneney et al. 2007; Mason et al. 2013). The adoption of risk-rated insurance premiums would also send a transparent risk signal to all parties; if this were successful in changing behaviour, then over time the population at risk to natural disasters should decrease (or at least not continue to rise in concert with increasing population and wealth). Any gains achieved here will put us in good stead for additional changes that a warming climate may eventually throw at us. Lastly, we add that these are complex issues with no easy answers and call for a strong dose of pragmatism by decision-makers.

References

American Academy of Actuaries, Actuarial Soundness Taskforce (2012) *Actuarial Soundness. A Public Policy Special Report.* American Academy of Actuaries, Washington, DC.

Barthel, F. and Neumayer, E. (2012) A trend analysis of normalized insured damage from natural disasters. *Climatic Change* 113, 215–237.

Bender, M.A., Knutson, T.R., Tuleya, R.E. et al. (2010) Modeled impact of anthropogenic warming on the frequency of intense Atlantic hurricanes. *Science* 327, 454–458.

Bouwer, L.M. (2011) Have disaster losses increased due to anthropogenic climate change? *Bulletin of the American Meteorological Society* 92, 39–46.

Burby, R.J. (2001) Flood insurance and floodplain management: the US experience. *Environmental Hazards* 3, 111–112.

Chen, K. and McAneney, J. (2010) *Bushfire Penetration into Urban Areas in Australia: A Spatial Analysis. Invited report prepared for 2009 Victorian Bushfire Royal Commission.* Risk Frontiers, Sydney.

Crompton, R.P. (2011) Normalising the Insurance Council of Australia Natural Disaster Event List: 1967–2011. Report prepared for the Insurance Council of Australia, Risk Frontiers, Sydney.

Crompton, R.P. and McAneney, K.J. (2008) Normalised Australian insured losses from meteorological hazards: 1967–2006. *Environmental Science Policy* 11, 371–378.

Crompton, R.P., McAneney, K.J. Chen, K., Pielke Jr, R.A. and Haynes, K. (2010) Influence of location, population, and climate on building damage and fatalities due to Australian bushfire: 1925-2009. *Weather, Climate and Society* 2, 300–310.

Crompton, R.P., Pielke Jr, R.A. and McAneney, K.J. (2011) Emergence timescales for detection of anthropogenic climate change in US tropical cyclone loss data. *Environmental Research Letters* 6, doi:10.1088/1748-9326/6/1/014003.

Emanuel, K. (2011) Global warming effects on U.S. hurricane damage. *Weather, Climate and Society* 3, 261–268.

Froot, K.A. (2001) The market for catastrophe risk: A clinical examination. *Journal of Financial Economics* 60, 529–571.

IPCC (2012) *Managing The Risks of Extreme Events and Disasters to Advance Climate Change Adaptation. A Special Report of Working Groups I and II of the Intergovernmental Panel on Climate Change.* Field, C.B., Barros, V., Stocker, T.F. et al. (eds) Cambridge University Press, Cambridge, UK, and New York, NY, USA.

Jaffe, D.M. and Russell, T. (2006) Should governments provide catastrophe insurance? *The Economists Voice* 2006, 1–8.

Jaffe, D. and Russell, T. (2013) The welfare economics of catastrophe losses and insurance. *The Geneva Papers* 38, 469–494.

Kunreuther, H. (1996) Mitigating disaster losses through insurance. *Journal of Risk and Uncertainty* 12, 171–187.

Kunreuther, H. (2006) Disaster mitigation and insurance: Learning from Katrina. *Annals of American Academy of Political and Social Science* 614, 208–227.

Mason, M., Haynes, K. and Walker, G. (2013) Cyclone Tracy and the road to improved wind resistant design. In: Boulter, S., Palutikof, J., Karoly, D.J. and Guitart, D. (eds) *Natural Disasters and Adaptation to Climate Change.* Cambridge University Press, Cambridge, pp. 87–94.

McAneney, J., Crompton, R. and Coates, L. (2007) *Financial Benefits Arising From Regulated Wind Loading Construction in Tropical Cyclone Prone Areas of Australia. Report Commissioned by the Australian Building Codes Board.* Risk Frontiers, Australia.

McAneney, J., Crompton, R., McAneney, D., Musulin, R., Walker, G. and Pielke Jr, R. (2013) *Market-based mechanisms for climate change adaptation: Assessing the potential for and limits to insurance and market-based mechanisms for encouraging climate change adaptation.* National Climate Change Adaptation Climate Change Facility, Gold Coast.

Pielke Jr., R.A. and Landsea, C.W. (1998) Normalized hurricane damages in the United States: 1925–95. *Weather and Forecasting* 13, 621–631.

Pielke Jr., R.A., Gratz, J., Landsea, C.W., Collins, D., Saunders, M. and Musulin, R. (2008) Normalized hurricane damage in the United States: 1900–2005. *Natural Hazards Review* 9, 29–42.

van den Honert, R. and McAneney, J. (2011) The 2011 Brisbane floods: Causes, impacts and implications. *Water* 3, 1149–1173.

5 Designing spatial adaptation planning instruments

ANDREW MACINTOSH[1], JAN MCDONALD[2]
AND ANITA FOERSTER[3]

[1]*ANU College of Law, Australian National University, Australia*
[2]*Faculty of Law, University of Tasmania, Australia*
[3]*The University of Melbourne, Australia*

5.1   Introduction

Spatial planning refers to a broad collection of methods and processes that aim to influence the spatial distribution of economic, social and environmental activities (European Commission 1997), and encompasses both formal land-use planning and other policy instruments. When well executed, spatial planning can be an indispensable tool for facilitating efficient and equitable adaptation (Burby and Nelson 1991; Productivity Commission 2013). Conversely, poor spatial planning can increase the social costs of climate change and result in particular groups bearing the costs of adaptation disproportionately (Scheraga and Grambsch 1998; Macintosh 2013). The risk of suboptimal outcomes is heightened by the complexities of adaptation decision-making, particularly the high levels of uncertainty and degree of contestation over values, objectives, property rights and governance structures.

Despite widespread acknowledgement of the role of spatial planning in adaptation, the focus of the existing policy science and administration literature is on the implementation of specific programs in particular countries or regions (Few et al. 2007; Macintosh 2013). The policy design aspects of spatial adaptation planning, concerning instrument choice, how instruments perform and interact under different circumstances and why and how instruments are selected (Linder and Peters 1984, 1989; Salamon 2001), have either been ignored or covered at the margins in implementation studies.

This chapter presents findings from an empirical evaluation of current spatial adaptation planning practice in Australia concerning climate-change-related coastal and bushfire hazards (Macintosh et al. 2013). The aim of the chapter is to help fill the policy design void in the spatial adaptation planning literature by: presenting a typology of policy instruments for adaptation; and outlining the key recommendations from the implementation analysis on the selection and design of the identified policy instruments.

5.2   Method

The research on spatial adaptation planning practice was conducted in four stages as follows.
1. A literature review on social and institutional barriers to, and the role of spatial planning in, climate change adaptation.
2. An analysis of existing legal and policy frameworks for planning and risk management in

Applied Studies in Climate Adaptation, First Edition. Edited by Jean P. Palutikof, Sarah L. Boulter, Jon Barnett and David Rissik.
© 2015 John Wiley & Sons, Ltd. Published 2015 by John Wiley & Sons, Ltd.

relation to coastal hazards and bushfire in all Australian jurisdictions.

3. The conduct of 50 semi-structured interviews across 13 state planning, emergency services and environmental protection agencies, 15 local governments (local government planners and elected officials) and 6 professional bodies to obtain information on the informal institutions and political dynamics surrounding the operation and application of relevant land-use planning and emergency service regulations.

4. A symposium involving 25 representatives from local, state and commonwealth planning agencies and representative bodies held in Melbourne in October 2012 to test the empirical findings and elicit further information from leading practitioners on the use of spatial planning tools to address adaptation challenges.

5.3   Typology of spatial planning instruments

Drawing on the literature review and empirical findings, we devised a seven-category instrument typology which is summarised in Table 5.1.

5.4   Institutional context for spatial adaptation planning

Instrument selection and design is strongly influenced by the institutional context in which policy choices are made. For spatial adaptation planning in Australia, the main institutional issues that are likely to affect instrument choice are:

- instrumental path dependency;
- federalism and the distribution of planning responsibilities; and
- property rights and liberal norms.

5.4.1   *Instrumental path dependency*

Instrument choice can exhibit a degree of path dependency that arises from the institutional context in which decisions are made (Hood 1986; Howlett 2009). Historically, the majority of

spatial planning in Australia has occurred via statutory land-use planning regimes which rely heavily on framing and regulatory instruments. The familiarity and institutional dynamics that arise from these arrangements bias the policy process in favour of these instruments, a fact evident from current Australian adaptation practice. To date, spatial adaptation planning responses have relied mostly on framing, fixed regulatory and broad information instruments.

5.4.2   *Federalism and the distribution of responsibilities*

Under Australia's federal system, there are three formal levels of government: the Commonwealth, the states and self-governing territories and local government. As a result of history and constitutional constraints on the powers of the Commonwealth, land-use planning is largely the formal responsibility of the states and territories, who then delegate many of the relevant functions to local government. The Commonwealth plays only a minor role.

Local governments are the 'front-line' planning service provider in most areas and, in that capacity, perform the majority of day-to-day planning functions. Notwithstanding this prominent role, the planning powers of local government are restricted and largely prescribed in land-use planning and local government statutes. They have limited capacity to impose taxes and charges, cannot create statutory liability shield instruments independently of state governments and their diminutive revenue powers hinder their use of voluntary and compulsory acquisition instruments. Even their use of framing, information and regulatory instruments is constrained by the planning powers of state governments. The discretion of local government can be narrowed further by planning appeal processes and the capacity of appeal tribunals to override their decisions.

State and territory governments face far fewer formal institutional restrictions on instrument choice than local governments. Their primary constraint stems from the Commonwealth's

Table 5.1 Typology of spatial planning instruments.

Instrument	Description	Examples
Framing instruments	Articulate overarching policy goals and objectives and outline how different regulatory and non-regulatory instruments can be used to achieve these objectives.	Objectives, principles and strategy clauses in planning legislation and/or state, regional and local planning policies.
Information instruments	Communicate information on climate hazards to landholders and the community.	Planning certificates; notations on title.
Regulatory instruments	Legally enforceable restrictions that dictate where, what and how land use and development occurs in order to reduce the severity of hazards, the harmful effects of hazards or hazard exposure. There are two subcategories of regulatory instruments: *Fixed*: based on the assumption that once lawfully commenced, an existing land use will be beyond the reach of the planning system and can continue indefinitely unless intensified, expanded or abandoned. *Flexible*: specifically to provide governments with powers to control land use and development, even after it has lawfully commenced.	*Fixed* instruments include: • zones and overlays (i.e. spatially explicit land use and development regulations); • standard permit requirements and approval conditions (e.g. setbacks and buffers); • codes and guidelines. *Flexible* instruments include: • for new development, qualified development or use rights (e.g. event-dependent and time-limited approvals); • for existing development, regulations that modify the rights and freedoms associated with the uses and development.
Compulsory acquisition instruments	Instruments that give governments powers to compulsorily acquire interests in land and other property, with or without compensation.	Property purchase; designation of acquisition land.
Voluntary instruments	Involve the use of positive incentives to control or influence where, what and how land use and development occurs. These instruments do not compel compliance or participation.	Financial inducements to undertake hazard mitigation activities; voluntary buy-back schemes; land swaps.
Taxes and charges	*Taxes* are a compulsory exaction of money by a public authority for public purposes, which is not a payment for services rendered. *Charges* are levies imposed by a public authority to cover the costs of providing particular goods or services.	*Taxes*: • elevated council rates imposed on particular land uses in high-risk areas; • broad-based taxes to raise disaster recovery funds. *Charges*: • coastal protection levies; • bushfire management service levies.
Liability shield instruments	Provide a partial or full exemption from legal liability to specified entities if they take a particular action, or fail to act in a particular way, in relation to climate hazards.	Statutory immunities from liability; developer indemnity agreements.

monopolisation of revenue powers and the distribution of economic activity, which leaves them dependent on inter-government transfers to address the associated vertical and horizontal fiscal imbalances. While facing constraints on their revenue powers, state governments have broad regulatory powers to deal with spatial adaptation issues. Importantly, there is no *constitutional* requirement for state governments to provide compensation if it acquires, extinguishes or otherwise interferes with any interest in property, including through the imposition of far-reaching regulatory requirements (HCA 1997, 1998; 2001).

In contrast to the situation in the states, the Commonwealth has broad revenue powers but its powers to acquire interests in property are constrained by section 51(xxxi) of the Australian Constitution. This provision arms the Commonwealth and, through it the territories, with the power to acquire property. However, it qualifies that power with the requirement that any acquisition must be on 'just terms'. Due to this, neither the Commonwealth nor territory governments can take title to, or assume possession and control of, land without providing compensation (HCA 1997). Far-reaching regulations that prevent a landholder from enjoying any of the normal benefits associated with the relevant interest in land could also trigger the constitutional guarantee of 'just terms' (Macintosh and Cunliffe 2012).

5.4.3 *Property rights and liberal norms*

Many Australian policy processes place considerable emphasis on secure 'property rights'. This reflects the influence of liberal ideology on Australia's institutions, particularly the notions that property is essential to freedom and that secure property interests promote the efficient allocation of resources. In planning systems, this emphasis is evident in the strong protections provided to 'existing use rights', or entitlements to continue to use land for a particular purpose that was lawful prior to the introduction of planning regulations that prohibit or regulate the use. All state and territory planning systems contain provisions that protect these interests. Similarly, while there is no constitutional impediment to states acquiring property without providing compensation, all states have statutes that guarantee the provision of 'just terms' compensation where interests in land are acquired by government agencies. The Tasmanian and

Western Australian planning regimes also provide compensation where land is set aside under planning regulations for a public purpose, and Queensland's planning regime provides compensation to the owners of interests in land where changes in planning provisions adversely affect the value of the land.

As a matter of law, the protections afforded to landholders under state laws are not absolute and can be altered or removed entirely by parliament. Technically, this provides policymakers with broad scope to introduce spatial planning measures. However, the discretion of policymakers is narrowed considerably by community norms regarding property.

5.5   Recommendations for instrument selection and design

Instrument selection and design choices are a product of the institutional context in which they are made and the power and preferences of policy actors. Policy design acknowledges this but seeks to encourage the structured and systematic consideration of alternatives and the consequences of different instrument choices, having regard to the context (Howlett and Lejano 2013). In this vein, we discuss six key recommendations on instrument choice and design for spatial adaptation planning in Australia in the following sections.

5.5.1 *Framing instruments and guidance*

Adaptation policy objectives are highly contested in Australia. For coastal and bushfire hazards there is disagreement about the role of the state, particularly whether governments should second-guess the risk preferences of landholders (intervene to stop people putting themselves in harm's way) and the extent to which governments should provide assistance if risks materialise. There is also disagreement over the freedom of property owners to defend their properties when doing so will impose costs on neighbouring landholders and the broader community, and whether (and to what extent) the state should notify property owners of potential risks.

Consistent with the findings from the policy science literature (Matland 1995), this conflict has resulted in vague objectives in the policy instruments deployed by governments. This has been especially prominent in the approach to coastal climate hazards, where framing instruments have generally not set clear overarching strategic objectives and decision-makers have been provided with minimal implementation guidance through other instruments. While the ambiguity may help minimise political costs, it has had adverse impacts on planning processes, including:

• delayed implementation of coastal climate hazard responses as local governments have waited for more specific state government direction;
• resource wastage due to local governments and other planning bodies undertaking information-gathering and assessment processes that could have been carried out more cost-effectively by state governments;
• inconsistencies in strategic and statutory decision-making within and between agencies;
• high levels of conflict and reliance on appeals processes; and
• increased transaction costs for landholders and developers.

Given the institutional context for spatial adaptation policy, particularly the polycentric governance arrangements that spread roles and responsibilities between multiple parties at a local and state level, there is a need for improved clarity about the objectives and principles that are intended to guide decision-making. This requires amendments to existing framing instruments. Local governments and other state planning agencies also need more detailed guidance on the implementation of regulatory, information and other coercive instruments.

5.5.2   *Using the available instruments*

As for example Linder and Peters (1990) have noted, the extent to which policymakers consider instrument options can profoundly affect outcomes. A prominent feature of current spatial adaptation planning practice in Australia is the heavy reliance on framing instruments, broad information instruments and fixed regulatory instruments. While there may not be a single optimal instrument or suite of instruments, the focus on these 'standard' land-use instruments is unnecessarily limiting the capacity of spatial planning to facilitate efficient and equitable adaptation. Three instruments that have been particularly underutilised are flexible regulatory instruments, targeted information instruments and liability shield instruments.

5.5.3   *Flexible regulatory instruments*

The dominant regulatory approach to climate hazards has centred on fixed regulatory instruments, particularly zones and overlays. Zoning is used to impose outright prohibitions on land use and development, or particular types of land use and development, in hazard-prone areas. Overlays are used to regulate development where the underlying assumption is that development should be allowed in an area providing it meets prescribed standards and conditions. These instruments allow for the targeted application of restrictions on a spatially explicit basis, thereby reducing ambiguity and the transaction costs associated with compliance.

The underlying principle behind fixed regulatory instruments is that, once a land use is lawfully commenced, the power or 'right' of government to stop or control that existing use via the planning system is lost. This is reflected in the rules governing existing uses; they are generally exempt or immune from new planning controls unless they are intensified, expanded or abandoned. The loss of regulatory powers after a use is lawfully commenced involves the transfer of an economic property right (the right to control the use of property) from government to the landholder (Coase 1960). If a government subsequently wants to stop an existing use, or alter the planning conditions that apply to it, it will usually be required to purchase that right back from the landholder by providing compensation.

A key deficiency associated with fixed regulatory instruments is their inflexibility. Once land uses have lawfully commenced, the regulatory powers are expended and they provide planning authorities with few options to shape land use. This is particularly problematic for climate hazards, the distribution, timing and magnitude of which are highly uncertain. Due to this uncertainty, there is a significant risk that fixed regulatory responses will later be judged to be an under- or overreaction.

Flexible regulatory instruments can be a cost-effective alternative to fixed responses because they allow for the use and enjoyment of land until such time as the relevant hazard materialises. This minimises the opportunity cost associated with regulatory responses and allows for the more effective management of the distributional issues associated with climate hazard policies. Despite their benefits, flexible regulatory instruments have rarely been used in Australia. This is partly because they are not well suited to hazards such as bushfire where the threat is acute and highly unpredictable. They are most applicable to threats such as coastal erosion and inundation, where the hazards are likely to develop incrementally and the changes are irreversible. Other reasons for their underutilisation include path dependency in the planning system, technical design challenges (e.g. how utilities provide reticulated services such as sewerage to temporary developments) and concerns about the willingness of future governments to exercise the options embodied in these instruments without providing compensation. Notwithstanding these issues, there remains considerable scope for the use of flexible regulatory instruments. They offer an efficient, equitable, ideologically consistent and legitimate response to the planning challenges associated with coastal climate hazards.

5.5.4 *Targeted information instruments*

Information instruments can be subcategorised on a number of grounds, one of which is scope: broad versus targeted. Broad information instruments seek to convey general information about hazards, mitigation strategies and/or management options (e.g. television advertisements, information brochures and general hazard maps). Targeted information instruments are designed to provide information at a property scale and directly influence decision-making surrounding its purchase or management (e.g. planning certificates provided at the point of sale, or notations on title).

Australian governments have expressed strong rhetorical support for the use of information instruments to encourage autonomous adaptation, possibly because of a liberal ideological preference for less-coercive instruments (Hood 1986). Despite this, the implementation of information instruments has been mixed. Broad information instruments have been widely used but there has been political resistance to the deployment of targeted information instruments that identify 'at risk' properties and provide detailed hazard information, particularly in relation to coastal hazards. Disclosure of coastal hazard information is generally limited to the details of relevant planning controls. This hesitancy appears to be a product of landholder and developer concerns about property price impacts and general scepticism towards climate change.

The underutilisation of information instruments is a significant impediment to efficient and equitable adaptive responses. It can accentuate the risks of information-related market failures, raise transaction costs and increase the legal exposure of planning authorities. Governments should review the use of information instruments, including their design. To this end, the available research suggests that: information should be kept simple; uncertainties should be specifically conveyed; information generation and delivery should be coordinated between the government and non-government entities involved; and the provision of information should be timed to coincide with relevant decisions (Wozniak et al. 2012; Macintosh et al. 2013).

5.5.5 *Liability shield instruments*

Planning practitioners frequently identify legal and appeal risks (formal legal liability and the costs associated with defending legal challenges and judicial and merits-review appeals) as a significant barrier to adaptation decision-making. Clear framing instruments supported by detailed and prescriptive codes and guidelines to assist in the application of regulatory instruments can play an important role in reducing conflict and appeals. However, these measures alone will not eliminate legal risks.

In many cases, the fear of litigation is exaggerated. A legal liability to compensate victims will only arise if planning authorities are found to have acted negligently in the conduct of their duties and the negligence contributed to the damage incurred. Provided planning authorities act reasonably, and with regard to the available information, the risk of future legal liability is likely to be small (McDonald 2007, 2011). Nonetheless, the presence of this risk can have a material effect on the way planning agencies perform their functions. In particular, it can lead to excessive risk aversion and approaches that increase the transaction costs associated with land use and development (HCA 2005). Similar problems can arise with private landholders. Due to a fear of liability, landholders may be unwilling to take steps to reduce hazard-related risks or to cooperate with others to address potential hazard impacts.

Liability shield instruments can be used to reduce the tendency for the fear of legal liability to lead to unwanted outcomes. These instruments provide a partial or full exemption from legal liability to specified entities if they take a particular action, or fail to act in a particular way, in relation to climate hazards. Typically, the intent in deploying these instruments is to stop people from unjustly pursuing governments or other third parties for legal compensation when hazard risks materialise.

New South Wales is currently the only Australian jurisdiction that has a broad statutory exemption from legal liability for local governments for actions taken and decisions made in respect of natural hazards. In other jurisdictions there has been some limited use of, and interest in, the use of individual indemnities from developers to address these issues. Due to the potential

for legal risks to pervert policy processes, and the transaction costs associated with individual indemnities, Australian governments should consider developing uniform liability shield instruments, preferably in the form provided in New South Wales (*Local Government Act 1993* (NSW), s. 733).

5.5.6   Monitoring and enforcement of conditions

Bushfire risks have typically been managed on a site-by-site basis in land-use planning processes via technical solutions imposed through conditions on development approval. The underlying assumption in these processes has been that development should generally proceed as long as siting, construction and other risk minimisation conditions are met. This historical practice is being reconsidered in some places following extreme bushfires in southeast Australia in 2009. The policy window created by these experiences has led to increased emphasis on the use of strategic planning to avoid locating new development (particularly sensitive land uses) in high risk areas and, in conjunction with this, a greater role for more stringent restrictions on land use. Despite this, bushfire risks are still mainly dealt with through approval conditions.

A concern with the heavy reliance on approval conditions is compliance and enforcement. In practice, the monitoring and enforcement of hazard-related approval conditions is low in many jurisdictions and, due to governance arrangements and fiscal imbalances, most local governments lack the resources to address this. The legacy effects of past regulatory decisions add to the monitoring and enforcement challenge faced by planning agencies. In many areas, past decisions have resulted in a patchwork of approval conditions; newer developments have been subject to more stringent requirements than apply elsewhere. To help overcome these issues, additional resources should be made available for compliance, for example, through intergovernmental transfers from federal and state governments and innovative use of taxes and charges. Approval conditions should also include self-monitoring and reporting requirements to ensure monitoring costs are borne mostly by the relevant land-holders, as commonly occurs with environmental approvals (Spencer 2001).

5.6 Conclusions

One of the gaps in the adaptation literature concerns the design of the spatial planning instruments that are used to promote adaptation and address climate hazards. The starting point for research on this issue is the classification of instrument types and identification of the characteristics of these instruments that are likely to influence design choices. The typology and analyses presented here are intended to address these issues and, in so doing, help encourage greater consideration of policy design in adaptation planning research. By placing greater emphasis on instruments and design, it is also hoped that this chapter may assist in structuring implementation research and facilitate improved comparative analyses of the policy experiences in different jurisdictions.

References

Burby, R. and Nelson, A. (1991) Local government and public adaptation to sea-level rise. *Journal of Urban Planning and Development* 117(4), 140–153.

Coase, R. (1960) The problem of social cost. *Journal of Law and Economics* 3, 144.

European Commission (1997) *The EU Compendium of Spatial Planning Systems and Policies*. European Commission, Brussels.

Few, R., Brown, K. and Tompkins, E. (2007) Climate change and coastal management decisions: insights from Christchurch Bay, UK. *Coastal Management* 35, 255–270.

High Court of Australia (1997) Newcrest Mining (WA) Ltd v Commonwealth. *Commonwealth Law Reports* 190, 513–663.

High Court of Australia (1998) Commonwealth v WMC Resources Ltd. *Commonwealth Law Reports* 194, 1–105.

High Court of Australia (2001) Durham Holdings Pty Ltd v New South Wales. *Commonwealth Law Reports* 205, 399–433.

High Court of Australia (2005) Bankstown City Council v Alamdo Holdings Pty Ltd. *Commonwealth Law Reports* 223, 660–678.

Hood, C. (1986) *The Tools of Government*. Chatham House Publishers, Chatham.

Howlett, M. (2009) Governance modes, policy regimes and operational plans: a multi-level nested model of policy instrument choice and policy design. *Policy Sciences* 42(1), 73–89.

Howlett, M. and Lejano, R. (2013) Tales from the crypt: The rise and fall (and rebirth?) of policy design. *Administration and Society* 45(3), 357–381.

Linder, S. and Peters, B. (1984) From social theory to policy design. *Journal of Public Policy* 4(3), 237–259.

Linder, S. and Peters, B. (1989) Instruments of government: perceptions and contexts. *Journal of Public Policy* 9(1), 35–38.

Linder, S. and Peters, B. (1990) Policy formulation and the challenge of conscious design. *Evaluation and Program Planning* 13, 303–311.

Macintosh, A. (2013) Coastal climate hazards and urban planning: how planning responses can lead to maladaptation. *Mitigation and Adaptation Strategies for Global Change* 18, 1035–1055.

Macintosh, A. and Cunliffe, J. (2012) The significance of ICM in the evolution of s 51(xxxi). *Environmental and Planning Law Journal* 29, 297–315.

Macintosh, A., Foerster, A. and McDonald, J. (2013) *Limp, Leap or Learn? Developing Legal Frameworks for Climate Change Adaptation Planning in Australia*. National Climate Change Adaptation Research Facility, Gold Coast.

Matland, R. (1995) Synthesizing the implementation literature: The ambiguity-conflict model of policy implementation. *Journal of Public Administration Research and Theory* 2, 145–174.

McDonald, J. (2007) The adaptation imperative: Managing the legal risks of climate change impacts. In: Bonyhady, T. and Christoff, P. (eds) *Climate Law in Australia*. Federation Press, Sydney, pp. 124–141.

McDonald, J. (2011) The role of law in adapting to climate change. *Wiley Interdisciplinary Reviews: Climate Change* 2, 283–295.

Productivity Commission (2013) *Barriers to Effective Climate Change Adaptation*. Commonwealth of Australia, Canberra.

Salamon, L. (2001) The new governance and the tools of public action: an introduction. *Fordham Urban Law Journal* 28, 1611–1674.

Scheraga, J. and Grambsch, A. (1998) Risks, opportunities, and adaptation to climate change. *Climate Research* 10, 85–95.

Spencer, D. (2001) The Shadow of the Rational Polluter: Rethinking the Role of Rational Actor Models in Environmental Law. *California Law Review* 89(4), 917–998.

Wozniak, K., Davidson, G. and Ankerson, T. (2012) *Florida's Coastal Hazard Disclosure Law: Property Owner's Perceptions of the Physical and Regulatory Environment*. University of Florida, Gainesville.

6 Public risk perceptions, understandings and responses to climate change

JOSEPH P. RESER, GRAHAM L. BRADLEY AND
MICHELLE C. ELLUL

Griffith Climate Change Response Program, Griffith University, Australia

6.1 Introduction

Australia has made a substantial investment in research into biophysical environmental changes and impacts, as well as research into climate change adaptation and mitigation considerations respecting human settlements and infrastructure (e.g. Steffen 2009; Department of Climate Change 2010). However, scant attention has been paid to establishing a national database and monitoring program addressing changes and impacts in the human landscape (Garnaut 2008; Weissbecker 2011; Reser et al. 2012b, c). More particularly, in the face of daunting climate change challenges, very modest consideration has been given to psychological and social considerations relating to the ways in which individuals understand, adaptively respond to and are impacted by the threat and unfolding impacts of global climate change (ISSC/UNESCO 2013).

This chapter briefly summarises research findings from a large Australian national research program addressing public risk perceptions, understandings and responses to the dual threats of climate change and natural disasters. We describe the approach taken, the constructs and measures addressed, the research objectives and the more psychological nature of the larger enterprise, which reflected in many respects the recommendations of the American Psychological Association's Taskforce on Psychology and Climate Change (Swim et al. 2011). Given the constraints of chapter length and book focus, we have placed particular emphasis on psychological adaptation as the most useful window through which to achieve a sense of the distinctive character of the research and findings. In particular, we attempt to give readers an appreciation of the underlying functional and mediating roles which psychological adaptation processes play with respect to other more familiar psychological parameters when addressing public risk perceptions, understanding and responses to climate change.

6.2 Methodology and procedures

The research was initiated in 2010 as a cross-national collaboration with Cardiff University's Understanding Risk Centre (Spence et al. 2010), and continued in 2011through 2012 as an Australia-specific program. The project's methodology, procedures and findings are described in two published research monographs (Reser et al. 2012b, c). The first survey was undertaken in mid-2010 ($n=3096$) and the second was conducted in mid-2011 ($n=4347$). The 2011 sample also included 1037 respondents who had taken part in 2010. Sampling procedures were identical in both years. First, we selected 35 geographic regions designated by post-codes from throughout Australia, with a view to sampling across a range of population, climatic and disaster-exposed parts of the nation. Second, potential respondents aged 15 years and over were randomly selected from their panel by a professional survey firm until an approximate gender-balanced quota of 50 respondents in 2010 and 80 respondents in 2011 had been obtained from each region. Precise response rates are not available. Detailed reporting of the representativeness of the final sample is provided in Reser et al. (2012b, c).

The anonymous online surveys included many multi-item, interval-level measures of individual difference variables, including knowledge of climate change science, belief in/acceptance of anthropogenic climate change, risk perception, psychological distress, perceived responsibility and willingness to act, psychological adaptation in the context of climate change, self- and collective efficacy and mitigation behaviours. Also explored were the nature and extent of participants' direct and indirect exposure to, and experiences of, climate change and natural disaster events and impacts. As detailed in Reser et al. (2012b, c), descriptive and inferential statistical analyses were performed to model the relationships between participants' experiences, cognitions, motivations, affective states and behavioural responses. These statistical findings were complemented by qualitative data given in response to a series of open-ended questions.

6.3   Overview of findings

When the individual survey item findings for 2011 are brought together, the clarity and strength of public views and sentiments becomes very clear. For example, 74% of respondents *personally thought* that climate change is occurring with 69% 'very' or 'fairly' *certain* that this was

happening, and 50% judged it as *already happening in Australia*. In addition, 42% reported it being a *serious problem right now*, 64% reported being *very* or *fairly concerned* about climate change, 43% reported that climate change was an *extremely* or *quite important issue* to them personally and 27% reported that they *think about climate change a lot*. Approximately 20% of respondents reported feeling at times *appreciable distress* at the prospects and implication of climate and its consequences. On the whole, respondents appear to feel that they themselves can and should be addressing this environmental threat (59%) and that the Australian government, state governments and corporate Australia should be doing the same. Well over one-half of respondents (61%) reported being prepared to greatly reduce their energy use to help tackle climate change (61%), and many are psychologically adapting to the threat of climate change and changing their behaviours and lifestyle with respect to reducing their own carbon footprint (Reser et al. 2012c).

6.3.1   Addressing and documenting psychological adaptation and change

A strong focus in the research was that of *psychological* adaptation to climate change, reflecting a long history of engagement within psychology on both intra-individual and behavioural adjustments and responses to experienced environmental change, threat and environmental stressors (e.g. Folkman 2011; Swim et al. 2011). This focus on adaptation, our related research questions, measures and findings and the nature and implications of this more psychological consideration of adaptation to climate change brings together and integrates much of our research efforts and output. The encompassing disciplinary investment within psychology is of particular relevance to interdisciplinary climate change science and the effective addressing and fostering of crucial climate change adaptation and mitigation responses on the part of individuals, communities, regional and national governments and policymakers (Swim et al. 2011; Reser et al. 2012a).

The nature of the *adaptation* construct and associated processes in the social and health sciences, and the conceptual perspective and set of assumptions underlying adaptation to climate change, would strongly suggest that such adaptation and adjustment takes place *within individuals and their psychological systems* as well as within communities, organisations and meta systems (e.g. Reser and Swim 2011; Reser et al. 2012a). Such adaptation also assumes interdependent interactions or *transactions* (Werner et al. 2002) between individuals and their physical and social environment and settings. Human ecological perspectives, environmental psychological perspectives and, indeed, much of social science presumes and addresses the nature and dynamics of human responses and adaptations to environmental threat and change (e.g. National Research Council 1992, 1999, 2010; Bell et al. 2001; Winkel et al. 2009; Lever-Tracy 2010). The present research program set out to measure and document important psychological and behavioural changes and impacts taking place in the human landscape in association with the threat and unfolding impacts of climate change and extreme weather events. This chapter, as with the preceding reports and research monographs associated with this research program (e.g. Reser et al. 2012b, c), examines possible changes relating to theoretically relevant variables, processes and impacts that might reasonably be associated with the processes of psychological adaptation to climate change.

Climate change science and Intergovernmental Panel on Climate Change (IPCC) definitions and specifications of adaptation in the context of climate change are very system-focused, whether by way of natural biophysical systems or human infrastructure, organisations and institutional systems. Across the non-psychological social science literature, and in health reports addressing the implications and impacts of climate change, adaptation continues to be defined in exclusive overt action and behaviour terms, institutional and structural terms or in medical model and physiological functioning terms, with no reference to psychological adaptation

and underlying psychological processes and adaptation costs and impacts (e.g. Hughes and McMichael 2011; Reser and Swim, 2011; Reser et al. 2012a).

This present, more psychological, research undertaking has addressed the very neglected construct and convergent processes of *psychological* adaptation as they relate to human risk perceptions and responses to the threat and unfolding physical environmental impacts of climate change and associated extreme weather events (e.g. Reser and Swim 2011; Reser et al. 2012a). Adaptation as a multi-faceted construct and convergent multi-level suite of organism–environment processes has been an integral and foundational construct and perspective within psychology and the social sciences since the inception of these respective disciplines (e.g. Piaget 1955; Pribram 1969; White 1974), yet the crucial relevance of these social-science-based perspectives to climate change adaptation processes is only beginning to be fully appreciated (e.g. Lever-Tracy 2010; Swim et al. 2011; Agrawal et al. 2012). Importantly, psychological adaptation also invokes and involves psychological *impacts* and both the benefits and costs of environmental changes and human adaptations.

What is 'psychological adaptation' to climate change? *Psychological* adaptation to climate change encompasses those intra-individual processes (e.g. risk appraisal, motivational responses, coping strategies and decision-making) relating to psychological responses, changes and adjustments to the *threat and implications of climate change* as well as to *direct experience with what are perceived to be the unfolding impacts* of climate change. More experientially and operationally speaking, psychological adaptation is conceptualised as changes and adjustments in one's thinking, feelings and general understanding of and responses to climate change, as well as changes in one's engagement with climate change as a topic, issue and/or concern which one follows, discusses, thinks about and is influenced by. Psychological adaptation also encompasses extra-individual behavioural responses and adjustments (e.g. community engagement and involvement, mitigation and information seeking) to the threat and perceived physical environmental impacts of climate change, which are typically mediated by intra-individual psychological processes and responses. The nature of psychological adaptation requires this brief clarification because such within-individual responses and changes are not always recognised as crucial aspects of climate change adaptation (e.g. Reser and Swim 2011; Reser et al. 2012a). Yet considerable policy attention is paid to how public perceptions, attitudes, values, motivations, decisions and understanding might be more effectively influenced and changed, leading to behavioural and lifestyle changes (Whitmarsh et al. 2011). All of these changes, whether or not influenced by strategic interventions or risk communications, constitute psychological adaptations to climate change.

No appropriate climate-change-specific scales that measure psychological adjustments, adaptations or impacts to the threat of climate change were available at the time of our first survey. While emerging social science and psychological research has been closely examining public climate change attitudes, perceptions, beliefs, intentions and behaviour (e.g. Brechin 2010; Leiserowitz et al. 2010, 2012; Brulle et al. 2012), this research has not, by and large, been framed in adaptation terms; the focus has been more strongly on mitigation and behavioural change. Perhaps the closest construct, set of processes and perspective available within psychology and currently being employed in the climate change arena is that of coping (e.g. Homburg et al. 2007; Folkman 2011; Van Zomeren et al. 2011; Reser et al. 2012a). However, coping typically relates more particularly to specific and acute stressors and situations rather than being a more ongoing response, state and set of processes for maintaining an acceptable and long-term transactional congruence between individual and environmental press and adaptive change (e.g. Lazarus and Launier 1978; Reser et al. 2012a). Convergent arguments suggest that the ongoing environmental stressor status of global climate change and the more encompassing nature and status of psychological adaptation processes make a climate-change-specific and sensitive measure of psychological adaptation a very strategic and possibly crucial contribution to the field.

6.4   Research findings

Psychological adaptation in the context of climate change was measured using 7 questionnaire items in 2010 and 11 items in 2011. All items addressed self-reported changes in respondents' thinking, feeling, understanding and acting in response to the threat and reality of climate change. As no prior study had developed or used a self-report measure of psychological adaptation to climate change, we developed original items based on the extant literature for the purpose of the current research. Findings reported in this chapter are based on just the 7 items that were common to both years' surveys. All items required a response on a 6-point (*strongly disagree* to *strongly agree*) Likert scale. Analyses confirmed that the items formed a reliable scale; Cronbach's alpha = 0.87 (2010) and 0.85 (2011). Responses were summed (possible range = 7–42), such that high scores indicate high levels of psychological adaptation to climate change.

Responses indicated that substantial numbers of our survey respondents were going through processes of psychological adaptation, with many participants reporting changes in their thinking, feelings and general responses to climate change. For example, 57% of respondents in 2010 and 56% in 2011 agreed with an item indicating that they had 'changed the way they think about the seriousness of environmental problems because of climate change', while 56% in 2010 and 50% in 2011 agreed that 'climate change has forced me to change the way I think about and view how we live in and use our natural environment in Australia'. Responses such as these provide insight into the kinds of adaptation processes that are taking place to assist Australians in their sense-making and adjustments to climate change.

Mean scores on this composite scale were 24.5 (2010) and 24.0 (2011) from a possible range of

7–42. In both years, psychological adaptation scores were higher for females than for males, for respondents who were university educated than for those who were school-only educated, for people born outside of Australia than for those born in Australia and for Green and Labor party voters than for respondents who intended to vote for the conservative parties. Age differences in psychological adaptation were relatively small and not consistent across the years of the survey. Scores did not vary systematically with residential status (urban vs. rural) or with income.

Psychological adaptation was hypothesised to be a pivotal construct in our studies of human understanding and responses to climate change. Consistent with this, scores on our adaptation scale were found to be significantly and positively correlated with: perceived residential vulnerability ($r=0.45$ (2010) and 0.40 (2011)); objective knowledge of climate change science ($r=0.45$ and 0.46); belief in/acceptance of climate change ($r=0.56$ and 0.57); concerns about climate change ($r=0.70$ and 0.71); climate change risk perception ($r=0.62$ and 0.63); climate change distress ($r=0.77$ and 0.75); perceptions of responsibility and willingness to act on climate change ($r=0.66$ and 0.68); and mitigation behaviours ($r=0.52$ and 0.57). In 2011, psychological adaptation was also found to vary in predictable ways with engagement in three types of coping behaviours. Specifically, psychological adaptation was positively correlated with self-reported tendencies to cope with climate change by seeking emotional support from other people ($r=0.61$) and by re-framing climate change concerns in more positive ways ($r=0.34$), but was negatively correlated with tendencies to cope by denying the existence of climate change ($r=-0.40$).

We performed a series of structural equation modelling analyses to gain a fuller appreciation of the role played by psychological adaptation processes in mediating the influence of other psychological variables on engagement in climate change mitigation behaviours. Figure 6.1 shows one of our better-fitting models. Of note is the finding that the effects on mitigation

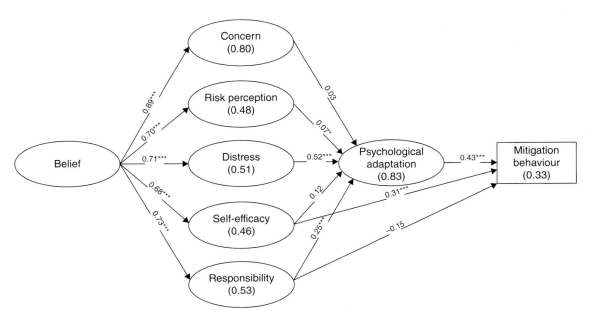

Figure 6.1 Model showing mediating role of psychological adaptation to climate change (standardised parameter estimates on lines, *** $P < 0.001$, * $P < 0.05$; percentage of variance provided in parentheses).

behaviours of several variables (e.g. risk perception, distress) are mediated by psychological adaptation, while only one other variable (self-efficacy) has a significant direct effect on behaviour. Perceived responsibility to act on climate change has two opposing effects on behaviour: a non-significant, negative direct path and a significant, positive, indirect path through adaptation. Most importantly, there is evidence of a strong effect of climate change belief/acceptance, through distress and psychological adaptation, to behaviour.

As a final illustration of our investigations into psychological adaptation, we conducted regression analyses using the repeat (2010–2011) sample of 1037 survey respondents to identify factors that predict changes in adaptation over a 1-year period. Included in a hierarchical model predicting 2011 levels of psychological adaptation were: (1) 2010 adaptation scores; (2) seven socio-demographic factors (gender, age, education, parental status, country of birth, income and voting intention); (3) three 'background' psychological factors (perceived residential vulnerability, objective knowledge and belief in climate change); and (4) the five proximate predictors shown in Figure 6.1 (concern, risk perception, distress, self-efficacy and perceived responsibility to act). Together, these predictors explained 60% of the variance in the criterion. In the final model, change in psychological adaptation was significantly predicted by education, perceived responsibility to act, distress and country of birth, with increases over time in adaptation associated with higher levels of education, higher perceived responsibility, higher distress and being born overseas.

6.5 Conclusions and implications

Psychological perspectives on climate change adaptation highlight a number of crucial but currently neglected aspects of adaptation and 'human dimensions of global change'. These include multi-level approaches and analytic frameworks that encompass individual and experience-focused levels of analysis, social psychological process responses to the threat and unfolding impacts of climate change and models, constructs and indicators relevant to assessing the *psychosocial* impacts of climate change. Psychological research on human response to global environmental change spans four decades, providing particularly helpful perspectives and insights on human adaptation and adjustment to environmental threat, natural and technological disasters and extreme and stressful environmental changes. However, this very relevant and extensive body of theoretical approaches, research findings and evidence-based applications continues to be a relatively unfamiliar disciplinary landscape in the environmental sciences. Of particular importance are the conceptual framing and theoretical elaboration of *psychological* adaptation processes across many different areas of psychology, seen to be important mediators of public risk perceptions and understanding, effective coping responses and resilience, overt behavioural adjustment and change and psychological and social impacts. This psychological window on climate change adaptation is arguably indispensable to genuinely multidisciplinary and interdisciplinary research and policy initiatives addressing the impacts of climate change. It may also provide the best disciplinary vantage point we have on important links between climate change adaptation and mitigation.

Our findings strongly suggest that psychological adaptation plays crucial interlinking and mediating roles with respect to other psychological variables, including behavioural engagement, and very likely physical and emotional wellbeing, reflecting the fact that psychological adaptation, essentially intra- and extra-individual adjustments to environmental press, both constitutes and fosters greater person–environment congruence (e.g. Bell et al. 2001). Importantly, this would suggest that psychological adaptation is a powerful mediator of individual-level behavioural adaptation and mitigation. Our findings reveal that psychological adaptation does not occur universally or evenly throughout the population. Rather, levels of adaptation tend to increase with education, knowledge of climate change science, climate change concern and other demographic and psychological variables. A particularly strong relationship was observed between distress and adaptation, suggesting that the latter is associated with some psychological costs; equally however there was a positive relationship between adaptation and motivation to take climate change action, suggesting that there are also benefits. Importantly, educational attainment was a strong unique predictor of increases over time in psychological adaptation. A direct causal effect of education on adaptation is plausible and, if true, has clear policy implications for assisting people in coming to terms with the reality of climate change.

Finally, and very importantly, behavioural engagement would appear to be not only a partially mediated outcome of psychological adaptation, but itself an integral part of psychological adaptation with respect to 'taking action', concretely 'doing something' and feeling that one is responsibly being part of the solution as well as the problem. Our qualitative findings very strongly point to just such psychological benefits and 'psychological significance' in the context of carbon-reducing behaviour engagements. Our structural equation modelling undertaken provides a coherent way of framing and understanding the interrelationships among these core variables and processes, and a template and platform for undertaking further research relating to specific variables and relationships. These findings confirm the importance of psychological adaptation to climate change, the psychometric sensitivity and utility of the measure and the multiple dividends of including such a parameter and measure in climate change survey research. The results provide evidence not only that psychological adaptation to climate change is taking place, but also that the diverse but convergent ways in which individuals are coming to terms with and adjusting to climate change constitute a crucially important component of climate change adaptation.

References

Agrawal, A., Orlove, B. and Ribot, J. (2012) Cool heads for a hot world: social sciences under a changing sky. *Global Environmental Change* 22, 329–331.

Bell, P.A., Greene, T.C., Fisher, J.D. et al. (2001) *Environmental Psychology*, 5th edition. Harcourt College Publishers, Fort Worth, TX.

Brechin, S.R. (2010) Public opinion: A cross national view. In: Lever-Tracy, C. (ed.) *Routledge Handbook of Climate Change and Society*. Routledge Taylor and Francis Group, New York, pp. 179–209.

Brulle, R.J., Carmichael, J. and Jenkins, J.C. (2012) Shifting public opinion on climate change: an empirical assessment of factors influencing concern over climate change in the US, 2002–2010. *Climate Change* 114, 169–188.

Department of Climate Change (2010) Australia's Fifth National Communication on Climate Change. A Report under the United Nations Framework Convention on Climate Change 2010. Commonwealth of Australia, Canberra.

Folkman, S. (ed.) (2011) *The Oxford Handbook of Stress and Coping*. Oxford University Press, New York.

Garnaut, R. (2008) *The Garnaut Climate Change Review: Final Report*. Cambridge University Press, Cambridge, UK.

Homburg, A., Stolberg, A. and Wagner, U. (2007) Coping with global environmental problems: development and first validation of scales. *Environment and Behavior* 39, 754–778.

Hughes, L. and McMichael, T. (2011) *The Critical Decade: Climate Change and Health*. Climate Commission Secretariat (Department of Climate Change and Energy Efficiency), Commonwealth of Australia, Canberra.

International Social Science Council and United Nations Educational, Scientific and Cultural Organization (2013) *World Social Science Report 2013: Changing Global Environments*. OECD Publishing and UNESCO Publishing, Paris, France.

Lazarus, R.S. and Launier, R. (1978) Stress-related transactions between person and environment. In: Pervin, L. A. and Lewis, M. (eds) *Perspectives in Interactional Psychology*. Plenum, New York, pp. 287–327.

Leiserowitz, A., Maibach, E. and Roser-Renouf, C. (2010) *Climate Change in the American Mind: Americans' Climate Change Beliefs, Attitudes, Policy Preferences, and Actions*. Yale University and George Mason University, New Haven, CT.

Leiserowitz, A., Maibach, E., Roser-Renouf, C. et al. (2012) *Global Warming's Six Americas March 2012 and November 2011*. Yale University and George Mason University, New Haven, CT.

Lever-Tracy, C. (2010) *Routledge Handbook of Climate Change and Society*. Routledge, London.

National Research Council (1992) *Global Environmental Change: Understanding the Human Dimensions*. National Academy Press, Washington, DC.

National Research Council (1999) *Global Environmental Change: Research Pathways for the Next Decade*. National Academy Press, Washington, DC.

National Research Council (2010) *Advancing the Science of Climate Change*. National Academy Press, Washington, DC.

Piaget, J. (1955) *The Child's Construction of Reality*. Routledge and Kegan Paul, London.

Pribram, K.H. (1969) *Adaptation: Selected Readings*. Penguin Books, Baltimore, MD.

Reser, J.P. and Swim, J.K. (2011) Adapting to and coping with the threat of climate change. *American Psychologist* 66, 277–289.

Reser, J.P., Bradley, G.L. and Ellul, M.C. (2012a) Coping with climate change: Bringing psychological adaptation in from the cold. In: Molinelli, B. and Grimalso, V. (eds) *Handbook of the Psychology of Coping: Psychology of Emotions, Motivations and Actions*. Nova Science Publishers, Hauppauge, NY, pp. 1–34.

Reser, J.P., Bradley, G.L., Glendon, A.I., Ellul, M.C. and Callaghan, R. (2012b) *Public Risk Perceptions, Understandings, and Responses to Climate Change and Natural Disasters in Australia and Great Britain*. National Climate Change Adaptation Research Facility, Gold Coast.

Reser, J.P., Bradley, G.L., Glendon, A.I., Ellul, M.C. and Callaghan, R. (2012c) *Public Risk Perceptions, Understandings, and Responses to Climate Change and Natural Disasters in Australia, 2010 and 2011*. National Climate Change Adaptation Research Facility, Gold Coast.

Spence, A., Venables, D. and Pidgeon, N. (2010) *Public Perceptions of Climate and Energy Futures in Britain: Summary Findings of a Survey Conducted from January to March 2010*. Understanding Risk Research, Cardiff University, Cardiff, UK.

Steffen, W. (2009) *Climate Change 2009: Faster Change and More Serious Risks*. Australian Government Department of Climate Change, Canberra.

Swim, J.K., Stern, P.C., Doherty, T.J. et al. (2011) Psychology's contributions to understandings and addressing global climate change. *American Psychologist* 66, 241–250.

Van Zomeren, M., Spears, R. and Leach, C.W. (2011) Experimental evidence for a dual pathway model analysis of coping with the climate crisis. *Journal of Environmental Psychology* 30, 339–346.

Weissbecker, I. (ed.) (2011) *Climate Change and Human Well-Being: Global Challenges and Opportunities*. Springer Publications, New York, NY.

Werner, C.M., Brown, B.B. and Altman, I. (2002) Transactionally oriented research: Examples and strategies. In: Bechtel, R.B. and Churchman, A. (eds) *Handbook of Environmental Psychology*. Wiley, New York, pp. 203–221.

White, R.W. (1974) Strategies of adaptation: An attempt at systematic description. In Coelho, G.V., Hamburg, D.A. and Adams, J.E. (eds) *Coping and Adaptation*. Basic Books, New York, pp. 47–68.

Whitmarsh, L., O'Neill, S. and Lorenzoni, I. (2011) *Engaging the Public With Climate Change: Behaviour Change and Communication*. Earthscan, London.

Winkel, G., Saegert, S. and Evans, G.W. (2009) An ecological perspective on theory methods, and analysis in environmental psychology. *Journal of Environmental Psychology* 29, 318–328.

7 Bridging the gap between researchers and decision-makers

DANIELLE C. VERDON-KIDD, ANTHONY S. KIEM
AND EMMA K. AUSTIN

School of Environmental and Life Sciences, University of Newcastle, Australia

7.1 Decision-making under uncertainty

Successful climate change adaptation outcomes are supported by decision-making that is informed by the best available climate science (e.g. Burton 1997; Sarewitz and Pielke Jr 1999; Patt and Dessai 2005; Power et al. 2005; Meinke et al. 2009) and also from previous lessons learnt on knowledge use in policymaking contexts (e.g. Gieryn 1983; Hoppe 1999, 2005; Guston 2001; Colebatch 2005, 2006). However, a fundamental gap exists between the information that climate science provides and the information that is practically useful for (and needed by) end-users and decision-makers looking to successfully adapt to climate variability and change (e.g. McNie 2007; Lemos et al. 2012; Kiem and Austin 2013a). In some cases the climate science, and associated information, is simply too 'uncertain' to be of any practical use, or the uncertainty is not adequately quantified and communicated. Further, as demonstrated by Kiem and Austin (2013a, b), even if the climate information does exist and the uncertainty is properly communicated, often decision-makers are unaware of it or cannot access it in a format they can readily use (i.e. the right information may exist but it is inaccessible to end-users due to time, expertise and/or technological constraints).

When carrying out research on human-induced climate change and its impacts, scientists have traditionally followed a linear pathway of activity starting from the specification of greenhouse-gas emissions and ending with impacts and possible response strategies. From the perspectives of those working in the adaptation field (i.e. natural resource managers, policymakers and decision-makers), the problem is that each step in this pathway has an associated uncertainty. More importantly these uncertainties compound at each step, so that by the time the stage of projecting climate change impacts at spatial scales relevant for decision-making is reached, the uncertainties have exploded (e.g. Jones 2000).

Several recent studies have reviewed the uncertainties associated with general circulation models (GCMs; e.g. Parry et al. 2007; Randall et al. 2007; Stainforth et al. 2007; Koutsoyiannis et al. 2008, 2009; Blöschl and Montanari 2010; Montanari et al. 2010) and the implications of those uncertainties for climate change impact assessments in Australia (e.g. Pitman and Perkins 2008; Verdon-Kidd and Kiem 2010; Kiem and Verdon-Kidd 2011). These studies demonstrate that, while climate models may represent the 'best available science' in terms of our understanding of global climate processes, the inherent uncertainties mean that climate model outputs

Applied Studies in Climate Adaptation, First Edition. Edited by Jean P. Palutikof, Sarah L. Boulter, Jon Barnett and David Rissik.
© 2015 John Wiley & Sons, Ltd. Published 2015 by John Wiley & Sons, Ltd.

do not necessarily represent useful information or meet the needs of end-users when planning for and making decisions about the future.

While uncertainties exist in relation to the generation of climate information, there is also uncertainty when attempting to define the differing and varied needs across the range of diverse end-users. End-user needs vary considerably as a result of location, sector, resources, existing knowledge, climate risks and the decision being made (e.g. Maraun et al. 2010; Kiem and Verdon-Kidd 2011). The capability, capacity to act (i.e. budget and time constraints), awareness of science information and attitudes of end-users also vary markedly across locations and sectors (e.g. Kiem et al. 2010a, b; Kiem and Austin 2013a, b). This means that, even if uncertainty in the science is dealt with, effort is also needed in addressing uncertainty in the way that science is viewed and implemented by end-users.

Climate science and its related uncertainties are often not the only consideration in making climate-sensitive decisions, or even the most important (Power et al. 2005). There is increasing support for the suggestion that the gap is not just between science and the decision-makers; rather, the decision has to be socially, politically, economically and environmentally acceptable for it to be implemented (e.g. Adger et al. 2005; Füssel 2007; Kiem and Austin 2013b). Even in a perfect world – where scientists provide useful information to end-users and end-users subsequently make robust climate change adaptation decisions based on that science – if the decision is not supported by society (i.e. individuals, communities, businesses, political groups, etc.) there will always be difficulty in getting that decision implemented (e.g. every time a desalinisation plant or reservoir is proposed, when water trading or allocation schemes are introduced or when sea-level inundation or flood management policies are changed).

Unfortunately, uncertainty surrounding climate science will not disappear (e.g. Kiem and Verdon-Kidd 2011) and improved decision-making under uncertainty and research aimed at

translating uncertainty into risk is required. Indeed, climate change adaptation is about providing decision-makers with the insight and tools needed to robustly and optimally deal with risk and the associated uncertainties. This chapter summarises the key findings of a project conducted for the Australian Government's National Climate Change Adaptation Research Facility (NCCARF), which aimed to investigate the gap between researchers and decision-makers with a particular focus on the issue of dealing with climate uncertainty in decision-making (www.nccarf.edu.au/publications/decision-making-under-uncertainty).

7.2   Assessing the gap between researchers and decision-makers

7.2.1   Online survey

An online survey was conducted to evaluate the gap between researchers' perceptions of what constitutes useful climate information (with a focus on the issue of uncertainty) and what decision-makers actually require to make robust decisions about climate change adaptation. The survey was specifically designed to answer the following questions.

1. How has climate information been used to date and from where is it sourced?
2. How do end-users and scientists rate the existing climate information (i.e. what are the strengths and weaknesses of the information)?
3. What do end-users understand uncertainty to mean and how well is uncertainty communicated?
4. What will be the major advances in climate modelling over the next 5–10 years, and will these advances reduce uncertainty?

The survey required participants to initially identify as 'providers' or 'end-users' of climate information. The survey then tailored subsequent questions according to this selection (see www.nccarf.edu.au/publications/decision-making-under-uncertainty for full survey questions and results). A total of 210 respondents participated

Figure 7.1 Understanding what uncertainty means in terms of climate information as reported by providers (left) and end-users (right).

in the survey with approximately 70% of respondents identifying themselves as end-users and approximately 30% as providers of climate information. Respondents worked within a range of Australian organisations, with government and university organisations particularly well represented. Various end-user sectors were also well represented (including primary industries, water management, human health, settlements and infrastructure, emergency management, social and economic dimensions, biodiversity and Indigenous communities).

One of the key focuses of the survey was to determine what providers and end-users understand uncertainty to mean and how they deal with uncertainty in climate information. There were clear commonalities between the two groups in their definition of uncertainty (Fig. 7.1) with the words 'models', 'future' and 'range' featuring in both sets of responses. However, there was also a clear distinction between the terminology of the two groups in terms of the word 'projections' versus 'predictions' (which have two very different meanings). Providers also used the terms 'accuracy' and 'error' when describing uncertainty, whereas end-users did not commonly use these words. Rather, end-users favoured terms such as 'confidence', 'quality', 'variability/variance' and

'spread'. Overall, the survey demonstrated that providers mainly think of uncertainty in terms of the ability of the model to accurately simulate reality; however, end-users think about uncertainty in a broader sense encompassing the range of model projections and variability between simulations. It was clear from subsequent questions that this difference in terminology contributes to the gap between scientists and decision-makers and the difficulties experienced in dealing with uncertainty in climate information.

The survey highlighted a general consensus that uncertainty in climate science is not well communicated to end-users (Fig. 7.2a), a result also supported by literature (e.g. Cash et al. 2003; Dow and Carbone 2007; Tang and Dessai 2012). Indeed, the majority of providers surveyed believe that uncertainty is poorly or very poorly communicated and end-users agree (less than 15% of end-users think communication of uncertainty is done well or very well). Figure 7.2b highlights the gap between provider and end-user perceptions around capabilities of end-users to deal with uncertain climate information. Providers appear to underestimate end-users' capacity to deal with uncertainty, or end-users overestimate their ability in this area. Either way, only a minority (less than 10%) of respondents thought that

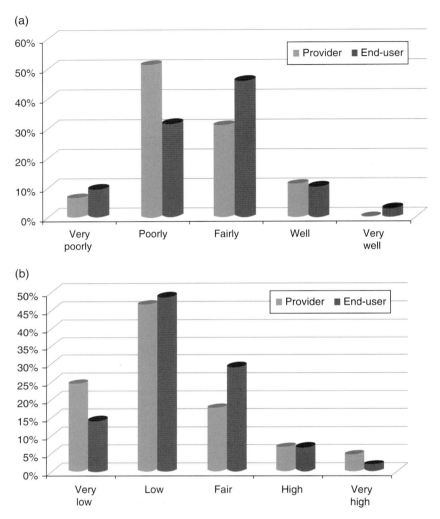

Figure 7.2 Provider and end-user perspectives on (a) how well uncertainty in climate information is communicated by the providers of that information; and (b) the capabilities of end-users to deal with uncertainty in climate information.

end-users were capable of dealing well with uncertainty in climate information.

A significant finding of the survey was that only 50% of providers believe that advances in science in the next 5–10 years will lead to a reduction in uncertainty (Fig. 7.3a), while 70% of end-users have expectations that uncertainty in climate information will reduce over this time period (Fig. 7.3b). This result demonstrates that there is a significant lack of communication around what end-users should realistically expect over the next 5–10 years in terms of improvements in the science. Perhaps the most striking result is the differing opinions among providers and end-users about the need to reduce current uncertainty; over 67% of end-users see this as

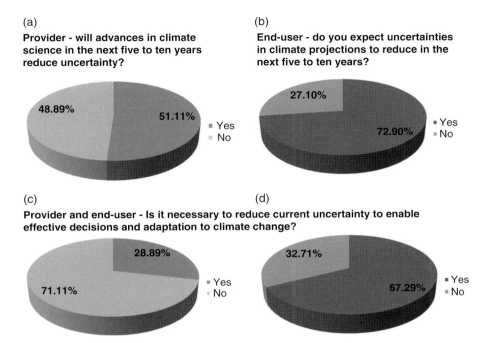

Figure 7.3 (a) Provider and (b) end-user views on advances in climate information and (c) provider and (d) end-user opinions on whether it is necessary to reduce current uncertainty to enable effective decisions and adaptation to climate change.

necessary (Fig. 7.3c) to make effective decisions, while less than 33% of providers believe this is required (Fig. 7.3d). This is an important finding given that some end-users may delay long-term adaptation planning while they wait for more certain climate projections that may never eventuate.

7.2.2 End-user and climate researcher workshop

Climate scientists (i.e. providers) and high-level decision-makers (i.e. end-users) were brought together for a two-day (12 and 13 April 2012) workshop in Canberra, Australia to discuss decision-making under uncertainty and the gap that currently exists between the information that climate science provides and the information that is practically useful for (and needed by) end-users and decision-makers. The objective of the

workshop was to give both providers of climate information and end-users 'a voice' where concerns, issues and beliefs could be raised and challenged in an organised forum. Workshop presenters and participants were carefully considered and selected in order to provide a comprehensive background on the state of current climate science, while also providing a succinct summary from the end-user perspective. Participants were selected because of their prominence in their particular field of climate science or their specific role as a decision-maker (e.g. representing a particular sector or level of government).

The workshop covered both end-users' experience in using climate information and the 'state of the science' (see www.nccarf.edu.au/publications/decision-making-under-uncertainty for the full workshop program and list of participants). A hypothetical case study exercise was also carried out (for further details see Kiem et al. 2014) which required all participants to engage in an interactive exercise that highlighted the information required and the different methods end-users rely on when forced to make decisions under uncertainty. This was followed by a wrap-up session where the key themes emerging from the workshop were discussed and given priority rankings.

At the start and conclusion of the workshop, participants were asked to reflect on:
• what they understood uncertainty to mean; and
• whether they thought it possible to effectively adapt and manage climate risk under uncertainty given the current tools and information available (if so why, if not why not).

Similar to the survey, end-users and decision-makers referred to uncertainty in terms of a range of possible outcomes and potential futures while responses from the scientists stressed the difference between uncertainty, precision and accuracy. By the end of the workshop scientist and end-user understanding of uncertainty did not change significantly; however from a climate scientist perspective, the importance of non-climate-related uncertainty was reinforced. Prior to the workshop end-users said that one thing that would make the workshop valuable is 'an appreciation that climate change adaptation is based on more than just climate'. Based on the following comments given by scientists after the workshop it seems that this was achieved:
• "The speaker mentioned that uncertainty of predictions was a problem when communicating with the community, and seemed to infer that it was scientists' problem. I am wondering if it is because as a community we are unaware of the level of uncertainty in which we exist? And as part of climate change conversations, we need to be educated as a community about uncertainty."
• "The workshop reinforced my view on uncertainty and reinforced that climate change uncertainty is not the only uncertainty and not always the most important."
• "I still tend to view uncertainty the same as before the workshop with respect to climate change projections, however I now recognise that there are political, social and demographic issues affecting decision-makers that may be of equal or greater magnitude than scientific uncertainty."

Participants were asked if they thought it was possible to effectively adapt and manage climate risk under uncertainty given the current tools and information available. End-users felt that although a range of information, processes and tools is already in use to inform decisions under current levels of uncertainty, these tools require improvement. Responses from scientists reiterated that the effectiveness of current information and tools is situation-dependent, indicating again that a more holistic approach is required.

To determine the key take-home messages and priority actions emerging from the workshop participants were presented with a list of issues that arose throughout the workshop and asked to vote on what they felt were the most important. Three primary issues that participants felt were the most important to address in order to bridge the gap and enable better decision-making under uncertainty were as follows.

1. *Improved communication and packaging of climate information.* Participants commented that this was not just more glossy brochures and presentations by climate scientists, rather the role of a 'knowledge broker' or 'boundary organisation' was identified to operate in the space between the two parties. The role of the 'knowledge broker' would be to package, translate (both ways) and transform climate information (e.g. Gieryn 1983; Guston 2001; Hoppe 2005; Kiem and Austin 2013a).

2. *A better understanding and quantification of baseline risk, natural variability and non-stationarity.* This issue arose on multiple occasions throughout the workshop and it is clear that guidance needs to be developed in order for end-users to integrate this into their climate impact assessment and adaptation processes, in particular how natural variability might change.

3. *Development of tools and methods to integrate between projections and decision-making.*

This issue is the focus of a parallel project funded by NCCARF, 'Understanding end-user decisions and the value of climate information under the risks and uncertainties of future climates' (refer to www.nccarf.edu.au/content/decisions-under-climate-risks and www.adaptation-decisions.com for further information).

7.3 Recommendations to help bridge the gap

Uncertainty in climate science can be a barrier to adaptation. As demonstrated by this project, this issue of uncertainty is multi-faceted with issues identified in terms of communication of uncertainty, misunderstanding of uncertainty and the lack of tools/methods to deal with uncertainty. The findings of the survey revealed that both providers and end-users of climate information felt that uncertainty was not well communicated, and this was noted as an area where significant improvements could be made in order to narrow the gap. In order to help narrow the gap, it is clear that further education of end-users is needed with respect to uncertainties in climate science, the limitations of the currently available information, the likely advances in the next 5–10 years and how to best use the available information (taking into account these uncertainties and limitations) to develop effective adaptation plans.

Improved communication and packaging of climate information was a key theme highlighted in the survey and workshop. The role of the 'knowledge broker' would be to package, translate (both from end-user to scientist and scientist to end-user) and transform climate information and also to educate decision-makers on climate change science and adaptation (e.g. via short courses, seminars and workshops, etc.). Importantly, education and communication of uncertainty need to be improved so that end-users are aware of (1) all the caveats and what can realistically be expected from climate science now and in the near future, and (2) how to properly use climate science insights and modelling outputs.

There is also a clear need to further develop tools and methods to integrate between projections and decision-making that properly account for uncertainty. The decision-support systems also need to account for non-climatic influences (e.g. social, environmental, political and economic) on what are essentially climate-driven problems that require solutions that address much more than just the climate. However, the development of decision-support tools is secondary to the need for improved education and communication, given that the delay in developing such tools may provide an excuse for end-users to procrastinate and delay the decision-making process.

Overall, the project identified that there is indeed a gap between end-user needs and science capability, particularly with respect to uncertainty, communication and packaging of climate information. This gap has been a barrier to successful climate change adaptation in the past. While it is unrealistic to think we could ever close the gap completely, based on the insights and recommendations provided in this chapter it may be possible to bridge the gap (or at least improve people's awareness of the gap).

References

Adger, W.N., Arnell, N.W. and Tompkins, E.L. (2005) Successful adaptation to climate change across scales. *Global Environmental Change* 15, 77–86.

Blöschl, G. and Montanari, A. (2010) Climate change impacts: throwing the dice? *Hydrological Processes* 24, 374–381.

Burton, I. (1997) Vulnerability and adaptive response in the context of climate and climate change. *Climatic Change* 36, 185–196.

Cash, D.W., Clark, W.C., Alcock, F. et al. (2003) Knowledge systems for sustainable development. *Proceedings of the National Academy of Sciences* 100, 8086–8091.

Colebatch, H.K. (2005) Policy analysis, policy practice and political science. *Australian Journal of Public Administration* 64(3), 14–23.

Colebatch, H.K. (2006) What work makes policy? *Policy Sciences* 39, 309–321.

Dow, K. and Carbone, G. (2007) Climate science and decision making. *Geography Compass* 1, 302–324.

Füssel, H.M. (2007) Adaptation planning for climate change: concepts, assessment approaches, and key lessons. *Sustainability Science* 2, 265–275.

Gieryn, T.F. (1983) Boundary-work and the demarcation of science from non-science: Strains and interests in professional ideologies of scientists. *American Sociological Review* 48, 781–795.

Guston, D.H. (2001) Boundary organizations in environmental policy and science: an introduction. *Science, Technology and Human Values* 26, 339–408.

Hoppe, R. (1999) Policy analysis, science and politics: from 'speaking truth to power' to 'making sense together'. *Science and Public Policy* 26(3), 201–210.

Hoppe, R. (2005) Rethinking the science-policy nexus: from knowledge utilization and science technology studies to types of boundary arrangements. *Poiesis and Praxis* 3, 199–215.

Jones, R.N. (2000) Managing uncertainty in climate change projections: Issues for impact assessment. *Climatic Change* 45, 403–419.

Kiem, A.S. and Verdon-Kidd, D.C. (2011) Steps towards 'useful' hydroclimatic scenarios for water resource management in the Murray-Darling Basin. *Water Resources Research* 47, W00G06.

Kiem, A.S. and Austin, E.K. (2013a) Disconnect between science and end-users as a barrier to climate change adaptation. *Climate Research* 58, 29–41.

Kiem, A.S. and Austin, E.K. (2013b) Drought and the future of rural communities: opportunities and challenges for climate change adaptation in regional Victoria, Australia. *Global Environmental Change* 23(5), 1307–1316.

Kiem, A.S., Askew, L.E., Sherval, M. et al. (2010a) *Drought and the Future of Rural Communities: Drought Impacts and Adaptation in Regional Victoria, Australia*. National Climate Change Adaptation Research Facility, Gold Coast.

Kiem, A.S., Verdon-Kidd, D.C., Boulter, S.L. and Palutikof, J.P. (2010b) *Learning From Experience: Historical Case Studies and Climate Change Adaptation*. National Climate Change Adaptation Research Facility, Gold Coast.

Kiem, A.S., Verdon-Kidd, D.C. and Austin, E.K. (2014) Bridging the gap between end user needs and science capability: decision making under uncertainty. *Climate Research*, doi: 10.3354/cr01243.

Koutsoyiannis, D., Efstratiadis, A., Mamassis, N. and Christofides, A. (2008) On the credibility of climate predictions. *Hydrological Sciences Journal* 53, 671–684.

Koutsoyiannis, D., Montanari, A., Lins, H.F. and Cohn T.A. (2009) Climate, hydrology and freshwater: towards an interactive incorporation of hydrological experience into climate research. Discussion of 'The implications of projected climate change for freshwater resources and their management'. *Hydrological Sciences Journal* 54, 394–405.

Lemos, M.C., Kirchhoff, C.J. and Ramprasad, V. (2012) Narrowing the climate information usability gap. *Nature Climate Change* 2, 789–794.

Maraun, D., Wetterhall, F., Ireson, A.M., et al. (2010) Precipitation downscaling under climate change: recent developments to bridge the gap between dynamical models and the end-user. *Reviews of Geophysics* 48, RG3003.

McNie, E.C. (2007) Reconciling the supply of scientific information with user demands: an analysis of the problem and review of the literature. *Environmental Science and Policy* 10, 17–38.

Meinke, H., Howden, S.M., Struik, P.C., Nelson, R., Rodriguez, D. and Chapman, S.C. (2009) Adaptation science for agriculture and natural resource management- urgency and theoretical basis. *Current Opinion in Environmental Sustainability* 1, 69–76.

Montanari, A., Blöschl, G., Sivapalan, M. and Savenije, H. (2010) Getting on target. Public Service Review. *Science and Technology* 7, 167–169.

Parry, M.L., Canziani, O.F., Palutikof, J.P., van der Linden, P.J. and Hanson, C.E. (2007) *Climate Change 2007: Impacts, Adaptation and Vulnerability. Contribution of Working Group II to the Fourth Assessment Report of the Intergovernmental Panel on Climate Change*. Cambridge University Press, Cambridge, UK.

Patt, A. and Dessai, S. (2005) Communicating uncertainty: lessons learned and suggestions for climate change assessment. *Comptes Rendus Geoscience* 337, 425–441.

Pitman, A.J. and Perkins, S.E. (2008) Regional projections of future seasonal and annual changes in rainfall and temperature over Australia based on skill-selected AR4 models. *Earth Interactions* 12, 1–50.

Power, S., Sadler, B. and Nicholls, N. (2005) The influence of climate science on water management in Western Australia: Lessons for climate scientists. *Bulletin of the American Meteorological Society* 86, 839–844.

Randall, D.A., Wood, R.A., Bony, S. et al. (2007) Climate models and their evaluation. In: Solomon, S., Qin, D., Manning, M. et al. (eds) *Climate Change 2007: The*

Physical Science Basis Contribution of Working Group I to the Fourth Assessment Report of the Intergovernmental Panel on Climate Change. Cambridge University Press, Cambridge, United Kingdom and New York, USA, pp. 589–662.

Sarewitz, D. and Pielke Jr, R. (1999) Prediction in science and policy. *Technology in Society* 21, 121–133.

Stainforth, D.A., Allen, M.R., Tredger, E.R. and Smith, L.A. (2007) Confidence, uncertainty and decision-support relevance in climate predictions. *Philosophical Transactions of the Royal Society* A365, 2145–2161.

Tang, S. and Dessai, S. (2012) Usable science? The UK Climate Projections 2009 and decision support for adaptation planning. *Weather, Climate and Society* 4, 300–313.

Verdon-Kidd, D.C. and Kiem, A.S. (2010) Quantifying drought risk in a non-stationary climate. *Journal of Hydrometeorology* 11, 1019–1031.

Section 2

Managing ecosystems under climate change

8 The challenge of biodiversity adaptation under climate change

JONATHAN T. OVERPECK

Institute of the Environment, University of Arizona, USA

8.1 The challenge

The risk of major biodiversity loss given continued human-driven climate change is real. Estimates of potential species loss are as high as possibly the Earth's sixth major mass extinction (Barnosky et al. 2011; NRC 2013); even though the most recent scientific consensus points to serious extinction risk, a firm estimate of potential species loss still eludes us (Leadley et al. 2010; Settele et al. 2014). Of course, there is more at risk than just the loss of species and diversity; maintaining the evolutionary and genetic flexibility of species to adapt to future stresses, climate and otherwise, is also an important objective (Hoffmann and Sgro 2011; Reed et al. 2011). Species and their genetic flexibility are the foundation of ecosystems that provide many key services to humans ranging from cleaning water and air, to providing renewable natural resources, to making up the composition of valued parks and preserves and much more. There is little doubt that humans have a strong need to avoid major ecosystem damage and biodiversity loss.

A key question for policymakers, conservation managers and humankind more generally is whether nature itself, or nature assisted by humans, can adapt to climate change.

Fortunately, we know nature by itself can adapt to some level of climate change. Climate has always changed on timescales of years to millions of years (Masson-Delmotte et al. 2014), and the Earth's biota has obviously persisted through much of this change relatively unscathed (Dawson et al. 2011; Settele et al. 2014). However, there have also been times when climate change has contributed to higher than usual extinction rates, all the way up to mass extinction (Barnosky et al. 2011; NRC 2013; Settele et al. 2014). We also know that most species on Earth have managed to adapt to c. 0.9°C global warming since 1901 (and greater change regionally; Hartmann et al. 2014; Settele et al. 2014), but that some species have been stressed substantially in the face of this recent climate change. This 0.9°C global warming could end up being small compared to changes in decades to come (Settele et al. 2014).

So how does the risk of biodiversity loss scale with the magnitude of human greenhouse gas emissions and the resulting global climate change? At present, all we know with confidence is that risk goes up with the magnitude of climate change (Settele et al. 2014). Can't we do better and quantify this relationship? Because extinction is forever, what level of extinction risk, and

Applied Studies in Climate Adaptation, First Edition. Edited by Jean P. Palutikof, Sarah L. Boulter, Jon Barnett and David Rissik.
© 2015 John Wiley & Sons, Ltd. Published 2015 by John Wiley & Sons, Ltd.

thus climate change, is acceptable? A more all-encompassing question remains: what level of anthropogenic climate change will trigger unacceptable levels of biodiversity loss?

Unfortunately, even if humankind is able to decide what constitutes 'unacceptable', estimates of how biodiversity loss will scale with future levels of greenhouse gas emissions and global climate change will probably never be secure enough to bet the Earth's biota on – at least not any time soon. The goal of this essay is to highlight why this is true (see also e.g. Stein et al. 2013; Settele et al. 2014) and to encourage great vigilance when assessing all of the serious threats biodiversity might face in any given region. If an overarching goal is to preclude mass extinction, it would be wise to err on the side of caution and fight more aggressively to reduce climate change and other threats to biodiversity.

8.2   The growing quiver of adaptation tools

The time to debate the value of adaptation is over; it is one of the two primary tools (the other being mitigation of greenhouse gas emissions) for dealing with the effects of climate change and other human stressors that are already threatening biodiversity (e.g. Staudinger et al. 2013; Settele et al. 2014). As a result, adaptation science and implementation will be defining human endeavours for the rest of the twenty-first century and beyond. The good news is that adaptation science has already come a long way (e.g. Dawson et al. 2011; Stein et al. 2013; Settele et al. 2014; see also Chapters 9–12), but the bad news is that much remains to be learned. The quiver of useful adaptation tools is growing, but will it be enough?

The four chapters that follow this essay (Chapters 9–12) collectively do a nice job of highlighting the science and application of biodiversity adaptation by describing efforts focused on the conservation of a number of species in Australia. Many aspects of biodiversity adaptation are summarised, including the daunting magnitude of the challenge associated with just conserving the focus species alone, and the implications of these papers go far beyond biodiversity conservation in Australia alone. In doing this, the much bigger financial challenge associated with conserving the complete biodiversity of an entire country is put in perspective ('a frightening sum', Garnett and Franklin 2014), as is the even larger challenge of conserving biodiversity in poorer countries and around the globe. Reading just these four chapters makes clear how expensive a successful global biodiversity adaptation effort could become, quite possibly more expensive than transforming the world's energy infrastructure sufficiently to avoid major climate change and biodiversity loss.

The following four chapters also offer a useful and succinct overview of adaptation strategies including the need to manage both climate and non-climate stresses, acting in the face of significant uncertainty and embracing continual learning and adaptive management. The case for expanded natural and social science research, as well as more extensive monitoring of the multiple stressors, is made. The inequity of perceived species value also comes though loud and clear; some charismatic species and ecosystems garner attention and investment, while many others are not on the current adaptation radar screen. Concern over this fact is heightened by poorly understood relationships among species and the fact that many species and other aspects of ecosystems go unstudied and unmonitored. Moreover, the papers highlight the key roles that local to national governance can – and, in some cases, must – play if adaptation efforts are to be successful. Collectively, these four chapters highlight a challenge that is as complex as it is potentially expensive.

8.3   Emerging threats to successful adaptation

Given space, it would be possible to generate a long list of issues that might threaten successful adaptation (e.g. Dawson et al. 2011; Stein et al.

2013; Rudd and Fleishman 2014; Settele et al. 2014, as well as the papers in this book). When combined with the 'forever' aspect of extinction, these issues should be dealt with better than might currently be possible. Some of the issues requiring more consideration are discussed in the Sections 8.3.1–8.3.4.

8.3.1   *Limits to climate models*

A primary climate science concern is that climate models – as good as they now are – do not agree on the magnitude and detail of climate change that might result from a given level of greenhouse gas forcing. Many adaptation efforts use multi-model ensemble averages to anticipate future climate stresses, when ranges and low-probability/high-risk changes might be just as important to biodiversity adaptation. Moreover, these same state-of-the-art models are not presently able to simulate all aspects of future climate that may be key to some biodiversity conservation efforts. Clever use of multi-model ensembles can provide improved estimates of the range of change that might occur in any given location (e.g. Dominguez et al. 2012; Vano et al. 2014) but some key variables, for example extremes such as drought and mega-drought, might not be well simulated by any model (Ault et al. 2013, 2014). Will adaptation strategies fare well if confronted by a drought that is much longer and hotter than ever seen before? Similarly, climate models seem unable to capture the full range of El Niño and La Niña behaviour that is apparent in instrumental observations (Flato et al. 2014). Even more troubling is the palaeoclimatic evidence that the range of ENSO behaviour in the instrumental record was eclipsed by larger, more prolonged El Niño–Southern Oscillation (ENSO) -related extremes in recent Earth history (e.g. Conroy et al. 2008, 2009). As a rule, wise adaptation plans will often need to have provision for much larger and longer climatic extremes than have been seen in the instrumental record or in state-of-the-art simulations of climate.

8.3.2   *The potential for abrupt change and surprise*

Another climate issue is abrupt change. Closely linked with the issue of 'climate surprises', there is a whole range of abrupt, tipping point or threshold climate system behaviour that is evident in palaeo-environmental records that is not fully understood, nor proven to be within the capability of our climate models to simulate (Overpeck and Cole 2006; Lenton et al. 2008; NRC 2013). One increasingly well-known issue is that of abrupt shifts between prolonged wet and dry periods, which is more likely than previously thought (Ault et al. 2013, 2014). Another is the abrupt sea-level rise that could take place at rates in excess of 1 m per century if ice-sheet collapse accelerates further. This could then drive unprecedented coastal storm impacts (Parris et al. 2012; NRC 2013; Church et al. 2014). Moreover, many multi-model projections of climate change are presented as if they will be steady and gradual through the coming century and beyond; they could in fact be characterised by abrupt shifts along the way, particularly at local scales critical for biodiversity conservation (Overpeck and Cole 2006; Higgins and Scheiter 2012).

8.3.3   *Historical evidence*

In some sense, it is comforting that the Earth's biota has dealt with the full range of climate system behaviour in the past and has persisted. That said, biota in many parts of the globe will soon be confronted by climate unlike any it has ever seen in its evolutionary history (Diffenbaugh and Field 2013; Settele et al. 2014). Uniquely high temperatures are likely in many places, as are higher rates of climate change than seen in millennia. The additional human-related stressors will also be uniquely challenging: unprecedented atmospheric carbon dioxide concentrations and other air and water pollution, depleted groundwater, fragmented landscapes, invasive species and predation. Great care must be taken to avoid assuming the past is a good guide for the future in terms of understanding how biota will respond to climate, and in terms of understanding the natural capacity of biota to deal with climate change.

8.3.4 *Future human behaviour*

As challenging as it may be to improve and compensate for climate model shortcomings, it is probably achievable. For example, careful model-based interdisciplinary scenario planning, coupled with clever use of instrumental and palaeoclimatic records (e.g. Ault et al. 2014), might allow conservation managers to bracket the full range of future climate threats. Biodiversity conservation plans could however conflict with other aspects of climate change adaptation and mitigation (e.g. renewable energy or biofuel deployment), and make the complexity of biodiversity conservation planning and implementation more daunting. Moreover, a bigger problem could be the inability to characterise future human behaviour, ranging from estimating future greenhouse gas emissions to assessing the stability of future governments and their commitments to biodiversity conservation. As Chapters 9–12 make clear, the cost of biodiversity adaptation on a global scale could be extremely expensive; further, it would have to be sustained for decades and longer to be successful. It seems risky to assume that the will to sustain an ever-expanding array of adaptation programs is guaranteed, even if the cost of failure could be the Earth's sixth major mass extinction event.

8.4   The bottom line

Managing and saving the Earth's biota in the face of rapid large climate change might turn out to be the largest challenge ever faced by humans. The stakes are high and the complexity of the problem immense. Thus far, and as the following chapters of this book attest, collaborations between scientists, resource managers and the public are already beginning to grapple with this complexity and to build capability. These chapters also support the growing consensus (e.g. Dawson et al. 2011; Stein

et al. 2013; Settele et al. 2014) that much more investment in science and practice will be needed to succeed. A key point, however, is that overconfidence in our ability to succeed could easily spell disaster for species and other biodiversity as they become threatened. Many uncertainties remain, and surprises will occur. The growing multitudes of biodiversity adaptation efforts around the globe should all include the provision for error, as well as the flexibility to face unexpected challenges. Failure comes at a high cost given that extinctions are forever.

Acknowledgements

I thank Jean Palutikof and anonymous reviewers for valuable feedback and suggestions. I also thank the Victorian Centre for Climate Change Adaptation Research for the funding support that made this paper possible.

References

Ault, T.R., Cole, J.E., Overpeck, J.T. et al. (2013) The continuum of hydroclimate variability in western North America during the last millennium, *Journal of Climate* 26, 5863–5878.

Ault, T.R. Cole, J.E., Overpeck, J.T., Pederson, G.T. and Meko, D.M. (2014) Assessing the risk of persistent drought using climate model simulations and paleoclimate data. *Journal of Climate,* published online 2014, doi: 10.1175/JCLI-D-12-00282.1.

Barnosky, A.D., Matzke, N., Tomiya, S. et al. (2011) Has the Earth's sixth mass extinction already arrived? *Nature* 471, 51–57.

Church, J.A., Clark, P.U., Cazenave, A. et al. (2014) Sea level change. In: Stocker, T.F., Qin, D., Plattner, G.K. et al. (eds) *Climate Change 2013: The Physical Science Basis. Contribution of Working Group I to the Fifth Assessment Report of the Intergovernmental Panel on Climate Change.* Cambridge University Press, Cambridge, in press.

Conroy, J. L., Overpeck, J.T., Cole, J.E., Shanahan, T.M. and Steinitz-Kannan, M. (2008) Holocene changes in eastern tropical Pacific climate inferred from a Galápagos lake sediment record. *Quaternary Science Reviews* 27, 1168–1180.

Conroy, J.L., Restrepo, A., Overpeck, J.T. et al. (2009) Unprecedented recent warming of surface temperatures in the eastern tropical Pacific Ocean. *Nature Geoscience* 2, 46–50.

Dawson, T.P., Jackson, S.T., House, J.I., Prentice, I.C. and Mace, G.M. (2011) Beyond predictions: Biodiversity conservation in a changing climate. *Science* 332, 53–58.

Diffenbaugh, N.S. and Field, C.B. (2013) Changes in ecologically critical terrestrial climate conditions. *Science* 341, 486–492.

Dominguez, F., Rivera, E., Lettenmaier, D.P. and Castro, C.L. (2012) Changes in winter precipitation extremes for the western United States under a warmer climate as simulated by regional climate models. *Geophysical Research Letters* 39, doi: 10.1029/2011gl050762.

Flato, G., Marotzke, J, Abiodun, B. et al. (2014) Evaluation of climate models. In: Stocker, T.F., Qin, D., Plattner, G.K. et al. (eds) *Climate Change 2013: The Physical Science Basis. Contribution of Working Group I to the Fifth Assessment Report of the Intergovernmental Panel on Climate Change.* Cambridge University Press, Cambridge.

Garnett, S. and Franklin, D. (eds) (2014) *Climate Change Adaptation Plan for Australian Birds.* CSIRO Publishing, Melbourne.

Hartmann, D. L., Klein Tank, A. M. G., Rusticucci, M. et al. (2014) Observations: Atmosphere and surface. In: Stocker, T.F., Qin, D., Plattner, G.K. et al. (eds) *Climate Change 2013: The Physical Science Basis. Contribution of Working Group I to the Fifth Assessment Report of the Intergovernmental Panel on Climate Change.* Cambridge University Press, Cambridge.

Higgins, S.I. and Scheiter, S. (2012) Atmospheric CO_2 forces abrupt vegetation shifts locally, but not globally. *Nature* 488, 209–212.

Hoffmann, A.A. and Sgro, C.M. (2011). Climate change and evolutionary adaptation. *Nature* 470, 479–485.

Leadley, P., Pereira, H.M., Alkemade, R. et al. (2010) *Biodiversity Scenarios: Projections of 21st Century Change in Biodiversity and Associated Ecosystem Services.* CBD Technical Series 50, Secretariat of the Convention on Biological Diversity, Montreal.

Lenton, T. M., Held, H., Kriegler, E. et al. (2008) Tipping elements in the Earth's climate system. *Proceedings of the National Academy of Sciences of the United States of America* 105, 1786–1793.

Masson-Delmotte, V., Schulz, M., Abe-Ouchi, A. et al. (2014) Information from Paleoclimate Archives. In: Stocker, T.F., Qin, D., Plattner, G.K. et al. (eds) *Climate Change 2013: The Physical Science Basis. Contribution of Working Group I to the Fifth Assessment Report of the Intergovernmental Panel on Climate Change.* Cambridge University Press, Cambridge.

NRC (2013) *Abrupt Impacts of Climate Change.* National Research Council, National Academy Press, Washington DC.

Overpeck, J.T. and Cole, J.E. (2006) Abrupt change in the Earth's climate system. *Annual Reviews of Environment and Resources* 31, 1–31.

Parris, A., Bromirski, P., Burkett V. et al. (2012) *Global Sea Level Rise Scenarios for the US National Climate Assessment.* NOAA Tech Memo OAR CPO-1. Silver Spring, MD.

Reed, T.E., Schindler, D.E. and Waples, R.S. (2011) Interacting effects of phenotypic plasticity and evolution on population persistence in a changing climate. *Conservation Biology* 25, 56–63.

Rudd, M.A. and Fleishman, E. (2014) Policymakers' and scientists' ranks of research priorities for resource-management policy. *BioScience*, published online 5 February 2014, doi: 10.1093/biosci/bit035.

Settele, J., Scholes, R., Betts, R. et al. (2014) Terrestrial and inland water systems. In Field, C., Barros, V., Mastrandrea, M. et al. (eds) *Climate Change 2014: Impacts, Adaptation and Vulnerability. Contribution of Working Group II to the IPCC Fifth Assessment.* Cambridge University Press, Cambridge.

Staudinger, M.D., Carter, S.L., Cross, M.S. et al. (2013) Biodiversity in a changing climate: a synthesis of current and projected trends in the US. *Frontiers in Ecology and the Environment* 11, 465–473.

Stein, B.A., Staudt, A., Cross, M.S. et al. (2013) Preparing for and managing change: climate adaptation for biodiversity and ecosystems. *Frontiers in Ecology and the Environment* 11, 502–510.

Vano, J.A. Udall, B., Cayan, D.R. et al. (2014) Understanding uncertainties in future Colorado River streamflow. *Bulletin of the American Meteorological Society* 95, 59–78.

9 Management options for bird conservation in the face of climate change

DONALD C. FRANKLIN[1], APRIL E. RESIDE[2]
AND STEPHEN T. GARNETT[1]

[1]*Research Institute for the Environment and Livelihoods, Charles Darwin University, Australia*
[2]*Centre for Tropical Biodiversity and Climate Change, James Cook University, Australia*

9.1   Introduction

The evidence is clear that recent climate change has already had an impact on birds and bird populations (Møller et al. 2010) as demonstrated both internationally (Thomas et al. 2006) and within Australia (e.g. Chambers and Keatley 2010; Chambers et al. 2011). Given that both climate change and the direct or indirect relationship between climate and bird ecology are pervasive, we should anticipate even greater impacts on the abundance and distribution of bird species with more extreme climate change in the foreseeable future (although there remains much uncertainty about the magnitude, direction and nature of these impacts; Pereira et al. 2010). Some of these impacts will doubtless be negative, threatening species with regional or global extinction (Maclean and Wilson 2011). Ecosystems may suffer loss of function as a result of the loss of species (Tomimatsu et al. 2013). Other bird species may benefit from change (Thomas et al. 2010). Still others may be impacted but in ways that do not affect populations. For example, mismatches between the timing of breeding and the timing of peak food availability induced by climate change may reduce reproductive success but sufficient young may still be produced to maintain the population at current levels (Chamberlain and Pearce-Higgins 2013; Reed et al. 2013). The response to these impacts may be gradual or abrupt, the latter being particularly unpredictable and the result of populations reaching demographic thresholds of stress.

For managers, the impacts of *ultimate* concern will mainly be regional or global extinction and loss of ecosystem function. More immediate concerns such as the spread of invasive or competitive species lead to ultimate concerns through their negative effects on species and ecosystems, and here we treat the spread of invasive species as one of many processes driving concern rather than the cause of concern itself. It is beyond the scope of this chapter for us to review such processes, which might include (but are far from restricted to): range shifts in response to shifts in climate envelopes (Parmesan and Yohe 2003); intolerance of extreme climatic events such as drought (e.g. MacNally et al. 2009); loss of habitat or resources (Cahill et al. 2013); mismatches in the timing of breeding or migration in relation to weather and resources (Thackeray et al. 2010); and changes to the balance of interspecific competition (Urban et al. 2012).

Rapid and uncertain change presents an enormous challenge for wildlife managers. If we are to avoid widespread extinction and loss of ecosystem function then we must equip ourselves with new ways of thinking and a new and/or reinforced toolkit of management actions. This chapter builds on the Australian context to explore how we might deal with the challenges likely to face birds as a result of climate change over the next half-century. We first introduce some general management principles, then detail options for action and finally briefly explore issues surrounding the timing of management responses. A more detailed outline of management options is provided by Franklin et al. (2014). A comprehensive analysis of climate change adaptation and Australian birds is found in Garnett et al. (2013) and Garnett and Franklin (2014).

9.2   The purpose of management in the face of climate change

As climate changes, those goals of conservation that are implicitly based on the assumption that climate is static must be set aside (Hannah et al. 2002). The pre-European state of Australian ecosystems, often held as an aspirational goal for conservation, will become increasingly irrelevant. Protected areas may end up conserving something quite different to that which was originally intended (Monzon et al. 2011). Some species may need to be moved to new locations, challenging concepts of 'native' and 'invasive' (Webber and Scott 2012). Fundamental values embodied in current legislation, such as subspecies being the basic units of conservation, will be undermined by both natural and assisted movements. There will also be new human dimensions. What levels of assisted colonisation will the public tolerate? Will a species or subspecies that can persist only in captivity because its environmental envelope has disappeared

continue to attract the necessary levels of expenditure for its survival?

9.3   General principles

Climate change is one among many threats (stressors) facing species, and the impact of climate change may be hidden by threats that seem more imminent. Negatively, the synergy of threats may well be worse than the sum of each (Brook et al. 2008); positively, it can mean that the mitigation of other threats is an appropriate (and more tractable) response to the threat posed by climate change (CCWAPWG 2009). This means that existing responses will need to be reinforced alongside the implementation of novel strategies (Lawler et al. 2010).

9.3.1   Strategies in the face of uncertainty

Types of uncertainty include unpredictable and often stochastic fluctuations in populations (Sinclair et al. 2006), ignorance about the relationship between a species' fundamental and realised climatic niche (Veloz et al. 2012), imprecision in climate projections at a local scale (Timbal et al. 2009), the likelihood that some climate components will change in abrupt steps as thresholds are breached (Lenton et al. 2008) and lack of a climatic benchmark (Head 2012). Possible management responses to ecological uncertainty include prioritising actions where current knowledge levels are high (Groves et al. 2012) and where detailed local projections of climate change and its impacts are not needed (Cross et al. 2012), and to practice adaptive management in a risk management framework (Conroy et al. 2011).

9.3.2   Three conceptual goals for conservation

Three conceptual goals for conservation are: avoiding the effects of climate change (resistance); promoting resilience to the impacts of climate change; and facilitating a response to it (Millar et al. 2007; Prober et al. 2012). Resistance

Table 9.1 Management options for the conservation of birds in a climate-change world (modified extensively from Mawdsley et al. 2009, see also Garnett et al. 2013, Shoo et al. 2013).

Management type	Practical options
Do nothing	–
Maintain and enhance habitat	Expand the protected area network; maintain and improve habitat quality; identify, protect and expand refugia; maintain and extend landscape connectivity; create new habitats
Facilitate the response of wild populations (intensive species management)	Assist colonisation by translocation; enhance the genetics of subspecies; manage other threatening processes (e.g. by predator control, habitat manipulation, captive breeding)
Preserve populations (the last resort)	Save species in captivity; store germplasm
Understand and prepare for what is happening	Monitor bird populations (general surveys e.g. Atlas; targeted species-specific monitoring); monitor habitats and threatening processes; investigate the ecology of species and communities; model habitat and climate envelopes in more detail; model management options

extends current management practice with climate change just another threat, and resilience combines current practice with new measures that increase population size or habitat area (Morecroft et al. 2012). The response goal acknowledges the need for inventive responses beyond current management.

9.3.3 Heirarchical adaptive management

A hierarchical approach to adaptive management is recommended in which steps increase in intensity, dollar cost, potential adverse environmental consequences and uncertainty as required (Table 9.1). The baseline is to do nothing. This should, however, only be considered where the consequences have been articulated, where monitoring is in place and where there is a commitment to institute management if circumstances change. As a second step, existing habitat can be maintained or enhanced by: expanding the protected area network; maintaining and improving habitat quality; identifying, protecting and expanding refugia; maintaining and extending landscape connectivity; and creating new habitats. Third, facilitating the response of wild populations through intensive species management would comprise actions such as: assisting colonisation by translocation; enhancing subspecies

genetics; or targeted management of other threatening processes by, for example, predator control, habitat manipulation or supplementary captive breeding. Finally, and as a last resort, populations can be preserved in captivity or as germplasm. Informing all this is monitoring and research to identify trends and understand the processes underpinning them.

9.4   Actions

9.4.1   Maintaining and enhancing habitat

Enhancing protected areas
For many species, the retention of habitat through a network of public and private protected areas on both land and on sea will be even more critical in the face of climate change (Hannah et al. 2007). This must include both intact and fragmented landscapes (Van Teeffelen et al. 2012) and must anticipate geographic shifts in climate if they are to be effective into the future (Hole et al. 2011). A coastal drift of climate space for many birds of south-eastern Australia (VanDerWal et al. 2013) and global sea-level rises (e.g. Thorne et al. 2012) presages conflict with some intensive land uses which requires anticipatory management. Some opportunities for reservation may arise in settled

districts where climate change renders farmland commercially unviable.

Enhancing habitat quality

An extensive body of research suggests that functional redundancy – in which more than one species serves a particular function within an ecosystem, e.g. dispersing fruit – and structural diversity provide strong support for species, so retaining or improving habitat quality will be an essential climate change adaptation strategy for many species. In an Australian context, weed and feral animal control and fire management will continue to be essential management goals, but may need to be intensified. Current prescriptions may well need to be tailored to new circumstances.

Refugia

A key management strategy will be the identification, protection and enhancement of places that will serve as climate refugia into the future. Refugia are places where the habitat and resources that a species needs are buffered against direct (e.g. temperature, moisture) and indirect (e.g. interactions, fire) impacts of climate change (Keppel et al. 2012). These might include, for example, mountain ranges, valleys, watercourses, springs and wetlands (Reside et al. in press). Refugia demonstrably facilitated the persistence of many species during the Ice Ages (e.g. Byrne 2008 for southern Australia). They need to be large enough for long enough to retain a genetically viable population (Ovaskainen 2002) and ideally be accessible for species to disperse to unaided. Three approaches can be taken to identify refugia (Ashcroft 2010): modelling of places where analogue climates will persist into the future independent of species (e.g. Mackey et al. 2012); identification of places where individual species are likely to persist; and identification of places where species have persisted in the past.

Ecological connectivity

Many bird species have already tracked geographical shifts in climate (Tingley et al. 2009; Brommer et al. 2012). For some, this movement will be prevented by a range of physical barriers. The 2000 km band of semi-arid vegetation may well prevent birds that inhabit the forests of south-western Australia from tracking their climate niche to forests in eastern Australia, for other species, however, barriers such as cleared land (e.g. Doerr et al. 2011), the ocean separating islands or even unsuitable habitat such as forests or rivers may operate at more local scales. The management response is to ensure that connectivity between current and potential habitat is retained or enhanced (Groves et al. 2012). Ecological connectivity also enhances the resilience of local populations through retention of metapopulation structure (Van Teeffelen et al. 2012). Planning for landscape corridors is well-advanced in Australia, enhancing the legacy of extensive tracts of uncleared habitat (DSEWPC 2012). In planning connectivity for birds, it is useful to characterise the species most likely to benefit from this form of management. Small species cross gaps less readily than large species (Lees and Peres 2009), and sedentary species and habitat specialists, especially those from closed vegetation, are often more reluctant to cross gaps than migrants, nomads or species of open habitats (Harris and Reed 2002). Birds with high dispersal capacity may not require connectivity (Mokany et al. 2013), whereas those with low dispersal capacity may not even be able to make use of corridors (Johst et al. 2011). With these qualifications in mind, investment in connectivity should be appraised carefully against other management options including assisted colonisation (see 'Assisted colonisation' in Section 9.4.2).

New habitats

Climates and communities with no previous analogue (Garcia-Lopez and Allue 2013 and Urban et al. 2012 respectively) may call for new habitats. Given time lags and expense, it is important to identify potential beneficiary species and have a clear understanding of the values underlying the management intent, especially where new habitats might be required to replace climate-change-induced degradation within existing natural areas. These issues have already received some appraisal with respect to plants (Booth and Williams 2012).

9.4.2   Intensive management

Assisted colonisation

Moving species to places they have never been recorded in anticipation of climate shifts is a logical but controversial extension to existing management practices for threatened species of creating insurance populations and restoring those that have been lost (Hoegh-Guldberg et al. 2008; Thomas 2011). Assisted colonisation may also be employed to retain or restore ecosystem function (Lunt et al. 2013), for example to ensure that bird-pollinated plants have a bird capable of pollinating them present in the ecosystem. The process entails risks to source populations, founder individuals and the receiving ecosystem that require careful appraisal (Schwartz et al. 2012). Issues relating to timing (McDonald-Madden et al. 2011), genetics (Weeks et al. 2011), policy (for Australia, Burbidge et al. 2011) and the mixed success of previous reintroductions (for Australia, Sheean et al. 2012) have been reviewed.

Enhanced genetics

Many Australian bird subspecies are likely to lose their climate space over coming decades, but the new climate within their current range will often become suitable for another subspecies (Garnett et al. 2013). Assuming that subspecies have specific adaptations to their current climate spaces, this suggests that the survival of the population is likely to become maladapted over time may be enhanced by genetic augmentation from the other subspecies (Thomas et al. 2013). Deliberate enhancement also brings risks of outbreeding depression and disease transmission, but these seem relatively minor (Weeks et al. 2011).

9.4.3   Preserving populations

Captivity

Populations of some bird species already exist only in captivity, though nearly always there is hope of reintroduction to the wild. With climate change, reintroduction may not be possible in the foreseeable future so long-term or permanent *ex situ* conservation must be among the potential management tools as an alternative to extinction (Pritchard et al. 2012). This raises substantial ethical, technical and economic questions which have been explored in the context of captive breeding to enhance the conservation of wild populations, including: (1) captive populations are rarely self-sufficient; (2) captive breeding is often expensive; (3) captive populations often drift into domestication and inbreeding; (4) disease is common; and (5) administrative continuity is rare (Snyder et al. 1996; Araki et al. 2007; Christie et al. 2012). The considerable cost calls for prioritisation (Joseph et al. 2009) and strategic use of private as well as public resources (Cannon 1996). Conflict may also arise between the need to maximise the number of founders needed for genetic viability of a captive populations and attempts to save a species in the wild (McDonald-Madden et al. 2011).

Germplasm and code

The technologies for storage of the germplasm, embryos, blood products, tissue and DNA of birds, probably the very last option for biodiversity conservation (Wildt et al. 1997) are developing rapidly (Glover and McGrew 2012). Both semen and primordial germ cells can now be stored cryogenically and used to fertilise host embryos (Wernery et al. 2010). The technology may be the only practical way to preserve the exact genetic make-up of original forms that, in some unforeseeable future, could be re-established in the wild. The recreation of organisms from their DNA code is currently in the realm of science fiction, but coding technologies have improved exponentially in the last decade.

9.4.4   Monitoring and research

Monitoring

The appropriate timing of intervention in the face of climate change relies heavily on our understanding of population numbers and trends as well as change to habitat and ecological and threatening processes

(McDonald-Madden et al. 2011). Ecological surprises are inevitable (Doak et al. 2008) and require early detection if the response is to be adequate. Monitoring of Australian birds is patchy at best and absent at worst (Lindenmayer et al. 2012), although atlas schemes have provided baselines for all but the rare species (Blakers et al. 1984; Barrett et al. 2002, 2003). Monitoring of habitats and threatening processes may in some circumstances serve as adequate proxies (Lindenmayer and Likens 2010), but this monitoring too is currently of limited extent.

Research

Monitoring may identify trends but more detailed research is required to identify their drivers and thus optimal responses. Research will usually be required to identify the contribution of climate change to observed declines (though this in itself may not be productive; Parmesan et al. 2013) along with the demographic and ecological processes underlying it. For example, declines may be a direct response to climate-induced changes in resources or an indirect response via changes to interspecific interactions. Field research on species and communities should be complemented by modelling of habitat, climate envelopes and management options; Harris et al. (2012) provides an Australian avian example of this.

9.5   Timing and continuity

For successful climate change adaptation, some actions must begin immediately while others can wait for the results of monitoring; some actions will be one-off while others will need to be ongoing. This yields a four-way matrix of action categories. Immediate one-off actions might include land purchases, assisted colonisation of species already seriously threatened by climate change, surveys of little-known species or subspecies to create a baseline for future comparisons and the identification of refugia. Ongoing actions starting now include monitoring and management and captive breeding for species already seriously threatened (Garnett et al. 2011). Future one-off and ongoing actions will primarily be those identified as necessary in response to monitored trends.

9.6   Conclusion

Managers are not helpless in the face of climate change. However, the range of options available is not large and it cannot be stated strongly enough that mitigation is by far the cheapest and most effective way to retain our biodiversity. Even though the Australian avifauna has been winnowed by a tough environmental history, the combination of pressure from climate change and other anthropomorphic environmental change will make retention of all – or even most – bird species (and subspecies) exceptionally challenging. Substantially greater funds will be needed and, even then, strong prioritisation of actions will be essential (Wilson et al. 2006; Joseph et al. 2009). Society will need to debate difficult decisions about how it values biodiversity.

References

Araki, H., Cooper, B. and Blouin, M.S. (2007) Genetic effects of captive breeding cause a rapid, cumulative fitness decline in the wild. *Science* 318, 100–103.

Ashcroft, M.B. (2010) Identifying refugia from climate change. *Journal of Biogeography* 37, 1407–1413.

Barrett, G., Silcocks, A. and Cunningham, R. (2002) *Australian Bird Atlas (1998–2001) Supplementary Report No. 1 - Comparison of Atlas 1 (1977–1981) and Atlas 2 (1998–2001)*. Birds Australia, Melbourne.

Barrett, G., Silcocks, A., Barry, S., Cunningham, R. and Poulter, R. (2003) *The New Atlas of Australian Birds*. Royal Australasian Ornithologists Union, Hawthorn East.

Blakers, M., Davies, S.J.J.F. and Reilly, P.N. (1984) *The Atlas of Australian Birds*. RAOU and Melbourne University Press, Melbourne.

Booth, T.H. and Williams, K.J. (2012) Developing biodiverse plantings suitable for changing climatic conditions 1: Underpinning scientific methods. *Ecological Management and Restoration* 13, 267–273.

Brommer, J.E., Lehikoinen, A. and Valkama, J. (2012) The breeding ranges of central European and Arctic bird species move poleward. *PLoS ONE* 7, e43648.

Brook, B.W., Sodhi, N.S. and Bradshaw, C.J.A. (2008) Synergies among extinction drivers under global change. *Trends in Ecology and Evolution* 23, 453–460.

Burbidge, A.A., Byrne, M., Coates, D. et al. (2011) Is Australia ready for assisted colonization? Policy changes required to facilitate translocations under climate change. *Pacific Conservation Biology* 17, 259–269.

Byrne, M. (2008) Evidence for multiple refugia at different time scales during Pleistocene climatic oscillations in southern Australia inferred from phylogeography. *Quaternary Science Reviews* 27, 2576–2585.

Cahill, A.E., Aiello-Lammens, M.E., Fisher-Reid, M.C. et al. (2013) How does climate change cause extinction? *Proceedings of the Royal Society B* 80, Art. no. 20121890.

Cannon, J.R. (1996) Whooping Crane recovery: A case study in public and private cooperation in the conservation of endangered species. *Conservation Biology* 10, 813–821.

Chamberlain, D. and Pearce-Higgins, J. (2013) Impacts of climate change on upland birds: complex interactions, compensatory mechanisms and the need for long-term data. *Ibis* 155, 451–455.

Chambers, L.E. and Keatley, M.R. (2010) Australian bird phenology: a search for climate signals. *Austral Ecology* 35, 969–979.

Chambers, L.E., Devney, C.A., Congdon, B.C., Dunlop, N., Woehler, E.J. and Dann, P. (2011) Observed and predicted effects of climate on Australian seabirds. *Emu* 111, 235–251.

Christie, M.R., Marinea, M.L., French, R.A. and Blouin, M.S. (2012) Genetic adaptation to captivity can occur in a single generation. *Proceedings of the National Academy of Sciences* 109, 238–242.

Climate Change Wildlife Action Plan Work Group (2009) *Voluntary Guidance for States to Incorporate Climate Change into State Wildlife Action Plans and Other Management Plans.* Association of Fish and Wildlife Agencies.

Conroy, M.J., Runge, M.C., Nichols, J.D., Stodola, K.W. and Cooper, R.J. (2011) Conservation in the face of climate change: The roles of alternative models, monitoring, and adaptation in confronting and reducing uncertainty. *Biological Conservation* 144, 1204–1213.

Cross, M.S., Zavaleta, E.S., Bachelet, D. et al. (2012) The Adaptation for Conservation Targets (ACT) framework: A tool for incorporating climate change into natural resource management. *Environmental Management* 50, 341–351.

Department of Sustainability Environment Water Population and Communities (2012) *National Wildlife Corridors Plan: A Framework for Landscape-scale Conservation 2012.* Department of Sustainability, Environment, Water, Population and Communities, Canberra.

Doak, D.F., Estes, J.A., Halpern, B.S. et al. (2008) Understanding and predicting ecological dynamics: Are major surprises inevitable? *Ecology* 89, 952–961.

Doerr, V.A.J., Doerr, E.D. and Davies, M.J. (2011) Dispersal behaviour of Brown Treecreepers predicts functional connectivity for several other woodland birds. *Emu* 111, 71–83.

Franklin, D.C., Reside, A.E. and Garnett, S.T. (2014) Conserving Australian bird populations in the face of climate change. In: Garnett, S.T. and Franklin, D.C. (eds.) *Climate Change Adaptation Plan for Australian Birds.* CSIRO, Collingwood, pp. 53–78.

Garcia-Lopez, J.M. and Allue, C. (2013) Modelling future no-analogue climate distributions: A world-wide phytoclimatic niche-based survey. *Global and Planetary Change* 101, 1–11.

Garnett, S.T. and Franklin, D.C. (eds) (2014) *Climate Change Adaptation Plan for Australian Birds.* CSIRO, Collingwood.

Garnett, S.T., Szabo, J.K. and Dutson, G. (2011) *The Action Plan for Australian Birds 2010.* CSIRO, Collingwood.

Garnett, S.T., Franklin, D.C., Ehmke, G. et al. (2013) *Climate Change Adaptation Strategies for Australian Birds.* National Climate Change Adaptation Research Facility, Gold Coast.

Glover, J.D. and McGrew, M.J. (2012) Primordial germ cell technologies for avian germplasm cryopreservation and investigating germ cell development. *Journal of Poultry Science* 49, 155–162.

Groves, C.R., Game, E.T., Anderson, M.G. et al. (2012) Incorporating climate change into systematic conservation planning. *Biodiversity and Conservation* 21, 1651–1671.

Hannah, L., Midgley, G.F., Lovejoy, T. et al. (2002) Conservation of biodiversity in a changing climate. *Conservation Biology* 16, 264–268.

Hannah, L., Midgley, G., Andelman, S. et al. (2007) Protected area needs in a changing climate. *Frontiers in Ecology and the Environment* 5, 131–138.

Harris, R.J. and Reed, J.M. (2002) Behavioral barriers to non-migratory movements of birds. *Annales Zoologici Fennici* 39, 275–290.

Harris, J.B.C., Fordham, D.A., Mooney, P.A. et al. 2012. Managing the long-term persistence of a rare cockatoo under climate change. *Journal of Applied Ecology* 49, 785–794.

Head, L. (2012) Decentring 1788: Beyond biotic nativeness. *Geographical Research* 50, 166–178.

Hoegh-Guldberg, O., Hughes, L., McIntyre, S. et al. (2008) Assisted colonization and rapid climate change. *Science* 321, 345–346.

Hole, D.G., Huntley, B., Arinaitwe, J. et al. (2011) Toward a management framework for networks of protected areas in the face of climate change. *Conservation Biology* 25, 305–315.

Johst, K., Drechsler, M., van Teeffelen, A.J.A. et al. (2011) Biodiversity conservation in dynamic landscapes: trade-offs between number, connectivity and turnover of habitat patches. *Journal of Applied Ecology* 48, 1227–1235.

Joseph, L.N., Maloney, R.F. and Possingham, H.P. (2009) Optimal allocation of resources among threatened species: a project prioritization protocol. *Conservation Biology* 23, 328–338.

Keppel, G., Van Niel, K.P., Wardell-Johnson, G.W. et al. (2012) Refugia: identifying and understanding safe havens for biodiversity under climate change. *Global Ecology and Biogeography* 21, 393–404.

Lawler, J.J., Tear, T.H., Pyke, C. et al. (2010) Resource management in a changing and uncertain climate. *Frontiers in Ecology and the Environment* 8, 35–43.

Lees, A.C. and Peres, C.A. (2009) Gap-crossing movements predict species occupancy in Amazonian forest fragments. *Oikos* 118, 280–290.

Lenton, T.M., Held, H., Kriegler, E. et al. (2008) Tipping elements in the Earth's climate system. *Proceeding of the National Academy of Sciences* 105, 1786–1793.

Lindenmayer, D.B. and Likens, G.E. (2010) *Effective Ecological Monitoring*. CSIRO, Melbourne.

Lindenmayer, D.B., Gibbons, P., Bourke, M. et al. (2012) Improving biodiversity monitoring in Australia. *Austral Ecology* 37, 285–294.

Lunt, I.D., Byrne, M., Hellmann, J.J. et al. (2013) Using assisted colonisation to conserve biodiversity and restore ecosystem function under climate change. *Biological Conservation* 157, 172–177.

Mackey, B., Berry, S., Hugh, S., Ferrier, S., Harwood, T.D. and Williams, K.J. (2012) Ecosystem greenspots: identifying potential drought, fire, and climate-change micro-refuges. *Ecological Applications* 22, 1852–1864.

Maclean, I.M.D. and Wilson, R.J. (2011) Recent ecological responses to climate change support predictions of high extinction risk. *Proceeding of the National Academy of Sciences* 108, 12337–12342.

Mac Nally, R., Bennett, A.F., Thomson, J.R. et al. (2009) Collapse of an avifauna: climate change appears to exacerbate habitat loss and degradation. *Diversity and Distributions* 15, 720–730.

Mawdsley, J.R., O'Malley, R. and Ojima, D.S. (2009) A review of climate-change adaptation strategies for wildlife management and biodiversity conservation. *Conservation Biology* 23, 1080–1089.

McDonald-Madden, E., Runge, M.C., Possingham, H.P. and Martin, T.G. (2011) Optimal timing for managed relocation of species faced with climate change. *Nature Climate Change* 1, 261–265.

Millar, C.I., Stephenson, N.L. and Stephens, S.L. (2007) Climate change and forests of the future: Managing in the face of uncertainty. *Ecological Applications* 17, 2145–2151.

Mokany, K., Harwood, T.D. and Ferrier, S. (2013) Comparing habitat configuration strategies for retaining biodiversity under climate change. *Journal of Applied Ecology* 50, 519–527.

Møller, A.P., Fiedler, W. and Berthold, P. (eds.) (2010) *Effects of Climate Change on Birds*. Oxford University Press, Oxford.

Monzón, J., Moyer-Horner, L. and Palamar, M.B. (2011) Climate change and species range dynamics in protected areas. *BioScience* 61, 752–761.

Morecroft, M.D., Crick, H.Q.P., Duffield, S.J. and Macgregor, N.A. (2012) Resilience to climate change: translating principles into practice. *Journal of Applied Ecology* 49, 547–551.

Ovaskainen, O. (2002) Long-term persistence of species and the SLOSS problem. *Journal of Theoretical Biology* 218, 419–433.

Parmesan, C. and Yohe, G. (2003) A globally coherent fingerprint of climate change impacts across natural systems. *Nature* 421, 37–41.

Parmesan, C., Burrows, M.T., Duarte, C.M. et al. (2013) Beyond climate change attribution in conservation and ecological research. *Ecology Letters* 16, 58–71.

Pereira, H.M., Leadley, P.W., Proença, V. et al. (2010) Scenarios for global biodiversity in the 21st century. *Science* 330, 1496–1501.

Pritchard, D.J., Fa, J.E., Oldfield, S. and Harrop, S.R. (2012) Bring the captive closer to the wild: redefining the role of ex situ conservation. *Oryx* 46, 18–23.

Prober, S.M., Thiele, K.R., Rundel, P.W. et al. (2012) Facilitating adaptation of biodiversity to climate change: a conceptual framework applied to the world's largest Mediterranean-climate woodland. *Climatic Change* 110, 227–248.

Reed, T.E., Jenouvrier, S. and Visser, M.E. (2013) Phenological mismatch strongly affects individual fitness but not population demography in a woodland passerine. *Journal of Animal Ecology* 82, 131–144.

Reside, A.E., Welbergen, J.A., Phillips, B.L., Wardell-Johnson, G., Keppel, G., Ferrier, S., Williams, S.E., Storlie, C. and VanDerWal, J. (in press) Characteristics of climate change refugia for Australian biodiversity. *Austral Ecology*.

Schwartz, M.W., Hellmann, J.J., McLachlan, J.M. et al. (2012) Managed relocation: Integrating the scientific, regulatory, and ethical challenges. *BioScience* 62, 732–743.

Sheean, V.A., Manning, A.D. and Lindenmayer, D.B. (2012) An assessment of scientific approaches towards species relocations in Australia. *Austral Ecology* 37, 204–215.

Shoo, L.P., Hoffmann, A.A., Pressey, R.L. et al. (2013) Making decisions to conserve species under climate change. *Climatic Change* 199, 239–246.

Sinclair, A.R.E., Fryxell, J.M. and Caughley, G. (2006) *Wildlife Ecology, Conservation and Management. Second Edition.* Blackwell Publishing, Malden, Massachussetts.

Snyder, N.F.R., Derrickson, S.R., Beissinger, S.R. et al. (1996) Limitations of captive breeding in endangered species recovery. *Conservation Biology* 10, 338–348.

Thackeray, S.J., Sparks, T.H., Frederiksen, M. et al. (2010) Trophic level asynchrony in rates of phenological change for marine, freshwater and terrestrial environments. *Global Change Biology* 16, 3304–3313.

Thomas, C.D. (2011) Translocation of species, climate change, and the end of trying to recreate past ecological communities. *Trends in Ecology and Evolution* 26, 216–221.

Thomas, C.D., Franco, A.M.A. and Hill, J.K. (2006) Range retractions and extinction in the face of climate warming. *Trends in Ecology and Evolution* 21, 415–416.

Thomas, C.D., Hill, J.K., Anderson, B.J. et al. (2010) A framework for assessing threats and benefits to species responding to climate change. *Methods in Ecology and Evolution* 2, 125–142.

Thomas, M.A., Roemer, G.W., Donlan, C.J. et al. (2013) Gene tweaking for conservation. *Nature* 501, 485–486.

Thorne, K.M., Takekawa, J.Y. and Elliott-Fisk, D.L. (2012) Ecological effects of climate change on salt marsh wildlife: A case study from a highly urbanized estuary. *Journal of Coastal Research* 28, 1477–1487.

Timbal, B., Fernandez, E. and Li, Z. (2009) Generalization of a statistical downscaling model to provide local climate change projections for Australia. *Environmental Modelling and Software* 24, 341–358.

Tingley, M.W., Monahan, W.B., Beissinger, S.R. and Moritz, C. (2009) Birds track their Grinnellian niche through a century of climate change. *Proceedings of the National Academy of Science, USA* 106, 19637–19643.

Tomimatsu, H., Sasaki, T., Kurokawa, H. et al. (2013) Sustaining ecosystem functions in a changing world: a call for an integrated approach. *Journal of Applied Ecology* 50, 1124–1130.

Urban, M.C., Tewksbury, J.J. and Sheldon, K.S. (2012) On a collision course: competition and dispersal differences create no-analogue communities and cause extinctions during climate change. *Proceedings of the Royal Society B: Biological Sciences* 279, 2072–2080.

Van Teeffelen, A.J.A., Vos, C.C. and Opdam, P. (2012) Species in a dynamic world: Consequences of habitat network dynamics on conservation planning. *Biological Conservation* 153, 239–253.

VanDerWal, J., Murphy, H.T., Kutt, A.S. et al. (2013) Focus on poleward shifts in species' distribution underestimates the fingerprint of climate change. *Nature Climate Change* 3, 239–243.

Veloz, S.D., Williams, J.W., Blois, J.L., He, F., Otto-Bliesner, B. and Liu, Z. (2012) No-analog climates and shifting realized niches during the late quaternary: implications for 21st-century predictions by species distribution models. *Global Change Biology* 18, 1698–1713.

Webber, B.L. and Scott, J.K. (2012) Rapid global change: implications for defining natives and aliens. *Global Ecology and Biogeography* 21, 305–311.

Weeks, A.R., Sgro, C.M., Young, A.G. et al. (2011) Assessing the benefits and risks of translocations in changing environments: a genetic perspective. *Evolutionary Applications* 4, 709–725.

Wernery, U., Liu, C., Baskar, V. et al. (2010) Primordial germ cell-mediated chimera technology produces viable pure-line Houbara bustard offspring: potential for repopulating an endangered species. *PLoS ONE* 5, e15824.

Wildt, D.E., Rall, W.F., Critser, J.K., Monfort, S.L. and Seal, U.S. (1997) Genome resource banks. *BioScience* 47, 689–698.

Wilson, K.A., McBride, M.F., Bode, M. and Possingham, H.P. (2006) Prioritizing global conservation efforts. *Nature* 440, 337–340.

10 Methods to prioritise adaptation options for iconic seabirds and marine mammals impacted by climate change

ALISTAIR J. HOBDAY[1], LYNDA E. CHAMBERS[2] AND JOHN P.Y. ARNOULD[3]

[1]*Climate Adaptation Flagship, CSIRO Marine and Atmospheric Research, Australia*
[2]*Centre for Australian Weather and Climate Research, Australia*
[3]*School of Life and Environmental Sciences, Deakin University, Australia*

10.1 Introduction

Climate change is already impacting a wide range of marine species around Australia. Changes in distribution, abundance, physiology and phenology have been documented at a range of lower trophic levels around Australia including phytoplankton (Thompson et al. 2009; Hallegraeff 2010), seaweeds (Wernberg et al. 2011), intertidal and subtidal invertebrates (Ling et al. 2009; Pitt et al. 2010) and coastal and pelagic fish (Figueira and Booth 2010; Last et al. 2011; McLeod et al. 2012). In the global warming hotspot of southeast Australia (Hobday and Pecl 2014), these changes are now apparent throughout most of the foodweb (Frusher et al. 2013). For Australia's iconic higher-trophic-level marine taxa, such as seabirds and marine mammals, there is less known regarding the responses of species to climate change (Chambers et al. 2011), although recent work is improving this situation (e.g. Dann and Chambers 2013; Schumann et al. 2013; Chambers et al. 2014).

Australia has a large number of marine mammals and seabirds, particularly when Australian Antarctic and Southern Ocean species are included: 110 species of seabird (Chambers et al. 2011) and 52 species of marine mammal (45 of which are cetaceans; Schumann et al. 2013). The marine mammals considered in this study include Sirenia (dugong) and Carnivora (seals and sea lions), but not cetaceans (whales and dolphins). Prominent seabird taxa include Phaethontiformes (tropicbirds); Procellariformes (petrels, prions, albatross, shearwaters); Sphenisciformes (penguins); Phalacrocoraciformes (frigatebirds, booby, cormorants) and Charadriiformes (e.g. terns, gulls, noddies). With the exception of dugong, we focused on taxa that breed in colonies. Colonial breeding enables long time-series to be collected on the same population at the same location and generally at single time points, which makes both intervention and monitoring of responses to adaptation actions easier compared to other taxa (e.g. whales). Around the Australian mainland a wide range of seabird species (including one species of penguin) form breeding colonies but there are just three seal and sea lion species that are regular breeders, while dugong access predictable feeding grounds.

Applied Studies in Climate Adaptation, First Edition. Edited by Jean P. Palutikof, Sarah L. Boulter, Jon Barnett and David Rissik.
© 2015 John Wiley & Sons, Ltd. Published 2015 by John Wiley & Sons, Ltd.

These iconic species are protected throughout Australia and in some cases are recovering from previous anthropogenic impacts including harvest (Arnould et al. 2003; Kirkwood et al. 2010; Alderman et al. 2011; Chambers et al. 2011; Croxall et al. 2012; Schumann et al. 2013). The conservation status of seabirds, one of the most threatened bird groups globally, is deteriorating, with Australia having the 4th highest number of species of conservation concern (Croxall et al. 2012). As a result of this special conservation status, a range of monitoring projects around Australia are documenting the recovery or otherwise of these species. Actions to reduce a range of non-climate threats such as fishing bycatch, land use changes, feral predators or coastal pollution have also been implemented (Garnett et al. 2011; DSEWPaC 2013). However, climate change is now seen an additional stressor for many of these species with a range of possible effects (Chambers et al. 2011, 2014; Table 10.1). Understanding the climate change impacts and developing adaptation options is important, although evidence of responses to climate variability and change and the functional processes driving these affects is limited for most species. Resolution of climate change impacts from other anthropogenic threats is needed for these species in order to implement appropriate and timely adaptive management responses. Managers report this as a major impediment to ongoing conservation management and planning in the face of climate variability and change.

While developing an improved understanding of the impacts of climate change on these species is important and will continue to be so, it is also necessary to begin to consider options for action. Here we report on recent work developing and prioritising a range of adaptation options for Australian seabirds (rather than shorebirds), seals and sea lions (rather than all marine mammals). We focus on the methods that can prioritise adaptation options that can be implemented by humans, in order to improve the coping ability of seabirds and marine mammals under a range of climate-related stressors.

10.2   Going beyond a shopping list of adaptation options

The published literature contains a range of adaptation options for large animals such as habitat modification, fire management and invasive species management that could be modified and applied to seabirds and marine mammals (e.g. Mawdsley et al. 2009; Dawson et al. 2011; Koehn et al. 2011). There is value in going further then just providing a range of options however, such that a clearer path to implementation is provided to decision-makers (*sensu* Mawdsley 2011). Providing a list of options is just the first step to adaptation implementation, and we suggest that three additional stages should follow the generation of a set of adaptation options (Table 10.2). In addition to generating options based on a vulnerability framework (see following section), three tools were developed to evaluate adaptation options. These tools allow (1) a rapid screening of cost-benefit-risk; (2) identification of potential institutional barriers in implementing adaptation options; and (3) assessment of the likely social acceptance of these options. Together, these evaluation tools allow relative prioritisation and identification of issues associated with each scenario that might need to be considered before attempting implementation.

10.3   Generating adaptation options

To generate a set of options, we first reviewed both the range of existing seabird and marine mammal adaptation options (e.g. Chambers et al. 2011) and approaches to generating adaptation options, including *ad hoc* and structured methods. We chose a structured approach, combining scenarios and a framework, to generate a range of adaptation options. The selected framework was the IPCC model of vulnerability to climate change (Fig. 10.1). We used this as a guide to generate a range of options that might be applicable under various climate scenarios leading to increased vulnerability for iconic marine species. Under this framework, vulnerability to climate

Table 10.1 Expected and observed responses of Australian marine mammals and seabirds to climate-related change. Source: Adapted from Chambers et al. 2014.

Climate variable	Projected climate change	Species response (expected)	Species response (observed)	Example
Precipitation, streamflow	Decrease (southern regions)	Changed prey species recruitment	No direct link observed	Freshwater input can be critical for the survival of some fish larvae (Santojanni et al. 2006; Dann and Chambers 2013)
Fire	Increase	Risk of extreme mortality events in affected colonies	Unexpected response for burrow nesting species (stay and burn)	Deaths in burrowing seabirds due to fire in grassland colonies (Renwick et al. 2007; Chambers et al. 2009)
Air temperature	Increase	Increased heat stress and mortality	Physiological limits known for some species but no clear signal observed	Prolonged exposure to temperatures above 35°C leads to heat stress and potentially death in Little Penguins *Eudyptula minor* (Dann and Chambers 2013)
		Reduction in breeding success Changed breeding timing	Limited evidence Some evidence	Little Penguin (Chambers et al. 2011) Winter breeding in cormorants in SE Australia (Taylor 2007)
Sea surface temperature (SST)	Increase	Reduced breeding participation and success	Some evidence for but also some contradictory	Increased SST associated with reduced breeding success in Great Barrier Reef (Chambers et al. 2011)
		Changed breeding timing	Some evidence but direction of effect varies by location and species	Interval between breeding seasons in Australian Sea Lions *Neophoca cinerea* may be extended when SST is warmer (Goldsworthy et al. 2009)
		Polewards range expansion	Some evidence in WA seabirds	Bridled Tern *Onychoprion anaethetus* in SW Australia (Dunlop and Surman 2012)
Ocean currents (including mixed layer depth and stratification)	Strengthen (at least in short term)	Changed foraging and breeding success	Some evidence	South Australian Current and upwelling influence pup production in Australian Fur Seals *Arctocephalus pusillus* (Gibbens and Arnould 2009)
Sea level	Increase	Breeding sites move with changing coastline	No quantitative link observed	–
Ocean acidification	Increase	Altered prey	No quantitative link observed	–
Winds, storms	Increase (frequency and intensity)	Increased mortality	Some evidence	Mass mortality of seabirds along Victorian coast following strong wind events (Chambers et al. 2011)
Cyclones	Increase in intensity	Reduced breeding success Changed breeding timing	Some evidence Some evidence	Tropical seabirds (Chambers et al. 2011) Tropical seabirds (Chambers et al. 2011)

change can be reduced by adaptation options that: (1) reduce exposure of the individuals/populations/species to the physical effects of climate change; (2) reduce the sensitivity of the individuals to the physical effects of climate change; and (3) increase the adaptive capacity of the individual/population/species to cope with the physical effects of climate change, such as decreasing the impact of other stressors.

The project team first generated a range ($n = 25$, 15 for seabirds and 10 for dugong and pinnipeds) of species-specific climate impact scenarios linked to a physical change on land or in the ocean based on known impacts (Table 10.1). We then used several expert workshops to challenge the participants (a total of 34 researchers, managers and policymakers) to propose multiple adaptation options for each of the 25 scenarios, initially guided by the existing examples we

Table 10.2 Stages in the development and prioritisation of adaptation options.

Stage	Responsible group	Tools to generate or assess options
1. Generate options	System or species experts	Vulnerability framework
2. Technically appropriate?	Science	Cost-benefit-risk
3. Institutionally possible?	Policy and management	Barriers analysis
4. Socially acceptable?	Citizens	Social acceptability

provided. Development of options in each of the three categories of the vulnerability framework was encouraged (Hobday et al. 2013).

After combining similar adaptation options for each scenario, a total of 198 options across these 25 scenarios remained: 156 for seabirds and 42 for marine mammals, representing an average of 10.4 adaptation options for each seabird scenario, and 4.2 for each marine mammal (see examples for each category in Table 10.3 and a full set in Hobday et al. 2013). These options were relatively evenly distributed between the three elements of the vulnerability framework: reduce exposure ($n = 63$), reduce sensitivity ($n = 64$) and increase adaptive capacity ($n = 71$). Rather than reporting a long list of options, we used a set of tools to evaluate a subset of potential adaptation options for a subset of all the scenarios (for seabirds $n = 17$ scenarios; for marine mammals $n = 8$) and tested these in workshops with researchers and managers (Table 10.4). As such, these options should be considered indicative; to make a real decision, a comprehensive set of options should be evaluated for a single scenario.

10.4   Evaluating options using a cost-benefit-risk framework

The first tool we developed is a simple 'cost-benefit-risk' (CBR) screening tool to evaluate each scenario-specific adaptation option against a number of semi-quantitative attributes. Criteria

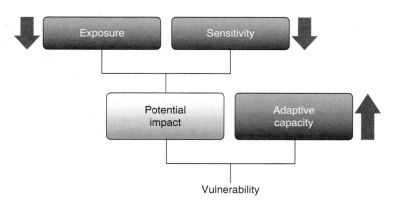

Figure 10.1 Vulnerability framework used to generate adaptation options. Options could reduce exposure, decrease sensitivity or increase adaptive capacity.

Table 10.3 Examples of adaptation options in each of three vulnerability categories. Without specifying the climate scenario, these options are intended to be illustrative only. Source: Upper and lower photographs by L. E. Chambers; middle photograph by Cheryl Reynolds/International Bird Rescue. Reproduced with permission. For colour details please see Plate 4.

Vulnerability category	Example adaptation option
Decrease exposure	Increase shading of burrows (photo); provide artificial nest boxes; translocate to new location; provide artificial haulouts; create rockpools for cooling; shade cloth for ground-nesting birds 
Decrease sensitivity	Supplemental feeding of chicks (photo); remove extra egg to increase survival of remaining chicks; chick rescue during extreme events; vaccinate chicks to disease; provide artificial cooling at nests 
Increase adaptive capacity	Add boardwalks to reduce human impact on colony nesting burrowing species (photo); cull competitors; reduce fishing effort (bycatch reduction); reduce feral predators; reduce competitor species 

Table 10.4 Summary of climate scenario adaptation option combinations evaluated for seabirds (S) or marine mammals (M). Note that several adaptation options were considered for some of the scenarios. Each option could reduce exposure (E), reduce sensitivity (S) or increase adaptive capacity (AC).

No.	Taxa	Climate scenario and resulting impact	Adaptation option	Vulnerability category
1	S	Increased air temperatures lead to decreased chick survival in burrow nesting birds	Decrease exposure via shade cloth over burrows	E
2	S	Increased air temperatures lead to decreased chick survival in burrow nesting birds	Shade burrows with re-vegetation to shrubby vegetation	E
3	S	Increased air temperatures lead to decreased chick survival in burrow nesting birds	Construct longer, deeper burrows that are cooler	E
4	S	Increased air temperatures lead to decreased chick survival in burrow nesting birds	Eliminate feral pest (e.g. foxes) on the island (e.g. size of Phillip Island)	AC
5	M	Rising air temperature leads to mortality in females pupping at isolated colonies, by increasing their time in water (exposure to predators), disease increase and disturbance to mother–pup bond	Build artificial rock pools to provide safe cooling areas	S
6	M	Rising air temperature leads to mortality in females pupping at isolated colonies by increasing their time in water (exposure to predators), disease increase and disturbance to mother–pup bond	Install shark deterrents (acoustic) off seal colonies	AC
7	S	Increased intensity of rainfall leads to flooding of burrows and chick mortality	Improve drainage around colony with agricultural drain	E
8	S	Increased intensity of rainfall leads to flooding of burrows and chick mortality	Remove chicks during extreme events and replace after event	S
9	M	Increased sea level and storm surge leads to overtopping of seal breeding colony and mortality of pups; overall population decline	Initiate island raising with dumping of very big rocks or concrete	E
10	S	Wind speed increases, lead to nesting failure of tree-nesting birds	Transition wind breaks from artificial structures to vegetation planting to replace artificial structures in time	E
11	M	Declining ocean productivity leads to declining participation of female seals in breeding	Artificial feeding of female seals during gestation period (when likely to be at the colony)	S
12	M	Declining ocean productivity leads to declining participation of female seals in breeding	Temporary closures of fisheries operating in the foraging range of the species	AC
13	M	Cyclone frequency increases and destruction of seagrass beds leads to starvation and death of dugongs following each cyclone	Relocate animals in affected areas to other locations	E
14	M	Increasing water temperatures leading to declines in dugong feeding areas (seagrass declines in some parts of the range)	Create strategic set-aside areas that reduce other stressors	AC

Table 10.4   *(Cont'd)*

No.	Taxa	Climate scenario and resulting impact	Adaptation option	Vulnerability category
15	M	Declining productivity of seagrass beds lead to starvation and mortality in some parts (c. 25%) of the dugong range	Initiate seagrass nurseries and outplanting to enhance natural production in these regions	S
16	S	A competitive bird species (e.g. silver gull or gannet) is favoured by climate change, arrives at the colony of a threatened species and begins to take over nesting sites	Cull competitor (e.g. firearms)	AC
17	S	Decreased foraging success of adults leads to chicks (*n*=2) fledging at lower weights and first-year survival declines.	Reduce brood size (e.g. from 2 to1) to increase condition and survival of remaining chick	S
18	S	Warmer weather leads to increased vegetation growth around burrows, leading to a fire risk and preventing birds from accessing the burrows	Burning of habitat in non-breeding season (i.e. when birds absent) to reduce vegetation overgrowth and fire risk during breeding season	S
19	S	Warmer weather leads to increased vegetation growth around burrows, leading to a fire risk and preventing birds from accessing the burrows	Introduce a grazing species to control vegetation (e.g. rabbit)	S
20	S	Decreased foraging success of adults leads to chicks fledging at lower weights and first-year survival declines	Decrease parasite loads in chicks via drenching	AC
21	S	A competitive bird species (e.g. silver gull or gannet) is favoured by climate change, arrives at the colony of a threatened species and begins to take over nesting sites	Provide alternative habitat for competitor, e.g. floating platform for gannets	S
22	S	Declining ocean productivity lead to declining fledging success of birds	Initiate fish farming to produce feed for marine species	AC
23	S	Warmer weather leads to increased vegetation growth around burrows, leading to a fire risk and preventing birds from accessing the burrows (e.g. kikuyu grass binds burrows, affecting shearwater and little penguin)	Reduce public access (manage human access) to reduce fire risk	S
24	S	Warming waters and a deepening thermocline lead to reduced foraging success and loss of northern colonies of seabirds (suitable areas elsewhere)	Translocate chicks to new location (assuming site fidelity)	E
25	S	Increased intensity of rainfall events leads to direct mortality of eggs and chicks of surface nesting seabirds (e.g. albatross, terns, gannets)	Corral chicks from crèche to under 'shelters' (e.g. crested tern)	S

Table 10.5 Cost-benefit-risk scoring criteria used to score attributes to assess each adaptation scenario.

Category	Attribute	Low (1)	Medium (2)	High (3)
Cost	Implementation cost	<$10,000	>$10,000 –<M$1	> M$1
	Ongoing cost (how many years of action are needed)	< 5 years	5–10 years	> 10 years
	Time to implement (lead time until action can begin)	Now	1–5 years	> 5 years
Benefit	Persistence of action	1 season	< 5 seasons	> 5 seasons
	Scale of benefit (at the scale action is applied)	Few individuals	Most of colony	Most of the population
	Benefit of action to target group	Minimal improvement	Partial solution	Solve problem
	Benefit of action to wider ecosystem	Low	Medium	High
Risk	Risk of action failing	< 33%	33–66%	>66%
	Risk of mal-adaptation (negative outcome on another strategy for target group)	< 33%	33–66%	>66%
	Risk of adverse impacts to wider (eco) system	< 33%	33–66%	>66%

for scoring each attribute for each adaptation option in the three categories – cost, benefit and risk – were developed (Table 10.5) and described in more detail in Hobday et al. (2013). This expert-based scoring approach is conceptually similar to other scoring systems used in ecological risk assessment (e.g. Chin et al. 2010; Hobday et al. 2011). Modification of these attributes or scoring criteria would be appropriate for other uses. The scoring cut-offs for each attribute were tested by the project team prior to use, and were intended to divide the results into approximate thirds. Each attribute was scored by workshop participants on the basis of their experience and the cut-offs as either low (1), medium (2) or high (3).

Scoring was carried out during workshops using *Turning Point©* software, which is embedded within a PowerPoint presentation and allows scoring using hand-held devices (similar to television remote controls). Each adaptation scenario was presented to the whole group along with the proposed adaptation option. The participants then individually scored each of the attributes using the hand-held devices and the results

were displayed to the group at the conclusion of the scoring for each attribute. Each adaptation option thus required scoring of 10 attributes. The agreement of scoring among participants was evaluated in a range of ways, but for illustration we show that there was agreement for the majority (93%) of scenario–attribute combinations presented here (Fig. 10.2).

Each adaptation option score for all participants was converted to a mean cost (average of the three cost attribute scores) and benefit (average of the four benefit attribute scores) score, and plotted with the three risk criteria scores averaged and represented on a plot as the relative size of the symbol (large symbols representing high risk; Fig. 10.3). This scoring identifies which adaptation options are of high cost and low benefit (lower left) and so might be discarded, and which are of high benefit and low cost (upper right) and might be rapidly implemented (depending on risk). Options that are low cost and low benefit might not be pursued, while those that are high cost but high benefit (upper right) deserve more detailed attention. An overall score can be calculated at the distance from the origin to

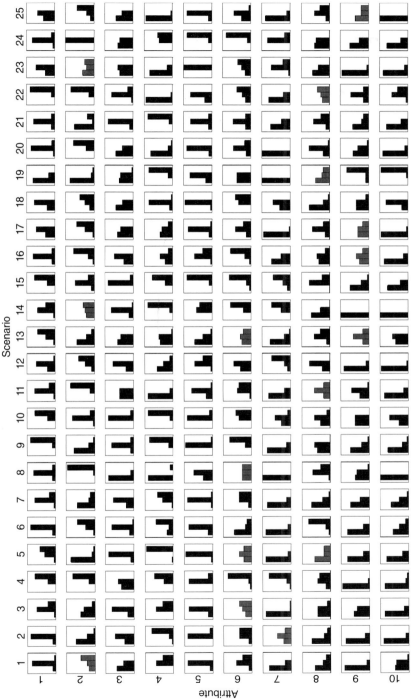

Figure 10.2 Consistency plot for the proposed seabird and mammal adaptation options. Adaptation scenarios are in the columns, with each row representing a different attribute. Each subplot is a histogram showing the number of responses scored as L, M and H (bars). Red subplots show where scores were distributed approximately uniformly across the options, indicating inconsistency. Of the 250 combinations, only 6.8% were scored inconsistently by the experts. For colour details please see Plate 5.

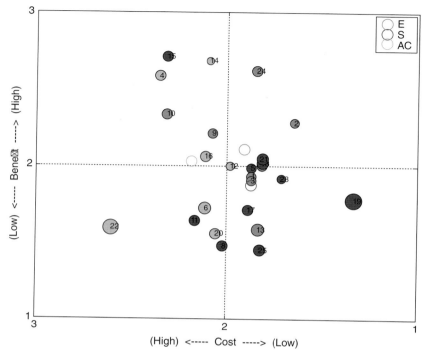

Figure 10.3 Summary cost-benefit-risk plot for 25 adaptation options (numbered as in Table 10.4), coloured by vulnerability category. Open circles represent the mean value for options that reduce exposure (E) or sensitivity (S) or increase adaptive capacity (AC). The size of the bubbles represents the risk score (small circles represent low risk, large circles higher risk). For colour details please see Plate 6.

the symbol, and thus options ranked by two of the dimensions (cost and benefit; see Table 10.6).

As an example, option 22 (provisioning foraging birds via development of aquaculture of prey species) was rated as highest cost and lowest benefit with a high degree of risk (Fig. 10.3). On the basis of this assessment, this option would not be worthwhile pursuing. In contrast, options 2 (growing vegetation to shade burrows) and 24 (translocating chicks to new locations) were rated as of high benefit and relatively low cost, and would be worthy of more detailed assessment or field trials. Overall, options to reduce exposure were rated higher (red) then options to reduce sensitivity (blue) and options to increase adaptive capacity which, although often of high benefit, were more costly and involved higher risk.

10.5   Evaluating barriers to implementing adaptation options

Even with technical merit, as assessed with the CBR tool, adaptation options can fail because of institutional problems with implementation. A second evaluation tool was therefore based on the conceptual framework on barriers to effective climate adaptation developed by Moser and Ekstrom (2010). They divide potential barriers into three stages (understanding, planning and managing) each with three elements (Fig. 10.4).

For each scenario-adaptation option combination, the nine elements in the barriers framework were scored on a Likert scale from 1 to 5, where 1 represented a likely barrier (strongly

Table 10.6 Summary of adaptation options as scored by the cost benefit and risk (CBR) barriers and social acceptability tools, sorted by average rank across all three tools. The highest scores for each tool are shown in pale grey, the lowest in mid grey and intermediate in dark grey.

Rank	Scenario	Average	CBR	Barriers	Social	Category
1	24	1.7	1	1	3	E
2	2	2.0	3	2	1	E
3	4	5.7	6	5	6	AC
4	1	7.3	14	6	2	E
4	7	7.3	10	8	4	E
4	10	7.3	13	4	5	E
7	21	8.0	7	10	7	S
7	23	8.0	8	7	9	S
9	14	8.3	2	16	7	AC
10	18	10.0	9	3	18	S
11	9	10.7	11	11	10	E
12	15	11.3	5	17	12	S
13	3	14.7	15	12	17	E
13	19	14.7	4	15	25	S
15	16	15.3	17	9	20	AC
16	8	16.0	23	14	11	S
17	25	16.3	20	13	16	S
18	5	16.7	12	23	15	S
19	12	17.0	16	21	14	AC
20	20	18.3	22	20	13	AC
21	13	19.7	19	18	22	E
22	6	21.3	21	24	19	AC
22	11	21.3	24	19	21	S
24	17	22.3	18	25	24	S
25	22	23.0	25	22	22	AC

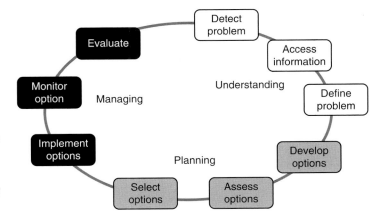

Figure 10.4 Barriers framework used to evaluate adaptation options. Source: Adapted from Moser and Ekstrom (2010). Reproduced with permission of Proceedings of the National Academy of Sciences.

agree) for the option and 5 (strongly disagree) represented no barrier. These questions were answered by workshop participants from the perspective of the likely agency charged with implementing the adaptation strategy. The specific questions were:

1. Detecting a signal will be a barrier for this adaptation strategy?

2. Gathering/using information will be a barrier for this adaptation strategy?

3. Defining the problem will be a barrier for this adaptation strategy?

4. Developing options will be a barrier for this adaptation strategy?

5. Managing the process will be a barrier for this adaptation strategy?

6. Selecting options will be a barrier for this adaptation strategy?

7. Implementation will be a barrier for this adaptation strategy?

8. Monitoring the outcomes will be a barrier for this adaptation strategy?

9. Evaluating effectiveness will be a barrier for this adaptation strategy?

Additional descriptions of each element were provided to participants and discussed prior to scoring the scenarios. The apparent barriers varied between the scenario-adaptation options (Fig. 10.5). For example, option 5 (building artificial rock pools) was considered to have barriers in the planning and evaluation stages, with an overall high barrier score. Option 17 (artificially reducing brood size) was also considered to have a range of barriers, while option 2 (planting vegetation to shade burrows) had few barriers. The scores for each element were then averaged for each of the three stages, and compared across scenarios (Fig. 10.6). In general the barrier scores were similar, and moderate, with a few options that did not score well. The eight adaptation options that would reduce exposure had the best (highest) mean barrier score of 3.3, while the decreasing sensitivity and increasing adaptive capacity options had similar but lower (worse) mean scores (2.93, $n = 10$, and 2.81, $n = 7$, respectively).

10.6   Evaluating social acceptability of adaptation options

Even with technical merit (CBR tool) and an understanding of the institutional barriers, adaptation options may not be acceptable to society at large or may be resisted by vocal opponents or groups. The final stage in evaluating the adaptation options was therefore designed to assess the social acceptability of each option. Awareness and identification of potentially contested options would be useful to managers charged with implementing adaptation options. Assessing social licence after some action is a difficult task, and attempts to assess it before the action has occurred are even more problematic. We believe considering the social acceptability of adaptation options is however valuable, even if the accuracy of the assessment cannot be evaluated.

We developed a short-list of issues that may be of concern when implementing a particular adaptation scenario in an expert workshop. This was designed to draw out the details of attributes to inform a more detailed assessment of social acceptability. After discussion, the group selected a set of 10 attributes to judge social acceptability in two categories – perceived impact and relationships – and then scored the total number of these attributes (i.e. between 0 and 10 attributes) that might be associated with each scenario. This represents an 'issues' count, where a low number of issues represent high acceptability and a high number represents low acceptability. In this approach, we assume that each attribute is equally important; however, weighting of attributes could allow differential importance to be included in an estimate of social acceptability.

Perceived 'impact' attributes are as follows:

1. Is the species threatened or protected? Such species can polarise stakeholders and intensify conflict around intervention.

2. Is the impacted species iconic, dangerous or cute?

3. Will other animals be killed or harmed or habitat modified? Ethical treatment of animals is important to society at large.

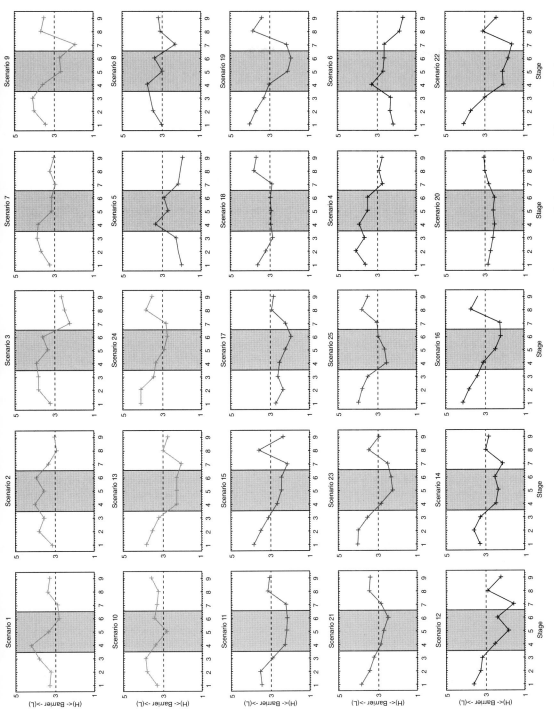

Figure 10.5 Barriers analysis for 25 scenario-adaptation options, for the nine elements in the Moser and Ekstrom (2010) framework. The scenarios are ordered by the vulnerability category where red lines are exposure reduction, blue lines sensitivity reduction and black lines adaptive capacity increase adaptation options. Higher scores on the *y*-axis represent lower barriers to implementation. For colour details please see Plate 7.

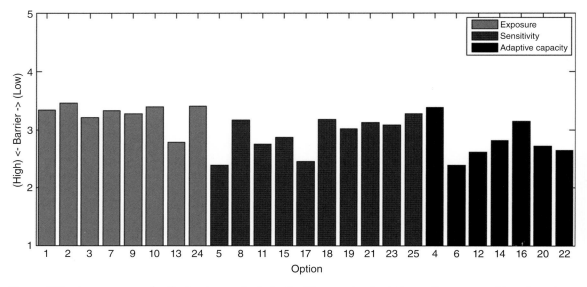

Figure 10.6 Average scores for the barriers analysis for the 25 scenario-adaptation options, grouped by adaptation options to reduce exposure, reduce sensitivity or increase adaptive capacity. Higher scores represent fewer barriers to implementation. For colour details please see Plate 8.

4. Contested action: is there a need for action, is there evidence for the need and are the benefits clear?

5. Will it impact on lifestyle, livelihood or recreation (e.g. closures, cat ownership, dogs on beaches)?

Relationship attributes are as follows:

6. Is there distrust for the proposing individual/organisation (e.g. government)?

7. Are community groups already engaged with a good history of engagements?

8. Is there a powerful or divergent interest groups involved?

9. Is it 'crazy' from public point of view (e.g. will it engage radio shock-jocks), e.g. use of public moneys (value for money).

10. Is the location for the activity in proximity to human settlement/recreation area? (e.g. pest control).

The scores for each scenario and the set of issues identified were then analysed to allow a ranking of 'preferred' and 'non-preferred' scenarios, in this case from an expert rather than general public perspective (Fig. 10.7). We note in

this case that we have used an expert group to rate the acceptability; a more robust use of the tool would be to canvass a group comprising members of the general public. This approach may not always be chosen by managers however, particularly around sensitive issues.

In the set of examples presented here, while the social acceptance for options that reduce exposure had a higher mean acceptability (3.0) than for options to reduce sensitivity (4.2) or to increase adaptive capacity (3.6), examples of high and low acceptability options occurred in all three categories. Here, option 2 (shading burrows via revegetation to reduce temperatures) was rated as the most acceptable, while option 19 (introduction of a grazing species to reduce vegetation cover) was the least.

10.7 Overall ranking of adaptation options

The scores from these three tools for each scenario were used to each generate a 'ranking' where a rank of 1 is more desirable than a rank of

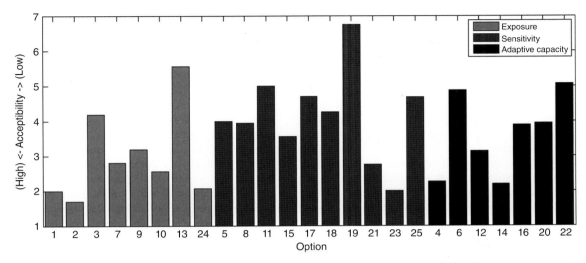

Figure 10.7 Relative social acceptability of the 25 adaptation options, scored as described in the text and grouped by adaptation options to reduce exposure, reduce sensitivity or increase adaptive capacity. Lower scores represent higher acceptability. For colour details please see Plate 9.

25 (Table 10.6). This quickly shows which options might be favoured overall. In general, options that were cost-effective also had few barriers and higher social acceptability. Option 24 (translocating chicks to more favourable regions) was the highest overall, rating highest in the CBR and barriers analysis and high with regard to social acceptability. This strategy has been implemented in a range of seabird species for conservation reasons (e.g. Priddel et al. 2006; Miskelly et al. 2009; Carlisle et al. 2012), and was not considered controversial by experts and managers. In contrast, option 22 (initiating fish farming to feed birds) was the poorest by all three measures. Of particular interest are options favoured by CBR analysis, but with high barriers or low social acceptability. While scientifically desirable based on CBR analysis, option 4 (eliminating feral pests) had both barriers and low potential social acceptability. In fact, eradication programs are often delayed by jurisdictional issues and social opposition, particularly where human activities may be impacted (e.g. Lord Howe Island; see http://www.theglobalmail.org/feature/100000-rats-42-tonnes-of-poison-and-

one-slightly-nervous-world-heritage-island/559/; Wilkinson and Priddel 2011).

Overall, these tools showed that for seabirds and marine mammals, adaptation options that reduced exposure had a higher average rank (8.8) than options to reduce sensitivity (14.5) or increase adaptive capacity (15.6). That said, within the range of options presented here, examples of favoured and poor choices existed within each of the vulnerability categories.

10.8   Conclusion

As for other taxa in Australia, climate is projected to have an impact on marine mammals and seabirds (Chambers et al. 2011; Schumann et al. 2013). Given the conservation status of these species, methods to assess the 'best' adaptation options are needed now. Policy intervention may be required to offer additional protection to species threatened under climate change, and we have shown that the design of adaptation options can be structured and evaluated. Fortunately there is a wide range of options for reducing vulnerability of colony-nesting seabirds; however, options seem more limited for marine mammals. In the examples presented here, we have not focused on particular adaptation options *per se*; rather, we have illustrated how these tools can be used to rank options. Some management responses for known climate threats can be simple. Monitoring to evaluate the effectiveness of adaptation options is also critical, and should be a focus in any adaptation experiments. Using these tools to examine options for specific taxa and situations is now underway (R. Alderman, pers. comm. 2014). Following on from specific application, testing some of the adaptation options in limited field trials would be a useful next step, further building the experience of researchers and managers charged with securing the status of these iconic species in the future.

Acknowledgements

This research contributed to the project 'Human adaptation options to increase resilience of conservation-dependent seabirds and marine mammals impacted by climate change', which was supported by NCCARF and funding from the FRDC–DCCEE on behalf of the Australian Government. We appreciate the support of many seabird and mammal researchers, managers and policy experts for participation in our workshops and discussions.

References

Alderman, R., Gales, R., Tuck, G. N. and Lebreton, D. (2011) Global population status of shy albatross and an assessment of colony-specific trends and drivers. *Wildlife Research* 38, 672–686.

Arnould, J.P.Y., Boyd, I.L. and Warneke, R.M. (2003) Historical dynamics of the Australian fur seal population: evidence of regulation by man? *Canadian Journal of Zoology* 81, 1428–1436.

Carlisle, N., Priddel, D. and Madeiros, J. (2012) Establishment of a new, secure colony of endangered Bermuda Petrel *Pterodroma cahow* by translocation of near-fledged nestlings. *Bird Conservation International* 22, 46–58.

Chambers, L.E., Renwick, L. and Dann, P. (2009) Climate, fire and Little Penguins, In: Steffen, W. (ed.) *Australia's Biodiversity and Climate Change.* CSIRO Publishing, Melbourne, p. 82.

Chambers, L. E., Devney, C.A., Congdon, B.C., Dunlop, N., Woehler, E. and Dann, P. (2011) Observed and predicted effects of climate on Australian seabirds. *Emu* 111, 235–253.

Chambers, L. E., Patterson, T. A., Hobday, A. J., Arnould, J. P. Y., Tuck, G. N., Wilcox C. and Dann P. (2014) Determining trends and environmental drivers from long-term marine mammal and bird data: Examples from Southern Australia. *Regional Environmental Change*, doi: 10.1007/s10113-014-0634-8.

Chin, A., Kyne, P.M., Walker, T.I. and McAuley, R. (2010) An integrated risk assessment for climate change: analysing the vulnerability of sharks and rays on Australia's Great Barrier Reef. *Global Change Biology* 16, 1936–1953.

Croxall, J.P., Butchart, S.H.M, Lascelles, B. et al. (2012) Seabird conservation status, threats and priority actions: a global assessment. *Bird Conservation International* 22, 1–34.

Dann, P. and Chambers, L.E. (2013) Ecological effects of climate change on Little Penguins *Eudyptula minor* and the potential economic impact on tourism. *Climate Research* 58, 67–79.

Dawson, T.P., Jackson, S.T., House, J.I., Prentice, I.C. and Mace, G.M. (2011) Beyond predictions: Biodiversity conservation in a changing climate. *Science* 332, 53–58.

Department of Sustainability Environment Water Population and Communities (2013) *Recovery Plan for the Australian Sea Lion* (Neophoca cinerea). Department of Sustainability Environment Water Population and Communities, Australia. Available at http://www.environment.gov.au/biodiversity/threatened/publications/recovery/n-cinerea/index.html (accessed 26 May 2014).

Dunlop, J.N. and Surman, C.A. (2012) Role of foraging ecology in the contrasting responses of two dark terns to a changing ocean climate. *Marine Ornithology* 40, 105–110.

Figueira, W.F. and Booth, D.J. (2010) Increasing ocean temperatures allow tropical fishes to survive over-winter in temperate waters. *Global Change Biology* 16, 506–516.

Frusher, S.D., Hobday, A.J., Jennings, S.M. et al. (2013) A short history of a marine hotspot: from anecdote to adaptation in south-east Australia. *Reviews in Fish Biology and Fisheries* 24(2), 593–611.

Garnett, S., Szabo, J. and Dutson, G. (2011) *The Action Plan for Australian Birds: 2010*. CSIRO Publishing, Melbourne.

Gibbens, J. and Arnould, J.P.Y. (2009) Interannual variation in pup production and the timing of breeding in benthic foraging Australian fur seals. *Marine Mammal Science* 25, 573–587.

Goldsworthy, S.D., McKenzie, J., Shaughnessy, P.D., McIntosh, R.R., Page, B. and Campbell, R. (2009) *An update of the report: Understanding the impediments to the growth of Australian Sea Lion populations*. South Australian Research and Development Institute Research Report Series No. 356. South Australian Research and Development Institute Research, Adelaide.

Hallegraeff, G.M. (2010) Ocean climate change, phytoplankton community responses, and harmful algal blooms: a formidable predictive challenge. *Journal of Phycology* 46, 220–235.

Hobday, A.J. and Pecl, G.T. (2014) Identification of global marine hotspots: sentinels for change and vanguards for adaptation action. *Reviews in Fish Biology and Fisheries* 24, 415–425.

Hobday, A.J., Smith, A.D.M., Stobutzki, I. et al. (2011) Ecological risk assessment for the effects of fishing. *Fisheries Research* 108, 372–384.

Hobday, A.J., Chambers, L.E., Arnould, J.P.Y. et al. (2013) *Developing Adaptation Options for Seabirds and Marine Mammals Impacted by Climate Change*. Final Report. FRDC-DCCEE Marine National Adaptation Research Project 2011/0533. Available at http://frdc.com.au/research/Documents/Final_reports/2010-533-DLD.pdf (accessed 12 June 2014).

Kirkwood, R., Pemberton, D., Gales, R. et al. (2010) Continued population recovery by Australian fur seals. *Marine and Freshwater Research* 61, 695–701.

Koehn, J.D., Hobday, A.J., Pratchett, M.S. and Gillanders, B.M. (2011) Climate change and Australian marine and freshwater environments, fishes and fisheries: Synthesis and options for adaptation. *Marine and Freshwater Research* 62, 1148–1164.

Last, P.R., White, W.T., Gledhill, D.C. et al. (2011) Long-term shifts in abundance and distribution of a temperate fish fauna: A response to climate change and fishing practices. *Global Ecology and Biogeography* 20, 58–72.

Ling, S.D., Johnson, C.R., Ridgway, K., Hobday, A.J and Haddon, M. (2009) Climate driven range extension of a sea urchin: Inferring future trends by analysis of recent population dynamics. *Global Change Biology* 15, 719–731.

Mawdsley, J.R. (2011) Design of conservation strategies for climate adaptation. *WIREs Climate Change* 2, 498–515.

Mawdsley, J.R., O'Malley, R. and Ojima, D.S. (2009) A review of climate-change adaptation strategies for wildlife management and biodiversity conservation. *Conservation Biology* 23, 1080–1089.

McLeod, D.J., Hobday, A.J., Lyle, J.M. and Welsford, D.C. (2012) A prey-related shift in abundance of small pelagic fish in eastern Tasmania? *ICES Journal of Marine Science* 69, 953–960.

Miskelly, C.M., Taylor, G.A., Gummer, H. and Williams, R. (2009) Translocations of eight species of burrow-nesting seabirds (genera *Pterodroma, Pelecanoides, Pachyptila* and *Puffinus*: Family Procellariidae). *Biological Conservation* 142, 1965–1980.

Moser, S.C. and Ekstrom, J. (2010) A framework to diagnose barriers to climate change adaptation. *Proceedings of the National Academy of Sciences* 107, 22026–22031.

Pitt, N.R., Poloczanska, E.S and Hobday, A.J. (2010) Climate-driven range changes in Tasmanian intertidal fauna. *Marine and Freshwater Research* 61, 963–970.

Priddel, D., Carlisle, N. and Wheeler, R. (2006) Establishment of a new breeding colony of Gould's petrel (*Pterodroma leucoptera leucoptera*) through the creation of artificial nesting habitat and the translocation of nestlings. *Biological Conservation* 128, 553–563.

Renwick, L., Dann, P. and Thompson, S. (2007) Effect of fire on Little Penguins at Seal Island, south-eastern Australia. In: Woehler, E.J. (ed.) *Abstracts of oral and poster presentations, 6th International Penguin Conference, Hobart, Australia, 3–7 September 2007*. Birds Tasmania, Tasmania.

Santojanni, A., Arneri, E., Bernardini, V., Cingolani, N., Di Marco, M. and Russo, A. (2006) Effects of environmental variables on recruitment of anchovy in the Adriatic Sea. *Climate Research* 31, 181–193.

Schumann, N., Gales, N.J., Harcourt, R.G. and Arnould, J.P.Y. (2013) Impacts of climate change on Australian marine mammals. *Australian Journal of Zoology* 61(2), 146–159.

Taylor, A. (2007) *Winter breeding in a temperate cormorant: the Black-faced Cormorant* Phalacrocorax fuscescens. BSc (Hons) Thesis, Deakin University, Burwood, VIC.

Thompson, P.A., Baird, M.E., Ingleton, T. et al. (2009) Long-term changes in temperate Australian coastal waters: implications for phytoplankton. *Marine Ecology Progress Series* 394, 1–19.

Wernberg, T., Russell, B.D., Thompson, M.S. and Doblin, M.A. (2011) Seaweed Communities in Retreat from Ocean Warming. *Current Biology* 21, 1–5.

Wilkinson, I.S. and Priddel, D. (2011) Rodent eradication on Lord Howe Island: Challenges posed by people, livestock, and threatened endemics. In: Veitch, C.R., Clout, M.N. and Towns, D.R. (eds) *Island Invasives: Eradication and Management*. IUCN, Gland, Switzerland, pp. 508–514.

11   Climate adaptation and adaptive management planning for the Macquarie Marshes: a wetland of international importance

GILAD BINO, RICHARD T. KINGSFORD AND KIM JENKINS

Australian Wetlands, Rivers and Landscapes Centre, University of New South Wales, Australia

11.1   Introduction

Human population growth and increasing economic development have severely degraded the world's river and wetland ecosystems (Lemly et al. 2000; Millennium Ecosystem Assessment 2005; Vörösmarty et al. 2010). Wetland biodiversity loss continues unabated with mounting global pressures facing freshwater aquatic ecosystems (Butchart et al. 2010). This is predominately driven through infrastructure development, land conversion, water withdrawal, pollution, over-harvesting and -exploitation and the introduction of invasive species (Dudgeon et al. 2006). Although exact global estimates are lacking, many of the world's wetlands have already been lost or severely degraded (Millennium Ecosystem Assessment 2005). Through the building of dams and appropriation of water, river development and water regulation impacts global freshwater ecosystems by diminishing the frequency and extent of available habitats, impacting on biodiversity (Vörösmarty et al. 2010). Already river regulation has led to major regime shifts in freshwater ecosystems (Gordon et al. 2008; Keith et al. 2013). Under climate change, unfavourable conditions are projected to exacerbate anthropogenic changes to the natural cycles, further affecting biophysical and ecological processes.

Wetlands within regulated river basins such such as the Macquarie Marshes (central New South Wales) will likely be affected by climate change more than those in free-flowing river basins as these have a diminished capacity to absorb further disturbances (Palmer et al. 2008). Increased temperatures and changes to hydrologic regimes may exceed tolerances of some aquatic biota (Palmer et al. 2008; Turak et al. 2011; Viers and Rheinheimer 2011). As responses of wetland ecosystems to environmental stressors (e.g. reduction in flooding events) are often non-linear (i.e. thresholds), loss of resilience and regime shifts can be expected (Catalan et al. 2009; Woodward et al. 2010; Bino et al. 2014). For example, changes in temperature regime are projected to shift the distribution and interactions of

Applied Studies in Climate Adaptation, First Edition. Edited by Jean P. Palutikof, Sarah L. Boulter, Jon Barnett and David Rissik.
© 2015 John Wiley & Sons, Ltd. Published 2015 by John Wiley & Sons, Ltd.

species and communities due to the thermal tolerances of freshwater organisms (Carpenter et al. 1992). Recorded historical increases in water temperatures have favoured thermophilic species (Chessman 2009).

Climate adaptation strategies will be essential within a management regime coping with already highly compromised freshwater ecosystems. Current management needs sufficient flexibility to incorporate and mitigate such impacts. Often management regimes are already struggling to cope with high variability, inadequate data, competing institutional objectives and many different stakeholders. Authorities responsible for the management of the Macquarie Marshes have also invested in more strategic methods of management, including strategic adaptive management (Kingsford et al. 2011). In this chapter, we highlight these challenges for the Macquarie Marshes, a Ramsar-listed wetland in the southeast of Australia, part of the highly developed Murray–Darling Basin (Kingsford 2000), and consider how management authorities might adapt to projected climate change given existing management processes.

11.2   The Macquarie Marshes: a Ramsar wetland

The Macquarie Marshes are an extensive, diverse and dynamic wetland system covering about 200,000 ha (Thomas et al. 2011), including the Macquarie Marshes Nature Reserve of about 20,000 ha (NPWS 2012). The area incorporates extensive areas of flood-dependent vegetation (Paijmans 1981; Shelly 2005), sustaining habitat for many species of fauna (Kingsford and Auld 2005), providing ecosystem services which support highly productive grazing and cropping industries and maintaining cultural and heritage values for Indigenous and non-Indigenous people (OEH 2010). The area is also one of the most important sites in Australia for breeding of colonial waterbirds in terms of population sizes, colony sizes, number of species and frequency of breeding (Kingsford and Thomas 1995; Kingsford

and Auld 2005; Bino et al. 2014). In 1986, the Macquarie Marshes were listed in the Ramsar Convention as a wetland of international importance (OEH 2012a). Once listed, signatory countries are expected to promote conservation and wise use and ensure that actions undertaken by either private or government do not lead to loss of biodiversity or diminish the ecological, hydrological, cultural or social values of the designated wetlands. In 2005, the maintenance of the ecological character of Ramsar-listed wetlands in particular became a critical objective where this includes the ecosystem components, processes, benefits and services that characterised the wetland when designated (Imhoff et al. 2004, Sekercioglu et al. 2004). The ecological character description provides a detailed assessment of the site's critical values (components, processes and benefits/services) and provides a benchmark for planning and management (Ramsar Convention, 2005). An avoidable loss of ecological character signals unsustainable use and management of the wetland which has led to the degradation of ecological, biological and hydrological processes.

11.3   Ecological character of the Macquarie Marshes

Over the past 50 years, river regulation and landscape modification have disrupted the natural cycles of flood and drought in the Macquarie Marshes (Kingsford and Thomas 1995). Since construction of Burrendong Dam in 1967 upstream of the Macquarie Marshes, water abstraction and release timing has severely altered natural flow frequency and duration (CSIRO 2008). The effects of this include: reduced median annual flows by about 40% (Ren and Kingsford 2011); increased average period between large flows (114%), critical for inundation; and reduced median flows by about 57% (CSIRO 2008). The frequency of small flows has also decreased (Jenkins et al. 2005) along with significant reductions in inundated areas (Kingsford and Thomas 1995; Ren et al. 2010; Thomas et al. 2011). Alterations to the natural flow regime have degraded the condition of many ecological assets of the Macquarie Marshes, including waterbirds (Kingsford and Thomas 1995; Kingsford and Johnson 1998; Kingsford and Auld 2005), fish (Rayner et al. 2009), invertebrates (Jenkins et al. 2009) and vegetation (Thomas et al. 2010).

Following a prolonged drought and continued degradation in the natural flow regime due to river regulation and water abstraction, a notification of likely change in ecological character of the Macquarie Marshes Ramsar site was submitted to the Ramsar Secretariat under Article 3.2 of the Ramsar Convention; this is essentially a formal acknowledgement that the ecological character had changed as a result of anthropogenic change. This represents a particularly significant challenge for management authorities and the commitment to maintain its ecological character. Projected climate change impacts of a drying climate and increasing temperatures (Herron et al. 2002; Vaze et al. 2011) are expected to further drive a decline in ecological character, necessitating the development of suitable management strategies.

11.4   Conservation management of wetlands

Conservation of biodiversity traditionally relies on establishing reserves aimed at minimising threatening processes (Lovejoy 2006). This has largely safeguarded refugia for biodiversity. However, managing freshwater ecosystems is often a more complex process as these ecosystems are heavily dependent on processes that originate far beyond the borders of a reserve. Consequently, there is a clear responsibility to manage threats and vital functions at appropriate scales. A key strategy for conservation management of freshwater systems is to maintain the quantity and quality of the natural water regimes, including the frequency and timing of flows (Arthington et al. 2009; Poff and Zimmerman 2010; Arthington 2012). This can be achieved by recovery of flow regimes, alteration of dam operations, management of protected areas and effective governance and adaptive management (Kingsford et al. 2011). For heavily regulated systems, recovery of flow regimes through procurement of environmental flows is a crucial conservation objective (Arthington et al. 2006). The aim is to mimic past flow and inundation regimes (e.g. timing, duration, drawdown, frequency). For example, ensuring environmental flows are realised in late winter and spring is vital to enable colonial waterbirds to build up their food reserves for breeding (Kingsford and Auld 2005). This is also true for river red gum forests, which are dependent on similar timing for germination (Bren 1988). Duration of inundation is also critical in providing sufficient time for waterbirds to fledge their young, as premature lowering of water levels may cause abandonment of young (Baker-Gabb 1985; Scott 1997: Kingsford and Auld 2005; Brandis et al. 2011). Also, prolonged inundation can deplete soil oxygen beyond tolerable levels for river red gums (Chesterfield 1986) and affect productivity of wetlands (Kingsford et al. 2004).

Australian governments have acquired increasing amounts of environmental water recently in the Murray–Darling Basin, exemplified by the Australian Government's acquisition of 1583 GL between 30 June 2009 and 30 April 2013 (http://www.environment.gov.au/ewater/about/index.html). For the Macquarie Marshes, governments now hold an allocation of more than 331 GL per year of environmental flows, driving strategic planning of environmental water allocation (MDBA 2012; OEH 2012a; SEWPAC 2012). With considerable uncertainty regarding the effects of developing climate change (CSIRO 2008; Jenkins et al. 2011), decisions also remain uncertain on how best to allocate water across the Macquarie Marshes in the long term to protect its ecological values. There is a need to establish management processes and practices which allow for management of such a complex ecosystem, including identification of values (a hierarchy of objectives linked to thresholds of change which identify appropriate management options for conserving freshwater species and ecosystems that may be near their climate limits; Kingsford et al. 2011). An important implementation approach is to establish a strategic management approach, recognising the different interacting elements.

11.5   Adaptive management

To address the change in ecological character, NSW's Office of Environment and Heritage has developed a response plan to guide an adaptive management and governance process to restore ecological structure and function. A plan adopting an adaptive management approach is driven by fostering structured links between science, policy and management (OEH 2012b). Following Nyberg (1998):

> Adaptive management is a systematic process for continually improving management policies and practices by learning from the outcomes of operational programs. Its most effective form, 'active' adaptive management employs management programs that are designed to experimentally compare selected policies or practices, by evaluating alternative hypotheses about the system being managed. (Nyberg 1998, p. 2)

We identified some essential elements for improving the basis for such an approach, including incorporation of climate adaptation strategies. This involved examining the type of information required for scenario planning and integration of objectives with further climate mitigation objectives. We followed a strategic adaptive management framework designed to illustrate the links between key values, objectives, management actions and monitoring (Kingsford et al. 2011; Kingsford and Biggs 2012). The strategic adaptive management framework has four major steps that cover adaptive governance, planning, management and evaluation with continuous feedbacks and iterative processes. Following this framework, we identified six key activities that would help mitigate impacts of projected climate change in the Macquarie Marshes. These key activities added to and interacted with all four main steps (Fig. 11.1).

11.5.1   Review and collate scientific information

Information relevant for management is critical, particularly that which allows for scenario planning and recognises cause and effect processes under management control. Ecological datasets can also inform development of fine-scale quantitative management objectives, forming the foundations for identifying thresholds for the ecosystem and responses to drivers, including climate change. We compiled available datasets of ecological values of the ecosystem as well as potential drivers of change (i.e. inundation and fire history), including temporal and spatial coverage. These focused on the key ecosystem components and ecological processes of the system. Available ecological databases formed the basis for developing an information platform for the Macquarie Marshes, modelling responses of key ecological assets and developing a comprehensive process model of the ecosystem to assist in identifying climate-change adaptation opportunities.

11.5.2   Climate change and hierarchy of objectives

Structured decision-making can provide insightful ways to help address the complexities involved with identifying and prioritising key conservation values and choosing alternative management strategies (Martin et al. 2009). Understanding linkages between management actions and ecological responses of values can be assisted through development of an objectives hierarchy. Within the context of structured decision-making, climate change adaptation strategies should be linked to three of the four high-level objectives already identified for the Macquarie Marshes (OEH 2012b). These include ecosystem, enabling and people objectives (excluding balancing objectives; Fig. 11.2). This objectives hierarchy can help understand and track progress of management, particularly when also informed by measurement of outcomes. This can be tracked through fine-scale ecosystem objectives focused on individual organisms,

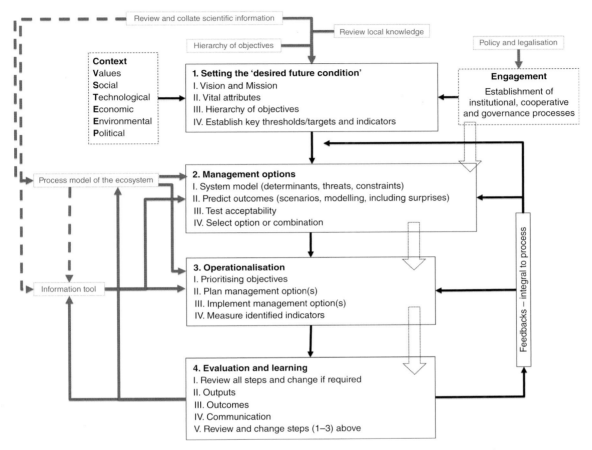

Figure 11.1 Integration of project activities within the strategic adaptive management framework. Source: Kingsford, Biggs & Pollard 2011. Reproduced with permission of Elsevier.

communities, processes or water management policies and planning. Investment in a hierarchy of objectives has occurred as part of an adaptive management plan for the Macquarie Marshes (OEH 2012b), but this does not explicitly incorporate climate adaptation. We developed climate adaptation objectives for this hierarchy (Fig. 11.2). These six climate adaptation objectives aim to separately address key aspects for the improvement of the Macquarie Marshes with clear responsibilities for different agencies (Table 11.1). **1.** The first is opportunities for adaptation in the management of the flow regime (Table 11.1; Fig. 11.2). Effective management of flows to the

Macquarie Marshes is achieved by delivery of available environmental flows on top of natural flows to maximise ecological outcomes for the system. Although climate change in the Macquarie Marshes will probably reduce flooding volumes and frequencies, past impacts of water regulation on loss of flooding may continue to overshadow those projected through climate change which will accentuate the drying impacts of river regulation. **2.** The single primary adaptation for restoring the Macquarie Marshes ecosystem remains return of adequate environmental water to restore the short and moderate inter-flood intervals. This can be achieved through increased water entitlements for

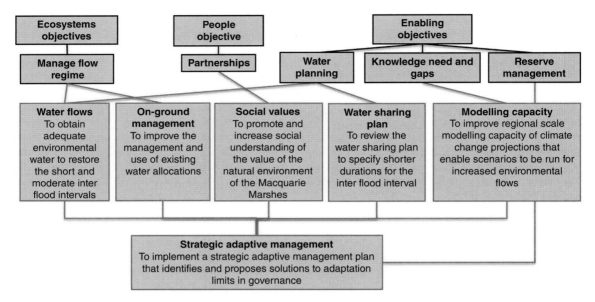

Figure 11.2 Integration of climate change adaptation strategies within existing objectives hierarchy presently developed by NSW Office of Environment and Heritage. Climate adaptations are all aimed at minimising the effects of climate change on the ecosystem through biophysical and behavioural adaptations.

Table 11.1 Six major adaptations that could be implemented to mitigate effects of climate change on the resilience of the Macquarie Marshes and its dependent biodiversity.

Measure	Adaptation	Responsibility
Flow regime	Obtaining adequate environmental water to restore the short and moderate inter flood intervals	State and federal governments
On-ground management	Improving the management and use of existing water allocations as well as to maximise the effectiveness of treatment and abatement activities	State government and other landowners
Social values	Promoting and increasing social understanding within the local and broader community of the value of the natural environment of the Macquarie Marshes	Communities and all tiers of government
Water-sharing plan	Reviewing the water-sharing plan to specify shorter durations for the inter flood intervals	NSW Office of Water
Modelling capacity	Improving regional-scale modelling capacity of climate change projections that enable ecological returns on environmental flows to be maximised	Research and management communities within and outside government
Strategic adaptive management	Implementing a strategic adaptive management with appropriate documentation that can be reviewed and used for decision-making	State and federal governments

the environment, involving reductions in extractive share for irrigation or through changes to practices of environmental flow release (Steinfeld 2013).

3. At a policy level, review of the water-sharing plan for the Macquarie River can improve environmental flow delivery with enhanced flexibility to meet ecosystem objectives.

4. Promoting social understanding (*people* objective) is also an important climate adaptation strategy focusing on building partnerships with local farmers, communities and government.

5. Further improving the modelling capacity to maximise ecological returns on environmental flows can only be attained through the support of science under *enabling* objectives.

6. Finally, progress can be made on adaptation through clear implementation of a strategic adaptive management framework, providing a transparent and coherent process of accountability (Kingsford et al. 2011).

Integration of the six climate change adaptation objectives demonstrated that climate adaptation would simply be an extension of implementing the strategic adaptive management plan for the Macquarie Marshes.

11.5.3 Process model of the ecosystem

Building a common understanding of system behaviour through a conceptual or process model is an important step for forming causal relationships among processes (Kingsford and Biggs 2012). Conceptual models can also be used to form the basis for identifying the core elements of an ecological character description, as part of the requirement under the Ramsar Convention (Davis and Brock 2008). We identified different wetland states and potential drivers related to water availability and climate change. Initial expert elicitation at a workshop helped sketch plausible models of cause and effect. Outcomes formed a basis for extending and formalising models to inform future management. We then developed a process model based on existing data and examined the responses of key indicators of the Macquarie Marshes to inundation and flow patterns. We explored alternative water

management strategies and identified maximal strategies for long-term management of the breeding of colonial waterbirds (Bino et al. 2014). For example, we found that management strategies could have a considerable effect on the likelihood of colonial waterbird breeding, increasing this by almost 50%. We also developed a quantitative state and transition models for key vegetation communities and a predictive capacity linking transition and persistence of vegetation communities to varying water allocations. Our approach highlighted the importance of semi-permanent vegetation, exhibiting a threshold response to water availability as an indicator for condition monitoring and assessment. Such an approach could extend to other organisms (e.g. frogs, invertebrates, epicormic growth of river red gums) to form a comprehensive ecosystem model driven by flow volumes. Much of this is highly dependent on knowledge of organisms and process and an understanding of cause and effect relationships that allow for predictive modelling.

11.5.4 Adaptive management information tool (AMIT)

Decision-making can be considerably improved with good information that can increasingly be used to develop predictive models, allowing the development of different scenarios for management (Kingsford and Biggs 2012). A critical constraint can be availability and access to scientific information. When available, high-quality datasets have the potential to support more informative decision-making and communicating of information to stakeholders (Michener and Jones 2012). Increased availability of information has raised standards for conservation and management requiring more scientific rigor (Fleishman et al. 2011). Consolidating multiple datasets to a single point of entry carries critical benefits for science (Reichman et al. 2011). A single dataset ensures data can be kept up to date and expanded to encompass all available data. Increased recognition of the value of data as a product of the scientific enterprise has led to increased emphasis on data stewardship, sharing and openness (Whitlock et al. 2010). Sharing of information is significantly more effectual in terms of resource use and can strengthen communication with the public on management outcomes. Importantly, key indicators can be jointly developed and used to improve understanding of responses to environmental variation or managerial actions. These returns can support better decision-making and guide future strategies of adaptation. Critical to adaptive management, continuous data collection and assessment forms the basis on which management can evaluate its actions and develop more efficient strategies for achieving desired objectives.

We developed an information platform that calls up data on biota, ecological processes and modelling into a spatiotemporal interface (http://test.amitweb.science.unsw.edu.au/). Google Earth network links were identified as a suitable mechanism as they provided fast and efficient distribution of data while allowing a degree of security to the project's original datasets. Central to the usability aspect of the information platform was the wide familiarity of users with the Google Earth interface. This novel and interactive search framework using the Google Earth KML (keyhole markup language) data structure allowed the efficient visualisation of point, polygon and continuous field data and can represent complex information structures required for model outcomes, scientific reports and additional project information. Our web interface also integrated modelled responses of flood-dependent ecosystem processes to determine adaptation opportunities to delivered flow regimes (e.g. environmental flow management), based on modelling (Ren et al. 2010). For adaptive management to succeed, management practices require constant feedback loops from data to planning. As monitoring continues, it will be important to continuously update the information platform and models to improve predictions and achieve management objectives.

11.5.5   Review local knowledge

It is not only scientific information that is critical. Many people living with complex social-ecological systems have deep insights about trajectories of

change and understanding of systems. We focused on documenting the local knowledge that graziers and government employees have gained from living and/or working in the Macquarie Marshes through the boom and bust cycles that characterise aridzone rivers and wetlands. The aim was to incorporate this into the conceptual model and potentially capture it within a framework of scientific data. We sought to record the ecological observations of graziers and government employees over decadal time scales across periods of drought, flood and changes in river management. We also wanted to understand how graziers might adapt to increased temperatures, reduced flooding and more frequent drought as predicted with climate change.

Landholders on the Macquarie Marshes have experienced extremes in these three variables and we were interested in their observations and strategies used under these conditions. We were interested in whether these practices may also help adapt to the possible consequences of climate change. We found that landholders are used to operating in a highly variable environment and many considered that this variability is well within the boundaries they anticipate with projected climate change. Grazing practices in the Macquarie Marshes use an adaptive approach to deal with water variability in the system. They identified the loss of flooding and variability as key impacts of regulation that reduced their resilience. Landholders have developed many practices to adapt to the loss of flooding due to regulation that they believe will enable them to adapt to current projections of climate change.

Collected information helped us develop and evaluate climate change adaptation strategies for floodplain wetlands and manage water for both agriculture and the environment. Climate change adaptation strategies involve complex trade-offs between the values different stakeholders associate with the ecosystem goods and services provided by floodplains and their wetlands. Climate change exacerbates the uncertainty associated with evaluating these trade-offs. By recording local knowledge, we ensured that these valuable memories of past and present events can be utilised in river and climate change planning. Interviews highlighted clear patterns of changes in the ecosystem over time. These included a decline in diversity and abundance of flood-dependent biota. Landholders also observed changes in the hydrological characteristics of the system including increased flow speed, erosion and changes in water condition.

11.5.6   Policy and legislation

Many different legislative and policy frameworks interact and focus on management of wetlands, particularly wetlands of international importance and their water supply (i.e. rivers) within a catchment. These include environmental and water legislation at the state and commonwealth level, as well as local and catchment planning frameworks. All commit to adaptive management implementation but generally fail to adequately integrate effort. Some of this is due to conflicting objectives and institutional power. We identified that alignment of the adaptive management approach was possible with current policies and processes, which could form a cohesive framework at different government and spatial scales. We particularly focused on recent changes to the governance of the Murray–Darling Basin implemented through the Water Act 2007 and its instrument the Murray–Darling Basin Plan, which ushered in a new era of water policy that aims to rehabilitate the basin's ecosystems and address overexploitation and mismanagement of water resources. Ultimately this could be the mechanism for driving a true collaborative commitment to adaptive management, but it would require a driving approach by the different organisations to make this happen. This could use all of the critical elements of strategic adaptive management (Kingsford et al. 2011; Kingsford and Biggs 2012) that are currently scattered and not well integrated. We conclude that governance, planning and policy, driven by legislative actions, requires considerably improved alignment to minimise redundancy and conflict to drive the necessary adaptive management framework while integrating climate adaptation.

11.6   The challenge

The ability to adapt in a changing environment is vital for effective management of the ecosystems. A strategic adaptive management framework provides a pathway through highly complex ecological and social processes. This requires the explicit development of a hierarchy of objectives to guide management, successively addressing relevant objectives. This is also partly dependent on increasing understanding of cause-and-effect relationships and capturing these in scientific tools (e.g. information platform, modelling). Without a clearly articulated strategic adaptive management framework, it is difficult to track whether management is effective, transparent and achieving its goals.

Adaptive management aims to provide a formal, systematic and rigorous framework while providing ongoing learning. Adaptive management requires a continuous synthesis of existing knowledge, exploration of alternative management actions and making explicit forecasts of ecological assets based on the best information, usually through modelling. This includes dealing with climate change effects and adapting to these.

Successful implementation of strategic adaptive management within the Macquarie Marshes will ultimately depend on champions within the environment agency and support from senior management. Other agencies involved in the sustainable management of the Macquarie Marshes also need to be involved, to avoid the problems of competing objectives and institutions. This requires investment in a jointly crafted vision, agreement of key values or assets and development of a hierarchy of objectives which can drive management, monitoring and reporting. Such a framework provides the flexibility to incorporate climate adaptation strategies specifically targeted to the resilience of a complex ecosystem such as the Macquarie Marshes.

References

Arthington, A.H. (2012) *Environmental Flows: Saving Rivers in the Third Millennium.* University of California Press, Berkeley.

Arthington, A.H., Bunn, S.E., Poff, N.L.R. et al. (2006) The challenge of providing environmental flow rules to sustain river ecosystems. *Ecological Applications* 16, 1311–1318.

Arthington, A.H., Naiman, R.J., McClain, M.E. et al. (2009) Preserving the biodiversity and ecological services of rivers: new challenges and research opportunities. *Freshwater Biology* 55, 1–16.

Baker-Gabb, D. (1985) Waterbird young perish. *Royal Australasian Ornithological Union Newsletter* 64, 1.

Bino, G., Steinfeld, C. and Kingsford, R.T. (2014) Maximising colonial waterbirds breeding events, using identified ecological thresholds and environmental flow management. *Ecological Applications* 24, 142–157.

Brandis, K., Kingsford, R., Ren, S. et al. (2011) Crisis water management and ibis breeding at Narran Lakes in arid Australia. *Environmental management* 48, 489–498.

Bren, L.J. (1988) Flooding characteristics of a riparian red gum forest. *Australian Forestry*, 51, 57–62.

Butchart, S.H.M., Walpole, M., Collen, B. et al. (2010) Global biodiversity: indicators of recent declines. *Science* 328, 1164–1168.

Carpenter, S.R., Fisher, S.G., Grimm, N.B. et al. (1992) Global change and freshwater ecosystems. *Annual Review of Ecology and Systematics* 23, 119–139.

Catalan, J., Barbieri, M.G., Bartumeus, F. et al. (2009) Ecological thresholds in European alpine lakes. *Freshwater Biology* 54, 2494–2517.

Chessman, B. (2009) Climatic changes and 13-year trends in stream macroinvertebrate assemblages in New South Wales, Australia. *Global Change Biology* 15, 2791–2802.

Chesterfield, E.A. (1986) Changes in the vegetation of the river red gum forest at Barmah, Victoria. *Australian Forestry* 49, 4–15.

CSIRO (2008) *Water Availability in the Macquarie-Castlereagh: A Report to the Australian Government from the CSIRO Murray-Darling Basin Sustainable Yields Project.* Canberra, CSIRO.

Davis, J. and Brock, M. (2008) Detecting unacceptable change in the ecological character of Ramsar wetlands. *Ecological Management and Restoration* 9, 26–32.

Dudgeon, D., Arthington, A.H., Gessner, M.O. et al. (2006) Freshwater biodiversity: importance, threats, status and conservation challenges. *Biological Reviews* 81, 163–182.

Fleishman, E., Blockstein, D.E., Hall, J.A. et al. 2011. Top 40 priorities for science to inform US conservation and management policy. *Bioscience* 61, 290–300.

Gordon, L.J., Peterson, G.D. and Bennett, E.M. (2008) Agricultural modifications of hydrological flows create ecological surprises. *Trends in Ecology and Evolution* 23, 211–219.

Herron, N., Davis, R. and Jones, R. (2002) The effects of large-scale afforestation and climate change on water allocation in the Macquarie River catchment, NSW, Australia. *Journal of Environmental Management* 65, 369–381.

Imhoff, M.L., Bounoua, L., Ricketts, T. et al. 2004. Global patterns in human consumption of net primary production. *Nature* 429, 870–873.

Jenkins, K.M., Boulton, A.J. and Ryder, D.S. (2005) A common parched future? Research and management of Australian arid-zone floodplain wetlands. *Hydrobiologia* 552, 57–73.

Jenkins, K.M., Kingsford, R.T. and Ryder, D. (2009) Developing indicators for floodplain wetlands: Managing water in agricultural landscapes. *Chiang Mai Journal of Science* 36, 224–235.

Jenkins, K.M., Kingsford, R.T., Wolfenden, B.J. et al. (2011) *Limits to Climate Change Adaptation in Floodplain Wetlands: The Macquarie Marshes.* National Climate Change Adaptation Research Facility, Gold Coast.

Keith, D.A., Rodríguez, J.P., Rodríguez-Clark, K.M. et al. (2013) Scientific foundations for an IUCN Red List of Ecosystems. *PloS one,* 8, e62111.

Kingsford, R. (2000) Ecological impacts of dams, water diversions and river management on floodplain wetlands in Australia. *Austral Ecology* 25, 109–127.

Kingsford, R. and Thomas, R. (1995) The Macquarie Marshes in arid Australia and their waterbirds: a 50-year history of decline. *Environmental Management* 19, 867–878.

Kingsford, R. and Johnson, W. (1998) Impact of water diversions on colonially-nesting waterbirds in the Macquarie Marshes of arid Australia. *Colonial Waterbirds* 21, 159–170.

Kingsford, R. and Auld, K. (2005) Waterbird breeding and environmental flow management in the Macquarie Marshes, arid Australia. *River Research and Applications* 21, 187–200.

Kingsford, R. and Biggs, H. (2012) Strategic Adaptive Management Guideline for Effective Conservation of Freshwater Ecosystems in and Around Protected Areas of the World. Sydney: IUCN WCPA Freshwater Taskforce, Australian Wetlands, Rivers, and Landscapes Centre.

Kingsford, R., Jenkins, K. and Porter, J. (2004) Imposed hydrological stability on lakes in arid Australia and effects on waterbirds. *Ecology* 85, 2478–2492.

Kingsford, R., Biggs, H. and Pollard, S. (2011) Strategic adaptive management in freshwater protected areas and their rivers. *Biological Conservation* 144, 1194–1203.

Lemly, A.D., Kingsford, R.T. and Thompson, J.R. (2000) Irrigated agriculture and wildlife conservation: conflict on a global scale. *Environmental Management* 25, 485–512.

Lovejoy, T.E. 2006. Protected areas: a prism for a changing world. *Trends in Ecology and Evolution* 21, 329–333.

Martin, J., Runge, M.C., Nichols, J.D. et al. (2009) Structured decision making as a conceptual framework to identify thresholds for conservation and management. *Ecological Applications* 19, 1079–1090.

Michener, W.H. and Jones, M.B. (2012) Ecoinformatics: supporting ecology as a data-intensive science. *Trends in Ecology and Evolution* 27(2), 85–93.

Millennium Ecosystem Assessment (2005) *Ecosystems and Human Well-Being: Wetlands and Water Synthesis.* World Resources Institute, Washington, DC.

Murray–Darling Basin Authority (2012) *Delivering A Healthy Working Basin: About the Draft Basin Plan.* Murray–Darling Basin Authority, Canberra.

National Parks and Wildlife Service (NSW) (2012) *Macquarie Marshes Nature Reserve.* Available at http://www.environment.nsw.gov.au/NationalParks/parkHome.aspx?id=N0449 (accessed 26 May 2014).

Nyberg, J.B. (1998) Statistics and the practice of adaptive management. In: Sit, V. and Taylor, B. (eds) *Statistical Methods for Adaptive Management Studies.* Land Management Handbook 42. BC Ministry of Forests, British Columbia.

Office of Enviroment and Heritage (2010) *Macquarie Marshes Adaptive Environmental Management Plan Synthesis of Information Projects and Actions.* Department of Environment, Climate Change and Water NSW, New South Wales.

Office of Enviroment and Heritage (2012a) *Ecological Character Description Macquarie Marshes Nature Reserve and U-block Components.* State of NSW, Office of Environment and Heritage, New South Wales.

Office of Enviroment and Heritage (2012b) *Macquarie Marshes Nature Reserve: Wetlands Adaptive Management Strategy.* In possession of the Office of Enviroment and Heritage, New South Wales.

Paijmans, K. (1981) The Macquarie Marshes of Inland Northern New South Wales, Australia. CSIRO Australian Division of Land Use Research Technical Paper no. 41. Commonwealth Scientific and Industrial Research Organisation, Melbourne.

Palmer, M.A., Reidy Liermann, C.A., Nilsson, C. et al. (2008) Climate change and the world's river basins:

anticipating management options. *Frontiers in Ecology and the Environment* 6, 81–89.

Poff, N.L. and Zimmerman, J.K.H. (2010) Ecological responses to altered flow regimes: a literature review to inform the science and management of environmental flows. *Freshwater Biology* 55, 194–205.

Ramsar (2005) Resolution IX.1 Annex A: A Conceptual Framework for the wise use of wetlands and the maintenance of their ecological character, In: *Ramsar, Wetlands and Water: Supporting Life, Sustaining livelihoods, 9th Meeting of the Conference of the Contracting Parties to the Convention of Wetlands,* Kampala, Uganda, 8–15 November 2005.

Rayner, T.S., Jenkins, K.M. and Kingsford, R.T. (2009) Small environmental flows, drought and the role of refugia for freshwater fish in the Macquarie Marshes, arid Australia. *Ecohydrology* 2, 440–453.

Reichman, O.J., Jones, M.B. and Schildhauer, M.P. (2011) Challenges and opportunities of open data in ecology. *Science* 331, 703–705.

Ren, S. and Kingsford, R.T. (2011) Statistically integrated flow and flood modelling compared to hydrologically integrated quantity and quality model for annual flows in the regulated Macquarie River in arid Australia. *Environmental management* 48, 177–188.

Ren, S., Kingsford, R.T. and Thomas, R.F. (2010) Modelling flow to and inundation of the Macquarie Marshes in arid Australia. *Environmetrics* 21, 549–561.

Scott, A. (1997) Relationships Between Waterbird Ecology and River Flows in the Murray Darling Basin. Technical Report 5/97. CSIRO Land and Water, Canberra.

Sekercioglu, C.H., Daily, G.C. and Ehrlich, P.R. 2004. Ecosystem consequences of bird declines. *Proceedings of the National Academy of Sciences of the United States of America* 101, 18042–18047.

SEWPaC (2012) *Commonwealth Environmental Water Holdings.* Department of Sustainability, Environment, Population and Communities: Water Programs, Canberra. Available at http://www.environment.gov.au/ewater/about/holdings.html (accessed 26 May 2014).

Shelly, D. (2005) *Flora and Fauna of the Macquarie Marshes Region,* Department of Infrastructure, Planning and Natural Resources, Dubbo.

Steinfeld, C. (2013) *Integrating Environmental Flows into Regulated River Systems.* PhD Thesis, University of New South Wales.

Thomas, R., Bowen, S., Simpson, S. et al. (2010) Inundation response of vegetation communities of the Macquarie Marshes in semi-arid Australia. In:

Saintilan, N. and Overton, I. (eds.) *Ecosystem Response Modelling in the Murray–Darling Basin.* CSIRO Publishing, Melbourne.

Thomas, R., Kingsford, R., Lu, Y. et al. (2011) Landsat mapping of annual inundation (1979–2006) of the Macquarie Marshes in semi-arid Australia. *International Journal of Remote Sensing* 32, 4545–4569.

Turak, E., Marchant, R., Barmuta, L. et al. (2011) River conservation in a changing world: invertebrate diversity and spatial prioritisation in south-eastern coastal Australia. *Marine and Freshwater Research* 62, 300–311.

Vaze, J., Davidson, A., Teng, J. et al. (2011) Impact of climate change on water availability in the Macquarie-Castlereagh River Basin in Australia. *Hydrological Processes* 25, 2597–2612.

Viers, J.H. and Rheinheimer, D.E. (2011) Freshwater conservation options for a changing climate in California's Sierra Nevada. *Marine and Freshwater Research* 62, 266–278.

Vörösmarty, C.J., McIntyre, P., Gessner, M.O. et al. (2010) Global threats to human water security and river biodiversity. *Nature* 467, 555–561.

Whitlock, M.C., McPeek, M.A., Rausher, M.D. et al. (2010) Data archiving. *The American Naturalist* 175, 145–146.

Woodward, G., Perkins, D.M. and Brown, L.E. (2010) Climate change and freshwater ecosystems: impacts across multiple levels of organization. *Philosophical Transactions of the Royal Society B: Biological Sciences* 365, 2093–2106.

12 Conservation of Australian plant-dwelling invertebrates in a changing climate

MELINDA L. MOIR

School of Plant Biology, University of Western Australia, Australia

12.1   Introduction

Anthropogenically induced global climate change is an insidious threat to biota in extensive regions of the world (e.g. Malcolm et al. 2006; Beaumont et al. 2011; Kingsford and Watson 2011; Bellard et al. 2012; Mokany et al. 2012). In combination with land clearing and fragmentation, climate change may be instigating the sixth mass extinction event in the Earth's history (Bellard et al. 2012).

Invertebrates comprise approximately 80% of global biodiversity. The number of species that we may lose from this component as a result of climate change is therefore likely to be very large (Dunn et al. 2009; Cardoso et al. 2011). One of the largest groups within the invertebrates comprises taxa that are closely dependent upon other species for their survival, for example herbivorous insects and their host plants. Approximately one-quarter of global terrestrial biodiversity is represented by plant-dwelling insects (Strong et al. 1984). In Australia, it is estimated that the total number of insect species ranges from 205,000 to 400,000 species (Cranston 2010). The potential for thousands of species to be extinguished through widespread disturbances such as a changing climate is therefore high (Moir et al. 2011, 2012a). This form of extinction is termed 'coextinction' as it occurs through the loss of the host or from some form of change in the population of the host (e.g. Koh et al. 2004; Moir et al. 2010; Colwell et al. 2012). An estimated 8–9% of all plant-dwelling insect species could be at risk of coextinction in certain regions such as heathland in the southwest of Australia (Moir and Leng 2013). Plant-dependent insects may be extinguished by a number of mechanisms associated with climate change including rising sea levels (New 2011; Andren and Cameron 2012), changes in temperature (Beaumont and Hughes 2002), mismatch in timing with host plant resources (Delucia et al. 2012) and increased competition or predation (Imbert et al. 2012). The number of species under threat is likely to be higher in areas that are geographically restricted but affected severely by climate change such as wetlands and mountain tops (for a review of these regions and climate change, see Hughes 2011).

12.2   Insect management under a changing climate

Given the numbers of species involved and the lack of knowledge on the majority of plant-dwelling insects, planning or implementing

Applied Studies in Climate Adaptation, First Edition. Edited by Jean P. Palutikof, Sarah L. Boulter, Jon Barnett and David Rissik.
© 2015 John Wiley & Sons, Ltd. Published 2015 by John Wiley & Sons, Ltd.

effective management options can be over-whelming. Perhaps because of the difficulty associated with managing these taxa, insects and invertebrates are rarely included in systematic conservation plans (Clark and May 2002; Pressey et al. 2003). To further compound this issue, Australian state and federal agencies charged with the conservation of all biodiversity demon-strate inequitable listing of vertebrate compared to invertebrate species, for example: South Australia will not list any invertebrate under their state legislation; Queensland does not have a single non-butterfly insect representative on their list; and the Federal government lists just 48 invertebrates as threatened compared with 336 vertebrates (12.5% of the total number of protected species; Department of the Environ-ment 2013). Insects represent only 3.6% (14 species) of all species listed by the Federal government (Department of the Environment 2013). This inequity becomes important when land managers consider which species require attention. Without the recognition that formal listing affords, few invertebrates are likely to be considered for conservation management or allocated resources towards their conservation under a changing climate.

When plant-dwelling insects are recognised as potentially threatened, a common misconception is that conserving the host plant will indirectly protect all dependent insect species reliant on that plant. Unfortunately, this is often not the case. Managing for the persistence of hosts alone under climate change may be insufficient to maintain populations of all their dependent species, just as the conservation of any species may not be assured through maintenance of its habitat alone. For example, natural enemy interactions may affect survival rates in the wild (Martin and Pullin 2004) and plant populations that have recently regenerated after fire may have lost their specialist herbivores (Taylor and Moir 2009).

Previously published frameworks can be used to determine whether an insect is threatened with extinction under a changing climate (e.g. Williams et al. 2008; Thomas et al. 2011; Shoo

et al. 2013). Despite the availability of such tools, Australian land managers indicate that lack of expertise in insect identification is the most important factor inhibiting the consideration of plant-dwelling insects in management planning (Moir and Leng 2013). As a consequence land managers currently struggle to determine which insect species inhabit their lands, much less iden-tify which are in need of conservation (Moir and Leng 2013). It is therefore unsurprising that conservation actions for plant-dwelling insects in Australia tend to occur when taxonomists and ecologists favour a particular group, which restricts a strategic and coordinated management approach from being taken. The exception is but-terflies, which are better known and invoke more public interest than any other insect group (Beaumont and Hughes 2002; New and Sands 2002, 2004). Employing dedicated conservation entomologists within the Australian federal and state governments could potentially bridge the interface between taxonomists, government conservation bodies, land managers and distur-bance ecologists, improving coordination and enabling a strategic approach to management (New 2012; Moir and Leng 2013).

12.3   Potential for adaptive management of threatened insects

Although many conservation programs aim to reduce the probabilities of extinction of plants (e.g. Millennium Seed Bank Project, Global Strategy for Plant Conservation), few consider the conservation of invertebrate species dependent upon those plant species for their survival such as herbivorous insects (Moir et al. 2012a). Typically the only insects that do receive attention in terms of assessing the response to climate change are butterflies (Lepidoptera) because they are charismatic and there usually exists good historical and contemporary datasets (New 2011; Wilson and Maclean 2011). Conservation actions to assist in threatened dependent insect species survival are required urgently (Moir et al. 2012b) for two reasons: (1) large numbers of host plant species are imminently threatened by climate change and other synergistic forces (e.g. Barrett et al. 2008; Yates et al. 2010), particularly habitat fragmentation (Mantyka-pringle et al. 2012); and (2) populations of host-specific insects may decline faster than that of their hosts because a minimum population size of host is required to sustain a viable population of insect, ensuring survival against genetic and environmental factors (e.g. Taylor and Moir 2009; Moir et al. 2010).

The adaptation management options for biota in the face of climate change often includes the following actions: *ex situ* conservation (seed banks, translocated populations, assisted migration); monitoring existing populations; managing other threats such as fire and competition from native and non-native competitors; and researching the impact of predict climate change on taxa (i.e. environmental tolerances). If the species is listed as threatened, then recovery plans are often developed to coordinate management actions which may include climate change adaptation management options. Most of these strategies assume that the ecology of the target organism is well-known, however (e.g. Thomas et al. 2011). This is not the case in regions such as Australia where the majority of invertebrates are undescribed and understudied. In such regions there are several preliminary steps that are required to determine which species are at threat and should be managed.

• *Biogeography*: surveys must be undertaken to determine rarity and distribution when known, and whether the species is likely to be threatened by climate change and any additional processes. A survey may show, for example, that an insect is restricted to a mountain summit, the climate of which is likely to dry with climate change.

• *Taxonomy*: a threatened invertebrate can be listed without a species name but a taxonomist must certify that it is a distinct species or subspecies. As management proceeds, genetic analysis may be required to determine whether different populations are genetically distinct, where translocated populations should be sourced from and to prevent inbreeding depressions.

• *Nomination*: successfully listing a species requires detailed background knowledge of the species, including successful and unsuccessful survey locations. Conservation resources are almost never directed towards a species that is not listed in Australia. This action therefore assists in future management and is often followed by development of recovery plans.

12.4   Case studies of management actions

The type of management and also the cost of different management strategies will vary greatly depending on the animal involved. The level of threat posed by climate change will also determine the urgency and cost of the management actions required. Some species may only require *in situ* management; for example, their habitat maintained without pollution or invasion by exotic diseases and species. Other habitats require active restoration or selectively planting particular host plants, and this may assist in the return of plant-dwelling insect populations (Littlewood et al. 2006; Sands 2008). Other *in situ* strategies include reducing habitat fragmentation by restoring 'corridors' to allow for natural movement of biota. A general framework (Fig. 12.1) for managing dependent insect species threatened with climate change is provided in Moir and Leng (2013). Examples are provided in the following sections of three *ex situ* adaptation management strategies employed to aid in the recovery of threatened plant-dwelling insects: reintroduction; introductions (also called assisted colonisation or migration); and captive management.

12.4.1   Reintroduction

Reintroductions are translocations of organisms back within their original range and, for plant-dwelling insects, reintroductions could encompass two strategies. Firstly, the host plant occurs

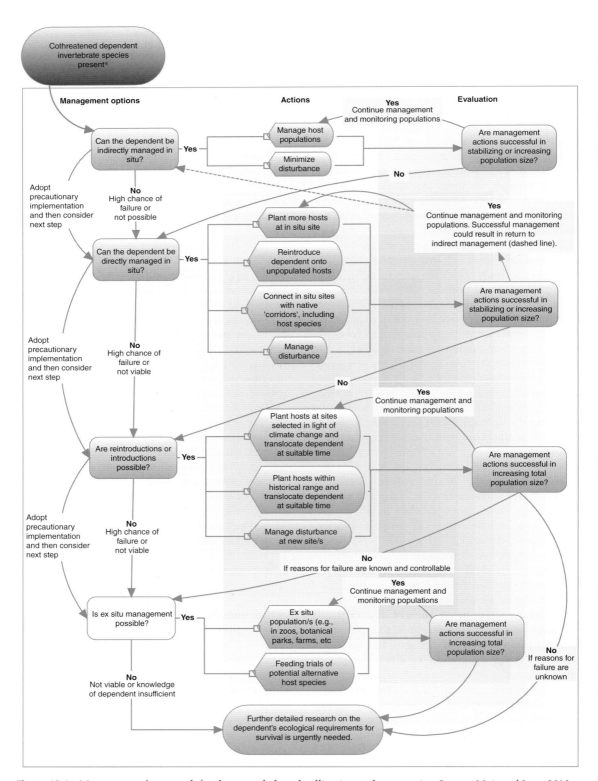

Figure 12.1 Management framework for threatened plant dwelling invertebrate species. Source: Moir and Leng 2013.

across the original range but the insect does not, and can be translocated onto the uninhabited host populations. Secondly, the host is grown within the original range and the insect then translocated onto the plants once they are of appropriate age and structure. Using plant-louse (Psylloidea) as examples, both strategies are expanded upon in the following.

Dependent species are rarely present on every population of their host. For example, the majority of plant species analysed in the southwest of Australia had insect assemblages differing significantly between plant populations (Moir and Leng 2013). This situation offers the opportunity to mitigate the risk of species extinction for the dependent insect by moving individuals onto unoccupied host populations. The advantages of such a strategy are that it is relatively cheap, easy and quick. There is no need to develop large enough populations of host, as generally the host population is of sufficient age to hold the dependent, particularly small, plant-dwelling insects. The disadvantages include not knowing whether the new location has suitable environmental parameters to facilitate dependent survival, and whether there are competitors, parasitoids or predators present that may decimate the threatened dependent insect. In addition, genetic factors should be considered (Allendorf et al. 2013), as numbers required to produce viable translocated populations (and minimise inbreeding) are debatable and often unknown for invertebrates. Finally, monitoring programs should be established to ensure that the translocation is successful in the long term (New 1994; Moir et al. 2012a).

An example of a reintroduction onto uninhabited host populations is that of the critically endangered (IUCN 2013) Vesk's plant-louse, *Acizzia veski* Taylor and Moir, in southwest Australia. Populations of the insect are expected to decline due to having a very narrow host breadth, host plants with small restricted populations and the plants being in danger of extinction through threats such as fire and climate change (Taylor and Moir 2009). *Acizzia veski* is currently found on a single population of *Acacia veronica*

(Maslin), a plant having a relatively low Priority 3 conservation status from the state government (http://www.dpaw.wa.gov.au/plants-and-animals/ threatened-species-and-communities for descriptions of conservation codes). Approximately 10 populations of *A. veronica* are present in the southwest. Through trial reintroductions, the plant-louse has been released into two uninhabited acacia stands (Moir and Leng 2013). It is too early to determine whether the initial translocations are a success in seeding plant-louse populations. However, the outlook is positive given the close proximity to transplant sites (within 30 km), the short amount of time it took to conduct the transplants (hours) and the lack of competitor and predator insect fauna on host plants at the new sites.

An example of a reintroduction of both host and insect populations is the translocation of the critically endangered (state government) plant-louse *Trioza barrettae* Taylor and Moir in southwest Australia. The plant-louse is expected to decline as it has a very narrow host breadth on a critically endangered host plant (Taylor and Moir 2013) The plant, the feather-leaf banksia (*Banksia brownii* R.Br.), is highly threatened by *Phytophthora* dieback as well as fire and climate change (Coates and McArthur 2010). This banksia is being grown in a seed orchard within what may have been the historical range of the plant. It may have once occurred throughout the area and only in recent history retracted to its current fragmented populations (Coates and McArthur 2010). Further, as the plant-louse is found on *B. brownii* stands over 80 km apart, it too may have been as widespread as its host. The plant-louse was translocated onto the host plants when the plants were approximately 7–8 years old. They were put into cages on the plant so that they could be monitored and not devoured by predators. At the time, only one translocation site was possible because of uncertainty of the genetic variability between plant-louse populations (Moir and Leng 2013). The host plant has shown differences between populations (Coates and McArthur 2010), and host plants from the location of the plant-louse source site were only translocated once. This translocation presents an opportunity to better understand the biology and ecology of these highly co-threatened, but understudied, insects. It should be noted that this species has only been named recently (Taylor and Moir 2013) and very little is known of its ecology.

In the Northern Hemisphere, particularly in Europe and North America, reintroductions have had a longer history (Shultz et al. 2008). Reintroduction as a management strategy to counteract the threat posed by climate change is still in its infancy, however (but see Pelini et al. 2009).

12.4.2 Introduction/assisted colonisation

An introduction, assisted migration or assisted colonisation of an organism is the establishment of an organism outside its historically known native range. In the case of plant-dwelling insects, either an alternative host plant must be identified and present in this new location or the host plant must also be introduced. As an example of the former, Martin and Pullin (2004) trialled the use of alternative host plants by the Large Copper butterfly, *Lycaena dispar batavus* (Oberthür), a rare European butterfly. They found that the butterfly could develop successfully on alternative hosts but that other factors, possibly competition and predation, reduced survival in the wild. In many cases however, the close association between an insect and plant species, or even a single plant population, has excluded the use of alternative hosts as an option (Pelini et al. 2010).

The mealybug *Pseudococcus markharveyi* Gullan provides an example of an introduction. It occurs naturally on mountain summits in southwest Australia on the critically endangered *Banksia montana* (George) (Gullan et al. 2013). The mealybug is not currently listed, but has been nominated at the state government level. Both species occur within a federally recognised Threatened Ecological Community (TEC) of flora species. The bug occurs on two populations of *B. montana*, although on one population it occurs only on a single plant. In addition to the low numbers of host plants with the mealybug occupying them, populations of *P. markharveyi* are expected to decline due to a combination of the following: (1) having a very narrow host breadth; (2) having low dispersal abilities; (3) host plants being rare and in danger of extinction; (4) host plants having small restricted populations; and (5) threats to host plants such as *Phytophthora* dieback, fire and climate change (Moir et al. 2012b; Gullan et al. 2013). *Banksia montana* has been successfully translocated and grown at a seed farm approximately 40 km to the south of these mountains. Translocation trials of the mealybug onto *B. montana* occurred recently, and after a year the mealybug populations are thriving within 40% of the bags (Fig. 12.2). Continued successful population growth would allow *P. markharveyi* to establish in this new location and decrease the chance of global extinction. As with the plant-lice described previously, this translocation also presents an opportunity to better understand the biology and ecology of these mealybugs and aid in establishing the first management protocols globally for the conservation of this group of insect.

12.4.3 Captivity

Maintaining populations of plant-dwelling insects in captivity, such as in zoos, is rarely a viable option. Such a strategy is costly to initiate and maintain. Furthermore, the vast majority of plant-dwelling insects are small and their ecological requirements poorly known, which means that few institutions have the expertise available to manage the insects (Daniels 2009). As such, examples of co-threatened insects managed in captivity are difficult to find.

One such Australian example is the giant Lord Howe Island phasmid, or stick insect, *Dryococelus australis* (Montrouzier). Thought extinct, it was rediscovered on Ball's Pyramid, an island off the coast of Lord Howe Island (Priddle et al. 2003). The threatening processes were predation by rats and lack of suitable habitat, but climate-related effects could have potentially contributed to its extinction through severe storms and resultant erosion of the steep slopes on Ball's Pyramid where the phasmid occurred. A successful

(a)

(b)

Figure 12.2 (a) Insect cages restrain the (b) threatened *Banksia montana* mealybug (*Pseudococcus markharveyi*) on critically endangered montane banksias (*Banksia montana*) at a translocation site in south-western Western Australia. Source: Photographs by M. Moir. For colour details please see Plate 10.

phasmid breeding program has been established at Melbourne Zoo (Carlile et al. 2009). Captive management of such poorly known and rare insects can be tenuous however; one of the two breeding pairs of Lord Howe Island phasmid died within a month of captivity, and the female of the second pair almost died, possibly due to the humidity in the enclosures (Carlile et al. 2009). There is currently no suitable habitat on Lord Howe Island due to the continued presence of rats but, as the island is larger and contains more habitat variability, it is hoped that the insect can be reintroduced in the future (Carlile et al. 2009).

Captive breeding is most successful when it occurs as a part of an integrated management plan, which often includes planned reintroductions and introductions. For example, the Australian Richmond bird-wing butterfly *Ornithoptera richmondia* (Gray) is threatened by numerous processes linked to climate change, including prolonged drought (Sands 2008). Integrated management across the species' range has included restoring the host plant into bush and gardens, community engagement such as workshops and school visits, removal of an invasive weed which reduces larval survival and careful captive breeding programs to reduce inbreeding

in certain small populations (for a full discussion see Sands 2008). Similarly, integrated management of the American Miami blue butterfly *Cyclargus thomasi bethunebakeri* (Comstock and Huntington), which includes captive breeding and reintroductions, has been successful to date (Daniels 2009). This success was predominantly due to continued monitoring which found that initial drought and cyclonic events severely impacted the populations; further supplementary reintroductions were found to be required (Daniels 2009).

Despite the success stories, the vast majority of threatened plant-dwelling insects are not currently receiving any management (if indeed they are even recognised as threatened). In general, invertebrates in captivity display the same characteristics; they are large and what the public consider either beautiful (e.g. tropical butterflies, lycaenid butterflies), weird (e.g. giant phasmids, giant rhinoceros beetles) or creepy (e.g. scorpions, tarantula spiders, giant cockroaches). Unfortunately most plant-dwelling insects do not fall into these categories. Currently, to be considered for captive management, these predominantly small insects may need to be on the brink of extinction with enthusiastic and vocal supporters calling for their rescue. Calls for such management have been made in Europe and North America. For example, potential future captive breeding and mass assisted migration of insects of Michigan prairie fen in North America (Landis et al. 2012) could provide novel methods for mass management and movement of entire insect communities from threatened habitats.

12.5 Summary

Plant-dwelling insect management with climate change is not currently a priority within many countries. Given that even widespread insect species are displaying negative impacts from a warming climate, it is unrealistic to expect that most plant-dwelling insects will survive if the habitat and host plant species are present. Within Australia, appointing state and federal conservation entomologists would assist with coordinating the identification, prioritisation and management of plant-dwelling insects that are threatened by multiple disturbances, including climate change and habitat fragmentation.

References

Allendorf, F.W., Aitken, S.N. and Luikart, G. (2013) *Conservation and the Genetics of Populations*. Wiley-Blackwell Publishing, West Sussex.

Andren, M. and Cameron, M.A. (2012) The distribution of the endangered Black Grass-dart butterfly, *Ocybadistes knightorum* (Lepidoptera: Hesperiidae). *Australian Zoology* 36, 159–168.

Barrett, S., Shearer, B.L., Crane, C.E. and Cochrane, A. (2008) An extinction-risk assessment tool for flora threatened by *Phytophthora cinnamomi*. *Australian Journal of Botany* 56, 477–486.

Beaumont, L.J. and Hughes, L. (2002) Potential changes in the distributions of latitudinally restricted Australian butterfly species in response to climate change. *Global Change Biology* 8, 954–971.

Beaumont, L.J., Pitman, A., Perkins, S., Zimmermann, N.E., Yoccoz, N.G. and Thuiller, W. (2011) Impacts of climate change on the world's most exceptional ecoregions. *Proceedings of the National Academy of Sciences* 108, 2306–2311.

Bellard, C., Bertelsmeier, C., Leadley, P., Thuiller, W. and Courchamp, F. (2012) Impacts of climate change on the future of biodiversity. *Ecology Letters* 15, 365–377.

Cardoso, P., Borges, P.A.V., Triantis, K.A., Ferrández, M.A. and Martín, J.L. (2011) Adapting the IUCN Red List criteria for invertebrates. *Biological Conservation* 144, 2432–2440.

Carlile, N., Priddel, D. and Honan, P. (2009) The recovery programme for the Lord Howe Island phasmid (*Dryococelus australis*) following its rediscovery. *Ecological Management and Restoration* 10, S124–S128.

Clark, J.A. and May, R.M. (2002) Taxonomic bias in conservation research. *Science* 297, 191–192.

Coates, D.J. and McArthur, S. (2010) Assessing the taxonomic status of *Banksia brownii* and patterns of genetic diversity in extinct and extant populations. Bankwest Landscope Project, Western Australian Department of Environment and Conservation, Perth.

Colwell, R.K., Dunn, R.R. and Harris, N.C. (2012) Coextinction and persistence of dependent species in a changing world. *Annual Review of Ecology, Evolution and Systematics* 43, 183–203.

Cranston, P.S. (2010) Insect biodiversity and conservation in Australasia. *Annual Review of Entomology* 55, 55–75.

Daniels, J.C. (2009) Cooperative conservation efforts to help recover an endangered south Florida butterfly. *Insect Conservation and Diversity* 2, 62–64.

Delucia, E.H., Nabity, P.D., Zavala, J.A. and Berenbaum, M.R. (2012) Climate change: resetting plant-insect interactions. *Plant Physiology* 160, 1677–1685.

Department of the Environment (2013) *EPBC Act List of Threatened Fauna*. Department of the Environment, Australian Federal Government, Canberra.

Dunn, R.R., Harris, N.C., Colwell, R.K., Koh, L.P. and Sodhi, N.S. (2009) The sixth mass coextinction: are most endangered species parasites and mutualists? *Proceedings of the Royal Society of London B* 276, 3037–3045.

Gullan, P.J., Moir, M.L. and Leng, M.C. (2013) New species of *Pseudococcus* from threatened *Banksia*. *Records of the Western Australian Museum* 28, 13–20.

Hughes, L. (2011) Climate change and Australia: key vulnerable regions. *Regional Environmental Change* 11, 189–195.

Imbert, C.E., Goussard, F. and Roques, A. (2012) Is the expansion of the pine processionary moth, due to global warming, impacting the endangered Spanish moon moth through an induced change in food quality? *Integrative Zoology* 7, 147–157.

International Union for the Conservation of Nature (IUCN) (2013) *Red List of Threatened Species*. Version 2013.1. IUCN Gland, Switzerland.

Kingsford, R.T. and Watson, J.E. (2011) Climate change in Oceania: a synthesis of biodiversity impacts and adaptations. *Pacific Conservation Biology* 17, 270.

Koh, L.P., Dunn, R.R., Sodhi, N.S., Colwell, R.K., Proctor, H.C. and Smith, V.S. (2004) Species coextinctions and the biodiversity crisis. *Science* 305, 1632–1634.

Landis, D.A., Fiedler, A.K., Hamm, C.A. et al. (2012) Insect conservation in Michigan prairie fen: addressing the challenge of global change. *Journal of Insect Conservation* 16, 131–142.

Littlewood, N.A., Dennis, P., Pakeman, R.J. et al. (2006) Moorland restoration aids the reassembly of associated phytophagous insects. *Biological Conservation* 132, 395–404.

Malcolm, J.R., Liu, C., Neilson, R.P., Hansen, L. and Hannah, L. (2006) Global warming and extinctions of endemic species from biodiversity hotspots. *Conservation Biology* 20, 538–548.

Mantyka-pringle, C.S., Martin, T.G. and Rhodes, J.R. (2012) Interactions between climate and habitat loss effects on biodiversity: A systematic review and meta-analysis. *Global Change Biology* 18, 1239–1252.

Martin, L.A. and Pullin, A.S. (2004) Host-plant specialisation and habitat restriction in an endangered insect, *Lycaena dispar batavus* (Lepidoptera: Lycaenidae) II. Larval survival on alternative host plants in the field. *European Journal of Entomology* 101, 57–62.

Moir, M.L. and Leng, M-C. (2013) *Developing Management Strategies to Combat Increased Coextinction Rates of Plant-Dwelling Insects through Global Climate Change*. National Climate Change Adaptation Research Facility, Gold Coast.

Moir, M.L., Vesk, P.A., Brennan, K.E.C., Keith, D.A., Hughes, L. and McCarthy, M.A. (2010) Current constraints and future directions in estimating coextinction. *Conservation Biology* 24, 682–690.

Moir, M.L., Vesk, P.A., Brennan, K.E.C., Keith, D.A., McCarthy, M.A. and Hughes, L. (2011) Identifying and managing cothreatened invertebrates through assessment of coextinction risk. *Conservation Biology* 25, 787–796.

Moir, M.L., Vesk, P.A., Brennan, K.E.C. et al. (2012a) Considering extinction of dependent species during translocation, ex situ conservation and assisted migration of threatened hosts. *Conservation Biology* 26, 199–207.

Moir, M.L., Vesk, P.A., Brennan, K.E.C. et al. (2012b) A preliminary assessment of changes in plant-dwelling insects when threatened plants are translocated. *Journal of Insect Conservation* 16, 367–377.

Mokany, K., Harwood, T.D., Williams, K.J. and Ferrier, S. (2012) Dynamic macroecology and the future for biodiversity. *Global Change Biology* 18, 3149–3159.

New, T.R. (1994) Needs and prospects for insect reintroductions for conservation in Australia. In: Serena, M. (ed.) *Reintroduction Biology of Australian and New Zealand Fauna*. Surrey Beatty and Sons, Chipping Norton, pp. 47–52.

New, T.R. (2011) *Butterfly Conservation in South-Eastern Australia: Progress and Prospects*. Springer, Dordrecht.

New, T.R. (2012) *Insect Conservation: Past, Present and Prospects*. Springer, Dordrecht.

New, T.R. and Sands, D.P.A. (2002) Narrow-range endemicity and conservation status: interpretations for Australian butterflies. *Invertebrate Systematics* 16, 665–670.

New, T.R. and Sands, D.P.A. (2004) Management of threatened insect species in Australia, with particular reference to butterflies. *Australian Journal of Entomology* 43, 258–270.

Pelini, S.L., Dzurisin, J.D.K., Prior, K.M. et al. (2009) Translocation experiments with butterflies reveal limits to enhancement of poleward populations under climate change. *Proceedings of the National Academy of Sciences* 106, 11160–11165.

Pelini, S.L., Keppel, J.A., Kelley, A.E. and Hellmann, J.J. (2010) Adaptation to host plants may prevent rapid insect responses to climate change. *Global Change Biology* 16, 2923–2929.

Pressey, R.L., Cowling, R.M. and Rouget, M. (2003) Formulating conservation targets for biodiversity pattern and process in the Cape Floristic Region, South Africa. *Biological Conservation* 112, 99–127.

Priddel, D., Carlile, N., Humphrey, M., Fellenberg, S. and Hiscox, D. (2003) Rediscovery of the 'extinct' Lord Howe Island stick-insect (*Dryococelus australis* (Montrouzier)) (Phasmatodea) and recommendations for its conservation. *Biodiversity and Conservation* 12, 1391–1403.

Sands, D. (2008) Conserving the Richmond Birdwing Butterfly over two decades: Where to next? *Ecological Management and Restoration* 9, 4–16.

Schultz, C.B., Russell, C. and Wynn, L. (2008) Restoration, reintroduction, and captive propagation for at-risk butterflies: a review of British and American conservation efforts. *Israel Journal of Ecology and Evolution* 54, 41–61.

Shoo, L., Hoffmann, A., Garnett, S. et al. (2013) Making decisions to conserve species under climate change. *Climatic Change* 119, 239–246.

Strong, D.R., Lawton, J.H. and Southwood, T.R.E. (1984) *Insects on Plants.* Blackwell Scientific Publications, Oxford.

Taylor, G.S. and Moir, M.L. (2009) In threat of co-extinction: new species of *Acizzia* Heslop-Harrison (Hemiptera: Psyllidae) from vulnerable species of *Acacia* and *Pultenaea*. *Zootaxa* 2249, 20–32.

Taylor, G.S. and Moir, M.L. (2013) Further evidence of the coextinction threat for jumping plant-bugs: three new cothreatened *Acizzia* (Psyllidae) and *Trioza* (Triozidae) from Western Australia. *Insect Systematics and Evolution*, published online November 2013, doi: 10.1163/1876312X-00002107.

Thomas, C.D., Hill, J.K., Anderson, B.J. et al. (2011) A framework for assessing threats and benefits to species responding to climate change. *Methods in Ecology and Evolution* 2, 125–142.

Williams, S.E., Shoo, L.P., Isaac, J.L., Hoffmann, A.A. and Langham, G. (2008) Towards an integrated framework for assessing the vulnerability of species to climate change. *PLoS Biology* 6, e325.

Wilson, R.J. and Maclean, I.M.D. (2011) Recent evidence for the climate change threat to Lepidoptera and other insects. *Journal of Insect Conservation* 15, 259–268.

Yates, C.J., Elith, J., Latimer, A.M. et al. (2010) Projecting climate change impacts on species distributions in megadiverse South African Cape and southwest Australian floristic regions: Opportunities and challenges. *Austral Ecology* 35, 374–391.

Section 3

Farming

13   Agricultural adaptations: social context and complexity

JOHN MORTON

Natural Resources Institute, University of Greenwich, UK

13.1   Introduction

For someone like me whose research interests lie predominantly in 'developing countries', Australia represents a fascinating case. The five chapters in this section (Chapters 14–18) all speak to the features of agriculture in the industrialised world: commercialised; highly competitive in a globalised world; flexible; backed by strong institutions for finance, resource allocation, export facilitation, insurance and technology transfer; and subject to sophisticated policy debates. Yet Australia is a largely tropical and subtropical country, and several features of Australian agriculture have resonances for some poorer countries of the tropics. These include the preponderance of family-owned and managed farms (Alston 2004), the heavy reliance on extensive livestock production, the sheer remoteness of many farming areas and the aridity, variability and proneness to extremes of the climate. For one of the world's most urbanised countries, agriculture and rural life loom remarkably large in the Australian self-image. The image of agriculture and rural life is one that acknowledges, even celebrates, the extreme climate as in the iconic poem *My Country* (by Dorothea McKellar in 1905):

I love a sunburnt country,
A land of sweeping plains,
Of ragged mountain ranges,
Of droughts and flooding rains.

(For a contemporary commentary in a climate change context, see Glenn Tamblyn on the Skeptical Science blog at http://www.skepticalscience.com/print.php?n=1296.)

Within this context, Australian agriculture is not just another industry. Alston (2004) surveys issues such as family farming, gender and ethnicity in Australian farming. Kingwell et al. (Chapter 15) relate successful adaptation to social characteristics of farm managers, including their support from their families, their community involvement and their ability to maintain a work-life balance. Hobday et al. (Chapter 16) mention in passing the particular status of Torres Straits Islanders in fisheries. The pioneering nature of the Australian Landcare movement in encouraging community action for the rural environment is well known and it has been promoted as a model for poorer countries (Hinchcliffe et al. 1995). The darker side of these linkages to broader society includes the now widely discussed linkage of drought to stress, mental health problems and suicide among male Australian farmers (Alston and Kent 2008). In the light of both positive and negative linkages it is no exaggeration that 'the centrality of the social', coined by

Applied Studies in Climate Adaptation, First Edition. Edited by Jean P. Palutikof, Sarah L. Boulter, Jon Barnett and David Rissik.
© 2015 John Wiley & Sons, Ltd. Published 2015 by John Wiley & Sons, Ltd.

Fairhead and Leach (2005) for the embeddedness of African farming in social relations, within and between households, is just as applicable to Australian agriculture.

Within these Australian and global contexts, I would like to pick up a few themes from the chapters in this section: definitions of adaptation; the conceptualisation of 'impact' and 'adaptation'; adaptation in complex public/private systems; adaptation within supply chains; and the idea of adapting to mitigation.

13.2   Definitions of adaptation

Discussions of adaptation, particularly of 'autonomous adaptation' (see the Glossary in IPCC 2007) by individuals, households or small enterprises, often proceed with a certain ambiguity between adaptations to climate variability and adaptations to climate change. 'Planned adaptation' is usually defined by its objectives, while autonomous adaptation is defined largely by its outcomes. Adaptations undertaken by farmers in response to climate variability and resource scarcity can be seen (in some cases argued, in some cases assumed) as being adaptive to climate change. Within this section, the two chapters on irrigation discuss the current fit between different models of irrigation water pricing under climate variability on the one hand and farmer requirements on the other; in particular, Wheeler et al. (Chapter 18) use their findings to argue for strengthened water markets as an adaptation to future climate change.

Kingwell et al. (Chapter 15) study general farm management and farm financial health, and their determinants over the course of a decade (2002–2011) that saw continuation of a warming and drying trend in south-western Australia, but equally saw other trends and stresses. Hertzler et al. (Chapter 14) more explicitly discuss adaptation to current and future climate change by using a modelling approach based on spatio-temporal analogues. My point here is not to prescribe any approach or declare any approach invalid, but to show the variety of ways in which current economic and agronomic decision-making can be studied in the context of adaptation to climate change.

13.3   Conceptualisation of 'impact' and 'adaptation'

An additional issue here is the methodological impossibility of disentangling 'impacts' from 'adaptation'. None of the chapters isolate the impacts of climate change, currently or in projections. Especially in contexts of high climate variability, impacts on agriculture stimulate short-term adaptations (and in some cases maladaptations), which in turn modify the system within which further impacts are experienced. Hertzler et al. (Chapter 14) in particular show how adaptation at farm level is a matter of constant management and adjustments. Water markets require farmers to make constant choices on utilising or selling water allocations in the face of fluctuations in supply and changes in price. From one point of view, adaptation is part of the noise, a glass through which climate impacts can be observed darkly, if at all. But the glass itself is what we most need to study.

13.4   Adaptation in complex systems

The two chapters on irrigated agriculture highlight the complex issues of adaptation in this sub-sector, characterised by hybrid institutions such as Goulburn–Murray Water. An effective monopoly supplier of water, GMW is a government-owned corporation, pricing water as an enterprise with capital and running costs but also subject to public policy choices. These policy choices cannot and should not be reduced to 'political interference'; they include environmental considerations, government's own transaction costs in establishing markets, and fundamental questions of equity. This is not to say that there is not also political interference of a more short-term and partisan nature, as well as simply the inadvertent impacts of other policies on pricing. In practice water pricing depends on a policy-determined estimation of the 'Regulatory Asset Base' (RAB) of the corporation, that is, the value of infrastructure that it is allowed to take into account in setting tariffs. The RAB excludes the considerable additional infrastructure gifted by government in recent years. Farmers make adaptive choices in the face not only of climate variability and change, but also of irreducible institutional complexity.

13.5 Adaptation within supply chains

Chapter 16 (Hobday et al.) is important not only because it broadens the coverage to fisheries and aquaculture, but because of its emphasis on supply chains. Far too little research has been done in either developing or developed countries on post-harvest aspects of agriculture and food systems and how they will be impacted by, and adapt to, climate change. For developing countries one of the rare examples, focusing on smallholder grain systems in Africa, is the article by Stathers et al. (2013). Despite the huge contrasts between the supply chains in question, there are important parallels both in the invisibility of post-harvest problems and solutions, even to knowledgeable actors in the supply chains, and the need for partnerships between stakeholders within and across supply chains to innovate and therefore adapt.

13.6 Adapting to mitigation

The climate impacts on which Hobday et al. (Chapter 16) focus are in fact not direct impacts, either pre- or post-harvest, but the possible impacts of emissions-reduction policies or of future consumer preferences for lower-carbon food on seafood supply chains, given the carbon emissions associated with capture, refrigeration and transport. The chapter reminds us that we are not only concerned with adaptation to climate variability or to climate change, or even to those in conjunction with economic trends and shocks, but also with adaptation to mitigation.

The agriculture and food sectors in Australia and internationally are changing in complex ways, of which climate change is only one. Other processes include globalisation of supply chains and of investment patterns in agriculture, demographic shifts in rural areas, an increasingly explicit international policy debate about the distinct questions of support for farming and support for rural communities, and also international responses to climate change such as emissions reduction, which have impacts of their own on agriculture. Adaptations to climate change will be made by families, farms, firms and governments, and can only be understood by researchers as a response to this nexus of change processes.

Acknowledgements

I am grateful to Mark Howden for introducing me to Dorothea McKellar's 'My Country'.

References

Alston, M. (2004) Who is down on the farm? Social aspects of Australian agriculture in the 21st century. *Agriculture and Human Values* 21, 37–46.

Alston, M. and Kent, J. (2008) The Big Dry: The link between rural masculinities and poor health outcomes for farming men. *Journal of Sociology* 33, 133–147.

Fairhead, J. and Leach, M. (2005) The centrality of the social in African farming. *IDS Bulletin* 36, 86–90.

Hinchcliffe, F., Guijt, I., Pretty, J. and Shah, P. (1995) New Horizons: The economic, social and environmental impacts of participatory watershed development. *International Institute for Environment and Development Gatekeeper Series No.50*. IIED, London.

IPCC (2007) *Climate Change 2007: Impacts, Adaptation and Vulnerability. Contribution of Working Group II to the Fourth Assessment Report of the Intergovernmental Panel on Climate Change.* Parry, M.L., Canziani, O.F., Palutikof, J.P., van der Linden, P.J. and Hanson, C.E. (eds) Cambridge University Press, Cambridge.

Stathers, T., Lamboll, R. and Mvumi, B. (2013) Postharvest agriculture in changing climates: its importance to African smallholder farmers. *Food Security* 5, 361–392.

14 Farmer decision-making under climate change: a real options analysis

GREG HERTZLER[1,7], TODD SANDERSON[1], TIM CAPON[2,8], PETER HAYMAN[3], ROSS KINGWELL[4,5], ANTHEA MCCLINTOCK[6], JASON CREAN[6] AND ALAN RANDALL[1]

[1]Faculty of Agriculture and Environment, University of Sydney, Australia
[2]CSIRO Land and Water, Australia
[3]South Australian Research and Development Institute, Australia
[4]Australian Export Grains Innovation Centre, Australia
[5]University of Western Australia, Australia
[6]NSW Department of Primary Industries, Australia
[7]Faculty of Natural and Agricultural Sciences, University of Pretoria, South Africa
[8]School of Economics, University of Sydney, Australia

14.1   Introduction

Agricultural practices are adaptations to prevailing conditions, including climatic, social and market conditions. Australian producers have improved their understanding of climate regimes over time and, by choosing particular crop varieties, production technologies and other inputs, they have adapted agricultural systems to a highly variable environment. However, a changing climate means that historical experience provides only a limited guide for future decisions. If we want to understand the options available for climate adaptation, we need to match existing scientific knowledge of the biophysical consequences of climate change with research into the likely responses of social and economic systems. We need to understand the decisions by communities and farmers as they respond to changes and uncertainties arising within the climatic, social and market systems. It is these decisions that will ultimately determine the nature of agricultural production and the resilience of rural communities.

Decisions can take various forms, such as (1) adjusting practices and technologies; (2) changing production systems; or (3) greater transformations, such as re-locating production (Howden et al. 2009; Rickards and Howden 2012), and can be viewed as decisions about alternative production regimes within the agricultural systems that will be affected by climate change. A switch from one regime to another can be irreversible or only partially reversible. Switching production regimes may require investments into production techniques (e.g. equipment or knowledge), as well as processing and infrastructure. Old technology may have a salvage value. Conversely, assets may become stranded with no possibility of recovering the investment. These complications throw up barriers to adaptation, with broader implications for rural communities and regional economies.

Applied Studies in Climate Adaptation, First Edition. Edited by Jean P. Palutikof, Sarah L. Boulter, Jon Barnett and David Rissik.
© 2015 John Wiley & Sons, Ltd. Published 2015 by John Wiley & Sons, Ltd.

To bridge the science of climate change impacts and the socio-economic realities of agricultural adaptation and transformation we need tools such as real options analysis, the modern analytical method for modelling the value of flexibility and the timing of action in decision-making under uncertainty (we follow common usage by defining 'uncertainty' as imperfect information and 'risk' as uncertain consequences; Dixit and Pindyck 1994; Copeland and Antikarov 2001). Previous research in the Australian context developed and demonstrated the appropriateness of real options as a framework to bridge this gap and to chart sequences of optimal climate change adaptations through time (Hertzler 2007). We apply the Real Options for Adaptive Decisions (ROADs) framework (Hertzler et al. 2013) to assess the likelihood of farmers in wheat-dominant agriculture transitioning to alternative production regimes.

Our three case study regions are in New South Wales, South Australia and Western Australia. We employ transects across space as an analogue for climate change (Hayman et al. 2010; Nidumolu et al. 2012). First, transects are defined to transcend rainfall isohyets and meteorological isopleths and capture a cross-section of wheat-dominant systems. Second, a set of possible agronomic systems are identified and the dynamics of these are estimated as stochastic differential processes. Finally, option values and regime transition thresholds are estimated using ROADs, and probabilities for transformations between production regimes are calculated. These transformation probabilities are a measure of the resilience of a particular agricultural system at a particular location.

14.2 Spatio-temporal analogues

The implications of climate change can be modelled by using a current spatial transect as an analogue for possible future changes (Hayman et al. 2010; Nidumolu et al. 2012). This allows us to model the adaptation and transformational processes that might occur in the future at one site by examining the nature of optimal decisions at another site where possible future conditions are currently observed. (Although spatial transects represent possible future conditions they cannot accurately represent likely future conditions. A current spatial transect is not the same as a temporal transect since the current environment at one location is very unlikely to become exactly the same as another. Temporal changes not captured in spatial changes include higher concentrations of CO_2, higher average temperatures and their interaction. Also, the set of future options for responding to a changing climate are unlikely to be the same as the currently available set of options. In the future, more profitable options for responding to climate challenges are likely to become available than are currently observed.) Accordingly, transects were identified across regions that represent the range of agricultural production conditions that characterise wheat-dominated agriculture in Australia, as shown in Figure 14.1 and detailed in Table 14.1.

The spatio-temporal analogues approach recognises that the location of a farm can be a good predictor of the prevailing farming activity. For instance, the transect in South Australia ranges from intensive cropping with a high proportion of relatively high-risk and high-return crops, through to an increasing proportion of cereals with lower inputs and then finally to grazing enterprises with opportunistic cropping (Hayman et al. 2010). This transect straddles Goyder's line, a line drawn through south-eastern South Australia in the 1860s by George Goyder, the Surveyor-General of South Australia, which demarcates the land suitable for extensive grazing to the north from the land suitable for cropping in the south.

Since we are interested in the location of thresholds for transformations to other agricultural systems, we must seek to understand both the wheat-dominant agricultural system and alternative systems such as extensive grazing. For our purposes we can think of wheat cropping and livestock grazing, in this case merino sheep production, as two alternative production regimes available to a farmer. Accordingly, the decision problems modelled for South Australia consist of switching into and out of wheat cropping and merino grazing. For New South Wales, additional regimes of wheat-dominant mixed farming and

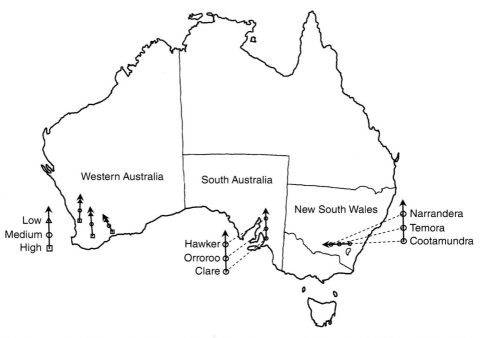

Figure 14.1 Agro-ecological zones in Western Australia and transects in New South Wales and South Australia.

Table 14.1 Descriptions of spatio-temporal analogues for each case study.

Case study state	Locations	Description
New South Wales	Cootamundra, Temora and Narrandera	This spatial transect is in the south of the state and represents a transition from high to low rainfall, starting with an average of 404 mm of annual growing season rainfall at Cootamundra, reduced by 18% to 331 mm at Temora, and reduced by a further 20% to 226 mm at Narrandera.
South Australia	Hawker, Orroroo and Clare	This spatial transect crosses Goyder's Line and represents a transition from relatively high and reliable rainfall near the south coast to desert conditions to the north. The average annual growing season rainfall is 486 mm at Clare, 225 mm at Orroroo and 201 mm at Hawker.
Western Australia	Agro-ecological zones L2, L3, M1, M2, M3, M4, M5, H4, H5	This spatio-temporal analogue is based on areas with a similar crop performance, agro-ecological zones that represent nine different combinations of rainfall level and length of growing season. Typical annual rainfall levels range from around 600 mm in the west to about 300 mm in the east. The length of the growing season is shortest in the north and longest in the south. The letters L, M and H are for low, medium and high rainfall and the numbers increase from north to south. For example, M1 is a zone of medium rainfall in the north and M5 is a zone of medium rainfall in the south.

sheep-dominant mixed farming are included in the analysis to better represent the production regimes available to farmers in this region. For Western Australia, the sequence of decision problems modelled is similar to that for the case study of South Australia except that a wheat-dominant cropping regime is modelled rather than a wheat-only cropping regime. The outputs of biophysical models such as APSIM (Agricultural Production Systems Simulator) and farm data provided by farm consultants were used to model the range of farm systems for the locations along each transect. This enabled us to characterise the dynamic and stochastic nature of the alternative production regimes available to farmers (Hertzler et al. 2013).

14.3   Real options analysis

Real options analysis extends traditional economic analyses of agricultural investment decisions based on cost-benefit analysis and net present values to better represent uncertainty. It does this by examining the trade-offs between acting sooner versus retaining the option to act later, by taking into account the value of flexibility and the value of new information that can help resolve uncertainty. Taking option values into account is especially important since some adaptation decisions can be costly to reverse or can even have irreversible consequences for farmers and rural communities. For instance, real options can be used to understand why farmers may decide to delay the adoption of new varieties of crops or new technologies. This hesitation to

adopt can exist because there are uncertainties about the impact of adoption on production and economic outcomes. For some farmers the risks may outweigh the apparent benefits predicted by the science. (Note that the use of the term 'farmer' does not denote any particular person. This study focuses upon changes in agricultural production on a representative hectare of land at a given location, for which the farmer is considered to act as a representative decision-maker. An actual farmer will have many objectives, including objectives about consumption and lifestyle. Because we are studying commercial farming however, consumption and lifestyle decisions are separable from production and investment decisions. Regardless of any other objectives, the farmer who survives on the land will produce and invest efficiently, given the thresholds and risks of climate change.)

As farmers transform agriculture in response to climate change, they will switch from one production regime to another. A production regime is defined by the activities undertaken within it. As an illustration, consider a sequence of four possible production regimes available to a farmer currently engaged in wheat-dominant cropping (Fig. 14.2). When conditions are uncertain and changing, there is a trade-off between switching immediately to a new regime versus retaining the option to switch later when new information might be available that reduces the uncertainty. ROADs is a framework to analyse this type of decision problem, by applying a real options analysis that decomposes complex decisions over time into choices among alternative regimes. In other words, ROADs implements

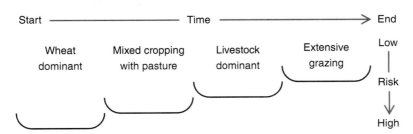

Figure 14.2   One of the many possible sequences of regime transitions with climate change.

the mathematics for determining the optimal adaptation and transformation pathways.

As the climate changes, wheat production may decline and the farmer may switch to a regime in which wheat is grown in some years and pasture in others. With extreme adverse climate change, this farmer might even switch to a regime of extensive grazing, a transformation that can be viewed as crossing a threshold. Within each of these broader regimes there is the possibility to adapt by making smaller changes to farming practices, such as adopting improved techniques or adoption of genetically modified crop varieties. However, even these adaptations may be both costly and risky, and farmers may be hesitant to switch immediately. The timing of regime switches depends on the uncertainties associated with the returns from the alternative regimes. A producer might choose to switch immediately or never, depending on how the climate is changing and the magnitude of the variability associated with that change. The value farmers place on these decisions is captured in the concept of option values, which is essentially the price a farmer is willing to pay for flexibility.

The ROADs framework models the timing of decisions by maximising the option values of retaining flexibility while minimising the costs of switching back after a mistaken decision. This allows us to determine the resilience of alternative agricultural production regimes to a warmer and drier climate. We calculate the thresholds, option values and expected times until the optimal switches among regimes, and we calculate the probabilities of transformation of wheat-dominant agriculture. The threshold is some observed state of the system, in this case some per hectare gross margin. Each regime has a gross margin threshold for entry and exit which may be crossed at any time, triggering a change between production regimes. The position of a regime threshold is the solution to a stochastic dynamic optimal stopping problem, which is solved in ROADs.

There are a number of ways to represent the variability (or stochasticity) in the data series that underpin the ROADs framework. In this study we employed an Ornstein–Uhlenbeck

process, which was defined over a series of gross margins derived from the simulated or observed data. The Ornstein–Uhlenbeck process is appropriate where it is reasonable to make the mathematical assumptions that (a) there is a tendency for the data series to revert to some mean (Doob 1942), and (b) that the convergence is linear, i.e. reverts towards the mean value at the same rate from higher or lower values. Most agronomic systems satisfy (a) over a sufficiently long period of time, although (b) is less well satisfied. After modelling the stochastic dynamic processes, we solve for the stochastic optimal choice of regimes using Itô stochastic control procedures (Hertzler 1991; Hertzler et al. 2013). Since the Ornstein–Uhlenbeck process possesses analytically defined probability and transition density functions, we can also estimate the probabilities of observing a transition between alternative regimes within a given period of time. These transition probabilities represent the chance that a farmer would cross a threshold that triggers the decision to change production regimes.

14.4   Results

The estimated transition probabilities that a farmer would shift to an alternative production regime within a given five year period are presented in Table 14.2 for each location along the spatial transects for South Australia, Western Australia and New South Wales.

In South Australia, if the climate becomes hotter and drier, Goyder's line would move south; Clare would become more like Orroroo, and Orroroo would become more like Hawker. To illustrate, the estimated probability of entering into a regime of wheat production at Clare under current climate conditions is 53% and the probability of a farmer transitioning to merino production or leaving agriculture altogether are both estimated as 0%. In contrast, the estimated probability of entering into wheat production at Hawker is currently 24%, the estimated probability of transitioning to merino production is 22% and the probability of exiting merino

Table 14.2 Comparative transformation probabilities (P_{trans}) for New South Wales, South Australia and Western Australia.

New South Wales			South Australia			Western Australia						
Decision	Location	P_{trans} (%)	Decision	Location	P_{trans} (%)	Decision	Zone	P_{trans} (%)	Zone	P_{trans} (%)	Zone	P_{trans} (%)
Entry into wheat-dominant cropping with possibility to exit	Cootamundra	45	Entry into wheat cropping with possibility to exit	Clare	53	Enter into wheat-dominant cropping with possibility to exit	H4	47	M1	52	L2	49
	Temora	27		Orroroo	39		H5	54	M2	54	L3	44
	Narrandera	12		Hawker	24				M3	51		
Exit from wheat-dominant cropping only to enter crop-dominant mixed farm	Cootamundra	6	Exit from wheat and enter merinos	Clare	0	Exit from wheat-dominant cropping to enter sheep-only farm	H4	13	M4	46	L2	5
	Temora	23							M5	52		
	Narrandera	38							M1	2		
Exit from crop-dominant mixed farm to enter sheep-dominant mixed farm	Cootamundra	3		Orroroo	9		H5	5	M2	1	L3	11
	Temora	24		Hawker	22				M3	6		
	Narrandera	44							M4	11		
Exit from sheep-dominant mixed farm to enter sheep-only farm	Cootamundra	27	Exit from merinos	Clare	0	Exit from sheep-only farm	H4	1	M5	3	L2	8
	Temora	50		Orroroo	1		H5	1	M1	9	L3	0
	Narrandera	58							M2	2		
Exit from sheep-only farm	Cootamundra	4		Hawker	5				M3	1		
	Temora	13							M4	2		
	Narrandera	35							M5	0		

production is 5%. If conditions at Clare or Orroroo become more like the conditions at Hawker with climate change, the landscape would change as wheat becomes less dominant, merinos are adopted on more farms and some farms leave agriculture.

Compare this with our model of wheat-dominant agriculture in Western Australia. The probability that the threshold is crossed which would trigger a farmer to enter into wheat production is high for all the agro-ecological zones considered (i.e. probabilities of 44–54%). Furthermore, the probability that a farmer currently producing wheat would transition to sheep production is low in all zones (i.e. probabilities of 1–13%). This suggests that even though climate change is expected to make the state hotter and drier, wheat will continue to dominate agriculture in all agro-ecological zones across the Western Australian wheat-belt since, based on current farm data, there are no serious alternatives to wheat.

Even though the transect in New South Wales may be less affected by climate change than Western Australia, the choice of production regimes in our model might be more sensitive to climate change. This is because farmers in our model of the New South Wales transect have a greater range of alternative production regimes to choose from. For example, at Cootamundra conditions are fairly reliable with a 45% probability that a farmer who was not already growing wheat would enter into wheat production within a 5-year period. However, in our model of the New South Wales transect, farmers can choose production systems with mixed crop and sheep production before switching to sheep-only production. While the probability at Cootamundra of exiting wheat-only production and entering crop-dominant mixed farming is estimated as 6%, if Cootamundra comes to resemble Temora then the probability of adopting a crop-dominant mixed farming regime would increase to 23%. Given additional mixed farming options, the probability of leaving wheat-only production with climate change is much higher for the New South Wales transect even if climate impacts New South Wales less than Western Australia.

Compared with Cootamundra and Temora, Narrandera has highly variable returns from agricultural production. Someone who was not already farming would only have a 12% probability of entering into wheat-only production at Narrandera. A farmer with a sheep-dominant mixed farming regime has a 58% probability of ending wheat production entirely and entering a sheep-only production regime. A farmer in this sheep-only production regime would have a 35% probability of leaving agriculture altogether. Clearly, if Cootamundra or Temora became more like Narrandera with climate change, the agricultural landscape would be transformed.

14.5   Discussion and conclusion

Wheat-dominant agriculture is important in Australia and has lessons for agriculture in general. Agriculture is a very diverse industry with many sectors, all dependent on the climate in complex ways. This research modelled a major determinant of the impact of climate change on agricultural productivity: the decisions made by farmers. We find that research into the impacts of climate change which does not consider farmers' decisions and the influence of uncertainty can be misleading.

For example, Western Australia will become hotter and drier, but is very unlikely to adapt its agriculture away from wheat, let alone transform into other production systems. The important role of wheat growing in Western Australia is also recognised in Chapter 15 (by Kingwell et al.). The Mediterranean climate and limited options for profitably growing other crops and livestock will ensure that farmers choose wheat. In this analysis, South Australia is more likely to adapt away from wheat as Goyder's Line moves south, mostly because sheep are a more viable option and wheat can become unprofitable and very risky. As a counterexample, New South Wales may be less subject to climate change but farmers have more options to choose from; they are more likely to transform their production systems into mixed farming enterprises as a buffer against climate variability.

The results presented in this chapter are not forecasts; they are predicated on the assumptions that space is a good analogue for climate change and that climate will change significantly. However, there are limits to the usefulness of using spatial transects as a proxy for time and our results are sensitive to assumptions about the profitability and production scale of wheat and sheep enterprises. If climate change is moderate different regions may take on characteristics of other regions, but will never become exactly like their analogues along a spatial transect. Even so, this analysis demonstrates that climate change will not translate directly into transformations of agriculture, but will instead affect the options available to farmers as they choose among alternative production regimes.

Climate adaptation can be costly, costly to reverse or effectively irreversible. By accounting for the value of flexibility in decision-making, real options analysis demonstrates that it can be rational for farmers to delay some actions in order to wait for new information that resolves uncertainty about the rate and extent of climate change. This is especially relevant for Australia, where high levels of rainfall variability over years and decades can mask the underlying climate change signal, making it difficult for farmers to identify trends and make plans for the future.

Unlike adaptation studies that focus primarily on the biophysical impacts of climate change, this study has focused on the role played by farmers' choices in mediating the impacts of climate change on agricultural production. By advancing the methodology of real options analysis to better characterise the stochastic returns from agricultural production, we have also demonstrated that the probability of transitions to alternative production regimes can be estimated using the ROADs approach. This contributes to our understanding of the likely timing of adaptation in the agricultural sector. Future research should investigate which agricultural industries will be most affected by climate change, by taking into account the central role of farmers' decisions.

References

Copeland, T. and Antikarov, V. (2001) *Real Options: A Practitioner's Guide*. Texere, New York.

Dixit, A.K. and Pindyck, R.S. (1994) *Investment under Uncertainty*. Princeton University Press, Princeton.

Doob, J.L. (1942) The Brownian movement and stochastic equations. *The Annals of Mathematics* 43 (Second Series), 351–369.

Hayman, P., Alexander, B., Nidumolu, U. and Wilhelm, N. (2010) *Using Spatial and Temporal Analogues to Consider Impacts and Adaptation to Climate Change in the South Australian Grain Belt*. 15th Australian Agronomy Conference: Lincoln, New Zealand, November 2010.

Hertzler, G. (1991) Dynamic decisions under risk: Application of Itô stochastic control in agriculture. *American Journal of Agricultural Economics* 73, 1126–1137.

Hertzler, G. (2007) Adapting to climate change and managing climate risks by using real options. *Australian Journal of Agricultural Research* 58, 985–992.

Hertzler, G., Sanderson, T., Capon, T. et al. (2013) *Will Primary Producers Continue to Adjust Practices and Technologies, Change Production Systems or Transform Their Industry: An Application of Real Options*. National Climate Change Adaptation Research Facility, Gold Coast.

Howden, S.M., Meinke, H. and Gifford, R.G. (2009) Grains. In: Stokes, C.J. and Howden, S.M. (eds) *Adapting Australian Agriculture to Climate Change*. CSIRO Publishing, pp. 21–48.

Nidumolu, U.B., Hayman, P.T., Howden, S.M., and Alexander, B.M. (2012) Re-evaluating the margin of the South Australian grain belt in a changing climate. *Climate Research* 51, 249–260.

Rickards, L. and Howden, S.M. (2012) Transformational adaptation: agriculture and climate change. *Crop and Pasture Science* 63 (3), 240–250.

15 Broadacre farmers adapting to a changing climate

ROSS KINGWELL[1,2], LUCY ANDERTON[3],
NAZRUL ISLAM[3], VILAPHONH XAYAVONG[3],
ANGELA WARDELL-JOHNSON[4], DAVID FELDMAN[3]
AND JANE SPEIJERS[3]

[1]Australian Export Grains Innovation Centre, Australia
[2]University of Western Australia, Australia
[3]Department of Agriculture and Food WA, Australia
[4]University of the Sunshine Coast, Australia

15.1 Introduction

Broadacre farm businesses in southern Australia are large mixed-enterprise dryland farms. Since the late 1990s these farms have experienced climate trends of higher average temperatures and lower precipitation (Hughes 2003; Asseng et al. 2011; Cai et al. 2012; Asseng and Pannell 2012; Cai and Cowan 2013). In south-western Australia grain farmers have experienced a proportionate decline in growing season rainfall greater than any other wheat-growing region in Australia (Hope et al. 2006; Asseng and Pannell 2012). Regarding these warming and drying trends in climate change in southern Australia, for the future there is far greater scientific confidence around the projected trends in warming than there is over rainfall changes (Hughes 2003; Rebbeck et al. 2007; Hennessy et al. 2007; Addai 2013).

These unfolding trends in climate, and the variability around these trends, pose business challenges for farmers in southern Australia. How financially successful these farmers can be will affect their adaptive capacity, and their ability to afford and employ adaptation responses. If farm businesses are financially weakened by the changing climate and have few prospects of financial viability, then their capacity to adapt to the adversity or opportunity associated with climate change will be highly constrained.

To explore these issues of the financial performance of farm businesses adapting to the challenges of a changing climate, this chapter reports on a study that tracked 249 farm businesses over a recent decade (2002–2011) in south-western Australia. This region has experienced a warming, drying trend in its climate in recent decades. We briefly outline these climatic trends and then draw on analyses of the farm data to answer two questions.

1. What characteristics of farm businesses and farm managers have allowed them to prosper during this period of climate adversity?
2. What adaptation strategies appear to have been successful in combating the changing climate?

Applied Studies in Climate Adaptation, First Edition. Edited by Jean P. Palutikof, Sarah L. Boulter, Jon Barnett and David Rissik.
© 2015 John Wiley & Sons, Ltd. Published 2015 by John Wiley & Sons, Ltd.

15.2   Study region

The study region is the broadacre farming region of south-western Australia (Fig. 15.1). The region has a Mediterranean-type climate, characterised by long, hot and dry summers and cool, wet winters. As indicated in Figure 15.1, average annual rainfall decreases rapidly in a north-easterly direction from the region's southwest corner. In most parts of the study region, around three-quarters of the average annual rainfall is received between April and October. Summer rainfall is highly variable, and is more common along the region's southern coast where growing seasons are longer. Compared to summer rainfall winter rainfall is greater and is much more reliable,

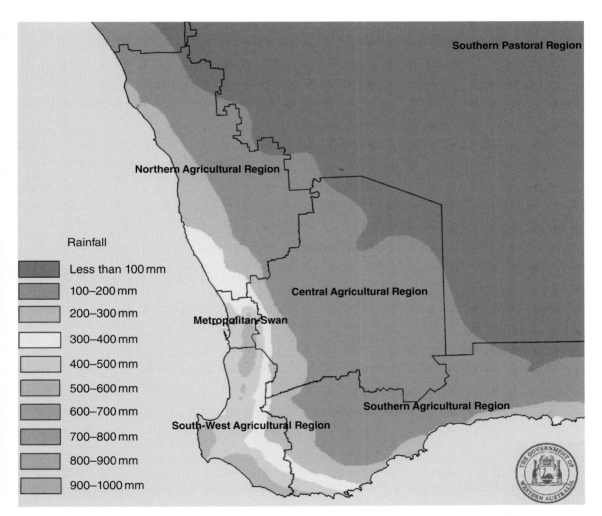

Figure 15.1   The study region of south-western Australia, its spatial pattern of average winter rainfall (based on rainfall from 1976 to 2007) and various agricultural regions. Source: DAFWA (2007).
© Western Australian Agricultural Authority 2007. Reproduced with permission. For colour details please see Plate 11.

making the region suitable for annual crops and pastures. The region's farming systems are rain-fed mixed enterprises. Wheat is by far the principal crop grown and sheep are the main live-stock enterprise.

Crops, primarily wheat (*Triticum aestivum* L.), barley (*Hordeum vulgare* L.) and canola (*Brassica napus* L.), are typically sown in late autumn through to early winter and crops are harvested in November/December. In some parts of the study region the occurrence of frost can greatly affect grain production and crop and pasture yields are positively correlated with growing season rain-fall. The main products from the farms are cereals, sheep and wool, most of which are exported (Kingwell and Pannell 2005; Doole et al. 2009).

The crops and pastures are grown mostly as opportunistic land-use sequences rather than in strict rotations. Canola now features in farming systems. Higher-yielding canola varieties and attractive prices have encouraged their greater adoption. Crop and pasture yields are regularly improved by the application of lime, phosphate and nitrogenous fertilisers. Pasture production varies by soil type and seasonal conditions.

Sheep are run on annual pastures during winter and spring. In summer months, livestock feed is mainly pasture residues and crop stubbles. In late summer through to early winter, sheep are supplementary fed to maintain their welfare. The sheep systems mainly involve merinos and include both wool and sheep-meat systems.

15.3   Farm business and managerial data

Data describing the farm businesses in the study region were supplied by agricultural consulting firms operating in the region. Farm business records of 249 farms were obtained for the years 2002–2011. Since the data come from farms able to afford agricultural consultants, they may not necessarily be truly representative of the wider farming community. The data nonetheless are unique longitudinal datasets that describe the farm production and financial records of each farm over the decade.

Complementing the physical and financial datasets of farm businesses were managerial data. These were questionnaire-based managerial assessments of their farmer clients provided by the consultants. Because the farmers have been clients of the particular consultancy firms for at least that decade, and because the farmers tend to retain the same consultant, often the consultant is well informed about the managerial environment and managerial characteristics of their farmer client.

15.4   Climate trends in the region

Since the early 1960s a warming trend has emerged (see Fig. 15.2) underlain by increases in minimum and maximum daily temperatures. Maximum daily temperatures have trended upwards in all seasons, but noticeably in winter. Minimum daily temperatures have trended upwards in all seasons, apart from winter. During the study period 2002–2011 the mean tempera-ture anomaly averaged +0.45 °C, based on the 30-year climatology of 1961–1990. In the study region plant growth benefits from warmer tem-peratures in winter, but heat in spring and summer can lessen grain-filling and hasten pas-ture senescence.

Figure 15.3 shows the pattern of trends in annual rainfall in Western Australia for the period 1970–2012. Since the 1970s the study region has experienced a downward trend in annual rainfall. The observed drying has been far greater than was projected in the late 1980s using the then best-available global climate models (Foster 2013).

Another important change in regional climate that has affected grain production in many parts of the study region has been an increased inci-dence of frost (GRDC 2012), especially late in the growing season when grain damage is more likely.

Besides some climatic adversity, farms in the study region have also faced pronounced price volatility, especially for grains. This volatility has been a global phenomenon (Kingwell 2012) that has greatly affected the profitability of grain pro-duction.

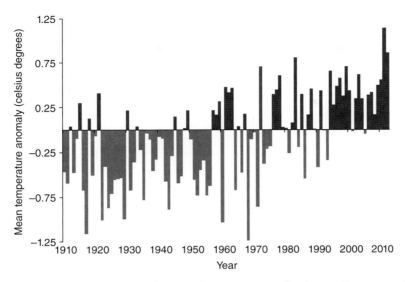

Figure 15.2   Annual mean temperature anomaly in south-western Australia. Source: Data sourced from the Bureau of Meteorology. For colour details please see Plate 12.

15.5   Research methods

Drawing on each farm's financial and physical records, a suite of farm performance financial indicators was derived, including measurement of business equity, operating profit per hectare, return on capital, debt to income ratio and income diversity. Generalised linear mixed effects models were used to fit a range of explanatory variables and interactions to these business indicators.

Besides these five indicators of farm performance, a novel set of additional descriptors of farm performance were created (Blackburn and Ashby 1995; Kingwell et al. 2013). These measures allowed farms to be classed as growing, strong, secure or less-secure businesses.

Complementing these classifications of farm types and farm performance was the derivation of estimates of the total factor productivity of each farm in each year using the approach outlined by O'Donnell (2010, 2012). For more details regarding the research methods and analyses interested readers are referred to Kingwell et al. (2013).

15.6   Key findings

The suite of analyses that were applied to the financial and socio-managerial datasets generated a rich set of findings; only key findings are reported here. Regression results revealed significant differences in operating surplus per hectare between agro-ecological zones (see Fig. 15.1) within the study region. Significant differences between the zones in their degree of enterprise diversity were also observed. The differences in operating surplus per hectare were mostly a product of differences in farm size and rainfall between the zones.

In the high-rainfall zones where farm sizes were smaller and crop and pasture yields were higher, operating surpluses per hectare were generally higher. By contrast, in low-rainfall zones where crop and pasture yields were lower and farm sizes larger then operating surpluses per hectare were smaller. The different environments in which the differently sized farm businesses operated led to significant differences in operating surpluses per hectare. Similarly, the observed significant differences in enterprise diversity

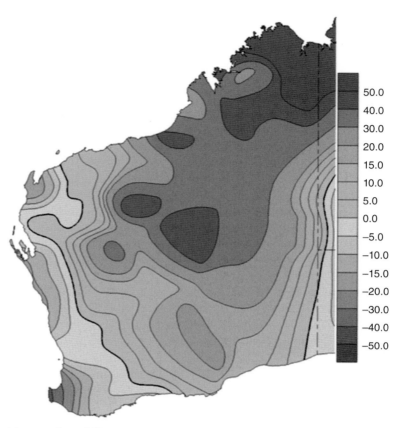

Figure 15.3 Trend in annual rainfall in Western Australia, 1970–2012 (mm/decade). Source: Reproduced with permission of the Bureau of Meteorology/Commonwealth of Australia. For colour details please see Plate 13.

between the zones were mostly due to differences in the physical environments and farm sizes within each zone. In the low-rainfall environments where farm sizes were greater, often farms were crop-dominant and they relied on cropping machinery with high work rates to provide economies of size. These farms had their income streams mostly dependent on wheat production, as wheat is highly adapted to these mostly flat low-rainfall environments. By contrast, farms in high-rainfall environments are smaller in size, often with steeper slopes and some areas prone to occasional water-logging. These farms are more likely to be mixed-enterprise farms with a smaller proportion of their farm area devoted to crops, yet

a greater diversity in their mix of crops. Canola and barley feature more in their cropping programs.

Some findings of concern were the significant decline in equity (as a percentage) and a significant increase in the debt to income ratio during the study period. Increasing indebtedness and associated debt-servicing were becoming important business challenges for some farms.

Further statistical analyses showed that the 'farm' effect was a main explanatory variable for many of the farm performance measures, indicating that often the financial performance of a farm has much to do with its unique characteristics such as farm location, growing season rainfall, wheat yield and farm size. Complementary analyses also revealed a strong positive correlation between farms' average operating surpluses and the variability of those surpluses. This is the same result found by Lawes and Kingwell (2012), indicating a positive risk reward structure in broadacre farming.

When the sample of farms was categorised into business performance categories, almost two-thirds (64%) were classed as growing or strong businesses and 15% were classed as less secure and so were financially vulnerable. Table 15.1 lists key characteristics of these different groups of farms. Many of the differences in characteristics between the groups were statistically significant. When the financially superior groups of growing and strong farms were compared to the weak group of less-secure farms, the more commercially successful farms tended to be larger in area, carried less debt per hectare, were slightly more crop dominant, had a much lower debt to income ratio, had slightly higher equity in percentage terms and generated similar livestock income per hectare but generated much higher crop income per hectare; overall, they generated much higher profits.

Farms classed as growing also had double the productivity growth of less-secure farms. Moreover, productivity growth was a principal contributor to profitability growth; the same finding as reported by Islam et al. (2014). Fuelling the results in Table 15.1 is the underlying merit of engaging in crop production, particularly wheat production, during that decade. The importance of successful wheat growing to farm performance is a finding consistent with that of Lawes and Kingwell (2012). The prime importance of wheat revenue, however, is not altogether surprising. Wheat revenue is the major source of income for almost all farm businesses in the region and so changes in wheat revenue can translate into changes in business performance.

The predominance of wheat production in farming systems in the region has served as a useful bulwark against the volatility and adversity in the region's changing climate. These findings suggest that as long as broadacre farmers in south-western Australia have ongoing access to improved wheat varieties and technologies that support the profitable growing of wheat, farmers are likely to be able to adapt to the changing climate. The biologically robust performance of wheat projected in the future (Ludwig et al. 2009; Asseng and Pannell 2012; Potgieter et al. 2012) should help underpin the future profitability of crop production. Moreover, the topography and climate over much of the study region favours wheat production and the wide adaptability of wheat further supports its preferred use by farmers. Farmers' enhanced dependence on wheat-growing as a principal source of farm income during the study period has proven to be a sensible adaptation strategy, with a majority of farmers increasing their farm equity in dollar terms and expanding the size of their operations.

Regarding socio-managerial characteristics, the managers of growing farms were found to display behaviours that were statistically significantly different from those managers of less-secure farm businesses. Managers of growing farms were better managers of farm labour, were more inclined to be involved in their local community, were more able to separate their home and office life, used more leasing of land, invested more in personal superannuation, were more likely to have made succession plans, invested more in Farm Management Deposits, owned more off-farm assets, used more farm business management tools (marketing strategies, decision support tools, GPS technologies, electronic paddock recording), participated more in training, had greater experience in farming and often displayed greater quality of care for their cropping gear.

15.7 Concluding remarks

Broadacre farmers in south-western Australia are experiencing a drying and warming trend in their climate. We examined almost 250 south-western Australian broadacre farm businesses and reported how they fared over a 10 year period (2002–2011). Most farms reduced their equity in percentage terms over that decade, but their farm equity in dollar terms improved as many

Table 15.1 Characteristics of farms in the four categories of farm business performance: averages for the period 2002–2011.

		Growing	Strong	Secure	Less secure
Gross farm income	$ × 10³	1577	1204	1071	791
Operating costs	$ × 10³	996	808	731	594
Operating surplus	$ × 10³	581	396	340	197
Profit	$ × 10³	273	138	115	−44
Personal expenses	$ × 10³	112	106	83	85
Interest payments	$	81	53	58	82
Machinery replacement	$ × 10³	115	100	84	74
Debt to income ratio	no.	0.99	1.05	1.35	1.64
Operating expenses as a % of gross farm income	%	69.5	73.1	79.3	91.9
Land owned	ha	3875	3422	3093	2739
Land operated	ha	3935	3502	3269	2660
Land value	$ × 10³	4686	4496	3557	3277
Farm assets	$ × 10³	6987	6202	4864	4608
Business assets	$ × 10³	7718	7049	5356	4986
Equity	$ × 10³	6431	5743	3963	3750
Equity as a %	%	82.4	82.2	75.6	76.7
Crop area	ha	2826	2313	2188	1770
Pasture area	ha	1110	1190	1081	890
Crop income as % of farm income	%	80	77	76	74
Crop income per ha	$/ha	464	427	403	379
Livestock income per ha	$/ha	250	201	295	255
Debt per ha	$/ha	375	393	429	515
Return on equity	%	11	8	10	6
Return on capital	%	5	3	4	−1
Growing season rainfall	mm	253	249	242	240

expanded their farm size. Most farms had starting equities greater than 80% that helped insulate them from the downsides of climate and price volatility during those years. The frequency and magnitude of poor seasons was insufficient to greatly jeopardise the profitability of most farm businesses. Rather, on average, farms prospered.

Farmers' increased dependence on wheat growing as a principal source of farm income was found to be a sensible adaptation strategy. Farm profitability improved, principally supported by productivity growth. Farm businesses that grew substantially were often better managed and achieved greater productivity growth.

Growing and strong farms formed almost two-thirds of the farm sample and, when compared to

less-secure farms (15% of the sample), they displayed a range of socio-managerial differences that supported their business performance. Typically, the managers of growing or strong farm businesses had a greater commitment to training, made greater use of modern technologies and were additionally blessed with a breadth and depth of experience and family support to engage in farming. These managers on average were more involved in their local community and expressed more care regarding their work–life balance. These important social and managerial dimensions that link to farm performance are often overlooked in agricultural and climate change policy. Hence, supporting and encouraging community engagement and management training and education may generate not only social rewards but, as suggested by this study's findings, may also promote farm business performance.

Lastly, because the unique characteristics of each farm business were found to be strongly associated with their performance, this suggests there is a legitimate role for personalised advice and support for individual farm businesses in order to further improve their performance and capacity to adapt to climate change.

References

Addai, D. (2013) *The Economics of Technological Innovation For Adaptation to Climate Change by Broadacre Farmers in Western Australia,* Unpublished PhD thesis, School of Agricultural and Resource Economics, University of Western Australia.

Asseng, S. and Pannell, D. (2012) Adapting dryland agriculture to climate change: Farming implications and research and development needs in Western Australia. *Climatic Change* 118, 167–181.

Asseng, S., Foster, I. and Turner, N.C. (2011) The impact of temperature variability on wheat yields. *Global Change Biology* 17, 997–1012.

Blackburn, A. and Ashby, R. (1995) *Financing Your Farm. 3rd Edition.* Australian Bankers Association, Melbourne.

Cai, W. and Cowan, T. (2013) Southeast Australia autumn rainfall reduction: A climate-change-induced poleward shift of ocean-atmosphere circulation. *Journal of Climate* 26, 189–205.

Cai, W., Cowan, T. and Thatcher, M. (2012) Rainfall reductions over Southern Hemisphere semi-arid regions: the role of subtropical dry zone expansion. *Nature (Scientific Reports)* 2, 702, doi: 10.1038/srep00702.

Department of Agriculture and Food, Western Australia (2007) *Climate Prediction Maps for Western Australia. Winter Rainfall.* Department of Agriculture and Food, Western Australia. Available at http://www.agric.wa.gov.au/objtwr/imported_assets/content/lwe/cli/winter_v2.pdf (accessed 27 May 2014).

Doole, G.J., Bathgate, A.D. and Robertson, M.J. (2009) Labour scarcity restricts the potential scale of grazed perennial plants in the Western Australian Wheatbelt. *Animal Production Science* 49, 883–893.

Foster, I. (2013) *Assessment of Climate Change Projections for WA: New Tools for Adaptation.* Selected paper for the 2013 Crop Updates, Crowne Complex, Burswood, 25–26 February, 2013. Government of Western Australia and GRDC, Perth.

GRDC (2012) Researchers probe warming climate frost puzzle. *GRDC Ground Cover.* Available at http://www.grdc.com.au/Media-Centre/Ground-Cover/Ground-Cover-Issue-101/Researchers-probe-warming-climate-frost-puzzle (accessed 27 May 2014).

Hennessy, K., Fitzharris, B., Bates, B. et al. (2007) Australia and New Zealand. Climate change 2007: Impacts, adaptation and vulnerability. In: Parry, M.L., Canziani, O.F., Palutikof, J.P., van der Linden, P.J. and Hanson, C.E. (eds) *Contribution of Working Group II to the Fourth Assessment Report of the Intergovernmental Panel on Climate Change.* Cambridge University Press, Cambridge, UK, pp. 507–540.

Hope, P., Drosdowsky, W. and Nicholls, N. (2006) Shifts in the synoptic systems influencing southwest Western Australia. *Climate Dynamics* 26, 751–764.

Hughes, L. (2003) Climate change and Australia: Trends, projections and impacts. *Austral Ecology* 28, 423–443.

Islam, N., Xayavong, V. and Kingwell, R. (2014) Broadacre farm productivity and profitability in south-western Australia. *Australian Journal of Resource and Agricultural Economics* 58, 1–24.

Kingwell, R. (2012) Revenue volatility faced by some of the world's major wheat producers. *Farm Policy Journal* 9, 23–33.

Kingwell, R. and Pannell, D. (2005) Economic trends and drivers affecting the wheatbelt of Western Australia to 2030. *Australian journal of Agriculture Research* 56, 553–561.

Kingwell, R., Anderton, L., Islam, N. et al. (2013) *Broadacre Farmers Adapting to a Changing Climate.* Final Report to National Climate Change Adaptation Research Facility, Gold Coast.

Lawes, R. and Kingwell, R. (2012) A longitudinal examination of business performance indicators for drought-affected farms. *Agricultural Systems* 106, 94–101.

Ludwig, F., Milroy, S. and Asseng, S. (2009) Impacts of recent climate change on wheat production systems in Western Australia. *Climate Change* 92, 495–115.

O'Donnell, C. (2010) Measuring and decomposing agricultural productivity and profitability changes. *Australian Journal of Agricultural and Resource Economics* 54, 527–560.

O'Donnell, C. J. (2012) Nonparametric estimates of the components of productivity and profitability change in U.S. agriculture. *American Journal of Agricultural Economics* 94, 873–890.

Potgieter, A., Meinke, H., Doherty, A. et al. (2012) Spatial impact of projected changes in rainfall and temperature on wheat yields in Australia. *Climatic Change* 117, 163–179.

Rebbeck, M., Dwyer, E., Bartetzko, M. and Williams, A. (2007) *A Guide to Climate Change and Adaptation in Agriculture in South Australia*. Primary Industries and Resources SA, Adelaide.

16 Growth opportunities for marine fisheries and aquaculture industries in a changing climate

ALISTAIR J. HOBDAY[1], RODRIGO H. BUSTAMANTE[1], ANNA FARMERY[2], AYSHA FLEMING[1], STEWART FRUSHER[2], BRIDGET S. GREEN[2], LILLY LIM-CAMACHO[3], JAMES INNES[1], SARAH JENNINGS[4], ANA NORMAN-LÓPEZ[1], SEAN PASCOE[1], GRETTA T. PECL[2], ÉVA E. PLAGÁNYI-LLOYD[1], PEGGY SCHROBBACK[5], OLIVER THEBAUD[1], LINDA THOMAS[1] AND E. INGRID VAN PUTTEN[1]

[1]*Climate Adaptation Flagship, CSIRO Marine and Atmospheric Research, Australia*
[2]*Institute for Marine and Antarctic Studies, University of Tasmania, Australia*
[3]*Climate Adaptation Flagship, CSIRO Earth System Sciences and Resource Engineering, Australia*
[4]*School of Economics and Finance, University of Tasmania, Australia*
[5]*School of Economics and Finance, Queensland University of Technology, Australia*

16.1 Introduction

Climate change is impacting the oceans around Australia with significant warming of ocean temperatures, observed on both the east and west coasts (Pearce and Feng 2007; Ridgway 2007; Lough and Hobday 2011). A range of other physical changes have also been documented, including to circulation and ocean chemistry (Lough and Hobday 2011; Poloczanska et al. 2012), and are projected to continue and even intensify in the future (Hobday and Lough 2011). Such changes are in turn impacting coastal marine ecosystems by altering the distribution, growth, recruitment and catch of exploited marine species (e.g. Pecl et al. 2011; Frusher et al. 2013) and their habitats (Ling 2008; Pratchett et al. 2011). In Australia, declines in lobster recruitment (Pecl et al. 2009), increases in abundance of tropical fish in southern waters (Last et al. 2011) and changes in growth of fished species (Neuheimer et al. 2011) have been reported. As a result, marine resource-based industries such as fishing and aquaculture are expected to experience both opportunities and challenges in coming years (Hobday et al. 2008; Norman-López et al. 2011; Doubleday et al. 2013). Given observed and projected climate-related changes, seafood sectors may need to adapt practices in order to maintain or enhance production to meet the needs of future populations (Rice and Garcia 2011; Merino et al. 2012; Bell et al. 2013). Adaptation is important as seafood plays a key role in regional food and economic security, supplying about 10% of world human calorific intake (Allison et al. 2009;

Applied Studies in Climate Adaptation, First Edition. Edited by Jean P. Palutikof, Sarah L. Boulter, Jon Barnett and David Rissik.
© 2015 John Wiley & Sons, Ltd. Published 2015 by John Wiley & Sons, Ltd.

Brander 2010), and in Australia is an important regional industry and employer (Hobday et al. 2008; Frusher et al. 2013).

16.2   Australia's seafood industry must take a supply chain view for effective adaptation

The response of wild fisheries and aquaculture sectors to climate change is an area of active investigation; however, the species and environment have received most attention (e.g. Cheung et al. 2010; Frusher et al. 2013; Salinger and Hobday 2013). Long-term shifts in target species and fisher activity have been reported from Australia and elsewhere (e.g. Nye et al. 2009; Caputi et al. 2010; Pinksy and Fogarty 2012), while climate-related extreme events such as heat waves, floods and cyclones also impact fisheries and aquaculture (Marshall et al. 2013; Pearce and Feng 2013). Planning responses to climate change relies on a solid biophysical understanding at the catch phase, yet this is not sufficient as the full range of opportunities and threats that will confront fisheries and aquaculture are not limited to biophysical changes at the catch phase (e.g. Marshall et al. 2013; van Putten et al. 2013). Consideration of the impacts of climate change along seafood supply chains, the steps a product takes from capture to consumer, is therefore vital to safeguard the ongoing supply of seafood from both wild fisheries and aquaculture.

Here we describe the results of an integrated consideration of the impacts of climate change across seafood supply chains. The two core objectives are:

1. to map supply chains for selected fisheries and aquaculture sectors as a basis for understanding potential impacts across their chains; and
2. to investigate possible climate-related impacts on these supply chains and develop adaptation approaches to overcome any barriers and take advantage of the opportunities to support continued growth of the Australian seafood sector.

Growth opportunities and adaptation by Australian seafood sectors in the face of climate change can be realised in several ways. First,

industry growth can result from improved environmental conditions, such that the level of sustainable biological production of particular species can increase. There are examples of species that will be advantaged in some regions of Australia (Brown et al. 2009; Hobday 2010; Fulton 2011); however, increased supply does not always mean increased growth for a fishery or aquaculture sector, as a market must exist for any additional seafood output. We do not discuss this first option further, as it is generally beyond the control of industry and management. Second, industry growth can also occur as a result of improving performance via: (1) decreasing operating costs as a result of increasing catch rates (e.g. more fish per trip); (2) increasing the value of catch as a result of increased seafood prices; or (3) reducing waste along the supply chain (Hamon et al. 2013). A third way that sectoral growth can occur is by increasing the value of existing production through value-adding or directing the product towards more profitable markets. Finally, growth can also occur by minimising vulnerability and instability in the supply chain by identifying critical elements and internal vulnerabilities that can be addressed by directed industry or government adaptation actions. Overall, we contend that growth and adaptation opportunities will be enhanced through increased awareness of markets and opportunities along the supply chain.

To explore these issues, we mapped supply chains for seven fishery and aquaculture sectors: southern rock lobster (Tasmanian sector), tropical rock lobster (Torres Strait fishery), western rock lobster, Sydney rock oyster, wild banana prawns (Northern Prawn Fishery sector), aquaculture prawns and Commonwealth trawl sector (CTS). These examples allowed comparison between competing wild products (e.g. lobster fisheries), wild and aquaculture products (prawns) and domestic and international markets (e.g. CTS and lobster) (Hobday et al. 2013). Here we provide examples of the ways in which growth in Australia's seafood sectors might be limited or facilitated in the face of climate change using social, economic and network analyses.

16.3 Seafood supply chains as a basis for adaptation planning

Supply chains are representations of the range of activities involved from the point of conception of a product, through different production phases, to final consumption (Kaplinsky and Morris 2001). Systematic management and strategic coordination of the supply chain has underpinned improvement of the long-term performance of individual companies and the supply chain as a whole in many industries (Mentzer et al. 2001).

Supply chains for our sectors were generated by identifying the steps in the chain in relation to product flows from fishers to consumers. Data, including the relative volume flow of product through alternative pathways, were derived from previous studies (e.g. Hamon et al. 2009; van Putten and Gardner 2010), supplemented with additional information collected through interviews with key representatives along the chain, and by fisheries-dependent data (Hobday et al. 2014). We used a nine-step supply chain template for consistency of comparisons between seafood sectors, which was customised for each sector (Fig. 16.1).

The resulting sector-specific supply chains were used to support a range of subsequent analyses including life cycle assessment (LCA) that can be used to help improve resource efficiency along the supply chain, critical element analysis to reduce shocks to the supply chain and economic analyses to understand alternative market options.

16.4 Growth through improved performance

To guide performance improvements that might lead to increased growth, supply chains were examined using LCA. This involved a compilation and evaluation of the environmental impact of a product using a systematic approach across the supply chain from producer to wholesaler (Hornborg et al. 2012). LCA methods are standardised through the International Organization for Standardization (ISO 14040:2006) to provide a structured format to compare a range of similar metrics for all stages of production, including broad analyses of resource dependency and emissions. For example, measures of greenhouse gas emissions can be used to develop tangible pathways for emissions reduction, which may also be an important adaptation strategy in a future carbon economy, or to encourage consumer acceptance of low-carbon seafood products, both of which can enhance industry performance.

Each LCA was undertaken using Simapro 7.3.3 ® software which models all processes at each supply chain stage across the life cycle of the product. Here we provide examples of two impact categories to characterise the environmental impact of different stages of each fishery supply chain – global warming potential (GWP) and cumulative energy demand (CED) – with additional categories and analyses described in Hobday et al. (2014) and Farmery et al. (2014). The GWP, measured as kilograms of CO_2-equivalent, is a commonly used impact category in LCA and includes greenhouse gases generated during the combustion of fossil fuels including oil, coal and natural gas. Greenhouse gas emission estimates were standardised per kilogram of fish. The second example of an environmental impact category, CED, describes the consumption of energy at each step in the supply chain. This measure of energy consumption (in units of megajoules per lower heating value, or MJ LHV) includes the direct uses and indirect energy consumption from fishing, processing and transport. Results of the LCA were aggregated from the nine-step supply chain into three stages to better allow comparison between the sectors. The first stage was 'capture' which corresponded to fishing activity including the gear, bait and fuel used in catching fish. The second stage, 'processor', corresponded to two supply chain stages from interim storage to fish receivers. The final stage, 'export', corresponded to the last six supply chain stages (from interim transport to consumers).

Results for the supply chains of two Australian lobster fisheries, the Tasmanian southern rock lobster (SRL) and the Torres Strait

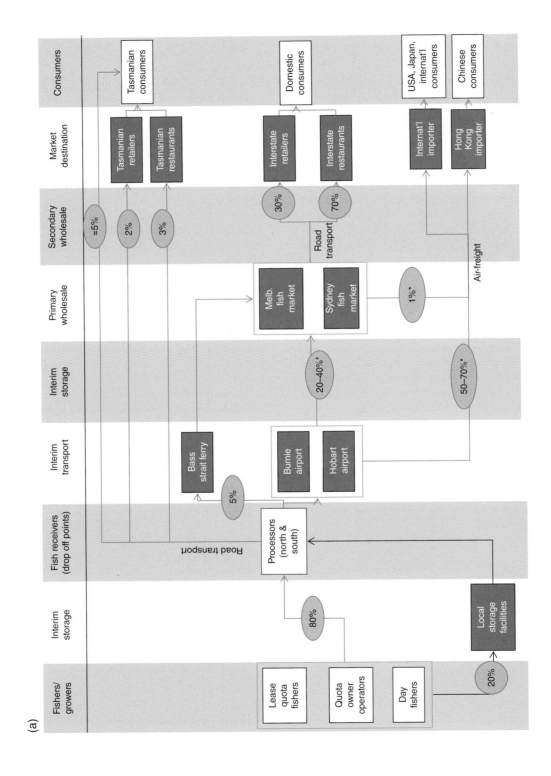

(a)

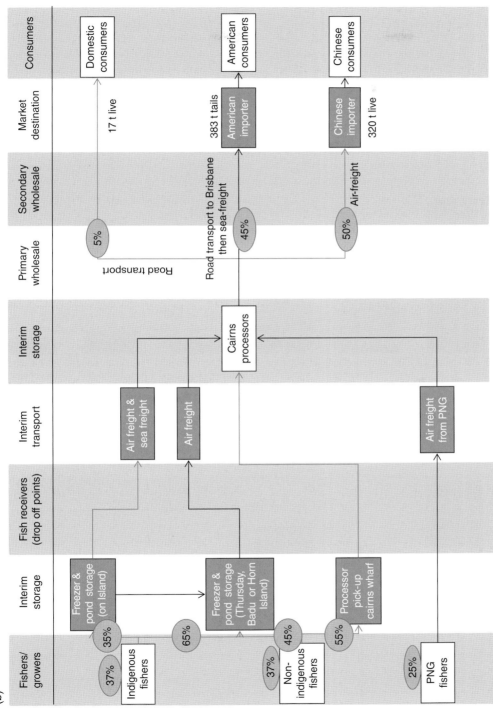

(b)

Figure 16.1 Example supply chains developed for (a) Tasmanian southern rock lobster fishery and (b) Torres Strait tropical rock lobster fishery. The approximate percentage of product along each pathway is shown in circles. For colour details please see Plate 14.

tropical rock lobster (TRL), show variation between fishery sub-sectors for each of the LCA categories (Fig. 16.2). GWP for the SRL was almost three times higher than the estimated sectoral average for the TRL (32 vs. 11.0 CO$_2$e kg^{-1} lobster). Over the whole TRL supply chain, the non-Indigenous sector for lobster tails had the lowest GWP (3.3 kg CO$_2$e kg^{-1} lobster) while the Indigenous commercial sector for live

lobster had the highest GWP (15.2 kg CO$_2$e kg^{-1} lobster) (Fig. 16.2a). In both the SRL and TRL, live fisheries more than 50% of GWP was accrued in the third stage of the aggregated supply chain. The capture and export of 1 kg of SRL consumed around three times more energy than 1 kg TRL (Fig. 16.2b). The capture and export components of the supply chain contributed a similar amount to the total cumulative

(a)

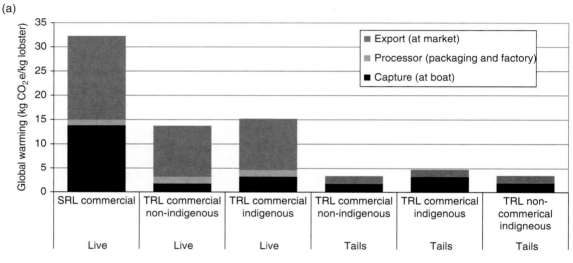

(b)

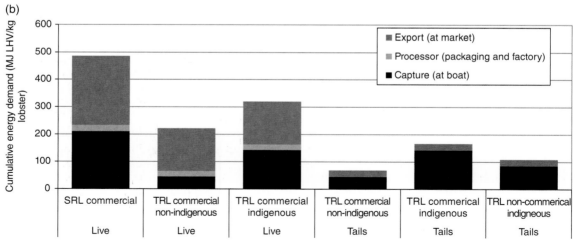

Figure 16.2 Life cycle assessment (LCA) result for southern rock lobster (SRL) and tropical rock lobster (TRL) fisheries. TRL is divided into 5 categories based on fisher typology and product type. (a) Global warming potential (kg CO$_2$e per kg lobster) and (b) cumulative energy demand (MJ LHV per kg lobster).

energy demand for the SRL fishery (210 and 253 MJ LHV kg^{-1}, lobster respectively).

If reducing greenhouse gas emissions becomes important to fishers or consumers, reducing the emissions generated at different steps of the supply chain could be a focus for adaptation strategies in its own right. (Note that here we consider the reduction of greenhouse gas emissions, while important as a mitigation strategy, is also an adaptation strategy when it enhances business performance regardless of the actual contribution to global emission reductions.) A number of adaptation options were considered, including those responding to changing abundance of the wild stocks (Hobday et al. 2014). To assess the relative benefit of such strategies to reduce GWP, two freight-related supply chain adaptation scenarios illustrate how LCA can be used to guide development and testing of adaptation options. One adaptation option considered for the TRL was a more direct route to markets, reducing air travel by 1700 km, while for the SRL a change from air to sea freight in the domestic and international transport phase was evaluated. While both scenarios raise other logistical and economic considerations, the LCA modelling explores the potential carbon reduction only. We compared the adaptation scenarios to the evaluated impact of maintaining the *status quo*.

The TRL fishery adaptation scenario of direct transport to markets in China reduced the carbon footprint for the commercial live TRL catch by 2 kg CO_2e kg^{-1} lobster (Fig. 16.3a). GWP was estimated to be 11 kg CO_2e kg^{-1} lobster for the commercial non-Indigenous sector (a total reduction of 490 t CO_2e per year across the total catch) and 13 kg CO_2e kg^{-1} lobster product for the commercial Indigenous sector (a total reduction of 109 t CO_2e per year across the total catch). Under the adaptation scenario for the SRL fishery, a 12% reduction in GWP was predicted (Fig. 16.3b) when sea-freighting lobster between domestic ports and then flying to the international destination. The GWP, based on fuel use, would be reduced by 99% (to 0.25 kg/CO_2e kg^{-1} lobster) if lobster were sea-freighted to the main international destination, assuming no extra emissions were incurred in keeping lobster alive for extended periods of time.

Even though the impact of the distance to markets on the environmental footprint is not currently a barrier to trading of seafood in these fisheries, carbon markets may influence future supply chains (Tang and Wang 2011). Airfreight currently represents the largest component of the overall environmental footprint in both fisheries and development of alternative transport routes or modes may be considered for future reduction in the greenhouse gas footprint (e.g. Tlusty and Lagueux 2009).

16.5   Growth through increasing prices

Standard economic theory predicts that as supply falls, prices will increase. Thus, some fishery sectors have considered that reducing supply (catch) could lead to enhanced profits and economic performance. In the face of declining biological production, such an approach may represent an adaptation strategy. However, such assumptions are rarely tested. We therefore used a range of economic analyses, including market integration analysis which is a useful tool for understanding the interrelation between different products, to assess market implications of potential adaptation options (Hobday et al. 2014; Norman-López et al. 2014; Schrobback et al. 2014). In particular, understanding the extent of the substitutability between different species and producer states would help them anticipate the price effects that climatic changes will have and to optimise supply to markets (Norman-López et al. 2014). For example, if different Australian seafood species are considered substitutes by international consumers, producers may be able to influence prices if they can coordinate supply. Conversely, if substitutability does not exist, exports of different species and/or producer states can be considered separate market segments, where changes in supply by one do not impact the price or revenue of others and the effect on their own price will be smaller than their change in supply (inflexible prices). Thus, when markets are integrated and products substitutable, managers in one fishery may need to consider harvesting and export activities in other related fisheries before implementing adaptation responses.

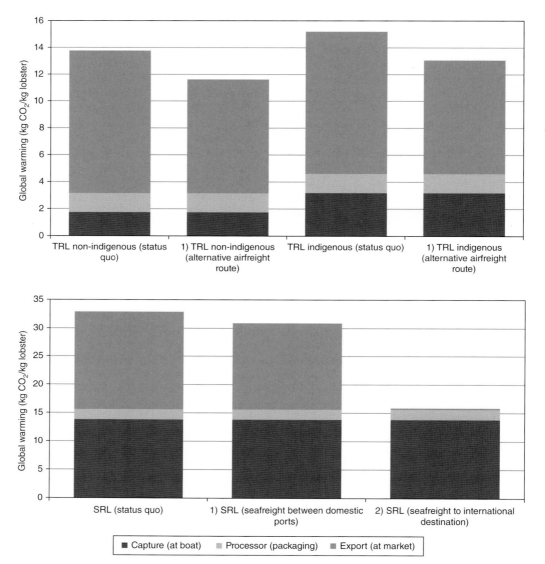

Figure 16.3 The LCA environmental metric 'global warming potential' for two lobster fisheries for potential climate change adaptation options: (a) commercial non-Indigenous and Indigenous TRL fishers comparing the current airfreight route to (1) the more direct airfreight route; and (b) SRL fishers comparing the current air freight route to (1) sea-freight between domestic ports (and then airfreight to international destination) and (2) sea-freight to the final international destination.

Given that temperate Australian rock lobster species are considered vulnerable to climate change, considerable thought is being given to the development of management strategies that can offset potential declines in total biomass, including a focus on enhancing the value of exports (Pecl et al. 2009). To determine if the international market for four Australian rock lobster species is integrated, and the possible price effects of changing supply, we used monthly export prices of live and whole fresh rock lobster to the Chinese (Hong Kong) market (the main market for Australian lobster at 75%, supplying 70% of Chinese imports; Norman-López et al. 2014) for the six different producer states, namely Queensland (QLD), Western Australia (WA), South Australia (SA), Victoria (VIC), Tasmania (TAS) and New South Wales (NSW). Prices were provided by Australian Bureau of Agriculture and Resource Economics and Sciences (ABARES) (Fig. 16.4).

Our results indicate a long-run export price relationship between the four Australian rock lobster species in the Chinese market (Norman-López et al. 2014). In other words, Chinese buyers regard live and whole fresh western rock lobster from WA, tropical rock lobster from QLD, eastern rock lobster from NSW and southern rock lobster from SA, VIC and TAS as substitutable and in the long run prices will have similar trends, although different values. The substitutability is particularly close for SRL between SA, VIC and TAS, so prices for products from these states will tend towards the same dollar value in the long run. Overall, changes in supply from one state will

impact not only its own prices but prices for all other Australian producer states. Such links between seafood sectors may restrict the range of adaptation options. For example, where competing products are seen as substitutes in the market, an adaptation strategy to reduce the catch of one species in order to boost price may not achieve the desired result. From a marketing point of view, producer states could coordinate total exports to the Chinese market in order to influence the market price they receive. If southern rock lobster exporters could work with western rock lobster exporters, they could have a stronger influence on the market price they all receive.

For the example considered here, the integrated nature of the Chinese export market for Australian lobster suggests that from an economic perspective the potential impacts of alternative fisheries management and development strategies cannot be considered in isolation for each state or species. In addition, impacts of external shocks affecting production in one state (e.g. climate change) can be expected to affect all Australian lobster fisheries. Hence, strategies to increase sector growth

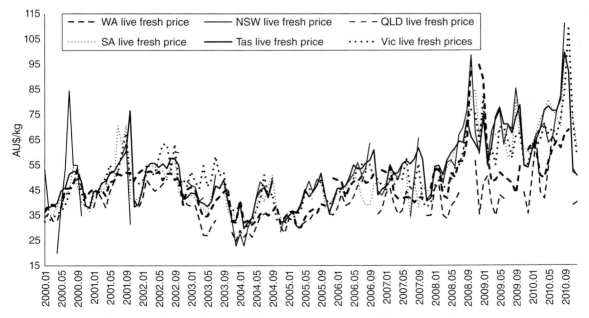

Figure 16.4 Monthly live and whole fresh rock lobster exports prices for different Australian producer states exporting to China (Hong Kong) from 1990 to 2010.

through decreasing supply to increase prices may not be an adaptation option that can be pursued by a single sector (Norman-López et al. 2014).

16.6   Growth through reducing vulnerability to shocks

Growth and profitability of seafood sectors can be enhanced if shocks to the production system are reduced. Unfortunately, this awareness often comes after an adverse event (Linnenluecke and Griffiths 2012; Marshall et al. 2013). To compare key features and critical elements in seafood supply chains under a changing climate, we developed a new quantitative metric analogous to indices used to analyse food webs and identify key species (Essington and Plaganyi 2012). Identification of key elements along the supply chain may assist in developing adaptation strategies to reduce anticipated future risks posed by climate change.

To identify critical supply chain elements – those with large throughput rates and greater connectivity – a Supply Chain Index (SCI) was developed (Hobday et al. 2014; Plaganyi et al. 2014). This single metric captures both the resilience and connectedness of a supply chain. The SCI for each element in the supply chain (j) with n nodes was calculated as the product of two variables – s_{ji} which represents the proportion of total product that receiver j receives from supplier i relative to all product flowing into that element j and p_{j}, which measures the proportion of the product that flows into receiver j – such that the multiplication of the two measures represents both connectance and magnitude of flow.

$$\mathrm{SCI}_{j} = \sum_{i=1}^{n} s_{ji} p_{j}^{2}$$

A lower overall score indicates greater clustering and connectivity in the supply chain, which may imply greater resilience to external shocks such as changes in the production and spatial distribution along the seafood production chain in response to changing climate. The SCI also provides information on the relative stability of different supply chains based on spread in the individual scores of the top few key elements, compared with a more critical dependence on a few key individual supply chain elements (Plaganyi et al. 2014).

By way of example, the distribution of critical elements for the Tasmanian SRL sector is shown in Figure 16.5. Most of the product from fishers is sent to processors located in the north and south of Tasmania, who then send the majority of the product to Australian mainland markets (primarily Sydney and Melbourne) and international destinations (primarily mainland China). The key elements identified by application of the SCI to the SRL supply chain are, respectively, the airports (Hobart and Burnie), the processors and Chinese consumers (Fig. 16.5). These elements are critical because of the volume of product that flows downstream from upstream suppliers.

Ensuring the resilience of key elements in a chain may be particularly important in maintaining the longer-term stability of the supply chain. On the other hand, an alternative way to strengthen the chain is to deflect or spread the criticality to other parts of the chain, such that the SCIj scores can help highlight the need to reduce risk associated with having a critical element. Hence for SRL, it might be fruitful to explore options for alternative Australian transport hubs, including Hobart airport, or alternative routes through other major cities, as a means of reducing the dependence on the key 'airport' node. Moreover, emphasis could be placed on supporting and building the resilience of other key elements such as the processors and Chinese consumers. For example, focusing effort on firmly establishing Chinese trade agreements may be one critical area providing scope for growth and building stability in the SRL supply chain (Hobday et al. 2014). Processors are also highlighted as important and hence the resilience of the chain can be strengthened by focusing on building the stability of this component. For example, contingency plans could be put in place to diversify in product types that are more versatile in terms of 'storability'

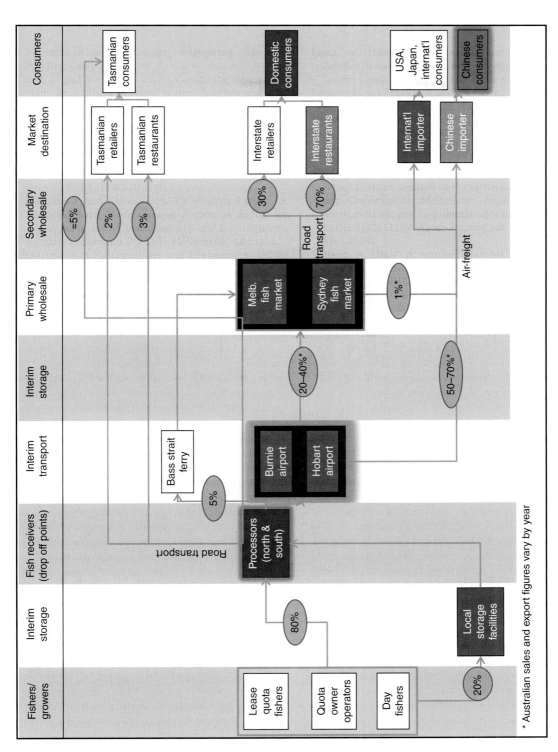

Figure 16.5 Tasmanian southern rock lobster supply chain with colour coding to highlight key elements identified using the supply chain index (SCI). For colour details please see Plate 15.

(for example, converting fresh product to frozen tails), thus making it possible to even out seasonal distribution of their product in anticipation of climate-driven environmental impacts (Plaganyi et al. 2014).

Comparison of SCI across supply chains showed that the top three key elements in each supply chain, as identified using the key supply indices for individual elements, differed across all the case studies, with the most common element being consumers (whether domestic or international). Key elements occur predominantly at the downstream end of the chain for TRL, WRL, banana prawn and prawn aquaculture supply chains in contrast to the case for SRL, oyster and the CTS (Fig. 16.6). These supply chains and the SCI also shed light on potential adaptation options, which can be examined through simulation and modification of the supply chain (Plaganyi et al. 2014).

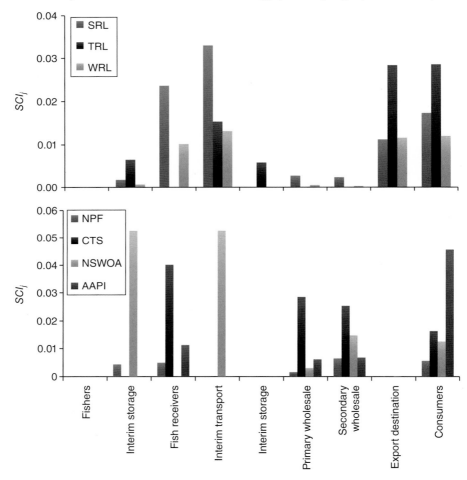

Figure 16.6 Plots of the standardised SCI_j metrics aggregated over different stages j of each wild seafood and aquaculture supply chain to compare the distribution of key stages in each of the supply chain case studies. Southern rock lobster (SRL), Western rock lobster (WRL), tropical rock lobster (TRL), northern prawn fishery (NPF), Commonwealth trawl fishery (CTS), Australian Aquaculture Prawn Industry (AAPI), New South Wales Oyster Aquaculture (NSWOA). For colour details please see Plate 16.

16.7 Stakeholder awareness of adaptation options

The information generated by the preceding analyses may not in itself lead to adaptation action. Adaptation across seafood supply chains will necessitate partnerships between the people involved in different elements in the chain; however, current connections are generally weak in terms of communication links within and between Australian fisheries and aquaculture supply chains. We undertook interviews with stakeholders at the beginning and end of this study, which generated insight into current awareness of climate change issues across supply chains. Initial interviews showed that, as for a range of other primary industries in Australia, awareness of marine aspects of climate change is high and potential impacts are well understood at the catch phase (Fleming et al. 2014). In some

seafood sectors, stakeholders reported that impacts and disruption to supply chains have already occurred. That said, climate change is not the highest priority for most stakeholders with other shorter-term priorities demanding attention such as rising fuel costs, labour shortages and legislative burden (Fleming et al. 2014).

A final set of stakeholder interviews evaluated the potential responses to, and the acceptability of, a range of adaptation options. Interviews showed that stakeholders were able to generate a range of adaptation options in response to a variety of possible scenarios. Interestingly, despite representation from people knowledgeable of all stages in the supply chain, the majority of adaptation actions were proposed for the catch stage, reinforcing the view that stakeholders need to be encouraged to take a holistic supply chain perspective to adaptation (Hobday et al. 2014; Fig. 16.7).

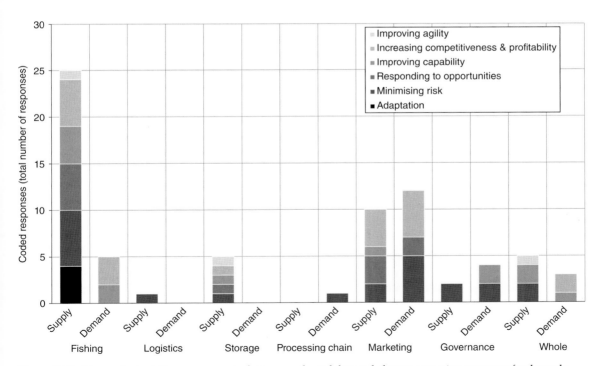

Figure 16.7 Types of adaptation responses to future supply and demand change scenarios across seafood supply chain stages. For colour details please see Plate 17.

16.8   Holistic adaptation across seafood supply chains and sectors

The approaches described here revealed a range of potential points for enhancing growth and profitability for Australian fisheries facing climate-related threats. Strengths and weaknesses in the supply chain were identified using critical element analysis and, together with the LCA, can further inform the development of adaptation options (Table 16.1). Economic analyses showed the influence of market factors on the price and profit adaptation options for selected sectors. Stakeholder interviews suggest that the seafood sector stakeholders could become more engaged in planning their future across the supply chain. Options that result in increased growth or efficiency will likely be more acceptable to industries, and may overcome barriers related to perceived importance of climate change relative to other issues. Importantly, development of options to overcome barriers or take advantage of opportunities should be undertaken in partnership with industry and managers, focusing on an agreed set of options and after defining 'supply-chain' related objectives for each fishery.

A number of further steps need to be taken before any of these alternatives can be implemented. The tools and outputs from this project can also be used to support detailed exploration of trade-offs between different adaptation options. For example, detailed feasibility and investment analyses need to be undertaken, accounting for the cost of developing the alternative transport infrastructure required to reduce a sector's carbon footprint. Developing future strategies that take advantage of opportunities identified in the LCA, critical metric and economic analyses is possible, particularly those that can address opportunities identified by several of the techniques. Such detailed consideration is a likely next step for the sectors we have worked with to date.

With regard to potential growth opportunities and adaptation options for Australian seafood sectors in the face of climate change, we have shown a range of possibilities for adaptation focus that are more in the control of industry than increasing the abundance of wild fish species. We have illustrated just a few of these options to illustrate how industry growth can occur as a result of improving efficiency, increasing the value of existing production through reducing waste along the supply

Table 16.1   Summary of approaches used to explore growth opportunities and adaptation potential for Australian seafood sectors.

Approach	Finding	Adaptation implication
Life cycle assessment	Environmental resource use and resulting emissions vary considerably between fisheries and stages in each supply chain.	Investigate options to reduce energy use for economic benefit and consumer advantage.
Market analysis	Similar seafood products that are seen as substitutes in the market will have co-varying prices.	Adaptation options need to be coordinated between similar seafood products to yield improved outcome.
Critical element analysis	Vulnerable elements in supply chains may be particularly susceptible to disruption under climate change.	Improving the resilience of supply chains will lead to improved growth, but may be inefficient in the short term.
Stakeholder perspectives	Climate awareness is high across the supply chain, but adaptation options are thought to be mostly at the production end of supply chains. Adapting to climate change is not seen as the highest priority.	Investigation of options higher up supply chains may allow future industry growth even if biological production is declining. Adaptation options that result in industry growth and/or increased profit should enhance uptake.

chain or by focusing on the most profitable markets (see Hobday et al. 2014) and by minimising vulnerability and instability in the supply chain by identifying critical elements.

Overall, we have developed a set of approaches to generate realistic adaptation management and policy options to enhance 'growth and opportunities' along seafood supply chains. These will directly benefit the adaptation efforts of the seven selected seafood sectors, and can be applied to additional sectors in future resulting in improved outcomes for primary industries in the face of a changing environment.

Acknowledgements

We greatly appreciate the contribution of stakeholders across the supply chains who participated in workshops, interviews and facilitated additional contacts. This research contributed to the project 'Growth opportunities and critical elements in the value chain for wild fisheries and aquaculture in a changing climate', which was supported by CSIRO, NCCARF and funding from the FRDC–DCCEE on behalf of the Australian Government.

References

Allison, E.H., Perry, A.L., Badjeck, M.-C. et al. (2009) Vulnerability of national economies to the impacts of climate change on fisheries. *Fish and Fisheries* 10, 173–196.

Bell, J.D., Ganachaud, A., Gehrke, P.C. et al. (2013) Mixed responses of tropical Pacific fisheries and aquaculture to climate change. *Nature Climate Change* 3, 591–599.

Brander, K. (2010) Impacts of climate change on fisheries. *Journal of Marine Systems* 79, 389–402.

Brown, C.J., Fulton, E.A., Hobday, A.J. et al. (2009) Effects of climate-driven primary production change on marine food webs: implications for fisheries and conservation. *Global Change Biology* 16, 1194–1212.

Caputi, N., Melville-Smith, R., de Lestang, S., Pearce, M. and Feng, M. (2010) The effect of climate change on the western rock lobster (*Panulirus cygnus*) fishery of Western Australia. *Canadian Journal of Fisheries and Aquatic Sciences* 67, 85–96.

Cheung, W.W.L., Lam, V.W.Y., Sarmiento, J.L. et al. (2010) Large-scale redistribution of maximum fisheries catch potential in the global ocean under climate change. *Global Change Biology* 16, 24–35.

Doubleday, Z.A., Clarke, S.M., Li, X. et al. (2013) Assessing the risk of climate change to aquaculture: a case study from south-east Australia. *Aquaculture Environment Interactions* 3, 163–175.

Essington, T.E. and Plaganyi, É.E. (2012) *Model and Data Adequacy for MSC Key Low Trophic Level (LTL) Species Designation and Criteria*. Marine Stewardship Council report, London.

Farmery, A., Gardner, C., Green, B.S. and Jennings, S. (2014) Managing fisheries for environmental performance: the effects of marine resource decision-making on the footprint of seafood. *Journal of Cleaner Production* 64(1), 368–376.

Fleming, A., Hobday, A.J., Farmery, A. et al. (2014) Climate change risks and adaptation options across Australian seafood supply chains–a preliminary assessment. *Climate Risk Management* 1, 39–50.

Frusher, S.D., Hobday, A.J., Jennings, S.M. et al. (2013) A short history of a marine hotspot–from anecdote to adaptation in south-east Australia. *Reviews in Fish Biology and Fisheries* 24(2), 593–611.

Fulton, E.A. (2011) Interesting times: winners, losers, and system shifts under climate change around Australia. *ICES Journal of Marine Science* 68, 1329–1342.

Hamon, K.G., Thébaud, O., Frusher, S. and Little, L.R. (2009) A retrospective analysis of the effects of adopting individual transferable quotas in the Tasmanian red rock lobster, *Jasus edwardsii*, fishery. *Aquatic Living Resources* 22, 549–558.

Hamon, K.G., Frusher, S.D., Little, L.R., Thébaud, O. and Punt, A.E. (2013) Adaptive behaviour of fishers to external perturbations: simulation of the Tasmanian rock lobster fishery. *Reviews in Fish Biology and Fisheries*, doi: 10.1007/s11160-013-9302-1.

Hobday, A.J. (2010) Ensemble analysis of the future distribution of large pelagic fishes in Australia. *Progress in Oceanography* 86, 291–301.

Hobday, A. J. and Lough, J. (2011) Projected climate change in Australian marine and freshwater environments. *Marine and Freshwater Research* 62, 1000–1014.

Hobday, A.J., Poloczanska, E.S. and Matear, R. (2008) *Implications of Climate Change for Australian Fisheries and Aquaculture: A Preliminary Assessment: Report to the Department of Climate Change, Canberra, Australia*. Commonwealth Government, Canberra.

Hobday, A. J., Bustamante, R. H., Farmery, A. et al. (2014) Growth opportunities and critical elements in the supply chain for wild fisheries and aquaculture in a changing climate. Final Report. FRDC-DCCEE Marine National Adaptation Program 2011/233.

Hornborg, S., Nilsson, P., Valentinsson, D. and Ziegler, F. (2012) Integrated environmental assessment of fisheries management: Swedish Nephrops trawl fisheries evaluated using a life cycle approach. *Marine Policy* 36, 1193–1201.

Kaplinsky, R. and Morris, M. (2001) *A Handbook for Value Chain Research.* Institute of Development Studies, Brighton, UK.

Last, P.R., White, W. T. Gledhill, D. C. et al. (2011) Long-term shifts in abundance and distribution of a temperate fish fauna: a response to climate change and fishing practices. *Global Ecology and Biogeography* 20, 58–72.

Ling, S.D. (2008) Range expansion of a habitat-modifying species leads to loss of taxonomic diversity: a new and impoverished reef state. *Oecologia* 156, 883–894.

Linnenluecke, M. and Griffiths, A. (2012) Assessing organizational resilience to climate and weather extremes: complexities and methodological pathways. *Climatic Change* 113(3–4), 933–947.

Lough, J.M. and Hobday, A.J. (2011) Observed climate change in Australian marine and freshwater environments. *Marine and Freshwater Research* 62, 984–999.

Marshall, N.A., Tobin, R.C., Marshall, P.A., Gooch, M. and Hobday, A.J. (2013) Social vulnerability of marine resource users to extreme weather events. *Ecosystems* 16, 797–809.

Mentzer, J.T., DeWitt, W., Keebler, J.S. et al. (2001) Defining supply chain management. *Journal of Business Logistics* 22(3), 1–25.

Merino, G., Barange, M., Blanchard, J. L. et al. (2012) Can marine fisheries and aquaculture meet fish demand from a growing human population in a changing climate? *Global Environmental Change* 22, 795–806.

Neuheimer, A. B., Thresher, R. E., Lyle J. M. and Semmens, J.M. (2011). Tolerance limit for fish growth exceeded by warming waters. *Nature Climate Change* 1, 110–113.

Norman-López, A., Pascoe, S. and Hobday, A. J. (2011) Potential economic impacts of climate change on Australian fisheries and the need for adaptive management. *Climate Change Economics* 2, 209–235.

Norman-López, A., Pascoe, S., Thebaud, O. et al. (2014) Price integration in the Australian rock lobster industry: implications for management and climate change adaptation. *Australian Journal of Agriculture and Resource Economics* 58(1), 43–59.

Nye, J.A., Link, J.S., Hare, J.A. and Overholtz, W. J. (2009) Changing spatial distribution of fish stocks in relation to climate and population size on the Northeast United States continental shelf. *Marine Ecology Progress Series* 393, 111–129.

Pearce, A. and Feng, M. (2007) Observations of warming on the Western Australian continental shelf. *Marine and Freshwater Research* 58, 914–920.

Pearce, A. F. and Feng, M. (2013) The rise and fall of the 'marine heat wave' off Western Australia during the summer of 2010/2011. *Journal of Marine Systems* 111–112, 139–156.

Pecl, G., Frusher, S., Gardner, C. et al. (2009) *The East Coast Tasmanian Rock Lobster Fishery–Vulnerability to Climate Change Impacts and Adaptation Response Options. Report to Department of Climate Change, Australia.* Commonwealth of Australia, Canberra.

Pecl, G. T., Ward, T., Doubleday, Z. et al. (2011) *Risk Assessment of Impacts of Climate Change for Key Marine Species in South Eastern Australia. Part 1: Fisheries and Aquaculture Risk Assessment.* Fisheries Research and Development Corporation, Project 2009/070.

Pinsky, M. and Fogarty, M. J. (2012) Lagged social-ecological responses to climate and range shifts in fisheries. *Climatic Change* 115(3), 883–891.

Plagányi, É. E., van Putten, I., Thébaud, O., Hobday, A. J., Innes, J. et al. (2014) A quantitative metric to identify critical elements within seafood supply networks. *PLoS ONE* 9(3), e91833. doi:10.1371/journal.pone.0091833.

Poloczanska, E.S., Hobday, A.J., Richardson, A.J. et al. (eds) (2012) *Marine Climate Change in Australia, Impacts and Adaptation Responses. 2012 Report Card.* CSIRO, Australia.

Pratchett, M.S., Bay, L.K., Gehrke, P. et al. (2011) Contribution of climate change to degradation and loss of critical fish habitats in Australian marine and freshwater environments. *Marine and Freshwater Research* 62, 1062–1081.

Rice, J.C. and Garcia, S.M. (2011) Fisheries, food security, climate change, and biodiversity: characteristics of the sector and perspectives on emerging issues. *ICES Journal of Marine Science* 68, 1343–1353.

Ridgway, K.R. (2007) Long-term trend and decadal variability of the southward penetration of the East Australian Current. *Geophysical Research Letters* 34, L13613.

Salinger, M.J. and Hobday, A.J. (2013) Safeguarding the future of oceanic fisheries under climate change depends on timely preparation. *Climatic Change,* doi: 10.1007/s10584-012-0609-z.

Schrobback, P., Pascoe, S. and Coglan, L. (2014) Impacts of introduced aquaculture species on markets for native marine aquaculture products: The case of edible oysters in Australia. *Aquaculture Economics and Management,* doi: 10.1080/13657305.2014.926465.

Tang, S. L. and Wang, J. W. (2011) The impacts of carbon quotas on supply chain management. In: Juan, S. (ed.) *Business and Economics Research.* Iacsit Press, Singapore.

Tlusty, M.F. and Lagueux, K. (2009). Isolines as a new tool to assess the energy costs of the production and distribution of multiple sources of seafood. *Journal of Cleaner Production* 17, 408–415.

van Putten, E. I. and Gardner, C. (2010) Lease quota fishing in a changing rock lobster industry. *Marine Policy* 34, 859–867.

van Putten, E. I., Jennings, S., Frusher, S. et al. (2013) Building blocks of economic resilience to climate change: A south east Australian fisheries example. *Regional Environmental Change,* doi: 10.1007/s10113-013-0456-0.

17 Water tariffs and farmer adaptation: the case of Goulburn–Murray Water, Victoria, Australia

BETHANY COOPER, LIN CRASE
AND NICHOLAS PAWSEY

Centre for Water Policy and Management, La Trobe University, Australia

17.1   Introduction

Many economists argue that the most effective way to facilitate human adaptation to a range of phenomena is to allow prices to signal the relative scarcity of resources/goods/services such that individuals and firms modify expectations and use. This response would align costs and prices so that (theoretically at least) less would be used of more costly (scarce) items and more of those less costly (scarce). In the case of water, for example, progressive increases in cost and setting prices to match would signal that users needed to exercise more caution in use and, where possible, substitute other resources (Johansson et al. 2002).

This broad approach to resource allocation problems only holds insofar as the resource/good/service in question is not subject to some form of market failure and as long as there are substitutes. It might come as no surprise then that water prices do not always reflect scarcity, with a range of reasons offered for a divergence between cost and price.

First, the provision of clean water is often considered a public good, especially from a human health perspective, and market provision would therefore result in under-provision relative to the optimum. Second, water is also related to a range of environmental amenities, which again have public good attributes or result in material spill-over effects. Third, water itself is a difficult resource to 'market'; establishing the property rights to make water a marketable product is no simple matter and assigning the resource through the intervention of the state might therefore be more straightforward and less costly. Fourth, there is a range of equity considerations that attend water, such as the notion that a minimum level of access is required to support life.

One of the consequences of these types of arguments is that water prices in many parts of the world are not remotely tied to scarcity or anticipated changes in availability (see, for example, Johansson et al. 2002). Moreover, even in a country such as Australia where there are robust institutions for isolating and protecting

Applied Studies in Climate Adaptation, First Edition. Edited by Jean P. Palutikof, Sarah L. Boulter, Jon Barnett and David Rissik.
© 2015 John Wiley & Sons, Ltd. Published 2015 by John Wiley & Sons, Ltd.

the 'public' aspects of water from the 'private' dimensions, differences in cost and price can be substantial.

In the context of climate change, the wedge between prices for water and costs has the potential to be especially problematic, unless accompanied by offsetting measures by the state. More specifically, the misalignment of prices and costs means that current water use cannot be optimal and, if these differences persist, there is a real danger that abrupt and serious adjustments in water use will be required. For example, if users face prices that are less than the cost of water they will be encouraged to have practices that are excessively water-intensive such that any increased scarcity will result in serious disruption.

This is not a problem limited to Australian irrigation; Wichelns (2010) details similar concerns in the US. Nonetheless, in the Australian milieu the problem is particularly relevant in the southern states of the Murray–Darling Basin (MDB): New South Wales, South Australia and Victoria. Each of these states has large irrigation sectors that account for a substantial amount of the value of agricultural production and a large proportion of water consumption. For example, in Victoria irrigated agriculture accounted for over 30% of agricultural output (ABS 2010) and most of this was produced in northern Victoria (i.e. part of the MDB). These industries are also highly exposed to potential changes in water availability due to climate change. In this context the Victorian Government's Department of Sustainability and Environment (2008) noted that 38% of all stream flows in the MDB were generated in northeast Victoria and yet water availability was predicted to decrease in the region by 5–50% by 2070. Against that background there are real risks that farmers will not adapt to match changes in water availability, especially if price or other signals are missing to encourage adjustments to scarcity.

In this chapter we provide details of a project which focused on water price reform in irrigation. The setting was northern Victoria where government policies had resulted in a significant mismatch between prices/water tariffs faced by

farmers and the costs of providing water. More specifically, the gifting of irrigation assets to irrigation districts as part of wider policy changes (e.g. Crase et al. 2012) means that water users are subsequently not obliged to pay tariffs that reflect the full cost. A question thus arises as to whether publicly funded infrastructure of this form distorts production choices, especially over the longer term, possibly exposing enterprises and communities to sudden future climatic shocks. A related question deals with those aspects of pricing reforms that can be undertaken now to improve the prospect of adaptation.

The remainder of this chapter is divided into four main parts. In Section 17.2 we provide an overview of the policy and regulatory environment that has led to the current mal-alignment of tariffs and costs. We describe an empirical approach for gauging the relative acceptability of tariff reforms from the perspective of irrigation farmers in Section 17.3. The key findings from this approach are presented in Section 17.4 before some brief policy considerations and concluding remarks are offered in Section 17.5.

17.2 Water prices, water tariffs and irrigation

While there are some advantages to having prices match scarcity and thus cost, achieving this alignment is difficult. There are a range of political and practical constraints that attend this task. For example, a water supplier is usually required to invest in 'lumpy' infrastructure (i.e. requiring large investment at irregular widely spaced intervals), so that adjusting prices to annual or seasonal water scarcity may not generate sufficient revenue to cover the long-run costs of infrastructure; rather, water will tend to be under-priced during dry years and over-priced in wet years (VAGO 2011). Similarly, the challenges of altering tariffs while maintaining certainty for customers can be non-trivial, especially when economic regulators set prices over 4–5-year planning cycles and customer bills arrive at infrequent intervals. Finally, increases to water tariffs are likely to be resisted politically, especially if there is a prospect of a small group gaining advantage by transferring modest costs to the general public (Horn 1995).

Tariff reform is thus a controversial topic in the water sector and this is probably no more apparent than in irrigation in Australia. The tariffs faced by irrigation farmers in northern Victoria represent a complex system of historical decisions, often made during times of relatively high water availability. The process for adjusting signals, via tariffs, is thus influenced by the legacy of earlier decisions; we do not start with a blank canvas. It is also worth noting that there are a range of other factors that bear on water prices in irrigation, not least the well-entrenched views about notions of food security and the sanctity of agrarian activity generally (Crase and Merton 2013). Complications also arise from the regulatory framework used for establishing prices.

Major adjustments occurred in the way water tariffs were struck in Australia when states, including Victoria, agreed to the National Water Initiative (NWI) in 2004. The basic tenet behind the NWI was that water tariffs should be cost reflective such that they did not distort incentives and that water resources would then move to the highest value use. By this time water markets had been established in most states so that water entitlements and allocations could be traded.

There is a necessary distinction that needs to be made between the way water markets operate to generate prices and the tariffs paid for using water. Water markets allow for the trade of perpetual shares that allow access to resource. The purchaser of a water entitlement can therefore expect a so-called 'allocation' that represents the volume of water available in a given timeframe (usually an irrigation season). Owners of entitlements can therefore trade allocations that attend those entitlements, along with the entitlements themselves. The prices paid for entitlements and allocations vary with the extent of competition for access, and generally represent the opportunity cost of the resource in alternative uses.

These are not the only prices/costs faced by water users, however. In addition to using the market to access water (or face the opportunity cost of holding water), users must then have the means to deploy water. Where farmers access water directly from a stream or aquifer, this will be the sum of any costs related to pumping (e.g. electricity, fuel) and payments to the state for so-called use rights. Clearly, if these costs are higher for some entitlements/allocation then the prices offered in the water market for those entitlements/allocations will be reduced to reflect the additional costs of deployment (i.e. this discounts the value of the entitlement or allocation in the water market). The corollary will also hold; if use costs are lower, farmers will be able to offer higher prices in the water markets and bid access away from other users.

Most irrigation water in northern Victoria is provided via communal irrigation networks and does not involve individuals pumping from streams or aquifers. Nonetheless, farmers must hold access right and pay tariffs to the owner of the irrigation network (in this case Goulburn–Murray Water, a state-owned irrigation corporation). Because an irrigation supplier such as Goulburn–Murray Water effectively holds a monopoly over the supply services, the tariffs charged to farmers are subject to economic regulation. In this instance, the Essential Services Commission (ESC) has responsibility for economic regulation and, at first sight, uses principles consistent with the NWI.

The starting point for the process of economic regulation is the 'building block approach', or what is sometimes called long-run marginal cost (LRMC) estimation. LRMC is a relatively static concept and involves estimating the short-run marginal cost (SRMC) and adding a margin to capture the average cost of bringing forward the next supply augmentation.

The nature of irrigation means there is a strong focus on capital costs; capital represents the main cost of running almost all water-supply businesses. Accordingly, establishing an appropriate method for estimating the costs related to capital has significant ramifications for water tariffs. In Victoria the notion of the regulatory asset base (RAB) is critical in this context. The RAB is the estimated value of infrastructure that can be legitimately related to water tariffs paid by end-users. This process involves establishing a return on capital (based on the estimated value of the RAB) to reflect the opportunity cost of the funds invested in the business, and a return of capital to reflect depreciation. These two costs are then added with projected operating costs to determine the notionally efficient revenue required to run the monopoly water supply business. Tariffs are then set with the aim of recouping only that revenue by estimating water use. The resulting fixed charges and volumetric charges paid by customers should not collectively exceed the revenue requirement of the water supplier, nor should it fall short of that requirement.

A major modification to water tariffs in communal irrigation occurred, almost inadvertently, as a result of other reforms in the water sector. Faced with a serious electoral challenge in 2007, the then Prime Minister John Howard hastily announced his government's National Plan for Water Security (Watson 2007). Developed in the midst of drought, the plan was to spend US$ 10 billion of public funds to address over-allocation in the MDB. A centrepiece of the plan was the commitment to fund irrigation infrastructure at taxpayer's expense on the proviso that this led to so-called water savings. Prime Minister Howard failed to secure re-election but the incoming government exceeded the infrastructure commitments in the plan through its *Our Water Future* manifesto.

An important element of this policy approach is that any infrastructure gifted by government is then not counted as part of the RAB. The upshot is that water tariffs do not cover the cost of the capital associated with such 'gifts', nor does the tariff provide for depreciation of the related assets. To gain some understanding of the impacts of these policy decisions, it is worth noting that the Commonwealth's contribution to the Northern Victorian Irrigation Renewal Project located in the irrigation districts controlled by Goulburn–Murray Water is estimated at about US$ 1 billion. Moreover, Pawsey and Crase (2013) estimate that tariffs in these irrigation areas

would need to increase by about 300% to recover costs in 2011, prior to full implementation of the gifted project. The point is that there is already a major divergence between tariffs and the cost of providing water to irrigation in this area and this has four impacts: (1) it encourages over-use of the resource in the current time period; (2) it distorts the water market and encourages water entitlement and allocations to be bid into the subsidised irrigation areas; (3) the failure to cover the cost of depreciation exposes the irrigation supplier and potentially the taxpayer to future shortfalls in revenue; and (4) the increased water intensity of production in these areas exposes industries to future shocks, should water become less plentiful as predicted by climate change modelling.

As we have already noted, tariff reform in water is notoriously problematic. In order to progress changes to tariffs such that they encourage adaptation to changed water availability, a useful starting point is to understand what farmers regard as 'acceptable' modifications to water tariffs and what is likely to be resisted more stridently. Armed with this knowledge, the policymaker could then focus attention and resources on matching the reform challenge.

17.3 How to discover what irrigation farmers want

The breakdown of the charges for water users in the Goulburn–Murray area alone makes the tariff structure complex to change. There are slightly more than 400 potential charges for customers. These can be broadly categorised into bulk water, entitlement storage, water delivery and drainage charges. The charges also differ across the six gravity irrigation districts. For instance, the charge for each delivery share presently ranges from $2700 to $4700 and, since these charges also determine termination fees (i.e. what has to be paid when a farmer 'disconnects' from the irrigation network), will also likely impact on incentives to adjust the scale of irrigation should climate change result in permanent reductions in water availability. The subsidised infrastructure modernisation also means that current charges do not reflect the service provided to irrigators, raising questions about the long-term sustainability of this approach and any attempts to improve the efficiency of the price signals so generated. For example, the level of service provided by the so-called 'service points' differs across the region; however, the current charge is not differentiated on the basis of quality of service.

Tariff reform involves a degree of coercion, inasmuch as there will be different preferences associated with alternative pricing structures depending on enterprise type and history of use. For example, a large water user will likely be worse off relative to a smaller water user should tariffs be recalibrated with greater emphasis on the volumetric component (other things being equal). Against that background it is important to understand the range and strength of preferences for different elements of a water tariff in order to understand the distributional effects of reform.

To try to gain some understanding of the current perceptions of water tariffs among farmers a survey technique known as best–worst scaling was deployed. A best–worst scaling experiment is a survey technique that presents participants with a range of features (in this case attributes of a tariff) and asks the respondents to rate the least- and most-preferred elements. The experiment is repeated continuously and the range of attributes varied such that data then provides a complete ranking (in this case, of tariff reform options). In this experiment, the *a priori* attributes for the best–worst scaling experiment were developed by reviewing the water pricing literature and through discussion with key stakeholders involved in rural water pricing in northern Victoria. To capture the perspective of customers of Goulburn–Murray Water, focus groups were conducted with the established Water Service Committees in each of the six districts in the Goulburn–Murray Irrigation District.

A four-part survey was subsequently developed. The first part included questions to capture the socio-demographics of the respondents, their water trading history and intentions and their attitudes towards the irrigation infrastructure modernisation scheme. The second part presented respondents with a range of questions to identify their level of understanding of the pricing process and the current tariff structure. The best–worst scaling experiment questions were presented in the third section of the survey, while the final part comprised questions about respondents' attitudes towards the environment, climate change and risk.

In total, eight attributes were identified to be included in the best–worst scaling experiment, described briefly in the following sections. The levels for each attribute are depicted in Table 17.1. Notably, a level is not related to level of importance; rather, it is a way to describe or express an attribute.

17.3.1　Service point fees

Currently, this charge is set at a standard rate of $250 per service point across all irrigation districts. The fee does not cover the full 'whole-of-life cost' of each outlet and the balance is recovered through higher infrastructure access fees. This proved to be a contentious issue and was therefore included as an attribute. This attribute was expressed in two levels, including the status quo. The second level was designed to capture preferences for an increase in the service point fee matched by a decrease in the infrastructure access fee.

17.3.2　Pricing strategy

Current tariff structure differentiates some charges across the different districts. Customers expressed varying opinions regarding any move towards a uniform tariff on this front. In the case of the experiment, this attribute was expressed in terms of the delivery share charge with two levels, including the status quo. The status quo was expressed as keeping charges for delivery share (ML/day) set by each irrigation district and the other level presented the option of adopting a single charge for delivery share (ML/day) across all irrigation districts.

Table 17.1 Best–worst attributes and levels.

Attribute	Level 1	Level 2	Level 3
Service point fees	Leave service point charge at a low fee of $250/service point per year; a high infrastructure access fee (Service Point Fees 1)	Increase the service point charge from $250 to $1000/ service point per year; lower the infrastructure access fee by reducing the cost of each delivery share by $600 (Service Point Fees 2)	
Pricing strategy	Keep charges for delivery share (ML/day) set by each irrigation district (Pricing Strategy 1)	Move to a single charge for delivery share (ML/day) across all irrigation districts (Pricing Strategy 2)	
Delivery share	Keep current volume at 270 ML per delivery share (Delivery Share 1)	Reduce volume of delivery share from 270 ML per delivery share to 150 ML per delivery share. (Delivery Share 2)	
Control	Keep current ratio of 90% fixed costs and 10% variable costs (Control 1)	Move to 100% fixed charges (currently 90% fixed costs:10% variable costs) (Control 2)	Move to 20% variable charges and 80% fixed charges (currently 90% fixed costs:10% variable costs) (Control 3)
Casual use fee	Keep casual use fees as they are. Typically, the fee is $80/ML (Casual Use Fees 1)	Increase casual use fees from $80/ML to $100/ML (Casual Use Fees 2)	
Termination fee	Keep the termination fee at 10 times the cost of each delivery share (Termination Fees 1)	Reduce the termination fee from 10 to 8 times the cost of each delivery share (Termination Fees 2)	
Payment instalment options	Keep the existing payment instalment options (i.e. upfront or with 4 instalments over 5 months) (Payment Instalment Options 1)	An increase in the number of payment instalment options that are available to you e.g. the option to pay once a month over 12 months (Payment Instalment Options 2)	
Breakdown of charges	Keep the existing breakdown of charges that appear on your bill (Breakdown of Charges 1)	Reduce the number of charges to simplify the bill (Breakdown of Charges 2)	

17.3.3 *Delivery share*

The infrastructure access fee is a fixed charge for the right to access the irrigation channel. Currently, it is charged per ML/day of delivery share held, irrespective of the volume of water owned or used. The delivery share charges are different for each of the six irrigation districts in the Goulburn–Murray Irrigation District (GMID). These fees are set with reference to the level of delivery share held (in ML/day) and vary between

$2700 and $4700 per delivery share. This variance reflects the historic decision taken regarding the levels of investment and density of service infrastructure in each district. Currently, the delivery share gives irrigators the right to access and use the capacity of the system. Each delivery share gives the right to 1 ML/day of delivery capacity. Holding more delivery shares provides priority access when the capacity of the channel is constrained. Currently, 270 ML is set as the upper limit that may be delivered for each ML/day of delivery share held (i.e. 1 ML/day over 270 days). Discussions regarding the Draft Tariff Strategy with the customers of Goulburn–Murray Water revealed this to be an important and controversial aspect of the tariff structure. The preferences of participants were mixed, with some arguing that the upper limit of the delivery share should be reduced to increase the scarcity of shares. Alternatively, others voiced concern about changing the rules associated with delivery shares as significant business decisions had already been made by customers. Delivery share charges were included as an attribute in the experiment with two levels, including the status quo. The second level offered the option to reduce the volume of delivery share from 270 ML to 150 ML per delivery share.

17.3.4   Control

The ratio of fixed and variable charges remained an important element of the tariff strategy. As mentioned previously, customers expressed various preferences for the ratio; some favoured a higher variable charge than the status quo and others a lower variable charge than the status quo. This attribute was included and expressed in three levels, including the status quo. The other two levels comprised an increase in fixed costs to 100% and a decrease in fixed costs to 80%.

17.3.5   Casual use fee

Casual use fees are levied when water usage exceeds a customer's entitlement. Currently, casual use fees are set to represent 150% of total

charges that would normally be paid. Typically, the fee is $80/ML representing about double the total delivery charges that would be paid at normal rates. Discussion with customers regarding the Draft Tariff Strategy also highlighted this as an important part of a reformed tariff structure. Thus, casual use fee was included as an attribute and expressed with two levels: the first is the status quo and the second represents an increase in the current casual use fee to $100/ML.

17.3.6   Termination fee

Currently, the termination fee is exclusively attached to the infrastructure access charge and is set at 10 times the cost of each delivery share. The focus groups that were conducted with the Water Service Committees stressed the role of the termination fee and its potential to lock irrigators into ongoing channel access, even though a more appropriate adaptation strategy might include dryland farming. This was therefore included as an attribute to identify preferences for reducing the termination fee. Notably, if the termination fee is reduced this will increase the cost for the remaining customers who will likely also be those that most value irrigation. (There are some equity issues that attend changes on this front, including whether those exiting an industry should pay to secure the future of others. The pecuniary nature of externalities when farmers exit irrigation has been a major driver of the government programs that led to the gifting of assets.) Two levels were used to express this attribute, including the status quo of the termination fee and a level for reducing the termination fee from 10 to 8 times the cost of each delivery share.

17.3.7   Payment instalment options

Respondents in focus groups identified a need to increase the variety of payment options that circumscribes their water bills. Currently, respondents receive one bill per year that covers fixed charges. This represents about 90% of water charges for most irrigators. Water users are given the option to pay this upfront or in four instalments over five months. In addition to the status quo, an option to increase the number of payment options to monthly instalments over 12 months was also included.

17.3.8 *Breakdown of charges*

Currently, there are up to 400 different potential charges in the GMID, arguably increasing the complexity of the bill. Although no consensus was reached among respondents with regards to the preferred breakdown of charges, it proved to be a contentious issue and was therefore included as an attribute. In this case, the status quo was expressed as 'Keep the existing breakdown of charges that appear on your bill' and the second level was described as 'Reduce the number of charges to simplify the bill'.

The survey was administered by mail to around 2000 customers of Goulburn–Murray Water, and the findings are briefly summarised in the following section.

17.4 What farmers favour in tariff reform

The response rate to the survey was around 14%. The final dataset consisted of around 200 respondents generating 6368 choices on which to formulate empirical models. The details of the empirical models are not reported here, but can be found in Cooper and Crase (2013).

The least-preferred tariff structure options identified by respondents are listed below, in order from those perceived as most egregious to those less egregious:
1. Increase the service point charge from $259 to $1000/service point per year, and lower the infrastructure access fee by reducing the cost of each delivery share by $600.
2. Keep the termination fee at 10 times the cost of each delivery share.
3. Move to 100% fixed charges (currently 90% fixed costs and 10% variable costs).
4. Move to a single charge for delivery share (ML/day) across all irrigation districts.

5. Leave service point charge at a low fee of $250/service point per year, and a high infrastructure access fee.

The preferred tariff structure options selected by respondents in order of most preferred are:
1. Reduce the termination fee from 10 to 8 times the cost of each delivery share.
2. Reduce the number of charges to simplify the bill.
3. Increase the number of payment instalment options that are available e.g. the option to pay once a month over 12 months.

The results for the best–worst scaling experiment highlighted several changes that would be strongly supported by farmers. First, a simplification of the tariff so that it communicated adequate information about water use, service and the underlying relationship to cost and water availability would be well regarded. While this was supported on theoretical grounds in the earlier work by Cummins and Associates (2008), the empirical data collected as part of this project arguably offers a more compelling argument for change. On the one hand, regulators have been keen to advocate that tariffs need to explicitly detail the basis of various charges. On the other hand, this detail comes at a cost to those who must interpret complex water bills. The data collected by this project suggests that adjustments in favour of a simplified and clear tariff are long overdue.

Second, there was strong support from the modelled data for an option to use multiple payment periods. This goes beyond the findings of Cummins and Associates (2008) and emphasises the importance of understanding the synergies between revenue management on the part of the water supply corporation and cash flow of farm businesses that access water. The support for this option among farmers may also highlight the extent to which the water suppliers' approach to revenue collection has not kept pace with revenue collection processes adopted by other suppliers of farm inputs.

Third, farmers were keen to see a reduction in the termination fees faced by irrigators. These findings have implications beyond the water supply corporation involved in this project.

Governments that are involved in gifting assets have been keen to promote this policy approach as a panacea for communal irrigation. Moreover, both state and federal governments have expressed a preference to gift assets and impose limits on the repurchase of water rights by government. The support of farmers for refurbished and gifted infrastructure has often been presumed in the public advocacy of this approach. However, farmers are clearly reluctant to be locked into infrastructure choices of this form, as evidenced by their desire to see a reduction in termination fees. This raises serious questions about the efficacy of a policy approach that claims to be supporting irrigation communities in the long run when there is evidence that it is counter to the motivations of individual irrigators to adapt and change.

17.5   Policy considerations and concluding remarks

The motivation for this work was the view that the wedge between irrigation water tariffs and the costs of providing water was increasing and this was likely to cause distortions in the way farmers adapt to changes in water availability. A further motivation was the fact that practical tariff reform was likely to be easier to achieve if changes were supported by farmers in general, or at least a sizable portion of the farming community.

The results for the best–worst scaling experiment showed several changes that would be strongly supported by farmers. First, a simplification of the tariff so that it communicated adequate information about water use, service and the underlying relationship to cost and water availability would be well regarded. Second, there was strong support for an option to use multiple payment periods. Third, farmers were keen to see a reduction in the termination fees faced by irrigators.

These findings are particularly important in the context of adaptation. In the case of the communicability attribute, it is self-evident that a charge that is understandable will help promote adjustment, provided that there is a relationship between resource availability and cost. Given the

oversight of the Essential Services Commission we have some grounds for feeling optimistic that this latter condition will be met.

The capacity to have multiple payments is also important from an adaptation perspective, since this allows farming enterprises to better manage cash flow during periods of change.

In the case of the support for reduced termination charges, there is a clear link between changes on this front and the scope for irrigation enterprises to scale down or exit industries completely should water become increasingly scarce or supplies less reliable over the longer term.

Importantly, these findings have been influential and impacted directly on the water supply corporation. In February 2013 the water corporation announced plans to overhaul its tariff structure, including actions to simplify pricing and facilitate staged payments. The arrangements for termination charges are also being reviewed.

The project also sheds light on the potential of sending mixed messages about adaptation when water buy-back is undertaken simultaneously with infrastructure upgrades. This adds weight to the call for adjusting tariffs to at least provide some incentives for adaptation behaviour and the requirement to limit the perception of being penalised for opting to reduce the extent of irrigation or leave the industry. While we acknowledge that tariff reform and clearer price signals are only one component of improving the adaptation by irrigation farmers, we recommend that any future gifting of irrigation infrastructure should be accompanied by a requirement on the part of irrigation companies to simultaneously reform tariffs. Material reductions in termination charges should be part of these reforms to ensure that mixed messages do not unduly slow farmer adaptation.

References

ABS (2010) *Agriculture and Industry Statistics*, Australian Bureau of Statistics, cat. No. 4610.0.55.008 released 29.11.2010.

Cooper, B. and Crase, L. (2013) *Leading Gifted Horses to Water: The Economics of Climate Adaptation in Government-Sponsored Irrigation in Victoria.*

National Climate Change Adaptation Research Facility, Gold Coast.

Crase, L. and Merton, E. (2013) Correcting misconceptions about links between water planning and food security in the Murray-Darling Basin. *Economic Papers* 32, 298–307.

Crase, L., O'Keefe, S. and Kinoshita, Y. (2012) Enhancing agri-environmental outcomes: Market-based approaches to water in Australia's Murray–Darling Basin. *Water Resources Research* 48, W09536, doi:10.1029/2012WR012140.

Cummins, T. and Associates (2008) *Pricing Signals for the Food Bowl Modernisation Project: A Report for the Department of Sustainability and Environment.* Frontier Economics Pty Ltd., Melbourne, Australia.

Department of Sustainability and Environment (2008) *Climate Change in the North East Region.* Victorian Government, Melbourne. Available at http://www.climatechange.vic.gov.au/__data/assets/pdf_file/0020/73118/NorthEast_WEB.pdf (accessed 28 May 2014).

Horn, M. (1995) *The Political Economy of Public Administration: Institutional Choice in the Public Sector.* Cambridge University Press, Cambridge.

Johansson, R., Tsur, Y., Roe, T., Doukkali, R. and Dinar, A. (2002) Pricing irrigation water: a review of theory and practice. *Water Policy* 4, 173–199.

Pawsey, N. and Crase, L. (2013) The mystique of water pricing and accounting. *Economic Papers* 32, 328–339.

VAGO (2011) *Victorian Auditor-General's Report: Water Entities: Results of the 2010–11 Audits.* Victorian Government, Melbourne. Available at http://www.audit.vic.gov.au/publications/20111109-Water-Entities/20111109-Water-Entities.pdf (accessed 28 May 2014).

Watson, A. (2007) A national plan for water security: Pluses and minuses. *Farm Policy Journal* 4, 1–10.

Wichelns, D. (2010) *Agricultural Water Pricing: United States.* OECD, Paris. Available at http://www.oecd.org/unitedstates/45016437.pdf (accessed 28 May 2014).

18 The role of water markets in helping irrigators adapt to water scarcity in the Murray–Darling Basin, Australia

SARAH ANN WHEELER, ADAM LOCH
AND JANE EDWARDS

CRMA School of Commerce, University of South Australia, Australia

18.1 Introduction

Australia has the largest water market in the world in the Murray–Darling Basin region, and serves as a leading example to other countries examining demand instruments to manage water scarcity. Over the last decade, different programs and incentives have reallocated water from consumptive (i.e. irrigation) users to the environment in Australia's southern Murray–Darling Basin (MDB), with a particular reliance on water markets as a reallocation instrument. The MDB water reallocation is globally unprecedented and, as the potential for environmental water trade increases, other nations are watching these reforms closely. Irrigator interest in utilising water markets to manage risks associated with water shortages and enhance farm viability is also growing. As such, water markets are likely to feature prominently among the strategies water users use to manage and adapt to the impacts of climate change. This chapter provides some details of water markets in the MDB and also canvasses some economic, social and environmental impacts of water reallocation through markets.

It concludes with some commentary on how to improve the effectiveness of water markets as an adaptation tool for irrigators.

18.2 The Murray–Darling Basin

Australia's Murray–Darling Basin (Fig. 18.1) in the country's southeast is an iconic area. It is crucial to agricultural production and the national economy (accounting for over one-third of Australia's food supply), its national ecological significance includes Ramsar-listed wetlands and the basin offers considerable amenity value due to recreation, tourism and cultural significance (MDBA 2012).

Because of its marked susceptibility to climate change, Australia's agricultural sector is expected to suffer more than any other. Less rainfall in the MDB during 1997–98 with corresponding inflow reductions saw storage levels decrease, and the advent of what is now known as the 'Millennium drought' (which ended in 2009–10). During this drought, water allocations to irrigators (allocations are a seasonal percentage of irrigators' long-term water ownership) declined sharply.

Applied Studies in Climate Adaptation, First Edition. Edited by Jean P. Palutikof, Sarah L. Boulter, Jon Barnett and David Rissik.
© 2015 John Wiley & Sons, Ltd. Published 2015 by John Wiley & Sons, Ltd.

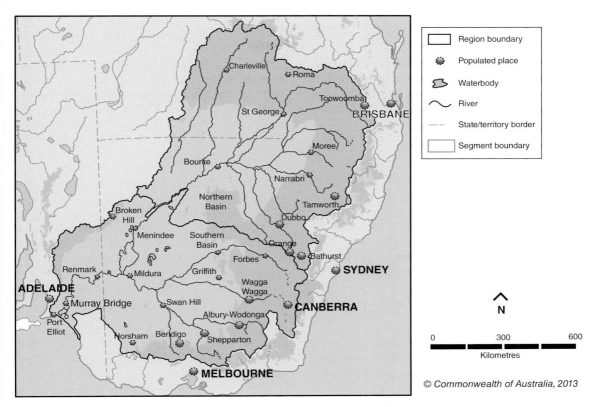

Figure 18.1 Contextual map of the Murray–Darling Basin. Source: Bureau of Meteorology, http://www.bom.gov. au/water/nwa/2010/mdb/physical.html.

Irrigators who had historically been used to receiving 100% of the water they owned suddenly found that their allocations were now sharply decreased in some drought years.

As a consequence of reduced allocations, the Millennium drought resulted in a 70% reduction in irrigation water use. But, the gross value of irrigated agricultural produce (sourced primarily from the basin) only fell by 14% (Prosser 2011) and this lower-than-expected drop in agricultural produce was the result of: (1) water trading between users; (2) greater utilisation of water entitlements; and (3) considerable farm adaptation. This illustrates the role markets can play in promoting adaptation and mitigating productivity losses. It also demonstrates that future water supply and

its efficient use will be increasingly important (Gunasekera et al. 2007).

Population growth and climate change projections predict increased risk to MDB water security. More frequent droughts are expected to occur in southern and south-eastern Australia (including the southern MDB). Further, droughts will be exacerbated by increased temperatures. Total surface water availability in the MDB is predicted to decline by 11% toward 2030 in a median scenario. These predictions have significant consequences for supply reliability of all water entitlements. Finally, an increase in extensive and prolonged flooding is expected, causing infrastructure damage and production/ environmental losses (CSIRO 2008).

The significant threat to environmental assets in the basin led to the Commonwealth developing the *Water for the Future* program, an AUD$12.9 billion program to address water scarcity issues. The Commonwealth also enacted the *Water Act* in 2007, which lead to the creation of the Murray–Darling Basin Authority and its objective to develop an integrated MDB Plan to balance and safeguard environmental, productive and social 'functions' of the system (MDBA 2012). For more detail on water reform in the MDB, see Loch et al. (2013).

18.3   Water user adaptation

Changes to climate conditions will require water users to adapt. Adaptation in response to perceived or actual climate change will entail either incremental or transformational responses. Transformational adaptation can be characterised as profound change, occurring when existing conditions cannot be maintained (Marshall et al. 2012; Park et al. 2012). It may entail exiting agriculture, relocating to a different area, selling all or part of a water entitlement and/or shifting to dry-land farming practices. On the other hand, incremental adaptation involves smaller, piecemeal change. Major decisions or access to significantly new information are generally not required (Park et al. 2012). Incremental options include adopting water-use efficiency improvements, employing deficit irrigation, trading water, using carry-over, investing in more water or land and/ or diversifying income from new commodity or off-farm sources (Loch et al. 2013).

One of the biggest instruments of adaptation by water users in Australia has been the use of water markets, described in the following section.

18.4   Water markets

Water trade has become a key risk-management tool for managing scarcity, particularly during severe droughts. Water markets are also being adopted globally (for example, in the United States, China, India, Spain, Brazil), and useful references for world issues include Grafton et al. (2011) and Maestu (2013). The adoption of water markets has been one of the most important adaptations by Australian irrigators in recent decades. We focus in this chapter only on the MDB in Australia, which has one of the world's largest and most sophisticated water markets. There are two broad water markets in Australia as follows.

1. Entitlement markets provide exclusive access to a water resource share within a specific area (predominantly in perpetuity); this is otherwise known as 'permanent trade'. Entitlements specify volumetric water allocations that can be extracted seasonally and put to beneficial use by entitlement holders. Allocations are announced at the beginning of each season as a percentage of the total entitlement, depending on seasonal conditions (NWC 2011a).

2. Water allocation markets (otherwise known as 'temporary trading') are water traded for seasonal use only. Diverse water allocation systems define pools of water available for consumptive use, and allocate water to entitlement owners.

The MDB comprised 94% of the volume of entitlements and allocations traded across Australia in 2011–12. Most of this trade (83% of volume traded) occurred in the southern MDB, which is hydrologically linked (NWC 2013).

Seasonal allocations are announced and updated based on stored water, expected inflows and other factors (NWC 2011a). However, water entitlements exist in both regulated systems (flows controlled through infrastructure to store and release water) and unregulated systems (opportunistic river or flood-based flows that may be harvested/extracted under approved circumstances). Regulated water entitlements have different reliability levels in the MDB (classified as high, general and low). Unregulated systems have no formal reliability and are usually determined by restrictions on extraction (Shi 2006).

Adoption and use of water markets in Australia has increased rapidly over the past three decades, although there is a long history. There are reports of irrigators first swapping (informal trading) water in the 1940s. Continual water scarcity, increased demand from irrigators and government water policy reform eventually led to the development of formal water markets in Australia. Temporary transfers of water were permitted in New South Wales during the droughts of 1966–67 and 1972–73, in Victoria during 1966–67 and in a restricted version during 1982–83 before its more general introduction in 1986–87 (NWC 2011b).

Irrigators adopted water allocation trade much faster than water entitlement trade. Consequently, water allocation trade grew rapidly while entitlement trading increased slowly until large-scale Commonwealth entry into the market during 2007–08 (as part of the *Water for the Future* program outlined previously). It is important to note that, although water markets were created by governments, they do not have to be used by irrigators; irrigators choose voluntarily whether they wish to buy or sell water. Indeed, many irrigators have *never* participated in the water market; by 2011 however, around two-thirds of all irrigators in the southern MDB had traded water (either entitlement or allocation) at least once (Wheeler et al. 2013).

18.4.1 Impacts of water markets

Although water scarcity resulted in the market reallocation of water between different users and uses (including urban uses) in a relatively effective and efficient manner, water markets have always been rather controversial. Since water trading was first introduced in Australia in the 1980s, there has generally been more concern about: entitlement trading than allocation trading; trading out of districts than trading within districts; and what has been termed speculative trading than trading between irrigators. However, surveys show that over time it is clear irrigators have become much more accepting of the benefits of water markets (Bjornlund et al. 2011).

Economists also suggest that considerable benefits have resulted from water markets in Australia (e.g. Qureshi et al. 2009, Wheeler et al. 2009, 2010, 2012; Wittwer 2011). Improved

understanding of trade behaviour, especially strategic behaviour which can lead to market failures, has further improved economic benefits; there is considerable work to be done in this area, however. Water trading has augmented the value of water entitlements and irrigator wealth (Young 2011), enabled rapid expansion in irrigated areas and helped irrigators avoid the worst impacts of the Millennium drought. However, it is also important to note that economists do not suggest that a market-led approach by itself will suffice; for success, a variety of reallocation instruments, government intervention and regulation are required. By way of example, Connell et al. (2005) suggest that the National Water Initiative (introduced in Australia in 2004) provided a misleading view that markets could self-regulate. Further, Bell and Quiggin (2008) termed the need for 'meta-governance' in water market oversight.

The overall benefits of water trading during scarcity have been calculated. Between 2006–07 and 2010–11 computable general equilibrium (CGE) modelling estimated that southern MDB production was AUD$845 million higher than it would have been without interregional water trading (NWC 2012). Further, from 2006–07 to 2010–11 southern MDB production was estimated to be AUD$4.3 billion higher than it would have been without on-farm reallocation of water between irrigation activities (NWC 2012). Because climate change is likely to exacerbate variability in water supply including periods of scarcity, it is likely that water markets will become an increasingly important adaptation tool. While the entitlement and allocation markets have different functions and outcomes, both have significant benefits for irrigators. For example, they promote flexibility as production can be changed relatively quickly. They also provide irrigators with clear opportunity–cost information for water and input costs relative to returns, thereby aiding productivity. They motivate producers to use water more efficiently and effectively. Trade can also aid farmers to re-organise farm finances, facilitate investment in on- and off-farm assets, reduce debt or exit agriculture (NWC 2012). Participation in the allocation market therefore promotes incremental adaptation by individual producers. In contrast, the entitlement market will facilitate transformational adaptation by helping irrigators make longer term decisions about their enterprises, that is, whether to restructure or exit (Wheeler et al. 2010).

18.4.2 Community impacts

Despite production advantages, there is community concern about negative social impacts arising from water trade. However, the available evidence does not generally support these fears. Community concerns over water transfers away from particular regions include: reduced community spending and reinvestment; population and employment losses; declining taxation bases; and loss of services and businesses. It is also feared that water transfers will drive regional production changes (e.g. conversion to dry-land farming) and legacy issues for remaining farmers (e.g. higher variable operating costs, stranded irrigation assets and/or pressure to rationalise marginal farms) (Australian Parliament 2011; EBC et al. 2011). Many of these issues are linked to ongoing structural change in agriculture as well as competition policy (Productivity Commission 2005). It is therefore difficult to disaggregate structural adjustment effects from water market impacts to establish causal relationships between trade and social change (Bjornlund et al. 2011, 2013; Wheeler et al. 2013). Some evidence even suggests socio-economic trends in rural communities were unchanged by water trading (Watson et al. 2007; NWC 2010a), and that there is variability in the capacity of rural communities to adapt to changing circumstances. It appears that water trading is only weakly linked to this adaptive capacity (NWC 2012).

18.4.3 Environmental impacts

There is ongoing debate over the environmental impact of water markets. Generally, trade has had beneficial impacts on river flows, lowered salinity via the downstream movement of water towards areas of greater scarcity and created a larger share of environmental water (Grafton 2011). Modelling suggests that the hydrologic and environmental impacts between 1998–99 and 2010–11 were small and mainly positive, as downstream water movement reduced summer flow stresses while leaving winter flow patterns unchanged (NWC 2010b, 2012). Connor et al. (2013) also suggest that allowing environmental water holder trade will have environmental benefits. However, large-scale activation of (previously under- or unused) water entitlement volumes after a cap was introduced in 1995 (i.e. restrictions on the licensing of further water diversions, thus freezing them at the 1993–94 extraction level) had significant negative environmental effects when these under-unused entitlements were traded between users. Previously, some of these under-utilised water entitlements provided beneficial water stocks in river systems, with attendant environmental advantages (Young 2013). Consequently, Young (2013) suggests governments should decide water allocation volumes, while markets then direct it to individual users.

Other common fears about negative environmental impacts from water trade include: (1) concentrating water use in areas with high water tables; (2) moving water to locations where use negatively affects water quality due to, for example, higher groundwater salinity levels; (3) moving water use upstream, resulting in reduced river flow between new and old extraction points; and (4) activating previously unused water, leaving less supply to support ecosystems (NWC 2012; Adamson and Loch 2013). Tisdell (2001) studied the potential impacts of water markets in the Border Rivers region of Queensland. He concluded that water entitlement trade might increase the differential between extractive demand and historical flow regimes, and water markets could limit water policy effectiveness in restoring natural flow regimes. In response to the concerns associated with water markets, state governments have instituted controls to limit potential harm. These include caps on permanent transfers out of regions and imposing factors such as exchange rates, tagged trade, salinity offsets, etc. (see Etchells et al. (2004) for more details).

18.4.4 *Future water market improvements*

Arguably, climate change and water management are two of the most important policy challenges facing Australia. For the MDB they are critical issues. Current water policy is well positioned to aid climate change impact management, especially for irrigators. Reforms to water pricing, entitlements and the development of markets also provide best-practice examples of adaptation.

Adaptation is influenced by farmers' willingness to adopt new strategies; these may be incremental (relatively common farming decisions) or transformative changes (representing a major change in livelihood, location or identity and much less commonly made). Successful policy should address both adaptation forms, and primarily focus on farmers' adaptive requirements to help them adjust to future water scarcity and sustainable management of resources. However, change cannot be expected from all farmers. Policy must also recognise that in light of environmental conditions, irrigation infrastructure costs and soil productivity, some irrigation districts in the MDB might not be viable in future (Productivity Commission 2010). Policy will therefore need to facilitate transformative and structural change. Small-block irrigated exit packages and the targeted buy-back of farms on inefficient delivery channel sections, no longer capable of being serviced or upgraded, are examples of relevant policies (Wheeler et al. 2012).

Policies and investment strategies related to climate change, energy and water are intertwined, with mitigation and adaptation in their sights. However, while adaptation and mitigation are usually portrayed as mutual, they can conflict. For example, adoption of efficient irrigation technologies to save water can increase energy use and carbon emissions. By contrast, carbon capture and storage is a mitigation measure that can thwart adaptation by increasing the use of, and competition for, water. Climate policy goals can also be traded off against other policy objectives, with positive or negative consequences. For example, promoting regional economic growth can lead to increased emissions. Mitigation and

adaptation strategies must rest on the same legal, financial, institutional and political footing to prevent serious imbalances between them (Moser 2012). To forestall conflict and incompatibility between policies targeting adaptation and mitigation, Loch et al. (2013) identified three areas of necessary water market reform as follows.

1. *Institutional reforms:* removing trade restrictions to promote more efficient transfers, which facilitate fluid farm adjustment and better groundwater regulation to avoid over-allocation. This includes expanding water trade products and markets (and cross-sector interaction) to improve approval procedures, while providing greater transparency in conflict of interest cases.

2. *Informational measures:* provide more robust and detailed market price signals. These increase use and prediction of future climatic information and improve seasonal water allocation announcements. Finally, they promote research into farm adaptive responses (capacity) across regions and industry sectors.

3. *Political reform:* regulate water brokers, raise market confidence and avoid catastrophic events (e.g. massive confidence loss). Consistent carry-over rules across states and districts and investigating the opportunity cost of infrastructure investments or alternative recovery programs would also be beneficial. Finally, farmer and rural community education to address climate change beliefs and attitudes that may improve adjustment and adaptation responses.

A better understanding of water trade behaviour, especially strategic trade issues that can lead to market failures, will assist in improving future water reallocation advantages. Many areas of further research in water markets are still required. For example, as Australian water markets are not fully mature, risk management potential is yet to be completely explored. Water markets in Australia have created wide-ranging economic, social and environmental impacts (some positive, some negative). There are also lessons to be learned about participation in other resource markets – for example, salinity trading, carbon farming or energy markets – to understand how benefits might translate to water markets (or vice versa). We suggest water markets have already been of net benefit for Australian irrigators. Future water trading is likely to be crucial in promoting new climate change adaptation.

18.5   Conclusions

This chapter has highlighted the many lessons to be learned – nationally and internationally – from southern MDB experiences during the Millennium drought. Access to water markets and political investments aided consumptive, social and environmental water use that avoided catastrophic outcomes. Climate change and water scarcity management are linked. Water users have little choice but to adapt. The cost of that adaptation at individual, regional and national levels – particularly to future water supply variability – can be mitigated by considering the existing advantages and future opportunities from water markets in Australia. Flexibility and adaptability are required to drive sustainable policy change for the future of the MDB. Water markets must therefore evolve to promote adaptive responses, and allow participants to play active roles in water reallocation. However, an individual's ability to adapt is governed by institutional, information and policy frameworks. The ability to respond sustainably to future challenges therefore needs a concerted effort across all jurisdictions. As such, links are required in policy, institutional and governance arrangements to deal with these issues.

Finally, the international significance of Australian water market operation needs to be understood in light of global developments. Climate change is already yielding less water at a time of increasing population demand for water and food supply and increasing industrial and urban demand. This places a premium on effectively and efficiently using water among stakeholders. Water markets are increasingly being used around the world to manage the challenge of a growing demand and a decreasing supply of water. However, the creation of effective, efficient and socially equitable markets is a complex process, implicating a range of factors. Optimal market operation requires adequate institutional

foundations such as legal clarity, supporting administrative structures, recognition of public interest, redressing market failures and responding to priority of use. Markets must be similarly structured to promote economic efficiency and environmental sustainability. If water markets are to deliver the hoped-for objectives, it is essential to assess the structures and fundamentals that provide institutional support and promote economic and environmental sustainability. Recent attempts to provide integrated and coordinated MDB water management may also provide a guide to market development that promotes environmental sustainability. In any case, Australian water markets are a useful guide for other jurisdictions to establish and expand institutions to optimally manage current and future challenges in balancing water supply and demand.

References

Adamson, D. and Loch, A. (2013) Possible negative feedbacks from 'gold-plating' irrigation infrastructure. *Agricultural Water Management*, published online October 2013, doi: 10.1016/j.agwat.2013.09.022.

Australian Parliament (2011) *Of Drought and Flooding Rains: Inquiry into the Impact of the Guide to the Murray–Darling Basin Plan*. Standing Committee on Regional Australia, Canberra.

Bell, S. and Quiggin, J. (2008) The metagovernance of markets: The politics of water management in Australia. *Environmental Politics* 17, 712–729.

Bjornlund, H., Wheeler, S. and Cheesman, J. (2011) Irrigators, water trading, the environment, and debt: Perspectives and realities of buying water entitlements for the environment. In: Grafton, Q. and Connell, D. (eds) *Basin Futures: Water reform in the Murray–Darling Basin*. Australian National University Press, Canberra, pp. 291–302.

Bjornlund, H., Wheeler, S. and Rossini, P. (2013) Water markets and their environmental, social and economic impact in Australia. In: Maestu, J. (ed.) *Water Trading and Global Water Scarcity: International Perspectives*. RFF Press/Taylor and Francis/Routledge, UK, pp. 68–93.

Connell D., Dovers, S. and Grafton, Q. (2005) A critical analysis of the National Water Initiative. *Australasian Journal of Natural Resources Law and Policy* 10, 81–107.

Connor, J., Franklin, B., Loch, A., Kirby, M. and Wheeler, S. (2013) Trading water to improve environmental flow outcomes. *Water Resources Research* 49, 4265–4276.

CSIRO (2008) *Water Availability in the Murrumbidgee: A Report to the Australian Government from the CSIRO Murray–Darling Basin Sustainable Yields Project.* CSIRO, Australia.

EBC, RMCG, MJA et al. (2011) *Community Impacts of the Guide to the Proposed Murray–Darling Basin Plan.* Murray–Darling Basin Authority, Canberra.

Etchells, T., Malano, H., McMahon, T. and James, B. (2004) Calculating exchange rates for water trading in the Murray-Darling Basin, Australia. *Water Resources Research* 40(12), W12505.

Grafton, Q. (2011) Economic costs and benefits of the proposed Basin Plan. In: Connell D, and Grafton, Q. (eds) *Basin Futures: Water Reform in the Murray–Darling Basin.* Australian National University E-press, Canberra.

Grafton, R.Q., Libecap, G., McGlennon, S., Landry, C. and O'Brien, B. (2011) An integrated assessment of water markets: A cross-country comparison. *Review of Environmental Economis and Policy* 5(2), 219–239.

Gunasekera, D., Kim, Y., Tulloh, C. and Ford, M. (2007) Climate change: impacts on Australian agriculture. *Agricultural Commodities* 14(4), 657–676.

Loch, A., Wheeler, S., Bjornlund, H. et al. (2013) *The Role of Water Markets in Climate Change Adaptation.* National Climate Change Adaptation Research Facility, Gold Coast.

Maestu, J. (2013) *Water Trading and Global Water Scarcity: International Experiences.* RFF Press, USA.

Marshall, N., Park, S., Adger, W. et al. (2012) Transformational capacity and the influence of place and identity. *Environmental Research Letters* 7, doi: 10.1088/1748-9326/7/3/034022.

Moser, S. (2012) Adaptation, mitigation and their disharmonious discontents: an essay. *Climate Change* 111, 165–175.

Murray–Darling Basin Authority (2012) *Proposed Basin Plan: A Revised Draft.* Murray–Darling Basin Authority, Canberra.

National Water Commission (2010a) *The Impacts of Water Trading in the Southern Murray–Darling Basin: An Economic, Social and Environmental Assessment.* National Water Commission, Canberra.

National Water Commission (2010b) *Australian Water Markets Report 2009–10.* National Water Commission, Canberra.

National Water Commission (2011a) *Australian Water Markets: Trends And Drivers, 2007–08 to 2009–10.* National Water Commission, Canberra.

National Water Commission (2011b). *Water Markets in Australia: A Short History.* National Water Commission, Canberra.

National Water Commission (2012) *Impacts of Water Trading in the Southern Murray–Darling Basin Between 2006–07 and 2010–11.* National Water Commission, Canberra.

National Water Commission (2013) *Australian Water Markets: Trends and Drivers 2007–08 to 2011–12.* National Water Commission, Canberra.

Park, S.E., Marshall, N.A., Jakku, E. et al. (2012) Informing adaptation responses to climate change through theories of transformation. *Global Environmental Change* 22, 115–126.

Productivity Commission (2005) *Review of National Competition Policy Reforms.* Productivity Commission, Canberra.

Productivity Commission (2010) *Market Mechanisms for Recovering Water in the Murray-Darling Basin, Final Report.* Productivity Commission, Canberra.

Prosser, I. (2011) *Water: Science and Solutions for Australia.* CSIRO Publishing, Victoria.

Qureshi, M.E., Shi, T., Qureshi, S.E. and Proctor, W. (2009) Removing barriers to facilitate efficient water markets in the Murray-Darling Basin of Australia. *Agricultural Water Management* 96(11), 1641–1651.

Shi, T. (2006) Simplifying complexity: Rationalising water entitlements in the Southern Connected River Murray System, Australia. *Agricultural Water Management* 86(3), 229–239.

Tisdell, J. (2001) The environmental impact of water markets: An Australian case-study. *Journal of Environmental Management* 62, 113–120.

Watson, A., Barclay, E. and Reeve, I. (2007) *The Economic and Social Impacts of Water Trading. Case Studies in the Victorian Murray Valley.* Rural Industries Research and Development Corporation, Murray-Darling Basin Commission and National Water Commission, Canberra.

Wheeler, S., Bjornlund, H., Shanahan, M. and Zou, A. (2009) Who trades water allocations? Evidence of the characteristics of early adopters in the Goulburn–Murray Irrigation District, Australia 1998–99. *Agricultural Economics* 40(6), 631–643.

Wheeler, S., Bjornlund, H., Zuo, A. and Shanahan, M. (2010) The changing profile of water traders in the Goulburn-Murray Irrigation District, Australia. *Agricultural Water Management* 97(4), 1333–1343.

Wheeler, S., Zuo, A., Bjornlund, H. and Lane-Miller, C. (2012) Selling the farm silver? Understanding water sales to the Australian government. *Environmental Resource Economics* 52(1), 133–154.

Wheeler S., Loch A., Zuo A. and Bjornlund H. (2013) Reviewing the adoption and impact of water markets in the Murray-Darling Basin, Australia. *Journal of Hydrology*, published online October 2013, doi: 10.1016/j.jhydrol.2013.09.019.

Wittwer, G. (2011) Confusing policy and catastrophe: Buybacks and drought in the Murray–Darling Basin. *Economic Papers* 30(3), 289–295.

Young, M. (2011) Water markets: a downstream perspective, In: Langford, J. and Briscoe, J. (eds) *The Australian Water Project Volume One, Crisis and Opportunity: Lessons of Australian Water Reform.* Committee for Economic Development of Australia, Melbourne, pp. 64–69.

Young, M. (2013) Trading into and out of trouble: Australia's water allocation and trading experience. In: Maestu, J. (ed.) *Water Trading and Global Water Scarcity: International Perspectives.* RFF Press/ Taylor and Francis/Routledge, UK, pp. 94–110.

Section 4
Coasts

19 Raising the seas, rising to greatness? Meeting the challenge of coastal climate change

SUSANNE C. MOSER

Susanne Moser Research and Consulting, USA

19.1 At the confluence of land and sea

Today, upwards of 40% of the world's population are living on the 5% of the world's land area that is within 100 km of the coast, exerting tremendous human pressures on coastal and marine ecosystems and resources (McGranahan et al. 2007). While many ocean-bordering countries have majorities of their populations in coastal areas, this is particularly true for Australia with 84% of its population living within 50 km from its nearly 26,000 km of coastline (Chapter 23, by Fletcher et al.). The reason for this high population concentration is that we all want so much from the coast: while some of us build and live there, others come to visit and recreate; we fish and farm, ship, transport and trade; we extract resources (oil, gas, minerals and water); we discharge our waste and station our troops there; and we grow our economies and tend our coastal cultures. Amidst all these benefits we derive from the coast, we also try to protect and save the remains of natural shorelines and ecosystems so that those who come after us may continue to enjoy them.

Yet the challenge to the sustainability of our coasts is profound. As ongoing research over decades has shown, all of these human activities are continuing and accelerating the loss of habitat and species, the declines in water and air quality, the hardening of the shoreline and the threats to human health and security (Agardy et al. 2005; Wong et al. 2005; UNEP 2007). Moreover, with continued influx of people to this narrow strip of land, these pressures are growing at the same time that climate variability and change, sea-level rise and related hazards are making human habitation in these areas less secure (Hinrichsen 2011; Nicholls et al. 2011). In fact, global sea-level rise already poses threats today; over the long term it will cause practically permanent alterations of our coasts. Together with storms, flooding, erosion, rising temperatures, reductions in sea ice, changes in freshwater runoff, increasing variability and climatic extremes and acidification of coastal waters (IPCC 2013), low-lying coastal areas and islands face the prospects of increasingly significant (and largely negative) ecological, social and economic impacts worldwide (Williams and Gutierrez 2009; Nicholls and Cazenave 2010; Moser et al. 2012). Yet in Australia as in other developed and developing nations, the beauty, benefits and wealth of coastal areas as well as the risks from climate change are not distributed equally, leading to significant differences in vulnerability

Applied Studies in Climate Adaptation, First Edition. Edited by Jean P. Palutikof, Sarah L. Boulter, Jon Barnett and David Rissik.
© 2015 John Wiley & Sons, Ltd. Published 2015 by John Wiley & Sons, Ltd.

and in the capacity to manage the growing risks
(e.g. Nicholls et al. 2007; Preston et al. 2008;
Martinich et al. 2013).

19.2   The challenges before us

The challenges before us are staggering indeed,
which is clearly the impression from the contributions to this section. Woodroffe et al. (Chapter 20)
lay the foundation for understanding and predicting
the physical risks Australians face from sea-level
rise and coastal climatic hazards. Driven by an
overriding goal to be relevant to practice, they
show how climate change considerations can be
integrated into familiar risk assessment approaches
and combined with state-of-the-art engineering
models to create more comprehensive assessments
of the changing risks faced at the coast.

Hadwen and Capon (Chapter 21) then add to
these the ecological and human complexities to
help us understand the vulnerabilities as well as
the capacities and limits of human and other-
than-human communities to adapt to the coming
changes. Similar to Woodroffe et al., Hadwen and
Capon offer a pragmatic step-by-step approach to
identifying key ecological vulnerabilities. They
then make a convincing case for a more integrated
approach to coastal zone management, one that
deliberately considers spatial, temporal and
sectoral interconnectivities between coastal ecosystems, human land and resource uses in order
to minimise the risk of maladaptation.

Ports and harbours are the special focus of
McEvoy and Mullett (Chapter 22), illustrating
how much of the Australian economy is
dependent not just on functional and well-
adapted coastal trade, transfer and transportation
infrastructure, but also on the protection of port
assets and on a well-prepared workforce that can
meet the twin challenges of near-term disaster
preparedness and long-term adaptation to
climate change. The authors clearly recognise
that climatic and non-climatic drivers influence
the business decisions of port authorities. They
make a strong case for adaptation as an iterative
social learning process in which researchers and

decision-makers engage in continual co-produc-
tion of relevant knowledge.

Fletcher et al. (Chapter 23) and Waters and
Barnett (Chapter 24) then take us into a range of
Australian coastal communities to explore the
economics, equity and institutional arrangements
for adaptation. Fletcher and colleagues set out
from the simple, if for many difficult-
to-accept fact, that coastal impacts of climate
change will affect some harder than others.
Accepting this premise of winners and losers from
climate change as much as from adaptation, the
onus is on governments to find a path to navigate
the tricky social and economic territory of
adaptation. Their study suggests that coastal residents and managers will often prefer familiar,
incremental and less legally risky adaptations over
novel, more disruptive, if maybe more effective
approaches over the long term. The price – quite
literally – for this preference is lower economic
efficiency. In other words, adaptation in the 'real
world' is not a rational, cost-minimising optimisation process, but instead a messier and more
expensive sequence of adjustments. Few economic
assessments to date are guided by this more realistic assumption. Likewise, few systems-thinking
scientists would recommend such a course of
action and few (if any) politicians to date are willing to acknowledge this more costly and likely
more disaster-prone and loss-ridden future.

What governance arrangement is best suited
to navigate these thorny trade-offs? Waters and
Barnett (Chapter 24) try to answer this question
by delineating – on the basis of detailed interviews with coastal managers and stakeholders in
two Australian communities – a preferred multi-
level governance approach to coastal adaptation.
While not suggesting unanimity in opinion, nor
general applicability to all coastal areas, they find
that local stakeholders have clear preferences for
the roles and responsibilities of different levels of
government, households and the private sector.
They see local governments ideally placed to
manage local public assets, regulate decisions
over private assets (which in turn are largely
implemented by individuals and businesses) and
co-ordinate public input into local planning,

while the state level is expected to play largely a co-ordinating role across locales and regions. The federal level in turn should preferably provide relevant information and generate and distribute the necessary financial resources for adaptation.

Of course, these preferences from the local perspective may not spell success for all interested parties when considering the many stakes nations have in their coastal areas: the economic assets and activities there, the infrastructure that connects coastal and inland areas and the ecological jewels and other commons. In highly interconnected national and global economies, the jury is still out on how to maximise the likelihood of successful adaptation for the greatest number of stakeholders at all levels while preserving the commons: the life support systems and biodiversity on which we all depend, and which many consider to have an intrinsic right to exist regardless of human needs and desires (Moore and Nelson 2011; Moser and Boykoff 2013).

19.3   Towards a human response equal to the test

Taken together, none of the findings in this section – whether on the multiple stresses and threats to coastal areas, the differential vulnerabilities and human–ecological complexities and uncertainties, the conflicts between public preferences for adaptation strategies and other ecological or economic goals or the preferred roles and responsibilities of different participants in governance – are entirely unique to Australia. Public views on desirable categories of adaptation actions and on the roles and responsibilities of different levels of government are also not unique to coastal adaptation, as a recent review of the adaptation literature on these questions found (Moser 2014).

This does not make them any less daunting, however. The clash of human development at the coastal fringe with the rising sea can easily create a sense of inevitable loss and paralysis. The future of the Australian coast (or any coast) is not hopeless, however.

Hopelessness as the ultimate failure to adapt is *not* an option when so many people, so much economic activity and so many ecological riches are at risk. It is therefore all the more important that coastal regions of the world learn from and with each other how to prepare for and minimise the negative consequences of climate change, and do so fast. Too much is at stake to reinvent the wheel countless times over, even as each coastal community and each coastal industry have to find locally contextualised and tailored, as well as regionally co-ordinated, answers to the shared challenges ahead.

American political and environmental scientist David Orr once said that it is our obligation as scientists and individuals at this time in the history of humans on Earth to tell the full truth of what we see happening to our planet, and thus to ourselves. But 'truth telling' in his mind does not mean merely listing a litany of (global) environmental crises to create a sense of doom. Instead, for him truth telling is a way of calling forth the best in humans: 'Telling the truth means that the people must be summoned to a level of extraordinary greatness appropriate to an extraordinarily dangerous time' (Orr 2011, p. 330).

In the contributions to this section, we find the increasingly sophisticated capacity of science to detect, explain and project forwards the unfolding impacts of climate change and sea-level rise on coastal areas (Chapters 20 and 21). We might see it as one measure of 'extraordinary greatness' just how mature that understanding has become over the past 30 or more years of climate change research. Maybe even more encouraging are the distinct advances reflected in these chapters on how to integrate not just physical and ecological knowledge, but also the social and economic sciences (Chapters 22–24).

What is more, each of the chapters expresses a willingness to engage with decision-makers, to learn from each other and to co-produce use-inspired and practice-relevant knowledge. This marks a step out of the academic comfort zone and, as such, a step closer to the extraordinary greatness demanded of us as scientists in this increasingly and extraordinarily challenging time. But this alone will not suffice.

Without a doubt, reliable, sophisticated, locally relevant and co-produced science as presented in these pages will help and may even be indispensable. Sharing data, tools and approaches up and down our coasts and across the seas will be extremely valuable. Even more beneficial will be greater co-operation across sectors and integration across scales of governance. The difficult recognition and public articulation of the coastal adaptation challenge lies ahead for most, however. The tough decisions and unpopular trade-offs will undoubtedly have to be made by a few and supported by many. The greatest of these challenges perhaps lies in collectively finding a path beyond our narrow short-sighted interests, a path towards a greater, common self-interest that ensures that our coasts continue to offer the riches that draw and sustain us. In this way, every inch of sea-level rise calls on each of us – in academia, in businesses and in public office – to rise to a human greatness more extraordinary than we have achieved to date.

References

Agardy, T., Alder, J., Dayton, P. et al. (2005) Coastal systems. In: Hassan, R.M., Scholes, R.J., and Ash, N. (eds) *Ecosystems and Human Well-Being: Current State and Trends: Findings of the Condition and Trends Working Group of the Millennium Ecosystem Assessment.* Island Press, Washington, DC, pp. 513–549.

Hinrichsen, D. (2011) *The Atlas of Coasts and Oceans: Ecosystems, Threatened Resources, Marine Conservation.* University of Chicago Press, Chicago, IL.

IPCC (2013) Summary for policymakers. In: Stocker, T.F., Qin, D., Plattner, G-K. et al. (eds) *Climate Change 2013: The Physical Science Basis. Contribution of Working Group I to the Fifth Assessment Report of the Intergovernmental Panel on Climate Change (IPCC).* Cambridge University Press, Cambridge, UK and New York, NY.

Martinich, J., Neumann, J., Ludwig, L. and Jantarasami, L. (2013) Risks of sea level rise to disadvantaged communities in the United States. *Mitigation and Adaptation Strategies for Global Change* 18(2), 169–185.

McGranahan, G., Balk D. and Anderson, B. (2007) The rising tide: Assessing the risks of climate change and human settlements in low elevation coastal zones. *Environment and Urbanization* 19, 17–37.

Moore, K.D. and Nelson, M.P. (2011) *Moral Ground: Ethical Action for a Planet in Peril.* Trinity University Press, San Antonio, Texas.

Moser, S.C. (2014) Communicating climate change adaptation: The art and science of public engagement when climate change comes home. *Wiley Interdisciplinary Reviews–Climate Change,* published online 14 February 2014, doi: 10.1002/wcc.276. 5(3): 337–358.

Moser, S.C. and Boykoff, M.T. (eds) (2013) *Successful Adaptation to Climate Change: Linking Science and Practice in a Rapidly Changing World.* Routledge, London.

Moser, S.C., Williams, S.J. and Boesch, D.F. (2012) Wicked challenges at land's end: Managing coastal vulnerability under climate change. *Annual Review of Environment and Resources* 37, 51–78.

Nicholls, R.J. and Cazenave, A. (2010) Sea-level rise and its impact on coastal zones. *Science* 328, 1517–1520.

Nicholls, R.J., Wong, P.P., Burkett, V. et al. (2007). Coastal systems and low-lying areas. In: Parry, M.M., Canziani, O.F., Palutikof, J.P. (eds) *Climate Change 2007: Vulnerability, Impacts and Adaptation, Contribution of Working Group II to the IPCC Fourth Assessment Report.* Cambridge University Press, Cambridge, pp. 315–356.

Nicholls, R.J., Marinova, N., Lowe, J.A. et al. (2011) Sea-level rise and its possible impacts given a 'beyond 4 degrees C world' in the twenty-first century. *Philosophical Transactions of the Royal Society A* 369, 161–181.

Orr, D.W. (2011) Hope (in a hotter time). In: Orr, D.W. (ed.) *Hope is an Imperative: The Essential David Orr.* Island Press, Washington, pp. 324–332.

Preston, B.L., Smith, T. and Brooke, C. (2008) *Mapping Climate Change Vulnerability in the Sydney Coastal Councils Group.* CSIRO Marine and Atmospheric Research, Aspendale.

United Nations Environment Program (2007) *Global Environmental Outlook 4: Environment for Development.* UNEP/Progress Press, Valletta, Malta.

Williams, S.J. and Gutierrez, B.T. (2009) Sea-level rise and coastal change: causes and implications for the future of coasts and low-lying regions. *Shore Beach* 77, 1–9.

Wong, P.P., Marone, E., Lana, P., Fortes, M., Moro, D. and Agard, J. (2005) Island systems. In: Hassan, R.M., Scholes, R.J. and Ash, N. (eds) *Ecosystems and Human Well-Being: Current State and Trends: Findings of the Condition and Trends Working Group of the Millennium Ecosystem Assessment.* Island Press, Washington, pp. 663–680.

20 A framework for modelling the risks of climate-change impacts on Australian coasts

COLIN WOODROFFE[1], DAVE CALLAGHAN[2],
PETER COWELL[3], DAVID WAINWRIGHT[2,4],
KERRYLEE ROGERS[1] AND ROSHANKA RANASINGHE[5,6]

[1]*School of Earth and Environmental Sciences, University of Wollongong, Australia*
[2]*School of Civil Engineering, University of Queensland, Australia*
[3]*School of Geosciences, University of Sydney, Australia*
[4]*Whitehead and Associates, Australia*
[5]*UNESCO-IHE, The Netherlands*
[6]*Research School of Earth Sciences, The Australian National University, Australia*

20.1 Introduction

The coast is one of the most dynamic components of the Australian landscape and is particularly vulnerable to the impacts of anticipated sea-level rise, including erosion during storms and inundation of low-lying areas. Sustainable management and strategic planning need to consider both long-term sedimentary processes that shape each section of coast and short-term event-driven processes that result in erosion and gradual recovery of beaches and foredunes. Unprecedented growth has led to billions of dollars worth of investment in coastal settlements and infrastructure, making it essential to consider the likely rates of recession of the coast.

In this chapter we describe a framework that integrates geomorphological and engineering understanding of coastal behaviour in a way that can be incorporated into coastal management decisions. We describe probabilistic assessment of a range of possible outcomes and their uncertainties, which can provide planners with greater flexibility than more deterministic approaches. The framework has been described in detail in our report on this NCCARF project, where its application was trialled on a data-rich case study site at Narrabeen Beach in northern Sydney and where we also outlined the potential for inclusion of an economic approach for risk assessment and development of suitable set-back guidelines (Woodroffe et al. 2012).

Understanding coasts and how they respond to hazards has often been based on a reactive paradigm after erosional events have occurred. Remedial hard-engineering protection works may be instigated on the basis of limited observations of processes, such as wave climate, to ameliorate future recurrence of such impacts. Although such strategies are built on sound engineering principles, they assume that boundary conditions are invariant and rarely incorporate consideration of geomorphological characteristics and longer-term geological processes. It is no longer realistic to consider sea level as stationary; it is apparent that the sea is rising at all tide-gauge stations around Australia (Church et al. 2006).

Applied Studies in Climate Adaptation, First Edition. Edited by Jean P. Palutikof, Sarah L. Boulter, Jon Barnett and David Rissik.
© 2015 John Wiley & Sons, Ltd. Published 2015 by John Wiley & Sons, Ltd.

Coasts around Australia have evolved over the past several thousand years during which sea level has been relatively stable, but statutory requirements now exist to incorporate sea-level rise into longer-term planning decisions, combining geologically informed trends with the likely extent of erosion by storms.

20.2   Variation in shoreline location

Quantifying impacts of sea-level rise on beaches and dunes is complex. Although it has been recognised for decades that a rise in sea level will translate into landward retreat, procedures to forecast shoreline positions at any particular date in the future are the subject of vigorous debate and provide little guidance to the most judicious setback line landward of which it might be prudent to allow buildings and infrastructure to be constructed.

The framework that we describe is based on policies and guidelines adopted by many agencies in Australia and New Zealand (Gordon et al. 1979; Gibb 1983) and applied routinely by consultants. It involves distinguishing between mean-trend and fluctuating constituents of coastal change, for which the terms *recession* and *erosion* have been adopted, respectively. In the case of a section of coast with a finite volume of sand (a closed coastal compartment) and with no variation of sea level (Fig. 20.1a), successive storms erode the beach and foredune with gradual recovery between events. When a sufficiently long sequence of storms has been observed a critical distance can be identified to which the largest storms cause erosion, which can be defined as a setback line for development of settlements and infrastructure. In many cases the boundary conditions are not static however, and there is a gradual recession (retreat) of this critical distance. This occurs in many coastal compartments because they are not 'closed', but experience net loss of sand through various pathways including loss into dune systems, estuaries, gradual by-passing around headlands or sequestration onto the inner continental shelf. Sediment budgets are a technique for considering sediment

pathways, inputs and exports from the coast (Rosati 2005). In some cases these pathways may actually supply sediment to a section of coast and the sediment budget, particularly at the mouth of large rivers, can be positive. However, recession is also likely to occur as a result of sea-level rise. This latter component has traditionally been treated using a simple heuristic called the 'Bruun Rule' (Bruun 1962), or a modified version of it, to assess the retreat of the shore as a response to sea-level rise.

Consequently characterising shoreline location for a section of coast involves discrimination of three components: the mean trend resulting from the overall sediment budget $\bar{R}_V$ (this long-term trend is generally one of recession); the trend associated with sea-level rise $\bar{R}_{SL}$ (recession); and fluctuating components E_D (erosion and recovery, termed storm cut, and temporarily marked by an erosional scarp cut into the dune face). A further allowance will need to be made for post-storm slumping and/or reduced foundation capacity (Nielsen et al. 1992) so that infrastructure is not sited too close to this eroded dune face, but this component can be addressed with adequate site investigation and treated deterministically.

20.3   Probabilistic approaches to modelling of coastal erosion

Many studies of the impact of sea-level rise on shorelines adopt a deterministic approach, particularly based on the Bruun Rule, by which a possible shoreline position for some time in the future is portrayed on a map. These lines provide potential setback lines, but they do not adequately show the considerable uncertainty which is inherent in calculation and portrayal of the three components of shoreline behaviour described above. The principal advantage of probabilistic modelling of coastal erosion is that it provides coastal managers with robust risk assessments that communicate the uncertainty associated with the input variables, making decision-making more transparent.

a) Stationary conditions

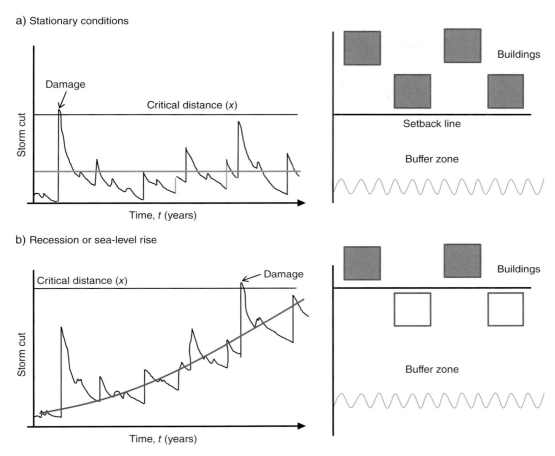

b) Recession or sea-level rise

Figure 20.1   Schematic representation of the pattern of storms resulting in storm-cut beach erosion and subsequent recovery over a period of time under (a) unchanged (stationary) boundary conditions, and (b) changed boundary conditions associated with long-term recession or retreat as a result of sea-level rise. Setback lines, based on the critical distance (*x*) are shown on the right. Source: Jongejan, Ranasinghe, Vrijling & Callaghan 2011. This figure was first published by Engineers Australia. Reprinted here with permission.

Although long-term coastal behaviour, short-term shoreline variability and sea-level rise have been considered important in coastal risk assessments for some time now, the derived output 'critical distance' for any prospective time in the future has invariably been presented as a deterministic line on a map without recourse to the levels of uncertainty that underlie selection of input parameters. This has most commonly been achieved using historical-change and sediment-budget analyses, erosion modelling and application of the Bruun Rule. The last of these is based on conservation of mass and presumes the translation of the beachface and shoreface profile upwards and landwards as determined by its gradient (Bruun 1962), producing one 'predicted' measure of shoreline retreat for any specified rise of mean sea level. Uncertainty is introduced wherever it is necessary to estimate appropriate input parameters. Probabilistic approaches account for much of that uncertainty. Repeated stochastic simulations that adopt values from

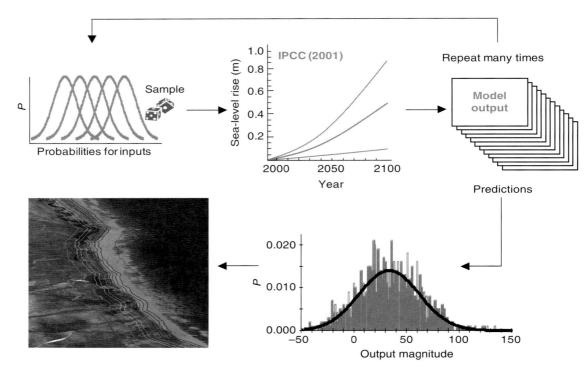

Figure 20.2   Schematic illustration of probabilistic simulation of critical distance. A probability distribution function (pdf) is produced for key variables spanning the likely range of variability; for example, the anticipated range of sea-level rise projections as produced by IPCC or other source of modelling (based on Cowell et al. 2006). The model is then run, repeatedly sampling stochastically from each of the input pdfs, producing a large number of predictions of varying magnitude. These can then be represented as potential setback lines to which probability values can be attached. For colour details please see Plate 18.

across the range of parameter domains quantify this uncertainty, producing an output probability distribution that forecasts the breadth of possible shoreline responses rather than predicting a specific shoreline retreat (Cowell et al. 2006).

This is shown schematically in Figure 20.2. Instead of a single value for key input variables (such as mean values), successive simulations are drawn stochastically from the full range of values considered possible for each of the parameters. In this schematic example, future sea-level rise is represented by the range of IPCC sea-level rise projections across the twenty-first century. In a similar manner, other constraints (such as shoreface morphology, beachface geometry, net sediment budget and sequestration in estuaries) can be varied simultaneously, and thousands of simulations aggregated into an output histogram that indicates likely shoreline movement. Examples of this procedure are described for Manly Beach by Cowell et al. (2006); it was shown that there is a small probability of the shoreline accreting seawards or eroding 100 m or more landwards, but a more likely forecast of recession of several tens of metres by 2100. These simulations produce a sequence of lines with modelled exceedance probabilities or risk contours (e.g. Fig. 20.2) that address the likelihood of shoreline retreat due to coastal erosion risks from storm events, incorporating both sediment-budget considerations and sea-level rise.

The strength of this conceptual framework is integration of best-practice engineering approaches with geologically informed assessments of past coastal behaviour, enabling managers and policymakers to incorporate estimated risk into adaptation options with greater confidence that the underlying risk assessment is transparently evidence-based. Alternative methods or models may be available to address each of these components. The framework approach does not prescribe particular models to adopt, but demonstrates the way in which different models might be convolved to produce outputs that express the range of shoreline locations when input parameters are varied across their potential range.

For example, E_D, also called the design erosion distance, may be set to a standardised level Average Recurrence Interval (ARI, or probability) such as the 100-year ARI, or to a theoretical maximum (probable maximum erosion) based on analysis of historical data. However, a more rigorous analysis is outlined by Callaghan et al. (2008) who developed the Joint Probability Method (JPM) by which they estimated storm erosion probabilities using 30 years of field validation measurements at Narrabeen Beach. Their approach involved temporal simulation of the dominant forcing parameters (wave height, period, direction, storm duration and surge effects), with a structural function that employs wave propagation, refraction and shoaling responses to offshore bathymetry (Fig. 20.3a). This was combined with an equilibrium-based convolution for storm-cut events separated by exponential beach accretion. Initial simulations used the analytical Kriebel and Dean (1993) storm erosion model which produced very large erosion estimates for rare low probability storms (Fig. 20.3b). Subsequent simulations, using semi-empirical SBEACH and more complex process-based XBEACH software, have produced similar erosion volumes although with less amplification during the largest events (Callaghan et al. 2013).

There are fewer longer-term geomorphological models from which to choose. The Shoreface Translation Model (STM) is a sediment-budget model that simulates morphodynamic attributes in a coastal cell through alongshore averaging (Cowell et al. 1995), accounting for geometric change of the cross-shore morphology through time. The STM and similar models (e.g. GeomBEST; Stolper et al. 2005; Moore et al. 2010) simulate redistribution of sediments kinematically based on geometric functions applied to beach, dune, shoreface and estuarine accommodation potential. Figure 20.3c shows the parameters used in a functional representation of the backshore through to the lower shoreface and inner continental shelf, within which cross-shore sediment exchanges are simulated.

The sea-level rise component is an important element of the modelling. At the simplest level it can be anticipated that all sandy shorelines will undergo long-term retreat as a consequence of sea-level rise. Determining the rate of retreat that can be ascribed to sea-level rise needs to be disentangled from observed shoreline changes that are a consequence of either short-term storm events or long-term sediment budget imbalances. Historical shoreline behaviour incorporates a component that is likely to be attributed to the slight rise of sea level observed over past decades. Determining the trajectory of change on the coast through monitoring, including field survey methods such as beach profile studies (McLean and Shen 2006), or photogrammetric analysis of aerial photographs, has generally not fully discriminated the relative contributions of short-term variability (storm-cut/recovery), longer-term trends (recession/progradation) and any component attributable to the gradual rise of sea level.

The sea-level rise component is factored deterministically by the Bruun Rule, but a wide range of predictions for the extent of retreat has been generated by such studies for the Sydney region of New South Wales (Ranasinghe and Stive 2009). In a preliminary application of a probabilistic approach termed the Probabilistic Coastal Recession (PCR) approach, Ranasinghe et al. (2012) used the wave impact erosion model presented by Larson et al. (2004) which relates impact force applied to the dune from swash zone

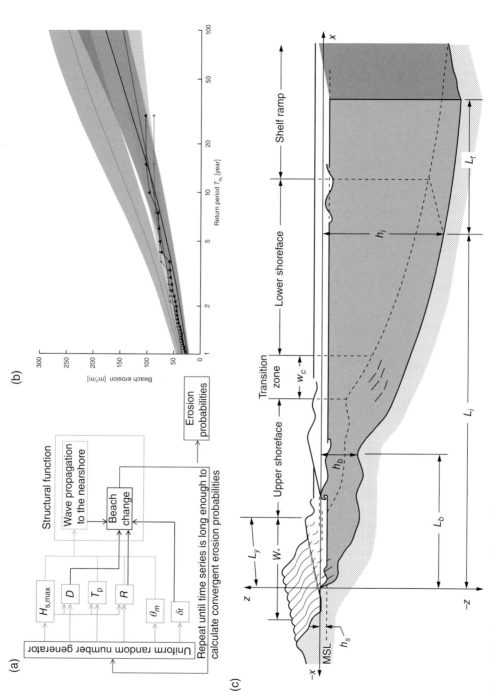

Figure 20.3 Components of the probabilistic modelling proposed in Woodroffe et al. (2012): (a) flowchart underpinning the stages of wave modelling and subsequent beach-change calculation adopted in the Joint Probability Method (JPM) of Callaghan et al. (2008) based on boot strapping of a synthetic sequence of storms and the structural function by which cross-shore beach change is calculated; (b) comparison of observed eroded sand volume above mean sea level (MSL) at Narrabeen Beach (empirical block estimates by block averaging (triangles) and consecutive volumes (circles)) and simulations (1000-year simulations repeated 2000 times) using Kriebel and Dean, SBEACH and XBEACH models (with 95% confidence margins shown) from Callaghan et al. (2013); and (c) geometric parameters used to characterise the shoreface and inner continental shelf in the Shoreface Translation Model (STM) following the procedures outlined by Cowell et al. (2006). For colour details please see Plate 19.

motion to determine the volume of sand eroded from the dune. Therein, the approach was formulated such that a non-stationary mean sea-level time series could be used as an input. Sea-level rise can also be incorporated in the STM component of the modelling of the shoreface and beach-dune morphology.

20.4   Modelling Narrabeen Beach and beyond

The repeated beach surveys at Narrabeen Beach represent an adequate dataset against which to compare simulation results. Accordingly a trial modelling exercise was undertaken in which geomorphic and engineering models were integrated for comparison with past behaviour of a profile in the centre of the Narrabeen coastal compartment (Woodroffe et al. 2012). Quantitative estimates of exchanges between significant sediment stores were determined on the basis of long-term trends (millennial to century scale) in sediment movement inferred from existing coring and stratigraphic data, together with limited radiocarbon dating of fossil shells. This provided the input to forecast future sediment transport and delivery, incorporating uncertainty by employing Monte Carlo modelling using the range of likely values for key parameters including rates of sea-level rise. In parallel, the storm-cut and recovery behaviour of the system was simulated using the PCR approach with stochastically generated 110-year time series of storms using the JPM methodology. The JPM-PCR method was able to generate estimates of recession, using the process-based dune impact model. The geomorphological component used the STM which incorporated response to the IPCC sea-level rise projections. Convolution of short-term variability and long-term trend probability density functions provided probabilistic estimates of risk exceedance, implying a 50% chance that the dune scarp will, by 2100, have retreated 15–25 m landward of its position in mid-1974 following the largest storms of the past half-century.

Despite the almost unparalleled dataset relating to past behaviour of Narrabeen Beach, the forecasts of future behaviour of this beach are constrained by several limitations. First, it has become apparent that the northern and southern ends of Narrabeen Beach undergo cycles of sand accretion and erosion (Harley et al. 2011). For this reason our simulations were applied to the centre of the beach where this phenomenon, termed beach rotation (Ranasinghe et al. 2004), is least apparent. Although the volume of sand in the sub-aerial beach is relatively well studied since the late 1970s, other components of the budget are less well constrained. Sand accumulates at the entrance of Narrabeen Lagoon at the northern end of the beach, demonstrating one obvious but still poorly estimated component of the sediment budget.

Still more significant on this urbanised beach is the impact of human interference. As a result of buildings and infrastructure being constructed on the active dune, there has been a history of attempted coastal protection works emplaced to try to stop erosion or in reaction to significant erosion events. Ad hoc seawalls have been built at various times and the effect that these might have had on the beach and the extent to which they may have interfered with its natural behaviour are largely unknown. A further human response has been periodic beach replenishment, involving the extraction of sand from the entrance to Narrabeen Lagoon and its return to the beachface. This replenishment has been supplemented by beach nourishment where sand has been brought in from outside the compartment (as much as 10,000 tonnes in 2003). This is an adaptation strategy not yet widely applied in Australia (though common in Europe and North America), but likely to become increasingly necessary in the future (for this beach in particular, and at other sites more broadly throughout Australia) in view of infrastructure threatened by storm erosion and the value of real estate along this beach. There remain other challenges to the wider adoption of this framework, particularly in terms of detailed estimates of the role of specific sediment sinks or sources such as estuaries in the modelling and the effectiveness with which adaptation measures, such as beach nourishment, can be represented.

20.5   Conclusion

The framework that we advocate for studying how open-coast sandy systems will behave in response to future sea-level rise involves integration of stochastic geomorphological modelling over longer timescales with similar probabilistic modelling of storm cut which together account for a large proportion of shoreline morphodynamics when compared with the component related directly to sea-level rise. The proposed framework enshrines many of the practices that are already adopted when producing forecasts of future coastal erosion and recession around Australia. Gaps and limitations can be identified however, even at data-rich sites such as Narrabeen Beach, relating to poorly constrained components of the sediment budget and including issues associated with longshore transport of sediment. It is likely that other boundary conditions have also been changed by human activities and may not remain constant in the future. For example, human activities have altered the rate of sediment supply down rivers (in some cases reduced by dams, in others increased by land-use change).

The effects of both direct (nourishment) and indirect (catchment land-use change) interventions will mean that future behaviour may increasingly deviate from that which can be inferred from the past. Further consideration of these aspects will be an important element of future research that can be directly incorporated into the design of adaptation measures to ameliorate the risks associated with sea-level rise.

References

Bruun, P. (1962) Sea-level rise as a cause of shore erosion. *American Society of Civil Engineering Proceedings, Journal of Waterways and Harbors Division* 88, 117–130.

Callaghan, D.P., Nielsen, P., Short, A. and Ranasinghe, R. (2008) Statistical simulation of wave climate and extreme beach erosion. *Coastal Engineering* 55, 375–390.

Callaghan, D.P., Ranasinghe, R. and Roelvink, D. (2013) Probabilistic estimation of storm erosion using analytical, semi-empirical, and process based storm erosion models. *Coastal Engineering* 82, 64–75.

Church, J.A., Hunter, J.R., McInnes, K.L. and White, N.J. (2006) Sea-level rise around the Australian coastline and the changing frequency of extreme sea-level events. *Australian Meteorological Magazine* 55, 253–260.

Cowell, P.J., Roy, P.S. and Jones, R.A. (1995) Simulation of large-scale coastal change using a morphological behaviour model. *Marine Geology* 126, 45–61.

Cowell, P.J., Thom, B.G., Jones, R., Everts, C.H. and Simanovic, D. (2006) Management of uncertainty in predicting climate-change impacts on beaches. *Journal of Coastal Research* 221, 232–245.

Gibb, J.G. (1983) Combatting coastal erosion by the technique of coastal hazard mapping. *New Zealand Engineering* 38, 15–19.

Gordon, A.D., Lord, D.B. and Nolan, M.W. (1979) *Byron Bay–Hastings Point Erosion Study*. Department of Public Works, New South Wales.

Harley, M.D., Turner, I.L., Short, A.D. and Ranasinghe, R. (2011) A reevaluation of coastal embayment rotation: The dominance of cross-shore versus alongshore sediment transport processes, Collaroy-Narrabeen Beach, southeast Australia. *Journal of Geophysical Research* 116, F04033.

Jongejan, R.B., Ranasinghe, R., Vrijling, J.K. and Callaghan, D. (2011) A risk-informed approach to coastal zone management. *Australian Journal of Civil Engineering* 9, 47–60.

Kriebel, D.L. and Dean, R.G. (1993) Convolution method for time-dependent beach-profile response. *Journal of Waterway, Port, Coastal and Ocean Engineering* 119, 204–226.

Larson, M., Erikson, L. and Hanson, H. (2004) An analytical model to predict dune erosion due to wave impact. *Coastal Engineering* 51, 675–696.

McLean, R. and Shen, J.-S. (2006) From foreshore to foredune: foredune development over the last 30 years at Moruya Beach, New South Wales, Australia. *Journal of Coastal Research* 22, 28–36.

Moore, L.J., List, J.H., Williams, S.J. and Stolper, D. (2010) Complexities in barrier island response to sea level rise: Insights from numerical model experiments, North Carolina Outer Banks. *Journal of Geophysical Research* 115(F03004), doi: 10.1029/2009JF001299.

Nielsen, A.F., Lord, D.B. and Poulos, H.G. (1992) Dune stability considerations for building foundations. *Civil Engineering Transactions, Institution of Engineers, Australia* CE34, 167–173.

Ranasinghe, R. and Stive, M.J.F. (2009) Rising seas and retreating coastlines. *Climatic Change* 97, 465–468.

Ranasinghe, R., McLoughlin, R., Short, A. and Symonds, G. (2004) The Southern Oscillation Index, wave climate, and beach rotation. *Marine Geology* 204, 273–287.

Ranasinghe, R., Callaghan, D.P. and Stive, M.J.F. (2012) Estimating coastal recession due to sea level rise: beyond the Bruun rule. *Climatic Change* 110, 561–574.

Rosati, J.D. (2005) Concepts in sediment budgets. *Journal of Coastal Research* 21, 307–322.

Stolper, D., List, J.H. and Thieler, E.R. (2005) Simulating the evolution of coastal morphology and stratigraphy with a new morphological-behaviour model (GEOMBEST). *Marine Geology* 218, 17–36.

Woodroffe, C., Cowell, P., Callaghan, D. et al. (2012) *A Model Framework for Assessing Risk and Adaptation to Climate Change on Australian Coasts*. National Climate Change Adaptation Research Facility, Gold Coast.

21 Navigating from climate change impacts to adaptation actions in coastal ecosystems

WADE L. HADWEN AND SAMANTHA J. CAPON

Australian Rivers Institute, Griffith University, Australia

21.1 Introduction

The world's coastal regions encompass an enormous diversity of ecosystems, supporting high levels of biodiversity and providing a wide range of critical ecosystem functions, goods and services. High degrees of exposure to rising sea levels and many extreme climatic events in addition to other climatic changes, along with strong sensitivities to these, suggest that coastal ecosystems are likely to be among the most vulnerable on the planet to deleterious impacts of climate change. Furthermore, the threats posed to natural ecosystems by climate change will be aggravated in most coastal regions due to the intensity of human activities in the coastal zone and the concentration of non-climatic threats which may also hamper the capacity of ecological systems to adapt autonomously. Planned adaptation to climate change is therefore critical to ensure the conservation of coastal biodiversity and protection of coastal ecosystem functions, goods and services.

Predicting the impacts of climate change on coastal ecosystems is fraught with uncertainty due to the inherent complexity and high degree of connectivity that occur within and between coastal ecosystems. Complex interactions and feedbacks with human communities, and their responses to climate change and its impacts, further muddy the waters. Indeed, it is precisely because of these human responses to climate change that adaptation for coastal ecosystems is so vital. Coastal regions are already hotspots for adaptation actions that focus predominantly on protecting human settlements (e.g. sea walls and barrages) and many of these actions are likely to have significant consequences for coastal species and ecosystems, further eroding their ability to cope with climate change. Where these impacts lead to the loss of critical ecosystem goods and services (e.g. provision of potable water or protection from storm surge), such narrowly focused adaptation actions may ultimately harm the very human communities they sought to benefit. Robust and holistic adaptation strategies that consider both human and ecological objectives across a range of possible futures are therefore urgently required to avoid escalating maladaptation in coastal regions.

In this chapter we provide an overview of the vulnerability of coastal ecosystems to climate change with respect to their exposure, sensitivity and adaptive capacity, emphasising the uncertainties involved in attempting to predict ecological impacts. We explore potential options for managed adaptation action that aim to conserve coastal biodiversity and protect ecosystem functions, goods and services. We conclude by briefly discussing factors influencing adaptation decisions for coastal ecosystems and present some guiding principles.

21.2 Ecological impacts of climate change in coastal ecosystems

21.2.1 *Exposure*

Due to their position at the interface of terrestrial and marine realms, coastal ecosystems are subject to particularly high levels of exposure to climatic changes. While the type and degree of exposure varies considerably among habitat types and between regions, coastal ecosystems are uniquely subject to the effects of sea-level rise in addition to increasing CO_2 concentrations, rising temperatures and altered precipitation. Coastal marine ecosystems, such as inshore coral reefs, are also exposed to changes in ocean pH, salinity levels and marine currents. In Australian coastal waters for example, ocean pH has fallen by around 0.1 since 1750 (Feely et al. 2004). Additionally, rising sea-surface temperatures and ocean salinity, driven by an intensification of the East Australian Current, have resulted in a shift in sea-surface climate of approximately 350 km between 1944 and 2002 (Ridgway and Hill 2009).

Terrestrial, estuarine and freshwater coastal ecosystems can be further exposed to changes in hydrological regimes (including precipitation, runoff, soil moisture and groundwater), wind and wave climate and fire patterns. Furthermore, coastal ecosystems are particularly exposed to many extreme climatic events (e.g. intense storms, floods and tropical cyclones), the frequency and intensity of which may shift as a result of climate change. In Australia for example, the number and magnitude of tropical cyclones in categories 3–5 are expected to rise as a result of climate change (Church et al. 2008).

21.2.2 *Sensitivity*

The structure and function of coastal ecosystems tend to be strongly controlled by the climate variables most subject to alteration under climate change. In marine habitats, for instance, the distribution of near-shore coral reefs is determined largely by the availability of shallow habitats (i.e. sea level) and sufficiently warm (i.e. >18°C) ocean temperatures (McClanahan et al. 2008) while macroalgal forests occur in cooler waters (Wernberg et al. 2009). Similarly, estuarine ecosystems are shaped predominantly by processes associated with either waves, tides or river flows, depending on their location (Bazairi et al. 2003). Coastal ecosystems at the terrestrial interface (e.g. sandy beaches, dune systems and rocky intertidal shores) are strongly influenced by wind, waves and near-shore currents in addition to terrestrial climatic processes (i.e. temperature, hydrologic and fire regimes). Consequently, coastal ecosystems across all of these realms are likely to be highly sensitive to the changes in climatic stimuli to which they are exposed.

Overall, rising air and ocean temperatures are expected to alter the distribution of coastal species and influence temperature-dependent ecological processes (e.g. primary production). For example, corals are extremely sensitive to temperature and high sea-surface temperatures can lead to coral-bleaching events, mass coral mortality and shifts in the composition of reef communities (e.g. the replacement of corals by algae; Hughes et al. 2007). Sea-level rise will redistribute shallow marine and intertidal habitats as well as causing the loss or transformation of many estuarine and coastal terrestrial and freshwater habitats due to saltwater intrusion. Increasing levels of erosion in many coastal habitats are also widely anticipated in response to climate change as a result of changes to dominant physical processes (i.e. waves, winds and river flows) as well as projected increases in the frequency of intense extreme events (e.g. storm surge and floods). In marine ecosystems the growth and survival of many organisms, especially those reliant on dissolved carbonate ions to build their shells or skeletons, will be further affected by ocean acidification (Hoegh-Guldberg et al. 2007). Table 21.1 provides a summary of likely climate change impacts expected in Australia's major coastal ecosystems.

Table 21.1 Selection of likely impacts of climate change in Australia's major coastal ecosystems. Source: Adapted from Hadwen et al. 2011.

	Coastal realm			
Climatic change	Marine	Terrestrial	Estuarine	Freshwater
Warming	Coral reef bleaching; coral mortality; replacement of corals by algae	Poleward migration of species' ranges; effects on primary production; altered fire regimes	Poleward migration of species' ranges; effects on primary production	Poleward migration of species' ranges; effects on primary production
Altered precipitation	Effects on water quality in marine environments	Effects on primary production; effects on growth and survival of coastal organisms; altered fire regimes	Changes to water quality; effects on primary production; effects on growth and survival of coastal organisms	Changes to water availability and quality; effects on primary production; effects on growth and survival of coastal organisms
Sea level rise	Redistribution (and loss) of shallow marine habitats	Redistribution (and loss) of intertidal habitats	Redistribution (and loss) of shallow marine habitats; saltwater intrusion	Redistribution (and loss) of shallow marine habitats; saltwater intrusion
Increased frequency of intense storm events	Effects on water quality in marine environments	Increased coastal erosion	Changes to terrestrial runoff inputs (quality and quantity)	Changes to terrestrial runoff inputs (quality and quantity)
Ocean acidification	Reduced growth and survival of many marine organisms; changes to marine food webs	Changes to inputs from marine food webs	Changes to inputs from marine food webs	
CO_2 enrichment	Increases in primary production for some algal species	Vegetation thickening and encroachment of grasslands by shrubs	Increases in primary production for some algal species	Increases in algal production by some species

The sensitivity of coastal ecosystems to climate change will be exacerbated by the high levels of connectivity that typically occur within and between coastal habitats and realms. For example, coastal food webs tend to be highly connected between marine, intertidal and estuarine habitats. Climate change impacts in marine ecosystems (e.g. declines in the abundance of calcifying organisms due to ocean acidification) are therefore likely to have indirect impacts on ecosystem structure and function in adjacent estuarine and intertidal habitats via food web connections. Similarly, many coral reef fish depend on estuarine habitats (e.g. mangroves) as

nursery habitats (Mumby et al. 2004). Climate change impacts on estuarine ecosystems (e.g. contraction of seagrass beds) may therefore have ramifications for neighbouring coral reefs.

Non-climatic stressors are likely to further intensify the sensitivity of coastal organisms and ecosystems to climate change (and vice versa). In the marine realm for example, the capacity of coral reefs to recover from periods of high ocean temperatures can be reduced by depleted fish stocks (Hughes et al. 2007) and poor water quality of terrestrial flood plumes as a result of agricultural and catchment management practices (McCulloch et al. 2003; Daley et al. 2008).

Invasive species are also attributed with declining condition and resilience of many coastal ecosystems. In southern Australian waters for example, large areas of macroalgal forest have been effectively converted to 'urchin barrens' (Valentine and Johnson 2005). Climate change may both limit the ability of macroalgae to recolonise following such invasions and increase the sensitivity of less degraded communities to the effects of sea urchin invasion.

21.2.3   *Adaptive capacity*

There is considerable potential for coastal species and ecosystems to adapt to changes in climatic stimuli, as has undoubtedly occurred on many occasions in the past. *In situ* adaptation pathways for species include acclimatisation, behavioural change and genetic adaptation. Shifts in the distribution of species may also be considered adaptive. At a community or ecosystem level, shifts in composition towards species more suited to new conditions can similarly be perceived as adaptive where ecosystem functions persist. Many coastal species exhibit significant adaptive capacity and indeed have evolved under highly variable conditions. For example, mangroves are widely considered to have high adaptive capacity as a result of their plasticity, dispersal capabilities and ability to adapt to local conditions (Alongi 2008; Proffitt and Travis 2010; Williamson et al. 2011). Furthermore, mangroves appear able to track changes in sea level with evidence from northern Australia suggesting that, over the last 40 years, mangroves in this region have both moved landwards and increased in abundance (Williamson et al. 2011). The key concern with respect to the adaptive capacity of coastal species and ecosystems is that the rate and extent of climate change are likely to outstrip natural adaptive processes in the absence of human intervention. For instance, rates of sea level rise greatly exceed rates of sediment accretion in mangrove forest and therefore net losses of this coastal ecosystem are expected in the future (Gilman et al. 2008). The intensity of human activities and non-climatic stressors in many coastal regions will likely limit the capacity of coastal species and ecosystems to adapt autonomously to climate change (Adger et al. 2005). For example, the landward migration of many estuarine and terrestrial coastal ecosystems (e.g. mangroves and salt marshes) in response to sea-level rise is significantly restrained by the presence of human developments in what is often referred to as the 'coastal squeeze'.

21.3   Managed adaptation options for coastal ecosystems

The disproportionately high value of coastal ecosystems, as well as their high vulnerability to deleterious climate change impacts, makes managed adaptation in the coastal zone a global priority; a wide range of on-ground and institutional approaches have been proposed and implemented (Klein et al. 2001; Abel et al. 2011). Many of these adaptation options focus solely on delivering human benefits (e.g. groynes, dredging and beach nourishment; El-Raey et al. 1999), while others are mainly intended to provide ecological benefits (e.g. dune revegetation, pollution management, etc.; Abel et al. 2011). More holistic approaches to adaptation (e.g. via ecological engineering) seek to protect human activities and developments in coastal regions by maintaining or enhancing ecosystem functions (Chapman and Underwood 2011).

There are a wide range of adaptation options available for coastal ecosystems with the aims of protecting or enhancing biodiversity and ecosystem goods and functions (Table 21.2). In many cases, options for adaptation involve adjusting current management actions to account for the risks associated with climate change. Other adaptation actions may be more novel. Managing existing threats and non-climatic stressors in order to reduce ecological sensitivity to climate change represents a major approach to adaptation (Table 21.2). Similarly, management to facilitate autonomous adaptation of species and ecosystems and build their resilience to climate change is widely acknowledged as a key adaptation

Table 21.2 Examples of on-ground adaptation options for Australia's coastal zone in Marine (M), Terrestrial (T), Estuarine (E) and Freshwater (F) realms. Source: Adapted from Hadwen et al. 2011.

Adaptation action	Relevant coastal ecosystems				Intended consequence	Unintended consequences	
	M	T	E	F		Ecological	Human
Management of existing stressors							
Environmental flow management	Y		Y	Y	Maintain the timing and magnitude of natural flow to rivers, estuaries and coastal waters to support species in all environments	May destabilise existing environments and species composition in some habitats	May take water from human users, creating conflict and limiting water resource development and associated industries
Control of invasive species and weeds	Y	Y	Y	Y	Remove non-native species to enable natives to recolonise and/or maintain healthy populations	Eradication measures (weed removal or trapping) may have impacts on native species	Will be costly in terms of money and human resources and may require management in human communities to reduce the risk of reintroduction of the invasive species
Predator exclusion	Y	Y	Y	Y	Provide organisms with relief from top-down pressures to enable populations to establish	May result in cascading flows through food webs which will favour some trophic levels over others and may result in negative consequences, such as algal blooms	Trapping and fencing to remove predators will be expensive to implement and maintain
Facilitation of autonomous adaptation/resilience building							
Protect refugia	Y	Y	Y	Y	Protect resistant populations or benign locations as source of persistence and recolonisation	May facilitate the establishment of invasive species	May reduce fishing opportunities and development options
Protect space for migration	Y	Y	Y	Y	Protect habitat for species that use multiple habitats	May facilitate the establishment of invasive species	May reduce fishing opportunities; reduce development options
Remove barriers to migration	Y	Y	Y	Y	Support species that move across realms and habitats within the coastal zone	May facilitate the establishment of invasive species	May reduce fishing opportunities and development options
Eco-engineering							
Artificial reefs	Y				Provide novel habitat for species persistence	May favour some species over others and promote patchy habitat with atypical species composition	Will be expensive to implement and maintain (storm damage, etc.) at any significant sort of scale

Option				Benefit	Negative consequences	Costs / positive outcomes
Re-seeding and/or planting	Y	Y	Y	Provide new habitat for vulnerable species, potentially in more favourable localities	Will require pre-planting treatment of recipient habitats which will disturb the pre-existing biota	Will be expensive (in terms of money and personnel) to implement and maintain to ensure long-term success
Ex situ adaptation						
Seed banks/zoos	Y	Y	Y	Preserve highly vulnerable species, especially threatened populations	Loss of diversity in the natural setting, consequences for food webs and community stability	Positive educational opportunities and increased opportunities for coastal development (threatened population having been removed)
Species relocation	Y	Y	Y	Move vulnerable species and populations to more suitable and protected environments	Loss of local habitat and ecosystem complexity; increased habitat complexity in recipient habitat (but with possible negative interactions with resident biota)	Increased developmental opportunities in damaged habitats; education and tourism opportunities in recipient habitat.
Responding to extreme events						
Fire management options	Y			Controlled burning to reduce risk of intense fire and very damaging events	Change in species composition and productivity with changed fire frequency and spatial pattern	Reduced risk of extreme fire events will enhance resilience and sustainability of human communities
Restrict access	Y	Y	Y	Protect vulnerable habitat or populations from the threat of extreme events	May reduce connectivity and spatial complexity of habitats	Loss of tourism, recreation and development opportunities
Microclimate management						
Install sprinkler systems	Y			Maintain conditions for vulnerable taxa and reduce fire risk	May promote conditions for non-native species	Includes costs or loss of water resources for alternative (human) uses
Artificial shading	Y	Y	Y	Produce and maintain conditions to protect vulnerable species	May favour some species over others and promote patchy habitat and vegetation types	Will be expensive to implement and maintain (storm damage, etc.) at any significant sort of scale

strategy for conservation and natural resources management (Steffen et al. 2009). Such strategies might involve strengthening currently protected area networks and their connectivity (Table 21.2), or may include more innovative approaches such as introducing spatial and temporal dynamics into reserve boundaries, e.g. designating protected areas in places expected to be significant under future climates. At the more extreme end of the spectrum, such adaptation strategies include managed retreat, i.e. the partial or complete removal of engineered structures (e.g. weirs or armouring) in order to provide space for species and ecosystems to adjust to changing conditions (Chapman and Underwood 2011). Managed retreat is particularly relevant to estuarine and terrestrial coastal ecosystems (e.g. mangroves and saltmarshes) in relation to sea-level rise (French 2006).

On-ground climate change actions in the coastal zone, as elsewhere, can be broadly classified as either 'hard' or 'soft' measures (McGlashan 2003). Hard engineering approaches involve the use of fixed structures to intervene in coastal processes, such as wave-driven erosion (Cooper and McKenna 2008), and typically aim to minimise local impacts by reducing exposure to climatic changes. Sea walls, groynes, rock armouring, offshore breakwaters, training walls, geotextile sandbags and artificial reefs are all examples of hard engineering adaptation options (Bacchiocchi and Airoldi 2003). Such actions have been implemented extensively in coastal regions of Australia (Abel et al. 2011) and around the world, typically to protect human settlements, infrastructure and investments although some (e.g. artificial reefs) have ecological objectives. In many cases, the ecological impacts of hard engineering actions are poorly known but can include reductions in sediment transport along coastlines (Gilman et al. 2008) and acceleration of beach erosion (Brown and McLachlan 2002). Hard engineering options are also associated with high opportunity costs and high risk of failure (Capon et al. 2013). Particularly in coastal areas, where hard-engineering measures may increase vulnerability to storms and

floods (Lucrezi et al. 2010) as well as reduce ecological resilience and adaptive capacity, hard-engineering adaptation options are likely to be maladaptive overall (Klein et al. 2001). Where such measures are unavoidable (e.g. due to political or social pressure) it is recommended that structures are designed with safety margins (e.g. dam heights determined in relation to projected flood levels) and reviewed regularly (Capon et al. 2013). Ecologically sensitive infrastructure design or retrofitting of existing infrastructure (e.g. inclusion of artificial habitat such as holes or caves) may also reduce perverse ecological outcomes of hard-engineering measures.

Soft adaptation measures include a range of actions that manipulate abiotic or biotic elements of ecosystems rather than employing man-made structures, e.g. beach nourishment and revegetation (French 2006). As with hard engineering approaches, some of these measures focus predominantly on providing benefits to people and may induce significant ecological impacts. For example, beach nourishment is widely used to address coastal erosion in areas of high tourism and recreational value and involves the movement of sand from low to high levels. While this process may benefit some coastal species by maintaining habitat, it can also lead to the deterioration of macrobenthos with considerable implications for beach food webs (Brown and McLachlan 2002). Many 'soft' adaptation actions are forms of ecosystem engineering, whereby particular species or species mixes are introduced or promoted to adjust ecosystem structure and function (Table 21.2). For example, stabilisation of fore dunes via the introduction of species such as Marram grass (*Ammophila arenaria*) is a management action widely employed around the world (Borsje et al. 2011). It should be noted that ecological engineering, as opposed to ecosystem engineering, may involve 'hard' and 'soft' measures, e.g. artificial reefs.

Where highly valued coastal species are considered to be extremely vulnerable to climate change impacts (e.g. facing likely extinction), *ex situ* adaptation actions may be deemed appropriate (Table 21.2). Such measures (e.g. species or genetic translocation) are highly interventional and therefore, like hard-engineering approaches, are associated with a high risk of failure and a high likelihood of unintended, perverse consequences.

21.4   Adaptation decisions

Current understanding of coastal climates and ecology leave little doubt that, in the absence of human interventions, the coastal zone is likely to be highly vulnerable to climate change impacts (Scavia et al. 2002). Understanding and predicting the magnitude, extent and, in some cases, even the direction of impacts in particular locations represents a major challenge for adaptation planning, however. In part, such high levels of uncertainty arise from a paucity of ecological knowledge, especially in relation to the role of climate and adaptive capacity, and highlight the need for urgent further ecological research. Capacity to project climate futures for coastal regions is also limited by the separation and lack of alignment between terrestrial and marine climate projections, which typically involve different parameters and spatial resolutions. Adaptation decision-making is further complicated by a lack of knowledge concerning the ecological impacts of adaptation actions themselves. Nevertheless, adaptation decisions for coastal regions around the world are urgently needed and must be developed despite these considerable uncertainties.

21.5   Guiding principles for adaptation decisions in the coastal zone

Adaptation strategies are all developed in relation to their own unique ecological, socio-economic and institutional context and are therefore dependent on the particular motives, opportunities and means for adaptation each situation entails. In light of the great challenges posed by first evaluating and then implementing climate change adaptation actions, there are seven

guiding principles that can provide clarity for decision-makers and reduce the risk of maladaptation, as follows.

1. Climate change adaptation approaches and actions require clearly stated goals and objectives to reduce the risk of establishing unrealistic expectations and implementing maladaptive strategies. Climate change adaptation planning should therefore include the setting of broad goals, e.g. 'keeping the system as it is' versus 'moving the system to a new state' as well as specific, targeted objectives concerning the desired ecological consequences of adaptation actions (e.g. conserving species, ecosystem function or ecosystem services).

2. Climate change adaptation planning and decision-making requires involvement from a wide range of stakeholders across environmental, social, economic and cultural sectors. Such integration is critical to developing appropriate goals and objectives for adaptation as well as identifying strategies which are likely to be successfully implemented and maintained and that involve minimal risks across sectors.

3. Climate change adaptation planning requires information to be made easily available and shared. This includes information regarding climate projections, physical, chemical and ecological parameters and social, economic and cultural components of the area of concern. Improved mechanisms of data collation, storage and delivery are also urgently required.

4. Climate change adaptation requires more integrated, flexible and dynamic legislative, policy and institutional frameworks which adequately reflect the temporal and spatial scales required by adaptation decision-making as well as allowing for adaptive management cycles.

5. Climate change adaptation requires a greater understanding and appreciation of connectivity within and among coastal ecosystems as well as between ecological and human systems, e.g. the high value of ecosystem services provided to human communities by coastal ecosystems.

6. Climate change adaptation actions will be implemented at local or regional scales, since these will determine which adaptation approaches

are (a) appropriate to address adaptation goals and objectives and (b) possible given the physical, ecological, social, economic and cultural features of the area of concern. However, larger scales require consideration since adaptation actions may have consequences for connectivity with ecological and human systems beyond this area (e.g. migrating species).

7. Climate change adaptation cannot be considered in isolation from existing non-climatic threats. Climate change adaptation should focus on coastal zone sustainability; explicit consideration of non-climatic threats is therefore essential, particularly as the impacts of many are likely to be exacerbated by climate change. As with other adaptation approaches, management of non-climatic threats should take into account the risks associated with their cost, efficacy and unintended ecological and human consequences.

Adherence to these guiding principles will ensure that climate change adaptation is considered in a holistic way, whereby coastal ecosystems and their values are explicitly incorporated into coastal zone plans not to the exclusion of human needs and concerns, but alongside them.

Acknowledgements

The authors wish to acknowledge the CERCCS project team for the ideas and material presented in this chapter along with the support of the project's steering committee and valuable contributions from the expert panel and stakeholder references groups as well as NCCARF.

References

Abel, N., Gorddard, R., Harman, B. et al. (2011) Sea level rise, coastal development and planned retreat: Analytical framework, governance principles and an Australian case study. *Environmental Science and Policy* 14(3), 279–288.

Adger, W. N., Hughes, T. P., Folke, C., Carpenter, S. R. and Rockstrom, J. (2005) Social-ecological resilience to coastal disasters. *Science* 309(5737), 1036–1039.

Alongi, D. (2008) Mangrove forests, Resilience, protection from tsunamis, and responses to global climate change. *Estuarine, Coastal and Shelf Science* 76, 1–13.

Bacchiocchi, F. and Airoldi, L. (2003) Distribution and dynamics of epibiota on hard structures for coastal protection. *Estuarine, Coastal and Shelf Science* 56, 1157–1166.

Bazairi, H., Bayed, A. Glemarec, M. and Hily, C. (2003) Spatial organisation of macrozoobenthic communities in response to environmental factors in a coastal lagoon of the NW African coast (Merja Zerga, Morocco). *Oceanologica Acta* 26(5–6), 457–471.

Borsje, B. W., van Wesenbeeck, B. K., Dekker, F. et al. (2011). How ecological engineering can serve in coastal protection. *Ecological Engineering* 37(2), 113–122.

Brown, A.C. and McLachlan, A. (2002) Sandy shore ecosystems and the threats facing them: some predictions for the year 2025. *Environmental Conservation* 29, 62–77.

Capon, S.J., Chambers, L.E., MacNally, R. et al. (2013) Riparian ecosystems in the 21st century: hotspots for climate change adaptation? *Ecosystems* 16, 359–381.

Chapman, M.G. and Underwood, A.J. (2011) Evaluation of ecological engineering of 'armoured' shorelines to improve their value as habitat. *Journal of Experimental Marine Biology and Ecology* 400, 302–313.

Church J., White N., Hunter J., McInnes K., Cowell P. and O'Farrell S. (2008) Sea-level rise. In: Newton, P. (ed.) *Transitions: Pathways towards Sustainable Urban Development in Australia*. CSIRO Publishing, Melbourne.

Cooper, J.A.G. and McKenna, J. (2008) Working with natural processes: the challenge for coastal protection strategies. *The Geographical Journal* 174(4), 315–331.

Daley, B., Griggs, P. and Marsh, H. (2008) Reconstructing reefs: qualitative research and the environmental history of the Great Barrier Reef, Australia. *Qualitative Research* 8, 584–615.

El-Raey, M., Dewidar, Kh. and El Hattab, M. (1999) Adaptation to the impacts of sea level rise in Egypt. *Climate Research* 12, 117–128.

Feely, R.A., Sabine, C.L., Lee, K. et al. (2004) Impact of anthropogenic CO_2 on the $CaCO_3$ system in the oceans. *Science* 305, 362–366.

French, J. (2006) Tidal marsh sedimentation and resilience to environmental change: Exploratory modelling of tidal, sea-level and sediment supply forcing in predominantly allochthonous systems. *Marine Geology* 235(1–4), 119–136.

Gilman, E.L., Ellison, J. Duke, N.C. and Field, C. (2008) Threats to mangroves from climate change and adaptation options: A review. *Aquatic Botany* 89, 237–250.

Hadwen, W.L., Capon, S.J., Kobashi, D., Poloczanska, E.S., Rochester, W., Martin, T.G., Bay, L.K., Pratchett, M.S., Green, J., Cook, B.D., Berry, A., Lalonde, A., Hall, A. and Fahey, S. (2011) *Climate Change Responses and Adaptation Pathways in Australian Coastal Ecosystems: Synthesis Report*. National Climate Change Adaptation Research Facility, Gold Coast, 359 pp.

Hoegh-Guldberg, O., Mumby, P.J., Hooten, A.J. et al. (2007) Coral reefs under rapid climate change and ocean acidification. *Science* 318(5857), 1737–1742.

Hughes, T.P., Rodrigues, M.J., Bellwood, D.R. et al. (2007) Phase shifts, herbivory, and the resilience of coral reefs to climate change. *Current Biology* 17, 360–365.

Klein, R.J.T., Nicholls, R.J., Ragoonaden, S., Capobianco, M., Aston, J and Buckley, E.N. (2001) Technological options for adaptation to climate change in coastal zones. *Journal of Coastal Research* 17(3), 531–543.

Lucrezi, S., Schlacher, T.A. and Robinson, W. (2010) Can storms and shore armouring exert additive effects on sandy-beach habitats and biota? *Marine and Freshwater Research* 61(9), 951–962.

McClanahan, T.R., Buddemeier, R.T., Hoegh-Goldberg, O. and Sammarco, P. (2008) Projecting the current trajectory for coral reefs. In: Polunin, N.V.C. (ed.) *Aquatic Ecosystems*. Cambridge University Press, Cambridge, pp. 242–262.

McCulloch, M., Fallon, S., Wyndham, T., Hendy, E., Lough, J. and Barnes, D. (2003) Coral record of increased sediment flux to the inner Great Barrier Reef since European settlement. *Nature* 421, 727–730.

McGlashan, D. (2003) Funding in integrated coastal zone management partnerships. *Marine Pollution Bulletin* 46, 393–396.

Mumby, P.J., Edwards, A.J., Arias-Gonzalez, J.E. et al. (2004) Mangroves enhance the biomass of coral reef fish communities in the Caribbean. *Nature* 427, 533–536.

Proffitt, C.E. and Travis, S.E. (2010) Red mangrove seedling survival, growth, and reproduction: effects of environment and maternal genotype. *Estuaries and Coasts* 33(4), 890–901.

Ridgway, K. and Hill, K. (2009) East Australia Current. In: Poloczanska, E. Hobday, A.J. and Richardson, A.J. (eds) *A Marine Climate Change Impacts and Adaptation Report Card for Australia 2009*. NCCARF Publication, Gold Coast, pp. 52–64.

Scavia, D., Field, J.C., Boesch, D.F. et al. (2002). Climate change impacts on US coastal and marine ecosystems. *Estuaries* 25(2), 149–164.

Steffen, W., Burbidge, A.A., Hughes, L. et al. (2009) *Australia's Biodiversity and Climate Change*. CSIRO Publishing, Collingwood, Victoria, Australia.

Valentine, J.P. and Johnson, C.R. (2005) Persistence of sea urchin (*Heliocidaris erythrogramma*) barrens on the east coast of Tasmania: inhibition of macroalgal recovery in the absence of high densities of sea urchins. *Botanica Marina* 48, 106–115.

Wernberg, T., Campbell, A., Coleman, M.A. et al. (2009) Chapter 11: Macroalgae and temperate rocky reefs. In: Poloczanska E.S., Hobday A.J. and Richardson A.J. (eds) *A Marine Climate Change Impacts and Adaptation Report Card for Australia 2009*. National Climate Change Adaptation Research Facility, Gold Coast.

Williamson, G.J., Boggs, G.S. and Bowman, D.M.J.S. (2011) Late 20th century mangrove encroachment in the coastal Australian monsoon tropics parallels the regional increase in woody biomass. *Regional Environmental Change* 11(1), 19–27.

22 Enhancing the resilience of seaports to a changing climate

DARRYN MCEVOY AND JANE MULLETT

RMIT University, Australia

22.1 Introduction

Seaports are critical components of the nation's infrastructure system, acting as vital 'hubs' that connect Australia to regional and global supply chains. The significance of seaports is emphasised by the recent National Ports Strategy, which states that 'ports and associated infrastructure are of the utmost economic and social importance to Australia' (Infrastructure Australia 2011, p. 6); as framed by an overarching aim to 'drive the development of efficient, sustainable ports and related freight logistics that together balance the needs of a growing Australian community and economy with the quality of life aspirations of the Australian people' (Infrastructure Australia 2011, p. 7). However, while there is considerable emphasis placed on the strategic importance of seaports, a changing climate will pose increasing challenges to the continuing successful operation of ports and their associated infrastructure over coming years and decades (Cox et al. 2012). Addressing this important agenda, the NCCARF-funded project on enhancing the resilience of Australian seaports to a changing climate (2010–12) was therefore timely. The research comprised a literature review, close engagement with case study ports and government agencies, the development of methodologies and tools and the publication of a series of reports and guidelines.

Evidence from the academic and grey literature review found that although the topic is relatively new with limited international studies on 'climate risks and seaports' to date, a knowledge base is slowly beginning to develop. Worthy of academic note is the international analysis of exposure to sea-level rise that was released by the World Bank and the OECD in 2008. This benchmark study – consisting of 136 ports of which 5 are Australian – mapped the port cities considered most vulnerable to climate change in 2070 (Nicholls et al. 2008). Analysis was based on a one-in-one-hundred-year (1:100) storm surge as the definitive water level from which to base calculations, with exposure of population and assets then estimated as a function of elevation against this water level (Hanson et al. 2011). Another international example which adopted an alternative 'actor-based' approach was the worldwide survey of Port Authorities undertaken by Becker et al. (2012) to elicit information on the sector's risk perceptions, the likely impacts of climate change on future international port operations and potential adaptation strategies. Perhaps the most pioneering example that was identified by the review was a comprehensive analysis of the Cartagena port facility, Colombia (Stenek et al. 2011). This was the only study identified by the literature review that had taken a system-based and integrated approach, explicitly considering both the functional and infrastructure assets of the case study port in its comprehensive analysis.

Grey documentation worthy of note includes a review of climate change adaptation measures available to seaports (International Association of Ports and Harbors 2011), and a growing number of climate risk assessments that have recently been carried out by ports in the UK (for example, see the Mersey Docks in Peel Ports Group 2011). Interestingly, this UK assessment activity was driven by national legislation, in this case the UK Climate Change Act 2008 which requires all major seaports (as well as other major infrastructure owners) to report to national government on their climate risk assessments and identified adaptation measures. These reports are openly available on the Government website, promoting transparency within the sector and supporting peer-to-peer learning. A final report provides an overview of the undertaking and observes that approaches to adaptation are nascent and variable across the ports sector (DEFRA 2012). The findings of the DEFRA report are remarkably consistent with those from the NCCARF research activity, as well as being supported by other contemporary Australian research (Nursey-Bray et al. 2013; Ng et al. 2013).

Whilst Australian seaports are not legally required to assess climate risks in such a way, there has been a marked shift in emphasis in recent times. For instance, adaptation is explicitly addressed in Infrastructure Australia's fourth annual report to the Council of Australian Governments. Here, adaptation is defined as 'assessing risks to infrastructure from extreme events, and understanding how asset management and the design and location of assets can be adapted in consideration of these risks' (Infrastructure Australia 2012, p. 21). Although less explicit about climate change, the recent National Ports Strategy recommends that ports' planning documentation, with a suggested minimum time horizon of 15–30 years, should consider external factors (both risks and opportunities) that may impact on the functioning of ports (Infrastructure Australia 2011).

22.2 An integrated assessment of vulnerabilities and future risks

At the current time, our understanding of the potential impacts of future climate change on seaport assets, workforce and operations and how best to adapt over time, is at an early stage. In order to address important knowledge gaps, this research was carried out by a multi-disciplinary team from two universities (RMIT University and the University of Queensland) over the period 2010–2012. Three seaports on the Eastern seaboard (Fig. 22.1), representative of different port types (container/bulk) and climatic regimes (uniform rainfall/wet summer, low winter rainfall), were selected to test and refine the assessment methodologies. Close engagement with the port authorities was integral to the research program, promoting co-generation of knowledge and ensuring that key deliverables were 'fit for purpose' for the seaports sector.

The research methodology was underpinned by an integrated assessment of current vulnerabilities and future climate risks, comprising quantitative, qualitative and participatory approaches. The resulting information, contextualised within the broader landscape of non-climate drivers, was then distilled to inform the assessment of vulnerabilities, risks and adaptation options for different 'elements at risk' within the port environs i.e. infrastructure assets, functional assets and operations and the workforce. An integrated assessment approach was necessary to adequately frame the key issues and interpret the complex mix of data and information that needed to be considered when conducting analysis in support of climate-resilient seaports.

Close liaison with climate information providers such as the Commonwealth Scientific and Industrial Research Organisation (CSIRO), the

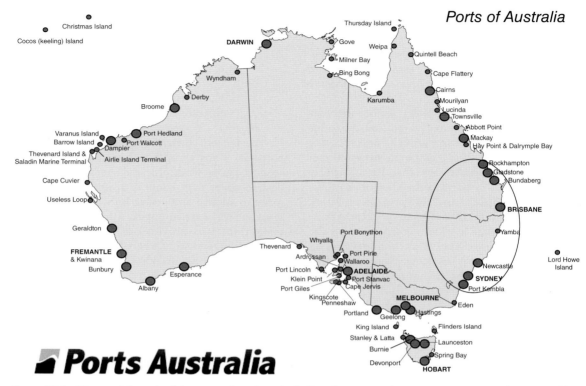

Figure 22.1 The spatial reach of the research project, including the case-study ports Port Botany and Port Kembla (now merged into NSW Ports) and Gladstone Ports Corporation Ltd. Source: Reproduced with permission of Ports Australia. For colour details please see Plate 20.

Bureau of Meteorology (BoM) and the Centre for Australian Weather and Climate Research (CAWCR) was critical to this process. Equally valuable was the commitment of the case study ports that provided 'bottom up' information on the identification and analysis of vulnerabilities, as well as stakeholder perceptions of future risk. This expert input helped to shape the research agenda in an iterative fashion as the project, and thinking, evolved.

The six main factors considered in the integrated assessment of current vulnerabilities and future risks were:
1. analysing ports as systems;
2. considering current (and past) climate/weather data;
3. interpreting future climate projections;

4. reconciling climate information with research needs (risk assessment and adaptation planning for infrastructure assets and functions);
5. compilation of climate information packs for each of the case study ports; and
6. contextualising climate data with non-climate drivers.

The sourcing and interpretation of future climate data for adaptation planning presented a significant challenge, ultimately requiring an iterative learning approach that sought to match the climate data currently available from the latest scientific global circulation models (GCMs) with the desired inputs for the engineering and logistical assessments (bridging the climate science–adaptation planning divide). It should be noted that there are a multitude of GCMs that can be applied to the Australian context, and selecting which of these to use in the project was a complex process. As highlighted by Clarke et al. (2011) there are currently projections from up to 24 GCMs, up to 6 emission scenarios, and around 12 climate variables, potentially resulting in many complex permutations.

The project's 'journey' of matching available future climate information to end-user requirements, and understanding and interpreting the processed data while dealing with the inherent uncertainties involved, formed a key part of the learning process. The suite of climate models ultimately used in the project resulted from the application of CSIRO's framework 'Climate Futures'. This enabled the choice of a set of possible futures (described as most likely; hotter and drier; and cooler and wetter) that encompassed a range of plausible modelled futures most appropriate for the geographical locations and for the likely climate risks (Table 22.1). This methodology helped guide the choice of a comprehensive and internally consistent set of models which provided salient data for the required climate variables.

From the project's experience, this framework is highly recommended for other Australian studies which require scenarios that are tailored to the needs of specific applications (reducing complexity by allowing users to select a smaller subset of climate models which are representative of different futures), and it has the potential to be of international relevance. The actual generation of different climate scenarios based on each of the representative models was performed using the online OzClim tool (www.csiro.au/ozclim/home.do), a publicly accessible tool designed to allow end-users to generate and explore scenarios up to 2100. This resource is now being updated to represent the new generation of probabilistic scenarios (CMIP5) discussed in the Fifth Assessment Report of the Intergovernmental Panel on Climate Change.

22.3 Functional resilience

To ensure a comprehensive system-wide analysis, the project assessed both functional vulnerabilities and the infrastructural resilience of the case study ports. Assessment of functional vulnerability involved considering the impact of extreme-weather-related events on functional assets, the movement of goods and the preparedness/adaptive capacity of the workforce. Due to the difficulty of forecasting how ports will operate in the distant future, analysis was framed by a relatively short time horizon (2030) with greater emphasis on the impacts of current-day extreme events and how port authorities are managing these risks today. Hence, a more immediate emphasis was used as a starting point before forecasting over a relatively short time period (addressing current-day adaptation deficits). This vulnerability-led approach also compensates for the still-emerging knowledge base on the likely frequency (average return period) and intensity of future extreme events, as well as being a time frame more relevant to port authority decision-making processes.

Three deliverables were produced as part of the analysis on functional assets. First, vulnerability matrices for key areas of all three case study ports were developed (resulting in a transferable methodology for other Australian seaports).

Second, two prototype models were built. The first of these was GIS-based (a tool for visualising key vulnerabilities); however, this approach proved to be both time and resource intensive

Table 22.1 Global circulation models used to represent different climate futures.

Climate Future	Sydney/Kembla	Gladstone
Most likely	Model MRI2.3.2	Model CSIRO Mk 3.5
Hotter, drier	Model CSIRO Mk 3.5	Model MRI2.3.2
Cooler, wetter	MIROC3.2-Medres	MIROC3.2-Medres
Warmer, wetter	MIROC3.2-Hires	MIROC3.2-Hires

and is therefore best suited to a more detailed assessment of risks that have been identified and ranked as priorities rather than as a first pass assessment. The second model was an agent-based simulation that modelled the throughput of containers in a port environment when perturbed by external climate-related stressors.

Attention was paid to the human dimension as the third element of investigation. Adaptive capacity was investigated through semi-structured interviews with ports staff from each of the ports (characterised by different management structures and operational arrangements), with specific reference to: (1) experiences of extreme weather events and how they were addressed; (2) personal perspectives on future climate change and implications for port operations; (3) existing procedures to deal with extreme weather conditions; and (4) their views on the vulnerability of seaports to extreme events both now and into the future. To consider issues of adaptive capacity, analysis was then framed according to skills and knowledge, systems and processes and organisational culture and norms. The analysis of the adaptive capacity of the workforce led to the development of a training manual aimed at different groups of actors involved with port operations.

22.4   Infrastructural resilience

For the purposes of this particular project, the engineering analysis concentrated on the long-term deterioration of infrastructure assets. As such, consideration of catastrophic failure brought about by the impact of a low-probability but high-consequence extreme event was outside the scope of this study. The research that was conducted consisted of several dimensions: structural asset identification; interpretation of climate data; long-term deterioration modelling taking account of changes to salient climatic variables; resilience matrices; a methodology for conducting life-cycle cost analysis; and the development of a software tool for use by port engineers.

A central component of the research was the development of deterioration models that simulated the effect of climate change on key infrastructure assets. Currently, there is a lack of models that explicitly include climate impacts when predicting deterioration over time. A numerical model was therefore developed, based on existing models extended through the inclusion of salient climate properties such as changing humidity, temperature, salinity, etc. Findings from the deterioration model, adapted to account for the changing exposure and sensitivity of different materials (concrete, steel, timber) to environmental parameters, predict that climate change will affect the timing of maintenance requirements by ports. Sensitivity analysis indicates concrete will be most impacted by temperature, marine timber by sea salinity and steel by relative humidity. These impacts point to important business implications, for example balancing future maintenance and capital budgets. A design and maintenance cost management methodology was also developed by the project to support decision-making in this regard.

22.5   Research findings

The multi-disciplinary research activity carried out by the project, evidenced through the integration of discrete work packages and reinforced by engagement with the case study ports, indicates that resilience to current-day climate variability is evident within the immediate port environment (at the level of individual organisations). This can be attributed to autonomous adaptation primarily as a result of a combination of regulatory and operational mechanisms such as Occupational Health and Safety (OHandS) requirements, organisational risk management strategies and incremental changes to practice brought about by the ports' experience of weather-related events, rather than as part of a conscious adaptation strategy (with the exception of sea-level rise and planning for raised berthing structures). However, important vulnerabilities were also identified: the seaward side of operations and the supply chain hinterland were found to be most affected by current climate variability (vulnerabilities which will intensify under a changing climate). It should also be noted that climate change may bring some benefits for individual ports; for example sea-level rise may allow the passage of ships with deeper draught, hence reducing the need for channel dredging in some ports.

Explicit assessment of future climate risks and adaptation planning for the longer-term was also less evident, again apart from consideration of sea-level rise. Low-probability high-impact events are also less well considered. Looking forward, although 'hard' infrastructure assets can be made more resilient by changing design and maintenance regimes (addressing sensitivity), land-use planning is likely to become even more important under a changing climate (changing levels of exposure, combined with developmental drivers). Functional resources (assets and the workforce) are likely to become increasingly vulnerable due to the likely increases in the frequency and intensity of extreme events. As such, consideration of future climate change impacts within current risk assessment and management processes would strengthen port resilience, with adaptation measures integrated as part and parcel of normal investment cycles or maintenance regimes. Analysis also needs to look beyond the immediate port environs to consider wider supply-chain issues (taking into account both internal and external trade routes). In this regard, adaptation is likely to require the promotion of flexibility and spare capacity within the system, ultimately going against the grain of business 'efficiency'.

Findings from across the comprehensive program of research were collated and analysed to inform adaptation guidance for climate resilient seaports. This guidance was embedded in a traditional risk management framework to best align with the sector's current risk management processes (involving step-by-step decision support). Throughout the project, several innovative adaptation options were identified including opportunities to improve logistics flow, manage infrastructural cycles and reduce potential occupational health and safety hazards. Building

adaptive capacity was also addressed in terms of awareness raising, skill development, data monitoring and collection and collaborative research. This recognises that adaptation actions are highly context specific.

22.6   Conclusions

Reconciling climate information for adaptation planning is not a simple process, and bridging the divide between the climate science and the needs of those responsible for implementing adaptation measures remains a challenging endeavour on both sides. It involves matching output from the evolving climate models with the information needs of different end-users in order to better understand risks and adaptation options at the local scale. One particular challenge is dealing with the complexity and uncertainty inherent in the climate data when making decisions. It was found that the information of greatest interest to port authorities – the future impact of extreme events on seaward and landward operations – is least well known at the current time. However, this uncertainty should not be seen as an impediment to consideration of potential impacts and implementation of low-regret adaptation responses in the short term.

Over the course of the project, a hybrid approach considering current-day vulnerabilities and future risks was developed and adopted for the climate assessment. This involved multi-actor dialogue and a strong participatory process to enable local and sectorial expertise to inform the framing of the 'problem', particularly from a bottom-up vulnerability perspective. This was then complemented by analysis of a range of possible futures described by the CSIRO 'Climate Futures' framework to ensure consistency across the application of scenarios. This framework was extremely useful in allowing an understanding of the range of possible future climates, including those of lower probability but high impact.

Given the complexities involved, the research findings pointed towards an important enabling role for Australia's Federal Government. There is a need for a trusted knowledge platform that not only allows stakeholders to access climate data but also assists with the interpretation of it to inform local risk assessments and adaptation planning, including guidance for decision-making under uncertainty. A national-level resource would also bring a nationally consistent approach to data and methodologies, which is currently lacking. It is also important to note that climate-related risks are only one set of risks facing current and future port operations; any evaluation therefore needs to be contextualised within a broader set of drivers. Authoritative national-level guidance on the application of non-climate scenarios would also help to provide a common framework for future analysis by industry and academics.

It is worth highlighting that ports do not operate in isolation. Due to their importance as operational hubs they have ongoing relationships with logistics providers and local authorities, so it would be highly beneficial for any assessments to be conducted in partnership to ensure that there is a co-ordinated approach, particularly regarding long-term planning of land use.

To conclude, the research also identified emerging opportunities for improved evidence-based decision-making through the provision of the necessary scientific information to allow ports to make more informed 'climate-related' decisions. The ongoing development of national-level port and freight strategies provides highly opportune policy platforms for integrating climate considerations into forward thinking and resilience strengthening activity for a critical infrastructure sector of the nation.

Acknowledgements

This chapter summarises some of the methodology and findings from the NCCARF-funded project 'Enhancing the resilience of seaports to a changing climate'. This was a collaborative effort involving many colleagues and their input to the research activity is fully recognised. Details for each of the individual work packages, and researchers involved, can be found from project reports downloadable from the NCCARF website.

References

Becker, A., Inoue, S., Fischer, M. and Schwegler, B. (2012) Climate change impacts on international seaports: Knowledge, perceptions, and planning efforts among port administrators *Climatic Change* 110, 5–29.

Clarke, J.M., Whetton, P.H. and Hennessy, K.J. (2011) Providing application-specific climate projections datasets: CSIRO's Climate Futures Framework. 19th International Congress on Modelling and Simulation, Perth, Australia, 12–16 December 2011. Available at http://www.mssanz.org.au/modsim2011/F5/clarke.pdf (accessed 12 June 2014).

Cox, R., McEvoy, D., Allen, G., Bones, A. and McKellar, R. (2012) *National Climate Change Adaptation Research Plan: Settlements and Infrastructure: Update Report*, National Climate Change Adaptation Research Facility, Gold Coast.

Department for the Environment, Food and Rural Affairs (2012) *Adapting to Climate Change: Helping Key Sectors to Adapt to Climate Change*. Report for the Adaptation Reporting Power, PB 13740, Department for the Environment, Food and Rural Affairs, London.

Hanson S., Nicholls, R. and Ranger, N. (2011) A global ranking of port cities with high exposure to climate extremes, *Climatic Change* 104, 89–111.

Infrastructure Australia (2011) *National Ports Strategy: Infrastructure for an Economically, Socially, and Environmentally Sustainable Future*. Infrastructure Australia and the National Transport Commission, Commonwealth Government, Canberra.

Infrastructure Australia (2012) *Australian Infrastructure: Progress and Action*. A report to the Council of Australian Governments, Infrastructure Australia, Canberra.

International Association of Ports and Harbors (2011) *Seaports and Climate Change: An Analysis of Adaptation Measures*. Port Planning and Development Committee, International Association of Ports and Harbours, Hamburg, Germany.

Ng, A.K.Y., Chen, S.L., Cahoon, S., Brooks B. and Yang, Z. (2013) Climate change and the adaptation strategies of ports: The Australian experiences. *Research in Transportation Business and Management* 8, 186–194.

Nicholls R. J., Hanson, S., Herweijer, C. et al. (2008) *Ranking Port Cities with High Exposure and Vulnerability to Climate Extremes: Exposure Estimates, Environment Working Papers no.1, OECD*. Organisation for Economic Co-operation and Development, France.

Nursey-Bray, M., Blackwell, B., Brooks, B. et al. (2013) Vulnerabilities and adaptation of ports to climate change, *Journal of Environmental Planning and Management* 56(7), 1021–1045.

Peel Ports Group (2011) *Mersey Docks and Harbour Company Ltd., Climate Change Adaptation Report, Report to DEFRA under the Adaptation Reporting Powers.* Peel Ports Group, Manchester.

Stenek V., Amado, J-C., Connell, R. et al. (2011) *Climate Risks and Business Ports: Terminal Martimo Muelle et Bosque.* World Bank, Cartagena, Colombia.

23 Equity, economic efficiency and institutional capacity in adapting coastal settlements

CAMERON S. FLETCHER[1], BRUCE M. TAYLOR[1],
ALICIA N. RAMBALDI[2], BEN P. HARMAN[1],
SONJA HEYENGA[1], K. RENUKA GANEGODAGE[2],
FELIX LIPKIN[1] AND RYAN R.J. MCALLISTER[1]

[1]CSIRO Ecosystem Sciences, Australia
[2]The School of Economics, University of Queensland, Australia

23.1 Introduction and background

Australia has one of the lowest population densities in the developed world (ABS 2012) and a long and sparsely populated coastline (Williams 2007). At the same time, 84% of the population are concentrated within 50 km of the coast (ABS 2003) and many coastal settlements are at risk of future inundation as a result of rising sea level (DCC 2009). Many decisions about how and whether to adapt to protect these communities are made at the local government level, reflecting this widespread distribution of risk across many small communities (Measham et al. 2011).

In both Australian and international coastal communities the spatial distribution of inundation risk and the benefits of adaptation may be very finely structured, affecting some properties more than others, and raising the question of what adaptations will provide the most equitable distribution of risk (McDonald 2007). Even when adaptation is economically sensible, communities may not have the capacity or desire to invest in adaptations in the short and medium term (Turner et al. 1995). This may be due to absolute financial constraints, but may also reflect a lack of community consensus about which adaptations to invest in (Buitelaar et al. 2007) because of the expected distribution of risk within the community or differing assessments of the future risk of coastal inundation.

The most appropriate adaptations for a given location also depend on institutional factors. Some adaptation options may be more acceptable to decision-makers because they reflect incremental changes to already familiar risk-management institutions, or because they are perceived to have few negative side effects or costs (Buitelaar et al. 2007). On the other hand, some potentially effective adaptations may be less popular if they represent a significant change in prior behaviours and institutions (Primmer 2011). Moreover, the scale at which adaptation decisions are made and find acceptance and the scale at which they are funded may directly affect the scope of adaptations considered, their equitability and their effectiveness (Fletcher et al. 2013).

Applied Studies in Climate Adaptation, First Edition. Edited by Jean P. Palutikof, Sarah L. Boulter, Jon Barnett and David Rissik.
© 2015 John Wiley & Sons, Ltd. Published 2015 by John Wiley & Sons, Ltd.

While the local scale at which adaptation decisions are made and funded in Australia partly reflects the scale of our particular circumstances, it also constrains the options that are available for adaptation now and into the future (Adger et al. 2005; Urwin and Jordan 2008). Although local governments are well placed to manage highly localised inundation risk with a variety of responses, making decisions at this level also risks poor economies of scale and adaptations mismatched to the environment in which they sit or the problem that they address. In particular, very large-scale adaptations are much less likely to be considered due to both the financial and institutional constraints faced by individual local governments. In contrast, European countries generally protect smaller and more densely populated areas; some of these include significant cities that have developed below sea level (e.g. Rotterdam) or in the path of expected sea-level rise (e.g. Venice). Many of these larger adaptation projects involve planning, co-ordination and funding from state and national governments. Harman et al. (2013) provide further discussion and detailed comparisons between international and Australian community risk profiles, governance arrangements and adaptation practices.

Involving larger scales of governance may help compensate for some of the limitations of managing adaptation at the local scale in Australia (Harman et al. 2013). Our institutional analysis suggested that where the scale of the problem exceeds institutional capacity, local governments are relying on insurance premiums and market signals to provide much of the impetus for adaptation without requiring a large-scale co-ordinated approach (Fletcher et al. 2013). Local governments are also hesitant to expose themselves to compensation risks by changing planning rules in ways that reduce the development rights of private property holders (Measham et al. 2011). These conditions suggest that while local decision-making is well suited to the structure of inundation risk in Australian coastal communities, there are perhaps situations in which new institutions that share risk to higher levels of governance may achieve some of the benefits of

the larger-scale decision-making used in other parts of the world (Harman et al. 2013).

To help inform this discussion, we performed complementary institutional and economic analyses of six coastal communities in three local government areas in Queensland, Australia. The next section describes the analysis of interview data conducted with key decision-makers in each case study area to determine how institutional factors were likely to enable or constrain the consideration and implementation of different adaptation options within broad 'protect', 'accommodate' and 'retreat' adaptation categories. The following section describes an economic analysis on the affordability, equitability and economic efficiency of three specific adaptation options within these categories: a seawall, changed minimum floor heights and retreat. We conclude with a short list of insights relating the Australian experience to the international context that may help inform further policy development to manage inundation risk in our coastal communities.

23.2   Assessing institutional capacity

Institutional capacity for adaptation at the local government scale is vital in Australia, because it is at this scale that many discussions around adaptation occur and decisions are made (Measham et al. 2011). To assess institutional capacity we analysed how local actors in our case study regions appraised different adaptation options to coastal inundation risk within their broader institutional context. We conducted two stages of semi-structured interviews in the Moreton Bay, Cairns and Sunshine Coast local government areas, designed to consider the organisational, structural and cultural dimensions of different adaptation strategies within the 'protect', 'accommodate' and 'retreat' categories of adaptation options (Fletcher et al. 2013).

Building an understanding of these issues is important because the institutional landscape has changed over the last decade as debates about adaptation strategies have shifted from the realm of the science–policy arena to the practitioner domain. This has been most notable in the identification of sea-level rise benchmarks to inform planning as part of legislation and guidance from state-level jurisdictions to local governments (Gurran et al. 2011). The Queensland State Government adopted a sea-level rise planning factor of 0.8 m by 2100 based on the IPCC findings (DERM 2012), and local governments prepared coastal hazard adaptation strategies under the provisions of the State Planning Policy 3/11: Coastal Protection (DERM 2012). The recent suspension of this state policy has effectively devolved responsibility for adaptation back to local authorities.

In the first stage of the analysis, 18 interviews were conducted that focused on: (1) the perceptions of current and future impacts of coastal inundation from storm surge; (2) the existing strategies in use for managing coastal inundation

risks; (3) the adequacy of those strategies; and (4) their social and political acceptability. Responses from these initial interviews were analysed for the different categories of adaptation options. In the second stage, these results were presented along with the economic modelling analysis to the original 18 and 6 additional local officers in each case study area in three group-based semi-structured discussions of the prospects of different responses, including the institutional context in which they operate. Figure 23.1 shows the number of favourable, unfavourable and neutral comments made about each adaptation option during the interviews grouped into protect, accommodate and retreat adaptation categories.

Across all interviews and discussions, seawalls were the most frequently mentioned single adaptation and accommodate actions the most frequently and positively talked about category. In both cases, analysis of the discussion revealed

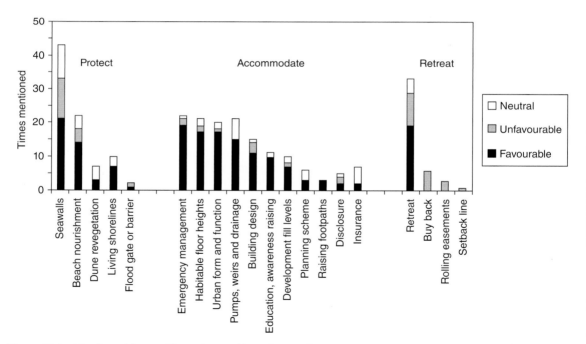

Figure 23.1 Number of favourable, unfavourable and neutral statements collated from interview transcripts across 18 personal interviews and three group interviews, involving a total of 24 respondents talking within their own case study areas about different adaptation options, grouped by category.

that familiarity with the adaptation was a key driver; decision-makers better understood and preferred incremental variations on established techniques for managing inundation risk and, vitally, felt that they would meet with greater acceptance both within local government and the broader community (Fletcher et al. 2013).

In contrast, while transformative changes such as retreat were talked about, they were often discussed as theoretical options for the future once sea levels had risen significantly rather than viable options for current consideration (Fletcher et al. 2013). This preference is not unexpected: previous work has shown that incremental responses that are able to be designed and implemented within the often-fragmented internal responsibilities of local government organisations are more likely to remain institutionally viable (Primmer 2011).

In addition to this preference for gradual incremental change, there is evidence that local governments are implicitly relying on 'outsourcing' to manage limitations in their own decision-making toolkits (Fletcher et al. 2013). For instance, many discussions focused on the importance of the insurance industry providing pricing signals through premiums to drive a market-based response to the equitable allocation of costs and benefits of managing storm surge. Outsourcing is a known institutional response where internal rules and practices are ill-equipped to address the spatial and temporal dimensions of risk sharing, for example. It is also a characteristic of multi-level environmental governance (Reed and Bruyneel 2010) previously observed in other Australian jurisdictions (Fletcher et al. 2013).

The results suggest that under existing institutional conditions local governments are unlikely to propose disruptive, unpopular or legally risky adaptations to manage coastal inundation risk even if they are likely to be effective. This suggests that gradual, incremental adaptations are likely to dominate discussions for the foreseeable future. Any major change to this pattern is likely only following episodes of community-led demand (Buitelaar et al. 2007) or direction from larger scales of governance (Reed and Bruyneel

2010), as seen in some high-profile adaptation projects internationally (Harman et al. 2013).

23.3   Economic analysis of affordability, equitability and efficiency

In parallel to the institutional analysis we created an economic model to calculate the distribution of the expected costs of storm surge inundation and the potential benefits of adaptation in terms of avoided costs. We considered the three most discussed economically model-able adaptation options within the 'protect' (seawall), 'accommodate' (changed minimum floor heights) and 'retreat' (retreat) categories. We investigated six case studies selected in collaboration with the Moreton Bay, Cairns and Sunshine Coast local governments. The case studies spanned a range of settlement types from the central business district of a coastal city to small coastal hamlets, and a range of inundation risks from highly exposed low-lying coastal strips to sheltered hamlets and canal estates.

The damage costs of inundation were calculated by estimating the depth of inundation on each property, and within each building, for an annual maximum storm surge event. The heights of these events were drawn from the observed extreme value distribution at each site (Gumbel 1958), with an offset to account for expected sea-level rise in future years. Sea-level rise was estimated using the global averaged SRES A1B sea-level rise scenario (Hunter 2010) with corrections for regional departures (CSIRO 2011), and generated sea-level rise of approximately 0.2 m by 2030 and 0.5 m by 2070 (Wang et al. 2010). These figures are broadly consistent with the rate of sea-level rise being factored into planning by local governments.

Inundation depths were converted to a dollar value of damage to infrastructure using published damage curves (Middelmann-Fernandes 2010). The devaluation of residential land relative to a fully protect property was estimated at $1.28\% + 5.45\%$ per metre of inundation during a 100-year Average Recurrence Interval event

(Rambaldi et al. 2013). This potential devaluation of land represents a vital component of the future impacts of sea-level rise on individual households as the land on which the family home rests represents the largest single asset of most Australians (Wilkins et al. 2009) and land values appreciate over time (Rambaldi et al. 2011). We estimated the potential avoided costs due to adaptation by implementing three types of adaptation within the model – seawalls, changed minimum floor heights and retreat – and rerunning it to compare with the unadapted case. The model was run a thousand times for each scenario to capture the impact of variation in the magnitude and timing of extreme events.

The expected per-property costs of adaptation, the affordability and the mean and median benefit-cost ratios of adapting each coastal community are presented in Table 23.1. A community-scale seawall adaptation was considered 'affordable' if a nominal $1000 annual levy, scaled to the local socio-economic conditions and contributed by every property in the case study, could completely fund construction by 2050. An adaptation was

considered 'economic' if the mean household benefit:cost ratio for each scenario averaged across all properties within the case study community was greater than one, indicating that across the community as a whole more damages were avoided than the cost of implementing the community-scale adaptation. An adaptation was considered 'equitable' if the median household benefit:cost ratio for each scenario was greater than 1, indicating that at least half the properties in the case study area achieved a net benefit from their contribution to a community-scale adaptation.

Affordability was primarily determined by the number of properties in each case study area and, to a much smaller degree, the socio-economic characteristics of the area. The mean benefit:cost ratio was greater than unity by 2100 for all case studies with a significant exposure to inundation risk (all except case study two). The median benefit:cost ratio provided drastically different results, however. In case studies three, five and six, although a seawall would have avoided more damages than it cost to construct, these benefits accrued to only a very small proportion of the

Table 23.1 Adaptation affordability, economy and equitability across 5 of the 6 case studies. Details for case study 2 are not shown because no property faced risk of inundation during an ARI 100-year event and only a small proportion were at risk by 2100, such that there was almost no economic benefit to households from adaptation. Case studies were chosen in conjunction with local council stakeholders.

Case study	Adaptation	Adaptation cost ($ × 10³/property)	Adaptation budget: cost	Mean benefit: cost	Median benefit: cost
1	Seawall	37	0.92	31.15	7.59
	Floor height	14	2.47	53.15	0.27
	Retreat	170	0.20	6.47	1.42
3	Seawall	34	0.87	11.22	0
	Floor height	7	4.08	35.61	0
	Retreat	59	0.50	6.43	0
4	Seawall	6	6.65	25.73	15.8
	Floor height	2	25.98	1.85	0.44
	Retreat	83	0.48	1.78	1.13
5	Seawall	9	5.46	4.09	0
	Floor height	1	40.12	0.65	0
	Retreat	56	0.92	0.71	0
6	Seawall	26	1.52	1.03	0
	Floor height	6	6.61	0.58	0
	Retreat	209	0.19	0.13	0

Table 23.2 A typology of settlement types, based around the economy, equitability and affordability of a community-scale seawall adaptation in each case study community.

Economic	Equitable	Affordable	Case study	Action
No	-	-	2	Do nothing
Yes	No	No	3	Retreat/household adaptation, e.g. raised floor heights
		Yes	5, 6	Household adaptation, e.g. raised floor heights
Yes	Yes	No	1	Funding from larger-scale government for community engineering, e.g. seawall
		Yes	4	Local council to fund community engineering, e.g. seawall

community with most achieving little to no direct benefit from their investment.

Based on these insights, Table 23.2 presents a typology of coastal settlements defined by their exposure to risk and the distribution of risk in the community, the potential benefits of adaptation and the potential capacity for adaptation in the community. The typology was determined by the suitability of a community-scale adaptation (in this case a seawall) to protect a community against future storm surge events around criteria of economic effectiveness, equitability and affordability. Results for the six case studies spanned five major categories. Case study two faced little risk of inundation, and there was no economic argument for adaptation to protect residential property. A traditional benefit:cost analysis of case studies three, five and six suggested that a community-scale adaptation such as a seawall would be economic. However, a more detailed analysis of the distribution of benefits showed that it was unlikely to be equitable, because almost all the benefit accrued to only a small proportion of the community. Case study four was a coastal Central Business District (CBD), and the high density of development and relatively uniform risk of inundation meant that a community-scale adaptation such as a seawall was economic, equitable and affordable. Case study one was a coastal suburb facing relatively uniform inundation risk such that a community-scale adaptation was economic and equitable. However, because of the relatively low density of

development at risk, construction of a seawall could not be affordably funded purely within the community. Cases like this suggest that co-funding or co-ordination from a higher level of governance, as is more common in many international settings, may be necessary in some situations to facilitate effective adaptation.

23.4 Conclusions

Our analysis and review of the international literature (Harman et al. 2013), along with further international comparisons in the full report (Fletcher et al. 2013), suggest four key findings relating the Australian experience to the international context that may help inform further policy development.

1. Australia has a unique distribution of risk due to its long coastline and relatively low population densities. This has led to well-developed institutions for making decisions about adaptations at local scales, but it also constrains the choice of adaptations to those which can be implemented and funded at the local scale.

2. Local governments 'outsource' some facets of the community adaptation process to other organisations such as the property market and the insurance sector. This is a rational and potentially effective approach, but can limit their ability to implement longer-term plans and may consequently expose communities to increased risk.

3. There will be cases where particular adaptation options (e.g. seawalls) are equitable and economically feasible, but not affordable at the local scale. Addressing this constraint means identifying and consistently implementing suitable cost-sharing models and risk-sharing policies between local and higher levels of government.

4. In the absence of mechanisms whereby local governments can share risk it is unlikely they will propose disruptive, unpopular or legally risky adaptations, even if incremental responses become suboptimal or increase risk over the long term. Lessons from the international context may provide important insights into how to most effectively combine state and national sponsorship with local implementation of coastal adaptation.

References

Adger, W.N., Arnell, N.W. and Tompkins, E.L. (2005) Successful adaptation to climate change across scales. *Global Environmental Change* 15, 77–86.

Australian Bureau of Statistics (ABS) (2003) *Census of Population and Housing: Population Growth and Distribution. Australia.* Australian Bureau of Statistics, Canberra.

Australian Bureau of Statistics (ABS) (2012) *Population clock (online).* Australian Bureau of Statistics. Available at http://www.abs.gov.au/AUSSTATS/abs@. nsf/Web+Pages/Population+Clock?opendocument (accessed 28 May 2014).

Buitelaar, E., Lagendijk, A. and Jacobs, W. (2007) A theory of institutional change: illustrated by Dutch city-provinces and Dutch land policy. *Environment and planning A* 39, 891.

CSIRO (2011) *Sea Level Rise: Understanding the past: Improving projections for the future.* Commonwealth Scientific and Industrial Research Organisation. Available at http://www.cmar.csiro.au/sealevel/sl_proj_regional.html (accessed 28 May 2014).

Department of Climate Change (2009) *Climate Change Risks to Australia's Coast: A First Pass National Assessment.* Department of Climate Change, Canberra, Australia.

Department of Environment and Resource Management (2012) *Queensland Coastal Plan.* Queensland Department of Environment and Resource Management. Available at http://www.ehp.qld.gov. au/coastalplan/pdf/qcp-web.pdf (accessed 28 May 2014).

Fletcher, C.S., Taylor, B.M. and Rambaldi, A.N. (2013) *Costs and Coasts: An Empirical Assessment of Physical and Institutional Climate Adaptation Pathways.* National Climate Change Adaptation Research Facility, Gold Coast.

Gumbel, E.J. (1958) *Statistics of Extremes.* Columbia University Press, New York.

Gurran, N., Norman, B., Gilbert, C. and Hamin, E. (2011) *Planning for Climate Change Adaptation in Coastal Australia: State of Practice. Report No. 4 for the National Sea Change Taskforce.* Faculty of Architecture, Design and Planning, University of Sydney, Sydney.

Harman, B.P., Heyenga, S., Taylor, B.M. and Fletcher, C.S. (2013) Global lessons for adapting coastal communities to protect against storm surge inundation. *Journal of Coastal Research*, published online 18 November 2013, doi: 10.2112/jcoastres-d-13-00095.1.

Hunter, J. (2010) Estimating sea-level extremes under conditions of uncertain sea-level rise. *Climatic Change* 99, 331–350.

McDonald, J. (2007) The adaptation imperative: managing the legal risks of climate change impacts. In: Bonyhady, T. and Christoff, P. (eds) *Climate Law in Australia.* The Federation Press, Sydney.

Measham, T.G., Preston, B.L. and Smith, T.F. (2011) Adapting to climate change through local municipal planning: barriers and challenges. *Mitigation and Adaptation Strategies for Global Change* 16, 889–909.

Middelmann-Fernandes, M. (2010) Flood damage estimation beyond stage–damage functions: an Australian example. *Journal of Flood Risk Management* 3, 88–96.

Primmer, E. (2011) Analysis of institutional adaptation: integration of biodiversity conservation into forestry. *Journal of Cleaner Production* 19, 1822–1832.

Rambaldi, A.N., McAllister, R.R.J., Collins, K. and Fletcher, C.S. (2011) *An Unobserved Components Approach to Separating Land from Structure in Property Prices: A Case Study for the City of Brisbane. School of Economics Discussion Paper.* University of Queensland, Brisbane.

Rambaldi, A.N., Fletcher, C.S., Collins, K. and McAllister, R.R.J. (2013) Housing Shadow Prices in an Inundation-prone Suburb. *Urban Studies* 50, 1889–1905.

Reed, M.G. and Bruyneel, S. (2010) Rescaling environmental governance, rethinking the state: A three-dimensional review. *Progress in Human Geography* 34, 646–653.

Turner, R.K., Adger, N. and Doktor, P. (1995) Assessing the economic costs of sea level rise. *Environment and Planning A* 27, 1777–1796.

Urwin, K. and Jordan, A. (2008) Does public policy support or undermine climate change adaptation? Exploring policy interplay across different scales of governance. *Global Environmental Change* 18, 180–191.

Wang, X., Stafford Smith, M., McAllister, R.R.J., Leitch, A., McFallan, S. and Meharg, S. (2010) *Coastal Inundation Under Climate Change: A Case Study in South East Queensland.* Commonwealth Scientific and Industrial Research Organisation, Canberra.

Wilkins, R., Warren, D. and Hahn, M. (2009) *Families, Incomes and Jobs, Volume 4: A Statistical Report on Waves 1 to 6 of the HILDA Survey.* Melbourne Institute of Applied Economic and Social Research, The University of Melbourne, Melbourne.

Williams, P. (2007) Government, people and politics. In: Thompson, S. (ed.) *Planning Australia: An Overview of Urban and Regional Planning.* Cambridge University Press, UK, pp. 29–48.

24   Who should do what? Public perceptions on responsibility for sea-level rise adaptation

ELISSA WATERS AND JON BARNETT

School of Geography, University of Melbourne, Australia

24.1   Introduction

As adaptation theory and practice evolve it is becoming clear that a key barrier to adaptation is the lack of clarity, at all scales, on the roles and responsibilities of different levels of government, public and private institutions and individuals (Tompkins and Adger 2005; Swart et al. 2009; Jantarasami et al. 2010; Beisbroek et al. 2011; Falaleeva et al. 2011; Productivity Commission 2011). However, assigning responsibility for adaptation is easier said than done given the heterogeneity of climate change risks, the range of adaptation activities associated with any given risk, the range of actors who may engage in such activities and the multiple scales over which adaptation occurs. Although an important determinant of effective adaptation, clarifying roles and responsibility for adaptation is therefore not easy.

In Australia, as in North America and Europe, substantial political effort has gone into negotiating the distribution of responsibility for adaptation (see Coastal Climate Change Advisory Committee 2010; Select Committee on Climate Change of the Council of Australian Governments 2012; Department of Sustainability Environment 2013). Nevertheless, there is persistent uncertainty around roles and responsibilities for adaptation to the risks posed by climate change, including those arising from sea-level rise (the focus of this chapter).

In Australia at present, responsibilities for coastal and sea-level rise planning reflect a complex mix of formal and informal management arrangements that are often shared and duplicated across levels of government, organisations and the public and private sectors (Lazarow et al. 2006). At a national level the Coastal Climate Change Advisory Committee has outlined these complexities, and has stated that the resolution of problems around responsibilities of the various coastal management actors and authorities is a matter of urgency (Coastal Climate Change Advisory Committee 2010). How this resolution might come about is still a matter of debate in both academic and public policy sectors. The situation is not helped by the absence of empirical research focused directly on this problem. Among the knowledge gaps there is a lack of evidence

Applied Studies in Climate Adaptation, First Edition. Edited by Jean P. Palutikof, Sarah L. Boulter, Jon Barnett and David Rissik.
© 2015 John Wiley & Sons, Ltd. Published 2015 by John Wiley & Sons, Ltd.

about the preferences of coastal people with respect to the distribution of responsibility for adaptation to sea-level rise. Such evidence is important for legitimate adaptation policies.

This chapter reports some key findings of a project that investigated community preferences for the distribution of responsibility for adaptation to sea-level rise (for the full report see Barnett et al. 2013). We draw on data from 80 hour-long interviews with coastal residents in two coastal communities. The interviews elicited preferences for the distribution of responsibilities for key adaptation tasks such as managing public and private assets and bearing the costs of adaptation. The findings reveal clear preferences for distributions of responsibilities at different scales. This information can inform efforts to allocate responsibility for adaptation to sea-level rise (Barnett et al. 2013).

24.2   Investigating community preferences for responsibility

The research project focused on two local government areas in Australia: Eurobodalla Shire in New South Wales and Mornington Peninsula Shire in Victoria. These two non-urban areas were chosen for two main comparative aspects: both had experienced the types of coastal problems that are analogous to those of sea-level rise; and both local governments were undertaking adaptation planning of some kind. There is therefore relevant experience with coastal changes and with adaptation policies at both locations.

In total, 80 semi-structured interviews were conducted across the two case study locations ($n = 37$ in Eurobodalla and $n = 43$ in Mornington Peninsula). A targeted sample was developed using a method of chain referral, which consisted of homeowners, business owners and coastal managers. The interviews, which each were approximately an hour in length, were designed to elicit considered opinions on coastal management, the risk of sea-level rise and responsibility for adaptation to sea-level rise. In particular we were interested in how respondents viewed the

appropriate distributions of responsibility for different tasks associated with adaptation, so the interview focused on five distinct responsibilities: providing information or creating knowledge; managing public assets; managing private assets; undertaking local planning; and bearing the costs of adaptation.

The following sections describe the responses to these five questions. The interviews were not closed questioned surveys but structured discussions in which participants developed their views through considered and iterative deliberation and conversation. To capture key findings, we coded and counted responses and show these as charts. The figures which follow show the total number of mentions of a particular actor or agency aggregated to each local government study site (total responses therefore do not sum to 100%). The value of this data is that it shows the overarching preferences for particular governance models. We also explain the logic behind these distributions of preferences for responsibility.

24.2.1   Preferences for responsibility for information provision

As shown in Figure 24.1, opinions on who should be responsible for providing information and creating knowledge on the risk of sea-level rise showed a clear preference for a role for the top two tiers of government. This was predominantly due to the scale and cost of information provision for adaptation, for which local governments were seen to be under resourced.

> I think it's a bit ridiculous to be asking the least resourced level of government to essentially go out and purchase advice that's so critical to how we manage the coast. (Interview 2, Manager, Eurobodalla)

Respondents also expressed significant concern about the credibility of information coming from local government due to the perceived influence of vested interests in local and state politics. Many believed state and local governments were more exposed to manipulation from external interest groups, particularly developers, whereas the federal

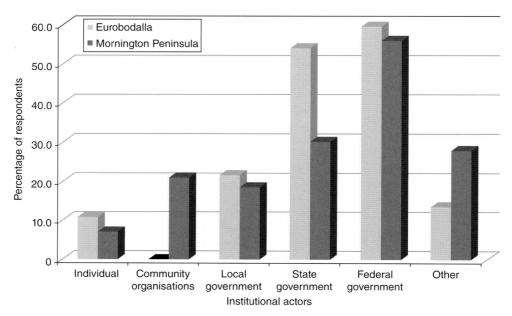

Figure 24.1 Preferences for responsibility for information provision on the risks of sea-level rise. Respondents were able to choose more than one category.

government was able to better maintain impartiality. In the words of a respondent:

> The federal government is the only one without a close vested interest, in my opinion. My experience here is that all coastal decisions are influenced by vested interests, whether it's on the state level or the local council level. (Interview 43, Home owner, Mornington Peninsula)

Overall, respondents saw information provision and knowledge creation as an important requirement for adaptation, and the majority identified state and federal governments as having the primary responsibility. Issues of capacity and trust explain people's preferences for the distribution of responsibilities.

24.2.2 Preferences for responsibility for managing public assets

Local Government was preferred as the primary responsible entity for decision-making for public assets (Figure 24.2). In most cases this was attributed to the belief that public asset management requires a sound knowledge of local environmental conditions and the specific coastal contexts. As one respondent explained:

> They're the only ones that know their local area and the situations, the winds, the tides, the weather events. The local government's the only one's got a real handle on what's happening up and down our coast. (Interview 19, Community Organisation Leader, Eurobodalla)

The interviewees indicated that this responsibility would need to be supported with funding and co-ordination from higher levels of government. The state government was the preferred higher-order actor responsible for consistency in decision-making across local government areas:

> Well I suppose it has to be done at state level and local government would have some responsibility. I think there needs to be uniformity along the coast. It's no good one little local government

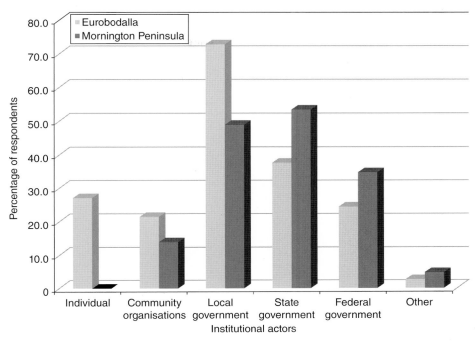

Figure 24.2 Preferences for responsibility for managing public assets for adaptation to sea-level rise. Respondents were able to choose more than one category.

doing one thing and another neighbouring one doing something different. (Interview 5, Home Owner, Eurobodalla)

Overall these results suggest a preference for a strong role for local government in the management of public assets for adaptation, which is an extension of their current roles in town planning and coastal management.

24.2.3 Preferences for responsibility for managing private assets

Figure 24.3 shows that preferences for responsibility for managing private assets favoured a combination of individual responsibility, but with government regulation to steer responses and ensure fairness. As explained by one respondent:

The individual property owner has to determine what's best and what they are willing to pay for and endure from the point of view of risk. But

they have to work within some sort of framework otherwise it will get out of hand. (Interview 37, Home Owner, Mornington Peninsula)

Interestingly, despite recent planning controversies in both Eurobodalla and Mornington Peninsula shires, most respondents were happy to see local governments continue to remain the primary agency responsible for regulating private actions. Unlike other discussions around responsibility, the management of private assets were seen to be the responsibility of only four groups (three levels of government and individuals); community organisations and 'other' options were almost entirely overlooked in responses.

24.2.4 Preferences for responsibility for local planning

Relative to the other four adaptation tasks which respondents were asked about, there was a preference for a distribution of responsibility for

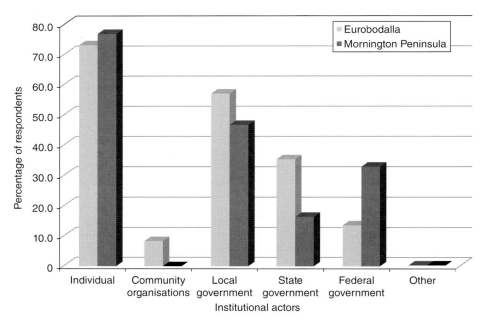

Figure 24.3 Preferences for responsibility for managing private assets for adaptation to sea-level rise. Respondents were able to choose more than one category.

local planning across multiple groups as shown in Figure 24.4. Very rarely did anyone see local planning as being the sole responsibility of any single agency or individual. People want to see governments collaborate on planning, but they also want to ensure the knowledge and skills of local groups and governments are influential. Somewhat like the question concerning information provision, there was also an element of desire to see higher levels of government provide oversight to check against policy capture at local levels, and to bolster resources where necessary.

Community organisations were identified as actors more often in this question than any other question. Much of the justification for this was based on a belief that community organisations are able to represent a range of local interests and are able to more effectively communicate local knowledge:

It will come from residents and community groups because of the need to incorporate local

knowledge, which I think is often overlooked. I think quite often the big view down from academia and others is that there is a set of operating principles which really has not a lot to do with the local variances. (Interview 51, Business Owner, Mornington Peninsula)

24.2.5 Preferences for responsibility for cost bearing for adaptation

As shown in Figure 24.5, the interviewees in both study sites had a preference for federal government to bear the costs of adaptation options. The main reason for this was the belief that the federal government is the only entity that has the capacity to raise the revenue necessary to cover the considerable costs of adaptation:

The federal government because they've got the coffers, they've got the GST, they've got the whole lot. I mean, it would have to come from federal government. (Interview 35, Manager, Mornington Peninsula)

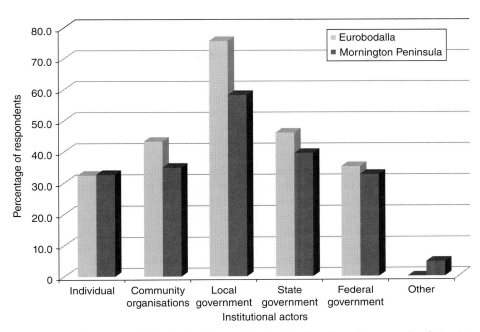

Figure 24.4 Preferences for responsibility for local strategic planning for adaptation to sea-level rise. Respondents were able to choose more than one category.

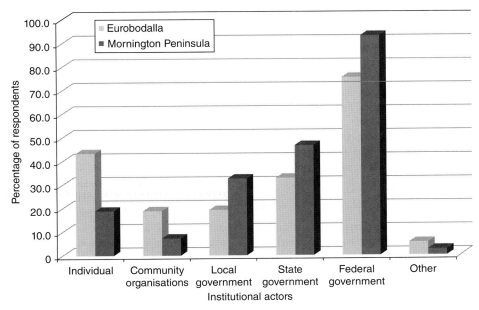

Figure 24.5 Preferences for responsibility for bearing the costs of adaptation to sea-level rise. Respondents were able to choose more than one category.

There was recognition by a majority of interviewees that the nature of federal taxation means that all residents of Australia will be contributing financially under this model. Many justified this position by characterising adaptation as a collective problem, for example:

> It's a collective, yes. At the end of the day everyone chips in. The same with our sewage system, our roads, everyone chips in. (Interview 27, Manager, Mornington Peninsula)

This question elicited very complex and sophisticated discussions about the nature of individual and collective responsibility, the practicalities of adaptation and issues of fairness and efficiency. It is important to note that this result does not reflect a simple desire from respondents to avoid paying for adaptation; indeed, as shown in Figure 24.5, individuals do accept the need to meet some of the costs.

24.3   Conclusion

This project sought to investigate community preferences for responsibility for adaptation to sea-level rise. Figure 24.6 outlines the majority preferences for different adaptation roles across the three tiers of government. Across the 80 respondents in the two case study locations there was a strong preference for a significant role for government in adaptation. Local government was seen to be best placed to manage public assets, regulate decision-making for private assets and lead and co-ordinate public input for local planning. Federal government was viewed as the most appropriate entity to take responsibility for information provision on the risks of sea-level rise and to bear the costs of adaptation. State governments, while not viewed as the primary responsible entity for any of these key tasks, was seen to have a role in co-ordinating adaptation actions across local government areas.

As noted in Section 24.1, there is limited agreement about the distribution of responsibility for adaptation to sea-level rise in Australia and therefore public policy and legislation on the issue remain unclear. This study helps to address one critical aspect of this uncertainty: we now have a better idea of public support for particular roles and responsibilities of different levels of government. Given that these data are from 80 interviews across two local government areas, the

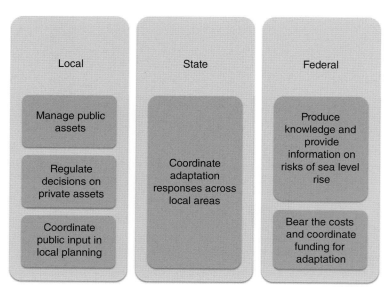

Figure 24.6   Summary of preferences for the role of different levels of government in adaptation to sea-level rise.

results cannot be said to represent the consensus of coastal residents. However, they clearly show that coastal residents in Australia want governments to engage on adaptation and that they are able and willing to have informed and considered discussions about adaptation governance. This supports a case for broader consultation with coastal communities as a way of encouraging public policy development on this issue.

While these results are limited to the context of Australia and sea-level rise, the international literature on scale and governance for adaptation can benefit from such situated and contextualised examples of how people actually conceive of ideal distributions of governance for adaptation at scale, and importantly what rationales drive those views. Our results therefore have broader significance than the simple construction of an ideal governance regime for a discrete climate hazard in Australia; they also inform global debate on scale and governance in climate change adaptation theory, which concerns the way adaptation responsibility can be best distributed and coordinated at vertical and horizontal scales (Adger 2001; Adger et al. 2005; Juhola and Westerhoff 2011). For example, the interviewees in this study attributed responsibility for the practical aspects of adaptation (decision-making, planning and asset management) to local authorities and groups, which supports emerging theory about the importance of local autonomy in adaptation decisions and outcomes (Few et al. 2007; Agrawal 2008). Further, our finding that coastal residents prefer the national government to be primarily responsible for information provision and cost bearing (supported by federal taxes) is important information for economic and political debates about the extent to which adaptation is or should be a collective responsibility to be managed nationally, or a more local matter with responsibility devolved to local governments and individuals at risk (Farber 2007; Macintosh 2012).

The interviews revealed a variety of reasons behind people's choices including assessments of the capacity of particular levels of government, the equity or fairness consequences associated with particular scalar configurations of responsibility and levels of familiarity with current governance arrangements. These reasons ground individual preferences in more fundamental debates on principles for adaptation governance, and have the potential to be used as platforms for further research on the distributive justice aspects of adaptation. Ultimately, in the vacuum of agreement about how to effectively distribute roles and responsibilities to deal with the multiple challenges of sea-level rise adaptation, these preferences represent a unique and valuable source of information.

References

Adger, N. (2001) Scales of governance and environmental justice for adaptation and mitigation of climate change. *Journal of International Development* 13(7), 921–931.

Adger, N., Arnell, N. and Tompkins, E. (2005) Successful adaptation to climate change across scales. *Global Environmental Change* 15(2), 77–86.

Agrawal, A. (2008) *The Role of Local Institutions in Adaptation to Climate Change*. The World Bank, Washington.

Barnett, J., Waters, E., Pendergast, S. and Puleston, A. (2013) *Barriers to Adaptation to Sea Level Rise*. National Climate Change Research Facility, Gold Coast.

Beisbroek, R., Klostermann, J., Termeer, C. and Kabat, P. (2011) Barriers to climate change adaptation in the Netherlands. *Climate Law* 2, 181–199.

Coastal Climate Change Advisory Committee (2010) *Issues and Options Papers Main Report*, The Victorian Government, Melbourne.

Department of Sustainability Environment (2013) *Victorian Climate Change Adaptation Plan*. The Victorian State Government, Victoria.

Falaleeva, M., O'Mahony, C., Gray, S., Desmond, M., Gaut, J. and Cummins, V. (2011) Towards climate adaptation and coastal governance in Ireland: Integrated architecture for effective management? *Marine Policy* 35(6), 784–793.

Farber, D. (2007) Adapting to climate change: Who should pay. *Journal of Land Use and Environmental Law* 23(1), 1–38.

Few, R., Brown, K. and Tompkins, E. (2007) Public participation and climate change adaptation: avoiding the illusion of inclusion. *Climate Policy* 7, 46–59.

Jantarasami, L., Lawler, J. and Thomas, C.W. (2010) Institutional barriers to climate change adaptation in US national parks and forests. *Ecology and Society* 15(4), 33.

Juhola, S. and Westerhoff, L. (2011) Challenges of adaptation to climate change across multiple scales: a case study of network governance in two European countries. *Environmental Science and Policy* 14(3), 239–247.

Lazarow, N., Souter, R., Fearon, R. and Dovers, S. (eds) (2006) *Coastal Management in Australia: Key Institutional and Governance Issues for Coastal Natural Resource Management.* Coastal CRC, Brisbane.

Macintosh, A. (2012) Coastal climate hazards and urban planning: how planning responses can lead to maladaptation. *Mitigation and Adaptation Strategies for Global Change* 18, 1035–55.

Productivity Commission (2011) *Barriers to Effective Climate Change Adaptation.* Commonwealth of Australia, Canberra.

Select Committee on Climate Change of the Council of Australian Governments (2012) *Roles and Responsibilities for Climate Change Adaptation.* The Australian Government, Canberra.

Swart, R., Biesbroek, R., Binnerup, S. et al. (2009) *Europe Adapts to Climate Change: Comparing National Adaptation Strategies.* PEER Report No 1. Partnership for European Environmental Research, Helsinki.

Tompkins, E.L. and Adger, N.W. (2005) Defining response capacity to enhance climate change policy. *Environmental Science and Policy* 8(6), 562–571.

Section 5

Building resilience among vulnerable groups

25   The 'turn to capacity' in vulnerability research

HALLIE EAKIN

School of Sustainability, Arizona State University, USA

Although societies globally continue to shun decisive action on greenhouse gas mitigation, climate change permeates the imagery of the twenty-first century. Glaciers melt, wildfires rage, shorelines are demolished by unprecedented floods and storm surges and harvests are lost to devastating drought: whatever our predilection for risk or ideological leanings, the climatic conditions we once considered part of the descriptive, stable geographies of human activity are undeniably changing at unprecedented rates and with unexpected consequences. It is in this context of social inertia and dramatic global change that the efforts of vulnerability analysis in the 1980s and 1990s to define and delimit climatic exposures and sensitivities have given way to a focus on capacity and, implicitly or explicitly, human agency. This focus typically emerges from the rationale that if we cannot eliminate the environmental risks that now consume us, we must develop new abilities to cope, live with and even master such risks in order to maintain (or possibly deliberately transform) life as we know it. While the 'turn to capacity' exemplifies a new realism in the face of climate change, it also echoes – at least in Western societies and those nations where neoliberal ideals have taken deep root – a prevalent belief in the individual, the entrepreneur, the local-level innovator and consensual communities as the sources of solutions to the inordinate social and environmental challenges of this century.

This focus on capacity as a solution is both inspiring and problematic for vulnerability analysts: prior decades of research have demonstrated that it is precisely limitations to individual capacities for social mobilisation and collective action, deficiencies in social capital and structural processes undermining trust, accountability and ultimately just and legitimate governance that characterise the social and political context of the world's most vulnerable populations (e.g. Hewitt 1983; Blaikie et al. 1994; Adger 2006). Rather than mobilising individual capacities, the conclusions of much of this research points to the need for fundamental structural change to address the social injustices and inequities that create and perpetuate the underlying social conditions of vulnerability.

By focusing on the capacities or lack of capacities in *a priori* vulnerable populations, we may in fact be pursuing only a partial line of inquiry. Understanding the production of vulnerability is as much about understanding what conditions, structures and resources have enabled some populations to escape the label of 'vulnerable' as it is about what attributes the 'vulnerable' have or do

Applied Studies in Climate Adaptation, First Edition. Edited by Jean P. Palutikof, Sarah L. Boulter, Jon Barnett and David Rissik.
© 2015 John Wiley & Sons, Ltd. Published 2015 by John Wiley & Sons, Ltd.

not have. In other words, vulnerability assessment must also ask how is individual capacity formed in conditions of extreme disenfranchisement and chronic exposure to stress, what enables individual capacity to generate abilities for collective action and what attributes enable populations to influence the conditions that perpetuate their disenfranchisement.

Capacity has typically been presented as the unique 'positive' attribute of vulnerability: the counterweight to exposure and sensitivity. The bulk of research on capacity has focused on its external dimensions: the material and immaterial resources and the assets and entitlements that predetermine the decision options available to an actor at any point in time to cope with losses and to anticipate future harm. Yet while these external dimensions are relatively more measureable and observable, they are less easily empirically linked to actual vulnerability reduction in causal chains. In other words, the functions these dimensions of capacity play in mediating exposure and sensitivity and ultimately reducing future harm is largely undemonstrated. In deference to Sen's theorisation of the instrumental roles of basic human freedoms in development (Sen 1999), we can look at capacity not in terms of its 'constitutive' elements (what individuals and households *have*, as end goals of vulnerability reduction) but rather as a source of influence and mobilisation. Capacity may provide an important conceptual structure for understanding the cross-scalar and networked nature of risk and loss in the twenty-first century. We have done a particularly poor job in understanding how *generic* capacities (those capacities most associated with Sen's conceptualisation of basic human freedoms) and *specific* capacities (those capacities associated specifically with climate risk reduction) interact and the possible trade-offs entailed in building these different types of capacity in different geographic contexts (Lemos et al. 2013).

While there is still far more work to do on the external political dimensions of capacity, we have just as much to learn about internal cognitive dimensions. Over the last decade, a number of scholars have invoked discussions of 1960s and 1970s on the cognitive and psychological aspects of vulnerability by illuminating those aspects of cognition, identity and perception that may be instrumental in both sensitivity to loss and ability to anticipate and manage future change. Identity, place attachment and willingness to change livelihood activities are once again considered important signifiers of vulnerability in the face of transformations in climatic conditions (Marshall 2010; Frank et al. 2011). Individuals must believe that they have the capacity to alter their vulnerability in order to act (Grothmann and Patt 2005). Self-efficacy matters.

Linking the internal cognitive attributes of capacity to the more instrumental and political dimensions of agency is one of the next challenges in vulnerability research. The concept of social learning and critical reflection and their association with collective action hold promise. Reflection and reflexivity lead not only to deeper understanding of the drivers of one's own vulnerability but also how, collectively, cognitive strategies and social behaviour may exacerbate vulnerability in the long run (Tschakert and Dietrich 2010). There are tensions here also: at what point does the fundamental optimism in the capacity of individuals to be empowered and emancipated from the risks that threaten to overwhelm them give way to frustration at their lack of control and influence over the structural processes that constrain their choice and limit opportunities? In contexts where failures in fundamental social contracts and the structural conditions of vulnerability have eroded individual internal as well as external capacities, is the emphasis on building local capacity misplaced and even misguided?

It is in this line of questioning that the old concerns over changing the structural determinants of vulnerability – the *ultimate* social causes of both chronic and emerging loss and harm – are resurfacing. Individual and community-level capacity building falls short of addressing the larger concerns of persistent global inequities in resource distribution and human welfare opportunity. While ecological resilience approaches have tended to assume the collective capacity to manage and alter system structure and function, vulnerability scholars have highlighted the insistent role of social position, power and privilege in influencing the normative outcomes of systemic transformation. Vulnerability scholars need to make better use of existing and emerging theories of the mechanisms of influence and control among disadvantaged and privileged populations in face of global environmental change; the politics of vulnerability and adaptation are likely to be far more instrumental in patterns of harm and loss than the environmental drivers that are the focus of so much of our attention (Manuel-Navarrete 2010). It is no easy thing to create capacities for social mobilisation and action among vulnerable actors, particularly in the face of entrenched interests and established formal and informal institutions that reproduce conditions of vulnerability (Wood 2003).

Finally, we can no longer assume that vulnerability is bounded in space and time in ways that are convenient for scientific analysis and policy intervention, but increasingly limited in explanatory power. Events of the last decade have brought home the fact that vulnerability is not easily constrained by the discrete units of analysis and the political, administrative and decision-making boundaries we've customarily used to organise social space and interactions such as sectors, households, cities and countries (Eakin 2010). Exposure to risk, experiences of stress and processes of social and environmental change are concurrent, spatially correlated and networked. Exposure and sensitivities to shocks are communicated via markets, tightly organised information and institutional networks, flows of virtual and material capital and movements of people and ideas (Eakin et al. 2009). Vulnerability is a networked and teleconnected phenomenon (Adger et al. 2009); climate signals are experienced directly but also indirectly, diffusely and synergistically with the diversity of stressors, stimuli and dynamics of the globalised economy and culture.

We have yet to understand adequately how vulnerabilities are produced, reproduced, amplified and transmitted in this interconnected world, and what specific capacities are associated with a population's or individual's ability to effectively negotiate risk in this context. Information and knowledge networks and communication systems (ICTs), social media and commercial ties all communicate risks and create the bridging capital that may be instrumental in creating conditions of social change and vulnerability reduction. As a result of these emergent networks governance is also changing, potentially enabling distal connections to be mobilised to reduce local vulnerabilities. Underlying the call for new social contracts (O'Brien et al. 2009; Pelling 2011) and renewed moral commitments across place and times is the recognition that, while perhaps inevitable, we are all complicit in the inequities in local and global vulnerability distribution. The 'turn to capacity' in vulnerability research may be the best way to mobilise this recognition for the structural and transformative changes that are so necessary to address vulnerability at its roots.

References

Adger, W.N. (2006) Vulnerability. *Global Environmental Change* 16, 268–281.

Adger, W.N., Eakin, H. and Winkles, A. (2009) Nested and networked vulnerabilities to global environmental change. *Frontiers in Ecology and the Environment* 7, 150–157.

Blakie, P., Cannon, T., Davis, I. and Wisner, B. (1994) *At Risk: Natural Hazards, People's Vulnerability and Disasters*. Routledge, London.

Eakin, H. (2010) What is vulnerable? In: Ingram, J., Ericksen, P.J. and Liverman, D. (eds) *Food Security and Global Environmental Change*. Earthscan, London.

Eakin, H., Winkels, A. and Sendzimir, J. (2009) Nested vulnerability: exploring cross-scale linkages and vulnerability teleconnections in Mexican and Vietnamese coffee systems. *Environmental Science and Policy* 12, 398–412.

Frank, E., Eakin, H. and Carr, D. (2011) Social identity, perception and motivation in adaptation to climate risk in the coffee sector of Chiapas, Mexico. *Global Environmental Change* 21, 66–76.

Grothmann, T. and Patt, A. (2005) Adaptive capacity and human cognition: The process of individual adaptation to climate change. *Global Environmental Change* 15, 199–213.

Hewitt, K. (1983) The idea of calamity in a techno-cratic age. In: Hewitt, K. (ed.) *Interpretations of Calamity*. Allen and Unwin, Winchester.

Lemos, M.C., Agarwal, A., Eakin, H., Nelson, D.R., Engle, N. and Johns, O. (2013) Building adaptive capacity to climate change in less developed countries. In: Asrar, G.R. and Hurrell, J.W. (eds) *Climate Science for Serving Society: Research, Modeling and Prediction Priorities*. Springer, Dordrecht.

Manuel-Navarrete, D. (2010) Power, realism, and the ideal of human emancipation in a climate of change. *Wiley Interdisciplinary Reviews-Climate Change* 1, 781–785.

Marshall, N.A. (2010) Understanding social resilience to climate variability in primary enterprises and industries. *Global Environmental Change* 20, 36–43.

O'Brien, K., Hayward, B. and Berkes, F. (2009) Rethinking social contracts: building resilience in a changing climate. *Ecology and Society* 14(2), 12.

Pelling, M. (2011) *Adaptation to Climate Change: From Resilience to Transformation*. Routledge, New York.

Sen, A. (1999) *Development as Freedom*. Knopf, New York.

Tschakert, P. and Dietrich, K.A. (2010) Anticipatory learning for climate change adaptation and resilience. *Ecology and Society*, 15(2), 11.

Wood, G. (2003). Staying secure, staying poor: the 'Faustian Bargain'. *World Development* 31, 455–471.

26 The limits to adaptation: a comparative analysis

JON BARNETT[1] AND JEAN P. PALUTIKOF[2]

[1] *School of Geography, University of Melbourne, Australia*
[2] *National Climate Change Adaptation Research Facility, Griffith University, Australia*

26.1 Introduction

Adaptation is essential to address the unavoidable impacts of climate change. Until recently, much of the research on adaptation avoided the question of what adaptation cannot achieve, implying by omission that adaptation can avoid all climate impacts. This is clearly not going to be the case for many systems, sectors and places at even modest rates of warming however, let alone at the more rapid rates of warming that seem almost inevitable (Meinshausen et al. 2009). The capacity of natural and human systems to adapt to climate change is limited by the severity of the climatic perturbation and/or by the nature of the social and ecological systems at risk (Dow et al. 2013a). In this chapter, we explore six case studies of climate change impacts in Australia, the scope and opportunity for adaptation, the possibility that limits to adaptation exist and their nature and causes.

Dow et al. define the 'limits to adaptation' as 'the point at which, despite adaptive action, an actor can no longer secure valued objectives from intolerable risk' (Dow et al. 2013b, p. 389). They consider that this point varies between actors and over time as knowledge and technology develop, and as attitudes to risk change. Adger et al. (2009) emphasise this mutability. They note that the discourse around adaptation limits is commonly constructed around three 'hard' exogenous limits or thresholds: ecological and physical, economic and technological. They propose a fourth mutable set of limits, stating that it is the values, perceptions, processes and power structures within society that impose limits more often than exogenous factors. Other authors, for example Moser and Ekstrom (2010), prefer to distinguish between 'limits' to adaptation which are absolute and 'barriers' which are malleable. Whatever nomenclature we use, once we introduce the concept of mutability the distinction between incremental and transformational adaptation has relevance. Where incremental adaptation has failed to overcome a limit, there remains the possibility that transformational change could provide options and strategies (Park et al. 2012; Klein et al. 2014).

The fourfold classification of adaptation limits proposed by Adger et al. (2009) is a useful basis for further exploration of their nature and characteristics. The first category of threshold or limit is ecological and physical. For example, there seems likely to be a threshold beyond which no amount of human action can avoid repeated and severe coral bleaching (Donner et al. 2005). Second, they identify economic thresholds which are where the costs of adaptation exceed the costs of impacts averted (that is, it is more expensive to adapt than it is to experience climate impacts). For example, while it seems the costs of protecting cities from sea-level rise are less than the costs of the impacts (Bigano et al. 2008), the same may not be said for protecting rural coastal settlements. Third, there are technological thresholds beyond which available technologies cannot avoid climate impacts. For example, there may be limits to engineering solutions to avoid flooding in certain places under extreme scenarios of change (Reeder et al. 2009).

The fourth threshold concerns the subjective nature of the limits to adaptation, and lies at the point where social groups judge adaptation actions to have failed. Because the social groups that may be part of adaptation processes are diverse, they will value things differently: what may be perceived as a successful adaptive response from one point of view may not be perceived the same way by others (Barnett and O'Neill 2010). Migration and relocation in response to climate change present examples of this: some populations faced with environmental disruption may make a positive choice to migrate over other possible adaptation strategies (Gemenne 2013) whereas others, such as Torres Strait Islanders in Australia (McNamara et al. 2012), find any suggestion that they might relocate in response to sea-level rise culturally unacceptable.

This conceptual framework of adaptation leads us to propose a relatively simple definition of the limits to adaptation. The advantage of such a definition is that it allows us to explore the existence, behaviour and outcome of adaptation limits through a case study approach, without placing excessive constraints on case study choice and methodological approach. We define a 'limit to adaptation' as the point or threshold at which actions to adapt cease to reduce vulnerability (Adger et al. 2009).

26.2   Six case studies of limits to adaptation in practice

Understanding the limits to adaptation is an emerging frontier of climate change research. It is important for decision-making about adaptation

for three reasons. First, it helps to determine which responses to climate change are both practicable and legitimate, and the time scales over which adaptation may be considered to be effective. Second, it helps to understand how people may respond to the damage to, or the loss of, things that are important to them, for which there may (in some cases) be substitutes or ameliorating policy measures. Third, it can help prioritise adaptation strategies, refine their intentions and identify communities that will be served by them.

Given this need to advance knowledge of the limits to adaptation, this chapter reports on a series of six linked research projects which investigated the possibilities of, and limits to, adaptation. The six case studies were selected purposefully. They are each well-studied places or systems where adaptation seems likely to reach its limits. They are all by and large iconic sites, that is, places of environmental or social significance (at least to Australians and in some cases to the world). Table 26.1 and Figure 26.1 describe and locate each study area.

Projects were conducted by multidisciplinary research teams. The main methods of data collection were desktop reviews of existing information about present and future climate and other drivers of risk and management responses in the case study areas; these were enhanced by interviews and workshops with key informants and stakeholders.

26.3   Key findings

Each study investigated the potential impacts of climate change in its respective study region, the potential for adaptation and the system characteristics that might impose limits on adaptation. The results of each study are summarised in Table 26.2. A synthesis of these results gives rise to five key findings, one of which does not relate to adaptation limits but is notable because it was a conclusion reached independently by all of the studies. The five key findings (discussed further after the list) are as follows:

1. adaptation goals can best be met through portfolios of adaptation strategies;

Table 26.1 The six case studies (full reports are available for download from www.nccarf.edu.au).

Study area	Report
The *Great Barrier Reef*, where climate change poses risks to the health of the reef and the businesses, communities and cultures that depend on it	Evans et al. 2012
Alpine areas, where climate change poses risks to the plants, animals, businesses, settlements and cultures that depend on winter snow	Morrison and Pickering 2012
The *Coorong and Lower lakes*, where climate change poses risks to the health of ecosystems and the businesses, communities and cultures that depend on them	Gross et al. 2012
The *Macquarie Marshes,* where climate change poses risks to the health of ecosystems and the businesses, communities and cultures that depend on them	Jenkins et al. 2012
The Torres Strait islands, where climate change poses risks to human settlements and cultural values	McNamara et al. 2012
Small inland communities in the Murray–Darling Basin, where climate change poses risks to the viability of small communities	Kiem and Austin 2012

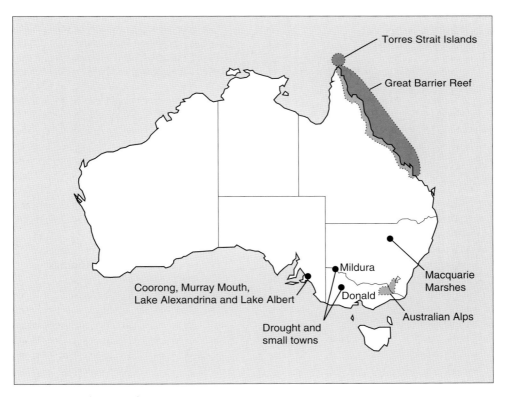

Figure 26.1 Location of case studies.

2. some barriers to adaptation are *de facto* limits;
3. socio-economic limits to adaptation may arise from processes that are distant in space and time;

4. limits to adaptation can arise through trade-offs in the way resources are allocated and places are managed; and

Table 26.2 Summary of results from six case studies.

Study region	Current stresses on ecosystem	Current stresses on communities	Impacts of future climate change	Potential adaptation strategies	Potential (non climatic) causes of limits to adaptation
Torres Strait Islands	N/a	Inundation is already occurring at high spring tides; out migration of young and well-educated	Sea-level rise and storm surge; changed weather patterns (seasonality, rainfall amounts and intensity, temperature variations); modified access to important traditional resources	Retreat, relocation, seawall construction	Cultural barriers to relocation and, to a lesser extent, retreat; physical barriers to retreat; governance failure (lack of confidence in government agencies); cost of sea defences
Great Barrier Reef	Climate change; declining water quality due to increasing pollutants and sediment in runoff; impacts from fishing; loss of habitat (coastal development)	Changes in markets; natural disasters; policy and management decisions	Ocean warming; ocean acidification; coral bleaching; increased flows of land-based pollutants; increased cyclone damage; increase in disease and pest outbreaks	Business management; stewardship; effort management; diversification; mobility of businesses	Lack of co-operation within and between sectors and with external stakeholders limits fisheries stewardship strategies; global market competition and preferences limit business management in fishing and tourism and limit mobility and diversification in fisheries; damaged reputations (through mis-communication or misunderstanding) limit business management in fishing and tourism
Macquarie Marshes	River regulation (reduction in flow volume by half; loss of flooding; loss of variability); climate extremes	Reduction in grazing productivity due to long unproductive dry spells; families moving away	Drying, especially increase in inter-flood interval (IFI). Possible loss of all floodplain with a 1–2 year IFI	Buy back and environmental flows to restore the short and moderate IFI floodplain; review of water sharing plans (WSPs); implementation of strategic adaptive management plans	Lack of capacity of river managers to learn from problems during droughts and implement WSPs; lack of willingness to change behaviour; intractable disagreements among stakeholders in water regulation regimes

Alpine	Fire; grazing by feral animals and livestock; weeds; impacts of tourism	Competition from other tourism destinations (national and international); stresses on local government from land use changes and demographic changes	Reduction in snowfall and in duration and depth of snow lying; loss of endemic species and communities; increase in invasive species; increase in wildfire number and intensity; decreased winter visitors	Technological solutions (snow making; super grooming); ecological/physical solutions (controlling/limiting invasive species; rehabilitating disturbed sites; restoring endemic communities and connectivity; off-site conservation; reducing soil erosion; controlling fires); economic solutions (diversifying tourism; increased real estate sales of vacation properties)	Technological solutions (cost, lack of water, environmental concerns for snowmaking); ecological solutions (strategies to deal with feral animals and weeds only slow their spread); lack of national recognition of the importance of the region for water supply to major urban centres
Coorong Wetlands	Water availability and drought; water sharing arrangements	Long-standing disagreements between stakeholder groups and governments regarding water sharing arrangements; socio-economic consequences of a degrading environment (e.g. impacts on fisheries and tourism)	Reduced freshwater inflow; increased salinity in Coorong and Lakes; more frequent drying of lakes; loss of species and communities (some culturally important) and formation of new ecosystems; effects on tourism and agriculture	Increase environmental inflows; engineering interventions such as weirs and channels to allow more active water management (social and governance issues around management goals, who would manage and risk that management strategies may prove to be overly narrow or mal-adaptive); raising barrages to protect against sea-level rise (an expensive option likely to be contested); better catchment management (re-vegetation, regulating water extraction, reducing erosion, controlling salinisation); improved long-term management structures (moving away from crisis response) based around a new regional institution which involves all local communities	Inadequate community engagement; parallel but unconnected planning initiatives; lack of trusted relationships between communities, governments and scientists; lack of management structures with long-term perspectives; intractable disagreements among stakeholders in water regulation regimes

(Continued)

Table 26.2 *(Cont'd)*

Study region	Current stresses on ecosystem	Current stresses on communities	Impacts of future climate change	Potential adaptation strategies	Potential (non climatic) causes of limits to adaptation
Drought in small inland towns	Climate extremes; increasing demand for water	Climate variability; uncertainty regarding future climate change; rural demographic changes; changes in the economic base (demise of the family farm, corporatisation of agriculture, deregulation, pressures from mining); social stresses (community changes, increasing debt levels); inadequate, misplaced or failed government support for agriculture; uncertainty around water markets and commodity prices	Warmer and drier conditions likely to lead to more rainfall droughts and more irrigation droughts; frequency of extreme rainfall events, both drought and flood, likely to increase	Development of water trading schemes which are climate resilient, leading to redistribution of industry, jobs and population	Social acceptability of changes in rural communities, especially loss of population; intractable disagreements among stakeholders in water regulation regimes

5. there is value in making adaptation goals, trade-offs and limits explicit.

The first key finding is *that adaptation goals can best be met through portfolios of adaptation strategies*. All projects identify such portfolios. In the Great Barrier Reef, for example, multiple strategies are identified that can help sustain tourism enterprises including a number that are not obviously related to climate change (e.g. business planning, marketing, currency devaluation and industry support packages) as well as diverse strategies designed to sustain the reef ecosystem. In Alpine regions, technological solutions include snow-making and super grooming, ecological solutions include controlling/limiting invasive species and rehabilitating disturbed sites and economic solutions include diversifying recreational activities. Because any single strategy may fail, avoiding the limits to adaptation might be possible though portfolios of multiple strategies. These multiple strategies must be well integrated to ensure they are not maladaptive.

The second key finding is that *some barriers to adaptation are* de facto *limits*. Most projects were unable to distinguish clearly between a barrier and a limit to adaptation; while in theory barriers are 'obstacles that can be overcome with concerted effort' (Moser and Ekstrom 2010, p. 22027), in practice many of the things that could be done with concerted effort were considered highly unlikely given, for example, prohibitive costs, community resistance or lack of precedence. From our six case studies, we found that the economic barrier to the strategy of coastal protection in the Torres Strait effectively constitutes a limit, as do the political and policy barriers to allocating sufficient water to the Macquarie Marshes wetlands. To say that a barrier is institutional – and therefore socially created – is not to say that it is ever likely to be overcome, even with concerted effort. In many cases the limits to adaptation arise from an inability of institutions to adjust, even if such adjustments seem possible in theory.

The third key finding is that *socio-economic limits to adaptation may arise from processes that are distant in space and time*. In the spatial sense, and taking the example of wetlands, upstream uses of water cause reductions in the flows necessary to sustain the ecological values of the wetlands and the social values they support. Local decision-makers seeking to find ways to adapt have little power over the institutions that create demand and influence the management of water in upstream areas. This is also true for inland towns and small farming communities, with farmers reporting frustration at their inability to influence the design of water markets and the effects of water markets and large institutional players in these markets on small communities.

The processes that limit adaptation are often global in scope. The limits to adaptation in tourism enterprises in the Great Barrier Reef and alpine areas, and in commercial fisheries in the Great Barrier Reef, arise in great part from increased competition in global markets and the appreciation of the Australian dollar, which puts pressure on costs and constrains investments in adaptation actions.

There are also limits arising from processes that operate over long time scales. Many of the institutions that create limits to adaptation exhibit path dependency. For example, if meaningful reductions in demand are to be achieved the institutions of water governance have to manage carefully the expectations of supply that tens of thousands of users have developed over past decades (Kenney et al. 2004; Tortajada and Joshi 2013). History also matters in the Torres Strait Islands. Many Indigenous people in these (and other mainland Indigenous) communities have strong negative associations with the idea of government involvement in resettlement, as in the past this has been the cause of enormous harm to individuals, families, communities and cultures: this makes them reluctant to consider resettlement as an adaptation option. It also leads to mistrust of governments, a perception not helped by repeated studies of the risks climate change poses to their islands with a repeated failure to invest in identified solutions.

The fourth key finding is that *limits to adaptation can arise through trade-offs in the way resources are allocated and places are managed.* The resources that are available for adaptation are finite. For example, in a drying climate there will be increased competition for water between the environment and users such as ski fields, irrigators and the public supply. Where water is scarce, decisions about its allocation will necessarily require trade-offs among demands.

Trade-offs arise through policy instruments. In creating markets as distributional mechanisms, the *economic* value of water is given preference because water is allocated to those who can afford to pay most for it, while other values of water – such as its *ecological* and *cultural* values – are traded off against its market value. Adaptation goals that rely on water are therefore also traded off: users that can pay for water can adapt while users that cannot pay, or cannot otherwise influence the allocation of water, face limits to adaptation.

Other examples of trade-off in adaptation and their selective effect on limits to adaptation include the Australian alps, where ski resorts may want to adapt by moving locations to more elevated areas but this would impinge on conservation areas. There is therefore a trade-off between the adaptation goal of conserving alpine ecosystems and ski-tourism in the alps. In the Great Barrier Reef there is also a trade-off between the ability of farming systems to adapt and the ability of reef ecosystems to adapt because increasing the resilience of the reef requires reducing land-based sources of sediment and pollution, much of which comes from agricultural practices.

The final key finding of these studies is that *there is value in making adaptation goals, trade-offs and limits explicit.* When adaptation reaches its limits, things that are valued will be lost. When those limits arise because of trade-offs in which the interests of some groups prevail over the interests of others, then adaptation becomes a matter of social justice. To take some examples from the case studies: small inland towns and wetlands, together with their environmental and cultural values, may lose if agricultural producers

adapt through consolidation of farms and economies of scale; reef ecosystems and the tourism and fisheries industries that depend on them could lose if farmers in the coastal hinterland adapt through increased application of fertilisers and pesticides; and winter tourism operators in the Alps may lose if endangered Alpine species (such as the Mountain Pygmy Possum) are supported to adapt by enlarging the area of protected land for their habitat.

The case studies show that most of the trade-offs that create limits to adaptation arise through institutional behaviours, for example: water markets are shaping the trade-offs between economic and environmental and cultural values; a commitment to high-input farming implies losses to the Great Barrier Reef; an unwillingness to invest public money in Indigenous affairs seems likely to imperil Indigenous settlements in the Torres Strait.

Difficult decisions about trade-offs between these things have not been necessary so far; most of the things that are important to many people both nearby (for example in local communities in the case study areas) and distant (for example people from across the country and around the world valuing Indigenous culture, the Great Barrier Reef and internationally protected wetlands) are sustained. Given current emissions trajectories and institutional behaviours, it seems however likely that choices between these things that are valued will be necessary. Theories of justice advise us that these choices about what to protect and what to let go should be made explicit, and is the subject of deliberation by stakeholders. In this way adaptation can arise through active (and admittedly sometimes difficult) choices rather than *de facto* institutional processes.

26.4 Conclusions

These projects suggest that there is utility in assessing the limits to adaptation actions. Most conventional guides to assessing vulnerability and adaptation to climate change fail to take the next step of thinking through what adaptation may not be able to achieve, what the drivers are of these limits and what their implications are for decision-making.

The projects each identify goals of adaptation and strategies to achieve them. They suggest that goals can be best met through identification and implementation of diversified portfolios of strategies, which provide the necessary flexibility to overcome barriers and limits at different thresholds of climate change.

Many of the limits to adaptation goals arise through exclusive allocations of resources and exclusive use of spaces and places. For example, demands for water are increasing while runoff seems likely to decrease and/or to become more variable under climate change: under these circumstances not all demands for water can be met. Identifying and discussing the trade-offs associated with adaptation decisions in advance can help focus attention on potential winners and losers from climate change adaptation, reveal public preferences with respect to acceptable and unacceptable losses from climate change, and stimulate thinking about changes that can be made now to avoid having to make these trade-offs.

Failure to identify these trade-offs and limits will mask the power of existing institutions and interests in adaptation processes, marginalise less powerful actors and may lead to climate change impacts that the public finds morally unacceptable.

References

Adger, W.N., Dessai, S., Goulden, M. et al. (2009) Are there social limits to adaptation to climate change? *Climatic Change* 93, 335–354.

Barnett, J. and O'Neill, S. (2010) Maladaptation. *Global Environmental Change* 20, 211–213.

Bigano, A., Bosello, F., Roson, R. and Tol, R. (2008) Economy-wide impacts of climate change: a joint analysis for sea level rise and tourism. *Mitigation and Adaptation Strategies for Global Change* 13, 765–791.

Donner, S.D., Skirving, W.J., Little, C.M., Oppenheimer, M. and Hoegh-Guldberg, O. (2005) Global assessment of coral bleaching and required rates of adaptation under climate change, *Global Change Biology* 11, 2251–2265.

Dow, K., Berkhout, F., Preston, B., Klein, R., Midgely, G. and Shaw, M. (2013a) Commentary: limits to adaptation, *Nature Climate Change* 3, 305–307.

Dow, K., Berkhout, F. and Preston B.L. (2013b) Limits to adaptation to climate change: a risk approach. *Current Opinion in Environmental Sustainability* 5, 384–391.

Evans, L.S., Fidelman, P., Hicks, C., Morgan, C., Perry, A. and Tobin, R. (2012) *Limits to Climate Change Adaptation in the Great Barrier Reef: Scoping Ecological and Social Limits*. National Climate Change Adaptation Research Facility, Gold Coast.

Gemenne, F. (2013) Migration doesn't have to be a failure to adapt: an escape from environmental determinism. In: Palutikof, J., Boulter S. L., Ash A.J. et al. (eds) *Climate Adaptation Futures*, 1st edition. Wiley-Blackwell, Chichester, pp. 235–241.

Gross, C., Pittock, J., Finlayson, M. and Geddes, M. (2012) *Climate Change Adaptation in the Cooring, Murray Mouth and Lakes Alexandrina and Albert*. National Climate Change Adaptation Research Facility, Gold Coast.

Jenkins, K.M., Kingsford, R.T., Wolfenden, B.J. et al. (2012) *Limits to Climate Change Adaptation in Floodplain Wetlands: The Macquarie Marshes*. National Climate Change Adaptation Research Facility, Gold Coast.

Kenney, D.S., Klein, R.A. and Clark, M.P. (2004) Use and effectiveness of municipal water restrictions during drought in Colorado. *Journal of the American Water Resources Association* 40, 77–87.

Kiem, A.S. and Austin, E.K. (2012). *Limits and Barriers to Climate Change Adaptation for Small Inland Communities Affected by Drought*. National Climate Change Adaptation Research Facility, Gold Coast.

Klein, R.J.T., Midgley, G.F., Preston, B.L. et al. (2014) Chapter 16: Adaptation opportunities, constraints and limits. In: Field, C., Barros, V., Mastrandrea, M. et al. (eds) *Climate Change 2014: Impacts, Adaptation and Vulnerability*. Contribution of Working Group II to the IPCC Fifth Assessment, Cambridge University Press, Cambridge.

McNamara, K.E., Smithers, S.G., Westoby, R. and Parnell, K. (2012) *Limits to Climate Change Adaptation for Low-Lying Communities in the Torres Strait*. National Climate Change Adaptation Research Facility, Gold Coast.

Meinshausen, M., Meinshausen, N., Hare, W. et al. (2009) Greenhouse-gas emission targets for limiting global warming to 2 °C. *Nature* 458, 1158–1162.

Morrison, C. and Pickering, C.M. (2012) *Climate Change Adaptation in the Australian Alps: Impacts. Strategies, Limits and Management.* National Climate Change Adaptation Research Facility, Gold Coast.

Moser, S.C. and Ekstrom, J.A. (2010) A framework to diagnose barriers to climate change adaptation. *Proceedings of the National Academy of Sciences* 107, 22026–22031.

Park, S.E., Marshall, N.A., Jakku, E. et al. (2012) Informing adaptation responses to climate change through theories of transformation. *Global Environmental Change* 22, 115–126.

Reeder, T., Wicks, J., Lovell, L. and Tarrant, O. (2009) Protecting London from tidal flooding: limits to engineering adaptation. In: Adger, W.N., Lorenzoni, I. and O'Brien, K.L. (eds) *Adapting to Climate Change: Thresholds, Values, Governance.* Cambridge University Press, Cambridge, pp. 54–63.

Tortajada, C. and Joshi, Y.K. (2013) Water demand management in Singapore: involving the public. *Water Resources Management* 27, 2729–2746.

27 Adaptation to extreme heat and climate change in culturally and linguistically diverse communities

ALANA HANSEN[1], MONIKA NITSCHKE[2] AND PENG BI[1]

[1]*Discipline of Public Health, University of Adelaide, Australia*
[2]*Department for Health and Ageing South Australia, Australia*

27.1 Background

Vast areas of the Australian continent experience periods of hot to very hot temperatures during summer, and extended hot spells are not uncommon. When heat events are extreme, resident populations can be impacted by adverse health effects ranging from marginal increases in morbidity to marked increases in mortality (Department of Human Services 2009; Nitschke et al. 2007, 2011; Tong et al. 2009). Heat-related deaths are thought to be largely preventable if individuals, populations and health systems are prepared for the potential impacts of hot weather. Heat-health action plans incorporating preventive advice on adaptive behaviours for populations at risk have shown promising reductions in heat-related morbidity and mortality (Matthies et al. 2008). Identifying and targeting those at risk in Australia and other heat-prone areas of the world is a challenge for public health authorities that may heighten with a shift to warmer temperatures.

Studies of heat-health impacts in the United States have shown that susceptibility can differ by race/ethnicity and is particularly higher in African Americans (Henschel et al. 1969; Klinenberg 1999; Whitman et al. 1997; Basu and Ostro 2008) with socio-economic disadvantage being a major contributing factor (Henschel et al. 1969; Betancourt et al. 2003). Ethnic groups around the world vary significantly on measures of preventable deaths partly due to behavioural and environmental risk factors (Kar et al. 2001). In terms of heat-susceptibility, literature regarding the influence of ethnicity is scarce although it is known that contributing factors can include poor living conditions, language barriers and increased heat exposure for those in outdoor employment (Hansen et al. 2013a).

While ethnicity can be a useful category for stratifying data in epidemiological studies, classifications are often subjective. In terms of differences in health outcomes, the concept of culture may be a more appropriate means of distinguishing social groups than ethnicity, as it captures the notion of how beliefs, customs and meanings are interpreted through a socially transmitted and learnt cultural lens (Ballard 2002;

Applied Studies in Climate Adaptation, First Edition. Edited by Jean P. Palutikof, Sarah L. Boulter, Jon Barnett and David Rissik.
© 2015 John Wiley & Sons, Ltd. Published 2015 by John Wiley & Sons, Ltd.

Egede 2006). Importantly, the term 'culture' sidesteps the ideology that one's genetic characteristics or birthplace can be a determinant of health outcomes and instead may better explain the influence of health beliefs, behaviours and practices.

Historically, Australian society has been shaped by immigration. As a consequence there is a high level of cultural and linguistic diversity in the population, with more than one-quarter of residents born in countries overseas (Australian Bureau of Statistics 2012). Reports of a higher than expected proportion of non-Australian-born people being hospitalised with heat-related conditions during a major South Australian heat wave (Zhang et al. 2012) suggest a disparity that requires further investigation; it is not known how culturally diverse groups, with different climate experiences and understandings of heat-related risk, adapt to Australia's hot summers. This is an important gap in public health knowledge that needs to be addressed to inform climate change adaptation strategies in preparedness for unprecedented heat waves and increasingly culturally diverse societies.

27.2   The study

A qualitative study was undertaken, the aim of which was to identify in culturally and linguistically diverse (CALD) communities socio-economic and linguistic factors affecting vulnerability to heat and climate change. The methodology involved face-to-face interviews and focus groups with 36 key informants during the warm months between December 2011 and April 2012. The study spanned three Australian cities: Adelaide, South Australia; Melbourne, Victoria; and Sydney, New South Wales. Respondents were employed in state and local government sectors, service provider organisations and migrant and refugee health services. Most were involved in liaison with people in refugee communities, new arrivals or long-established migrants. Many were first- or

second-generation European, Asian, Middle Eastern or African immigrants themselves and were able to share their professional and/or personal experiences on the research topic. Snowball sampling also resulted in the recruitment of members of a Vietnamese community group and a family of recently arrived Bhutanese refugees.

Questions for respondents related to experiences with extreme heat, factors contributing to vulnerability, adaptive behaviours and knowledge of heat-health warnings, as outlined by Hansen et al. (2013b). A thematic analysis of recorded transcripts from each city involved both deductive and inductive methods (Moretti et al. 2011). In this chapter, we present an overview of study findings and supporting literature and discuss the implications of future environmental and demographic changes.

27.3   Findings

Our findings showed that while being of a particular cultural background does not itself confer vulnerability to heat, at-risk subgroups exist within CALD communities. Whereas some respondents thought extreme heat was not a major issue for communities with which they were familiar, others were of the opposite opinion; on three occasions in our study people from Africa and Bhutan spoke passionately about issues facing those in new and emerging communities and the need for information to assist adaptation.

Those said to be at risk include older migrants, as well as new arrivals and refugees who lack acculturation and acclimatisation to the uniquely hot conditions in Australia. Asylum seekers, low income earners, the homeless, those with poor English skills, the socially isolated, mothers with babies, young children and people with chronic illnesses and/or physical or mental disabilities are also vulnerable. The underpinning reasons why some people in CALD communities face difficulties during periods of extreme heat are interrelated and multi-factorial.

27.3.1 Socio-economic issues

Discussions with service providers revealed that newly arrived migrants and refugees who are unable to find employment are reliant on welfare payments and often live in poor-quality rental accommodation that lacks air conditioning, insulation and at times even fans. During hot weather there is therefore limited opportunity to modify the indoor thermal environment. Housing characteristics can contribute to vulnerability (Maller and Strengers 2011) and studies have shown that in hot weather, temperatures inside thermally inefficient homes can be higher than outdoors (Strengers 2008), increasing the likelihood of heat stress for inhabitants.

The use of air conditioning clearly acts as a protective measure against heat stress (Naughton et al. 2002). Respondents however reported that, for many families, the very high cost of electricity associated with operating energy-hungry air conditioners constitutes a major barrier to usage. This was said to be particularly so in South Australia where the cost of power was a topic raised in each interview. Concerns about energy costs are shared by many in the population, including the frugal first-generation post-WWII migrants whose advancing age increases risks of poor heat-related health outcomes and low income earners in new and emerging communities whose financial disadvantage can be compounded by a strong sense of responsibility to send monetary support to overseas kin (Colic-Peisker 2009).

27.3.2 Cultural factors and norms

Wearing heavy dark-coloured garments can be part of cultural and religious norms, and may not align with the need to dress lightly during very hot weather. Respondents recounted that on hot days overdressing can be an issue for elders and mothers who unknowingly dress their children in warm clothes. For others however, there is a rapid transition to wearing lighter garments. The risk of dehydration can present in people whose fluid intake is inadequate for reasons that include a lack of awareness, a dislike or distrust of tap water or being unable to drink during daytime for the period of Ramadan. For some cultures cold food is not palatable; however, cooking may increase indoor temperatures on hot days. Seeking assistance can be problematic for those wary of authorities due to previous experiences in their home countries, and immigrants and refugees can face a range of barriers when accessing health care as identified in other Australian studies (Murray and Skull 2005; Sheikh-Mohammed et al. 2006). Finally, physical characteristics render some migrants more 'visible' than others and as a result there is the potential to encounter street discrimination in public places (Colic-Peisker 2009). This fact may underpin our findings that dark-skinned people can feel ill at ease spending any length of time in publicly cooled spaces such as shopping centres.

27.3.3 Health issues

Underlying health status can be a determinant of heat-susceptibility (O'Neill et al. 2009). Refugees can have a range of chronic health conditions, nutritional deficiencies and mental health issues which can affect vulnerability (Burnett and Peel 2001; Benson et al. 2013), and older people can have multiple co-morbidities associated with declining physical and mental health, contributing to the risk of heat-related morbidity and mortality.

A previous study showed that being unfamiliar with the harsh conditions in Australia can be a predisposing factor for deaths attributed to high environmental temperatures (Green et al. 2001). However, our data revealed that Australians can make inaccurate assumptions about people from hot countries and their ability to cope with heat, despite the conditions and environment being potentially quite different from homeland regions. For example, respondents spoke of the uniquely dry heat of south-eastern Australia and the discomfort felt by some new arrivals. Furthermore, the heat often fails to abate overnight during heat waves, unlike regions where cool breezes contribute to an evening drop in temperature. The sun in Australia causes the skin to burn easily, yet wearing sunscreen can be a foreign concept for many. When asked about the health effects of very hot weather, a Bhutanese refugee family in our study spoke about headaches, feeling lazy, itchy skin rashes and sunburn, together with the lack of acclimatisation and underlying health problems in the community.

27.3.4 Social isolation and language barriers

Instances were reported of individuals and families of CALD backgrounds being socially and/or linguistically isolated in neighbourhoods. This can occur even though social capital can be strong within CALD communities. In particular, respondents mentioned isolation occurring in people lacking proficiency in English and asylum seekers and families without social networks. Language barriers and low literacy levels disadvantage those seeking information or health care, and can hinder attempts to find help and articulate concerns (Sheikh-Mohammed et al. 2006).

Social isolation and advanced age are both significant risk factors for heat-related mortality (Naughton et al. 2002). Our study revealed that many older (e.g. European) migrants live alone, some do not speak English and occasionally some have no close relatives. Respondents experienced with aged care related how people who immigrated from Europe in the mid-1900s are now experiencing ageing-associated problems including, for some, reversion to their primary language and culture even though they may have spoken English for decades (Schmid and Keijzer 2009). This can result in isolation from neighbours and family. Additionally, people who resettle in Australia late in life from non-English-speaking countries can find English acquisition particularly difficult. A respondent from Africa spoke of older refugees feeling isolated even within the family bounds as the young, having spent their childhood in refugee camps, may not speak the traditional dialects of their elders. Furthermore, the aged and new arrivals can feel isolated if unable to access cooler places outside the home due to a lack of transport options.

Additionally, as recognised in our study and elsewhere (Arcury et al. 2010), language and literacy barriers in migrant and 'guest' workers employed in thermally challenging situations (e.g. the agricultural industry) can pose a safety issue if there is a poor understanding of local language and occupational safety information.

27.3.5 Adaptive and maladaptive behaviours

Migrants generally adapt well in Australia by virtue of experiential resilience and necessity. Younger people with English skills have a higher capacity to adapt than older generations, as do those who are not faced with the barrier of financial disadvantage. The risk of heat stress can be moderated by simple harm minimisation behaviours; however, these are not necessarily intuitive to those new to Australia's summers and our findings showed that, for new arrivals, information about appropriate heat-adaptive behaviours can be lacking.

Study respondents reported that for those lacking access to air conditioning, confinement in a hot house can be 'emotionally disturbing' and even 'tormenting'. Many find it oppressive if unable to have sunlight and fresh air in the home, and hence open doors and windows in the heat of the day in an attempt to cool the house (and in doing so, inadvertently let in hot, not cool, air). This practice also occurs elsewhere and can have health consequences. For example, an epidemiological study in France showed that during a major heat wave the behaviour of opening windows in the afternoon was associated with an increased risk of death (Vandentorren et al. 2006). By contrast, as raised in our study and cited in the international literature, opening windows and doors after dark to ventilate the house can elicit security fears in some residents, thus constituting a barrier to ventilating their homes (Klinenberg 1999).

Social capital can augment adaptive capacity, and the strong community spirit encompassing family and religious connections within CALD communities engenders support for others during the heat, particularly for older people who are cared for at home by the family. Conversely, reports suggest strong bonding networks may not by themselves always reduce the vulnerability of the elderly during heat waves (Wolf et al. 2010). For example, as mentioned by one respondent in our study, there may be a disincentive for an older person to use air conditioning during the day because of the potential financial stress on the family budget.

The popular Australian practice of visiting beaches and swimming pools on hot days is not part of the cultural norm for many migrants and has, on occasion, resulted in mishaps and drownings. Study respondents spoke of adaptation in terms of migrants availing themselves of swimming lessons (Leo 2012) and of innovative adaptation by a Muslim Australian woman who has designed culturally appropriate swimwear for women (www.burquini.com).

In other examples of positive cultural adaptation to the climate, older Italian and Greek migrants often establish significant shady gardens in their backyards; and Asian females sensibly use umbrellas to protect their skin from the sun, unlike the majority of Australian-born women. Some respondents mentioned behaviours undertaken by people in their home countries during hot weather that may not be possible in Australia, such as sitting by streams under shady trees in Africa or using bamboo sleeping mats on hot nights in China. Indeed, it was suggested by a community worker that Australia could learn from the climate experiences of people from other countries.

27.3.6 Implications for climate change at the international level

Globally, climate change-driven increases in the occurrence of heat events, along with demographic changes in the proportion of ethnic minorities and ageing migrants within populations, will see a potential escalation in the size of vulnerable subgroups. Hence, in terms of climate change adaptation, there will be a greater need for cultural competence in governance and decision-making, risk communication and in health and emergency services. This will be necessary both to enable pre-emptive action plans to prepare for the challenges of climate change and as crises arise.

At the structural level, better-quality thermally efficient housing is required for tenants including migrants. However, simply installing air conditioning in homes is costly and has been viewed by some as a short-sighted option that is contributing to climate change. More focus therefore needs to be placed on alternative adaptation strategies such as passive cooling strategies (Maller and Strengers 2011). There is growing need for guidance on maintaining thermal comfort and personal well-being in the heat without incurring high energy bills. Respondents indicated that this information needs to be available to low-income migrants as well as refugees who may be unfamiliar with cost-effective use of electrical appliances.

Climate change adaptation also requires preparedness for extraordinary and unprecedented events. Suggestions from this study included greater availability of free drinking water during heat waves or, for those seeking refuge outside of the home, cooled spaces where members of culturally diverse communities can feel at ease.

Different levels of understanding about climate change exist within CALD communities and the various means of educating groups about imminent warming temperatures and climate risks need to be considered. Adaptation involving community engagement has been identified previously as a means of bolstering a community's resilience to climate stressors (Ebi and Semenza 2008). Respondents in our study suggested that information sessions about environmental change and extreme heat could be organised for community groups and headed by bi-cultural community engagement officers to relay information in a culturally competent manner. Put into perspective however, these sessions may not be well attended as issues regarding climate change and heat risk are not major issues for those facing greater pressures associated with resettlement and financial stress.

Those at risk will be particularly vulnerable to climate change if unable to receive messages which allow informed decision-making in extreme weather and emergencies. Risk communication about emergencies such as bushfires, storms, floods and heat waves therefore needs to be inclusive of those who speak a language different to that of the majority. We found there are unmet needs in this regard and that migrant groups should be considered in policies such as heat-health action plans and other weather-related emergency contingencies. Respondents stressed that warnings need to be in a multilingual format as exclusion from access to monolingual weather reports and heat-health information can lead to a lower uptake of adaptive behaviour messages (Uejio et al. 2011) and therefore increased vulnerability. Formatting and disseminating such emergency messages has its challenges; there is no 'one size fits all' approach universally suited to a culturally diverse target audience with a range of literacy levels. Stakeholders stressed that messages need to be simple and communicated via one or more channels which may include written, oral, aural and visual means depending on which is most suited to the needs of individual cultural groups. Communication channels suggested by respondents included ethnic radio and television, articles in community newspapers, emergency phone lines, websites and social media; new and innovative methods of engagement should also be considered. In particular, community leaders can play an important role in one-to-one communication and small group information sessions. Information via the young at schools and technical training colleges can also be useful in getting messages to the extended family.

Importantly, a co-ordinated approach to climate change risk communication and health promotion is required which takes into account circumstances within communities including their language and needs, and with a focus on interpersonal communication (Boughtwood et al. 2012). Multicultural health-promotion interventions should involve prior community engagement to assess appropriateness and translatability, and subsequent evaluation to gauge effectiveness in promoting behaviour change. In this way the

nexus between cultural factors and health issues can be more thoroughly addressed, with consonant messages fashioned for discrete cultural groups and their respective behaviours, beliefs and inherent potential for adaptation.

As climate change affects whole populations regardless of nationality or cultural background, a holistic approach to adaptation is required. Stronger partnerships between government, policymakers and community organisations will aid in breaking down cross-cultural barriers at the organisational level. As an example, a multidisciplinary working group consisting of representatives from emergency services, academia and government organisations was recently established in South Australia to help facilitate better dissemination of heat-health messages in languages other than English.

27.4   Conclusion

This study builds on international evidence that points to a disparity in the risk of heat-related illness in people of different ethnic/cultural backgrounds (Klinenberg 1999; O'Neill et al. 2003; Uejio et al. 2011) and attempts to disentangle the underlying complexities in an Australian context. With a relatively small sample size, findings are not intended to be generalisable or representative of the broader, vastly heterogeneous CALD population in Australia as individuals/groups can differ markedly in terms of risk factors. For instance, immigrants who arrive as part of the skilled migration program are competent in English and have unemployment rates below the national rate (Australian Government Department of Immigration and Citizenship 2010). They are therefore less likely to encounter barriers to thermal comfort and climate change adaptation to the same degree as financially disadvantaged immigrants. Additionally, we have framed cultural diversity in the context of migrancy, and have not addressed the important public health issue of vulnerability to heat and climate change in Indigenous Australians who experience intergenerational economic and social disadvantage, chronic health conditions, inadequate housing and problems accessing culturally appropriate health care, especially in remote areas of Australia (Green et al. 2009; Hansen et al. 2013a).

In summary, it is recognised that people experiencing inequality and poverty are likely be affected most by the impacts of climate change (Australian Council of Social Service 2013). This includes disadvantaged members of CALD communities, particularly new arrivals and refugees. In this study we have found that overseas-born people generally acculturate well and certain cultural norms and life experiences can be beneficial to the resettlement process in Australia. Nevertheless, within CALD communities there are minorities and high-risk groups likely to be poor and underserved (Kar et al. 2001). For these, socio-cultural issues and prior traumatic experiences can increase susceptibility to the health impacts of extreme heat and climate change. Risk in the vulnerable can be compounded in hot weather by language barriers, underlying physical and mental health issues, socio-economic disadvantage and poor-quality housing, lack of acclimatisation to local conditions, cultural factors, limited fluid intake, social isolation and language/literacy barriers to accessing health care and warning messages. Recently arrived migrants and humanitarian arrivals on low incomes who are unable to speak the local language, as well as isolated and older migrants who lack access to a temperature-controlled environment, are of particular concern. It follows that for non-English-speaking older people in new and emerging communities the risk may be multiplied.

Nevertheless, stakeholders in our study were of the opinion that with equal opportunity and access to risk communication information and appropriate health care, potential ethnic disparities in vulnerability to extreme heat and climate change will be minimised. Furthermore, encouraging a more socially inclusive and integrative society will likely strengthen adaptive capacity for communities worldwide to collectively address climate change adaptation (Ebi and Semenza 2008).

With rapidly growing global multiculturalism and more people from CALD backgrounds becoming part of the ageing, heat-sensitive population, there are complex challenges ahead in terms of addressing the public health impacts of environmental change. Health communication strategies can no longer assume target audiences are culturally homogenous (Kar et al. 2001) and monolingual. A suite of socially inclusive communication tools is therefore required that takes into consideration people of diverse cultural backgrounds, language skills and literacy levels. Furthermore, formulation of cross-cultural adaptation policies could benefit from community consultation with immigrants to incorporate experiential knowledge of adaptive responses to hot weather in other countries, and to better understand health beliefs and practices within different cultural contexts. With a culturally diverse global population it is vital there be a collaborative approach in addressing the health impacts of climate change on vulnerable populations. Importantly, more participatory action research on barriers to adaptation within minority groups is required so that forward-looking policies will be relevant to the changing face of multicultural societies.

References

Arcury, TA., Estrada, J.M. and Quandt, S.A. (2010) Overcoming language and literacy barriers in safety and health training of agricultural workers. *Journal of Agromedicine* 15, 236–248.

Australian Bureau of Statistics (2012) Migration, Australia, 2010–11, Canberra. Available at http://www.abs.gov.au/ausstats/abs@.nsf/Products/840748 89D69E738CCA257A5A00120A69?opendocument (accessed 29 May 2014).

Australian Council of Social Service (2013) Extreme Weather, Climate Change and the Community Sector, ACOSS Paper 197. ACOSS, Strawberry Hills NSW.

Australian Government Department of Immigration and Citizenship (2010) How New Migrants Fare: Analysis of The Continuous Survey of Australia's Migrants, Commonwealth of Australia. Available at http://www.immi.gov.au/media/publications/research/_pdf/csam-results-2010.pdf (accessed 29 May 2014).

Ballard, R. (2002) Race, ethnicity and culture. In: M. Holborn (ed.) *New Directions in Sociology*. Causeway, Ormskirk.

Basu, R. and Ostro, B.D. (2008) A multicounty analysis identifying the populations vulnerable to mortality associated with high ambient temperature in California. *American Journal of Epidemiology* 168, 632–637.

Benson, J., Phillips, C., Kay, M. et al. (2013) Low Vitamin B12 levels among newly-arrived refugees from Bhutan, Iran and Afghanistan: A multicentre Australian Study. *PLoS One* 8, e57998.

Betancourt, J.R., Green, A.R., Carrillo, J.E. and Ananeh-Firempong, O. 2nd (2003) Defining cultural competence: a practical framework for addressing racial/ethnic disparities in health and health care. *Public Health Reports* 118, 293–302.

Boughtwood, D., Shanley, C., Adams, J. et al. (2012) Dementia information for culturally and linguistically diverse communities: sources, access and considerations for effective practice. *Australian Journal of Primary Health* 18, 190–196.

Burnett, A. and Peel, M. (2001) Asylum seekers and refugees in Britain: Health needs of asylum seekers and refugees. *British Medical Journal* 322, 544–547.

Colic-Peisker, V. (2009) Visibility, settlement success and life satisfaction in three refugee communities in Australia. *Ethnicities* 9, 175–199.

Department of Human Services (2009) January 2009 Heatwave in Victoria: An Assessment of Health Impacts, Victorian Government Department of Human Services, Melbourne.

Ebi, K.L. and Semenza, J.C. (2008) Community-based adaptation to the health impacts of climate change. *American Journal of Preventive Medicine* 35, 501–507.

Egede, L.E. (2006) Race, ethnicity, culture, and disparities in health care. *Journal of General Internal Medicine* 21, 667–669.

Green, D., King, U. and Morrison, J. (2009) Disproportionate burdens: the multidimensional impacts of climate change on the health of Indigenous Australians. *Medical Journal of Australia* 190, 4–5.

Green, H., Gilbert, J., James, R. and Byard, R.W. (2001) An analysis of factors contributing to a series of deaths caused by exposure to high environmental temperatures. *American Journal of Forensic Medicine and Pathology* 22, 196–199.

Hansen, A., Bi, L., Saniotis, A. and Nitschke, M. (2013a) Vulnerability to extreme heat and climate change: is ethnicity a factor? *Global Health Action* 6, 21364.

Hansen, A., Bi, P., Saniotis, A. et al. (2013b) *Extreme Heat and Climate Change: Adaptation in Culturally and Linguistically Diverse (CALD) Communities*. National Climate Change Adaptation Research Facility, Gold Coast.

Henschel, A., Burton, L.L., Margolies, L. and Smith, J.E. (1969) An analysis of the heat deaths in St. Louis during July, 1966. *American Journal of Public Health and the Nation's Health* 59, 2232–2242.

Kar, S.B., Alcalay, R. and Alex, S. (2001) *Health Communication: A Multicultural Perspective*. Sage Publications, Inc, Thousand Oaks, California.

Klinenberg, E. (1999) Denaturalizing disaster: A social autopsy of the 1995 Chicago heat wave. *Theory and Society* 28, 239–295.

Leo, J. (2012) Schools and Surf Life Saving SA band together to teach surf safety to new arrivals. AdelaideNow. Available at http://www.adelaidenow.com.au/news/south-australia/schools-and-surf-life-saving-sa-teach-surf-safety-to-new-arrivals/story-e6frea83-1226516973421 (accessed 29 May 2014).

Maller, C.J. and Strengers, Y. (2011) Housing, heat stress and health in a changing climate: promoting the adaptive capacity of vulnerable households, a suggested way forward. *Health Promotion International* 26, 492–498.

Matthies, F., Bickler, G., Cardeñosa Marín, N. and Hales, S. (2008) Heat-health action plans. WHO Regional Office for Europe, Copenhagen (Denmark).

Moretti, F., van Vliet, L., Bensing, J. et al. (2011) A standardized approach to qualitative content analysis of focus group discussions from different countries. *Patient Education and Counseling* 82, 420–428.

Murray, S.B. and Skull, S.A. (2005), Hurdles to health: immigrant and refugee health care in Australia. *Australian Health Review* 29, 25–29.

Naughton, M.P., Henderson, A., Mirabelli, M.C. et al. (2002) Heat-related mortality during a 1999 heat wave in Chicago. *American Journal of Preventive Medicine* 22, 221–227.

Nitschke, M., Tucker, G. and Bi, P. (2007) Morbidity and mortality during heatwaves in metropolitan Adelaide. *Medical Journal of Australia* 187, 662–665.

Nitschke, M., Tucker, G., Hansen, A., Williams, S., Zhang, Y. and Bi, P. (2011) Impact of two recent extreme heat episodes on morbidity and mortality in Adelaide, South Australia: a case-series analysis. *Environmental Health* 10, 42.

O'Neill, M.S., Zanobetti, A. and Schwartz, J. (2003) Modifiers of the temperature and mortality association in seven US Cities. *American Journal of Epidemiology* 157, 1074–1082.

O'Neill, M.S., Carter, R., Kish, J.K. et al. (2009) Preventing heat-related morbidity and mortality: new approaches in a changing climate. *Maturitas* 64, 98–103.

Schmid, M.S. and Keijzer, M. (2009) First language attrition and reversion among older migrants. *International Journal of the Sociology of Language* 200, 83–101.

Sheikh-Mohammed, M., Macintyre, C.R., Wood, N.J., Leask, J. and Isaacs, D. (2006) Barriers to access to health care for newly resettled sub-Saharan refugees in Australia. *Medical Journal of Australia* 185, 594–597.

Strengers, Y. (2008) Comfort expectations: the impact of demand-management strategies in Australia. *Building Research and Information* 36, 381–391.

Tong, S., Ren, C. and Becker, N. (2009) Excess deaths during the 2004 heatwave in Brisbane, Australia. *International Journal of Biometeorology* 54, 393–400.

Uejio, C.K., Wilhelmi, O.V., Golden, J.S., Mills, D.M., Gulino, S.P. and Samenow, J.P. (2011) Intra-urban societal vulnerability to extreme heat: The role of heat exposure and the built environment, socioeconomics, and neighborhood stability. *Health and Place* 17, 498–507.

Vandentorren, S., Bretin, P., Zeghnoun, A. et al. (2006) August 2003 heat wave in France: risk factors for death of elderly people living at home. *European Journal of Public Health* 16, 583–591.

Whitman, S., Good, G., Donoghue, E.R., Benbow, N., Shou, W. and Mou, S. (1997) Mortality in Chicago attributed to the July 1995 heat wave. *American Journal of Public Health* 87, 1515–1518.

Wolf, J., Adger, W.N., Lorenzoni, I., Abrahamson, V. and Raine, R. (2010) Social capital, individual responses to heat waves and climate change adaptation: An empirical study of two UK cities. *Global Environmental Change* 20, 44–52.

Zhang, Y., Nitschke, M. and Bi, P. (2012) Risk factors for direct heat-related hospitalization during the 2009 Adelaide heatwave: A case crossover study. *Science of the Total Environment* 442, 1–5.

28 Experiences of resettled refugees during the 2011 Queensland floods

IGNACIO CORREA-VELEZ[1], CELIA MCMICHAEL[2], SANDRA M. GIFFORD[3] AND AUGUSTINE CONTEH[1]

[1]School of Public Health and Social Work, Queensland University of Technology, Australia
[2]School of Social Sciences, La Trobe University, Australia
[3]Swinburne Institute for Social Research, Swinburne University of Technology, Australia

28.1   Introduction

With consensus among scientists that anthropogenic climate change is occurring, increasing attention is focused on understanding the connections between climate change and migration (Bardsley and Hugo 2010; McMichael et al. 2012). A central assumption is that climate change will contribute to substantial increases in human population movement in coming decades (Tacoli 2009; Foresight 2011; Scheffran et al. 2011; Black et al. 2012). Most research to date has focused on the potential magnitude and pathways for migration (Black et al. 2011), the significant barriers to migration for people affected by environmental hazards and the capacities of communities to adapt to environmental changes (Kniveton et al. 2008; Black et al. 2012). Relatively little consideration has been given to those who have been forcibly displaced for reasons other than environmental catastrophes and then find themselves confronted with climate-related disasters in their new resettlement setting.

With projected climatic changes it is expected that refugees and other forced migrants will increasingly spend protracted amounts of time in transit countries or will resettle in locations that experience ecological vulnerability. In these situations, climate change is not a 'push factor' but rather a context within which existing migration and settlement patterns occur. Although refugees make up only 7% of the world's 232 million international migrants, they are among the most vulnerable with nearly 9 out of every 10 refugees living in developing countries in 2013 (United Nations 2013). Only 1% of those recognised as a refugee by the UNHCR will be granted protection through resettlement in another country (UNHCR 2013).

Australia is one of 26 UNHCR resettlement countries and ranks third in the number of refugees it accepts each year (UNHCR 2013). During 2010–11, a total of 13,799 humanitarian visas were granted by Australia. Over the last five years, most arrivals have come from Southeast Asia, the Middle East and Africa (Department of Immigration and Citizenship 2011). While

Applied Studies in Climate Adaptation, First Edition. Edited by Jean P. Palutikof, Sarah L. Boulter, Jon Barnett and David Rissik.
© 2015 John Wiley & Sons, Ltd. Published 2015 by John Wiley & Sons, Ltd.

Australia – and other countries offering resettlement – provides security from persecution, refugees must respond to new challenges in the settlement context that are likely to impact on their health and wellbeing such as learning a new language, gaining employment, engaging with education and integration into the host society (Porter and Haslam 2005; Colic-Peisker 2009).

On 10 January 2011 a flash flood hit the Toowoomba region in Southeast Queensland, Australia, causing the loss of 23 lives and severe damage to 18,000 properties (van den Honert and McAneney 2011). On 13 January, the Brisbane and Bremer rivers surged flooding in many Brisbane and Ipswich suburbs, forcing the evacuation of 20,000 people. Concern about the impact of this event on resettled refugees was first raised by the Multicultural Development Association (MDA), the peak settlement service to refugee migrants in Queensland. A submission to the Queensland Floods Commission Inquiry 2011 by MDA reported that the floods displaced about 70 refugee client families and that 30 families had ongoing complex needs at the time of the submission (MDA 2011). Most of the clients reported 'feelings of uncertainty, fear, isolation and helplessness during the flood crisis' (MDA 2011, p. 5). Despite these feelings, many clients approached MDA with offers to help others during the flood clean up. MDA caseworkers reported that many clients were experiencing difficulties adjusting to new housing provided for them and that the trauma due to the floods was 'further compounded for many by re-traumatisation from previous experiences in their home countries as well as settlement issues already present in adjusting to life in Australia' (MDA 2011, p. 6).

What is the impact of extreme weather events on populations who have already experienced forced displacement? Do their previous experiences make them better equipped to deal with environmental disasters? In what follows, we report on the impact of the 2011 Queensland floods among 141 refugee migrant men who were interviewed before (first half of 2010) and after (between September 2012 and April 2013) the floods. The paper compares pre- and post-disaster

measures of health and wellbeing between unexposed and exposed participants. It also examines the impact of the floods on participants' relationships with their neighbours and how their past refugee experiences impact their capacity to cope with environmental disasters.

28.2   Methods

The findings reported in this chapter are derived from a follow-up of a cohort of men from refugee backgrounds who participated in the 2008–10 SettleMEN project (Correa-Velez and Gifford 2011). SettleMEN was a longitudinal investigation of health and settlement among a group of 233 refugee migrant men living in Southeast Queensland. SettleMEN participants were interviewed four times between 2008 and 2010 at intervals of 6 months. The SettleMEN study yielded pre-disaster measures of health and settlement among this cohort of men. The 2012–13 follow-up study offered a rare opportunity to investigate the impact of an environmental disaster on the health of a resettled refugee population.

The follow-up study used a mixed-method approach. A survey was administered in person by peer interviewers to 141 study participants in their preferred language. Peer interviewers were trained in research skills and the ethical conduct of research. The survey consisted of a combination of quantitative and open-ended questions, including: socio-demographic characteristics; educational and occupational indicators; health measures (e.g. WHOQOL-BREF questionnaire which assesses the physical, psychological, social relationships, and environment wellbeing domains) (WHO 1996); family and social support; living arrangements, experiences living in Australia, social exclusion; and impact of the floods. Flood impact questions were adapted from previously published disaster research (Ginexi et al. 2000; Queensland Health 2011). Below we focus on the flood impact findings and on the changes in pre- and post-disaster measures of subjective health and wellbeing. Two focus group discussions (one in Brisbane and one in

Toowoomba) were conducted to explore in more detail participants' experiences during the floods. A total of 25 refugee migrant men participated in the focus groups.

SPSS (IBM SPSS v21) was used to analyse the quantitative data. Descriptive statistics were used to summarise demographic and flood impact variables. Linear Mixed Models (West et al. 2006) were used to compare changes in pre- and post-disaster measures of subjective health and wellbeing between unexposed and exposed participants (controlling for age, time in Australia, region of birth and area of settlement). Thematic analysis guided the interpretation of the qualitative data. Transcripts of the survey open-ended questions and focus groups were analysed for major themes and sub-themes (Patton 2002). Pseudonyms are used throughout.

28.3   Results

28.3.1   Participants' characteristics

Table 28.1 summarises the demographic and pre-migration characteristics of the 141 participants. Participants were born in nine different countries, with 106 (75%) born in Africa (the majority in South Sudan followed by Rwanda, the Democratic Republic of Congo, Burundi, Congo, Liberia and Tanzania), 23 (16%) in the Middle East (Iraq) and the remaining 12 (9%) in Southeast Asia (Burma). Respondents had been in Australia for an average of 6 years.

28.3.2   Exposure to and impact of the floods

Figure 28.1 illustrates the impact of the floods on participants by region of settlement. Overall, 65 (46%) participants had to temporarily evacuate their homes and 59 (42%) reported that there was floodwater in their houses. A total of 106 (76%) respondents reported that close family or friends from the same ethnic background had to be evacuated during the floods. Compared to participants living in Brisbane, men living in the Toowoomba region were more likely to have been exposed to

Table 28.1 Demographic and pre-migration characteristics of participants (n = 141).

Age in years: mean ± SD (range)	36.2 ± 8.2 (22–62)
Time in Australia in years: mean ± SD (range)	6 ± 1.4 (4–8)
Place of residence during 2011 Queensland floods	
Greater Brisbane	81 (57%)
Toowoomba region	60 (43%)
Region of birth	
Africa	106 (75%)
Middle East	23 (16%)
Southeast Asia	12 (9%)
Religion	
Christian	111 (79%)
Muslim	30 (21%)
Highest educational level completed in Australia	
None/secondary education	13 (9%)
English course	53 (38%)
TAFE/college/trade (other than English course)	55 (39%)
University degree	20 (14%)
Marital status	
Never married	59 (42%)
Married/de facto	78 (55%)
Separated/divorced	4 (3%)
Current employment status	
Employed full-time	46 (33%)
Employed part-time or casual	62 (44%)
Unemployed	32 (23%)
Weekly income	
< AU\$ 399	30 (22%)
AU\$ 400–799	88 (63%)
> AU\$ 800	21 (15%)
How well participant speaks English	
Well	105 (74%)
Not well	36 (26%)

and impacted by the floods in terms of damage to property, disruption to their work, loss of income, illness and injury.

28.3.3   Comparing pre- and post-disaster measures of wellbeing

When comparing unexposed with exposed participants (and controlling for age, time in Australia, region of birth and area of settlement), the Linear

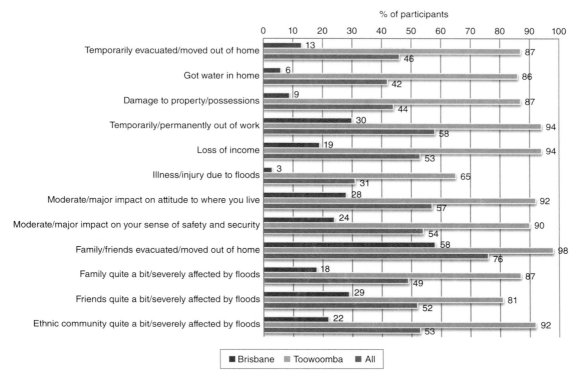

Figure 28.1 Exposure to and impact of the floods on participants by region of settlement. Source: Adapted from Ginexi et al. 2000 and Queensland Health 2011.

Mixed Models reported no significant changes in pre- and post-disaster measures of quality of life ($ß = -0.28$; S.E. = 0.18; $P = 0.12$), subjective health status ($ß = 0.01$; S.E. = 0.19; $P = 0.98$) and wellbeing in the physical domain ($ß = -0.07$; S.E. = 0.49; $P = 0.88$), psychological domain ($ß = 0.28$; S.E. = 0.47; $P = 0.55$), social relationships domain ($ß = -0.30$; S.E. = 0.59; $P = 0.61$) and environment domain ($ß = 0.09$; S.E. = 0.45; $P = 0.84$).

28.3.4   *Changes in relationships with neighbours*

A key question of this study is whether an environmental disaster strengthens social networks with neighbours; there is evidence that, in times of disaster, bonding networks are intensified (Aldrich 2012). Of the 91 participants who provided qualitative data on their interaction with neighbours, 40 (44%) reported that they continued to have no or very limited interaction, while 36 (40%) reported having good, friendly interactions with their neighbours prior to the floods and that there were no changes after the floods. Importantly, for 15 (16%) respondents, providing assistance to their neighbours during the floods had a positive impact on their relationship:

> Due to the assistance I rendered to my neighbour's children, our interaction has improved. Joseph (32 years, Sudan)
>
> [My neighbours]…began to greet me because I assisted them in the floods. Louis (39 years, Sudan)

While a small number of studies have demonstrated that migrants draw upon familial and ethnic networks in post-disaster evacuation, recovery and resilience (Leong et al. 2007), the findings from this study demonstrate that the floods encouraged some refugee migrant men to also seek and provide support to broader neighbourhood and community members.

28.3.5 Previous refugee experience and the Queensland floods

For a number of respondents, refugee experiences of war and trauma could not be compared to that of environmental disasters:

> I don't think that my past experience as a refugee or having witnessed persecution in my home country will help me cope better with the floods since the situation is quite different in their nature. Zay (47 years, Burma)

However, 64% of participants reported that their previous refugee experiences helped them to better cope during the floods. Many respondents felt that their prior experiences as refugees were substantially more challenging and this increased their sense of personal resilience both during and after the floods:

> As a refugee you cope with many problems in regard to food, water even medical but during the flood, I did not fear worse. As I was used to worse in refugee camp I coped well during the floods. Patrick (47 years, Burma)
>
> During the civil war in Darfur-Sudan, we sometimes missed and go without food for three days. These experiences have helped my wife and I to cope better during the flood. Simon (45 years, Sudan)

Many participants compared their experience of the floods with that of their children, who had not experienced traumatic events prior to the 2011 floods:

> Although we lost many of our belongings due to flood water we were not affected very much in comparison to our children who were seriously affected. It is because they have never witnessed

such an event in their life time. Chol (38 years, Sudan)

During the floods, a number of men described how they used strategies and skills learned during their refugee experience:

> I was able to apply strategies used before as refugee in dealing with difficult situations such as relaxation approaches to monitor positive mental state. Paul (29 years, Sudan)
>
> It reminded me about life in camps and war… how to evacuate… keeping calm, courageous and following instructions on how to evacuate the area we were living in. Of course my previous experience helped me a lot to cope with the floods. Compared to Australians I was very much less stressful than they were. Alfred (38 years, Burundi)

The skills and capacities that these men developed through their past adverse experiences provided important personal resources for helping them cope with the floods in the settlement context.

28.4 Discussion

This paper has described the impact of the 2011 Queensland floods on a cohort of 141 refugee migrant men residing in Southeast Queensland. Overall, we found that the floods had a considerable impact on this group of men, their families and communities in terms of being forced to evacuate their homes, work disruption, loss of income and personal belongings. The percentage of our participants who were forced to evacuate their homes (46%) is almost twice the percentage of Queensland adult males (24%) who moved out of their homes due to the floods (Queensland Health 2011). However, we found no significant differences in pre- and post-disaster changes in health and wellbeing measures between the unexposed and exposed participants. Importantly, many of these men reported that their previous refugee experience helped them to cope better during and after the floods; for some, providing assistance to others during the floods impacted positively on their relationship with their neighbours.

To our knowledge, this is the first study that has investigated the impact of an environmental disaster on former refugees living in Australia. A recent study by Osman and colleagues (Osman et al. 2012) assessed how 72 former refugees (55% males) coped after the September 2010 earthquakes in Christchurch, New Zealand. They found that three-quarters of participants reported coping well despite most of them not experiencing an earthquake before and less than 20% receiving support from mainstream agencies. Similar to Osman and colleagues' research, in our study participants with young children stated that the disaster was particularly stressful for their children because they had not experienced such events before.

Previous research has shown that individuals who experience traumatic events commonly report a re-evaluation of their purpose in life, improved family and community relationships and a greater sense of spirituality and resilience (Schaefer and Moos 1998; Tedeschi and Calhoun 2004; Hutchinson and Dorsett 2012). While resilience has been traditionally conceptualised as an individual attribute in the field of psychology (the capacity of an individual to bounce back; O'Malley 2011), in the hazard and disaster literature resilience is understood as 'the sustained ability of a community to withstand and recover from adversity' (Chandra et al. 2011, p. xiii). Resilience is therefore developed through interactions between the individual, their family and community and the social institutions (Mason and Pulverenti 2013). Communities are 'actors that respond to adversity' (Chaskin 2008, p.66) or threats from the environment, and this is particularly important for potentially marginalised communities such as former refugees in a resettlement country.

Successful long-term settlement depends not only on the capability of refugees to maintain the bonding links of the immediate family and ethnic community, but also on their ability to develop bridging links with their new neighbourhoods and other members of the society (Fielding and Anderson 2008; Loizos 2000). However, previous

analysis of social capital among this cohort of refugee men found a significant decrease in bridging relationships over time (Correa-Velez and Gifford 2011). Importantly, while many of the participants indicated that their interaction with neighbours was not altered or improved by experiences during and after the floods, some reported that their experiences had a positive impact on their interaction with neighbours. For them, facing an environmental disaster created an opportunity to greet and be greeted and to help and be helped by those living in close proximity. Building community spirit and acting collectively at the neighbourhood level can be important strategies for adapting to disasters (Adger 2003; Nix-Stevenson 2013).

Our paper has a number of limitations. First, the time period between the floods (January 2011) and post-disaster data collection (September 2012– April 2013) may have influenced participants' recollection of their flood experiences. Second, although there were no statistically significant differences between unexposed and exposed participants in terms of pre- and post-disaster changes of health measures two years after the floods, our study could not assess whether participants' levels of health and wellbeing decreased immediately after the floods. Third, given that the original cohort was recruited using a non-probabilistic sampling technique, our findings may not be representative of the entire population of refugee migrant men who experienced the floods. Finally, this research only involved men and there is clearly a need to hear about the experiences of women from refugee backgrounds.

Nevertheless, our study has shown how the previous experiences of being a refugee are perceived by many participants as resources they can draw on to better cope with environmental disasters. This underscores the strengths and capabilities of people with refugee backgrounds, including the survival skills developed during their refugee experience, enabling them to cope with and adapt to extreme weather events and other environmental disasters. During environmental disasters, risk and vulnerability are not distributed evenly but are shaped by factors such as socio-economic disadvantage, race and ethnicity and gender (Nix-Stevenson 2013). A central component of disaster risk reduction is to identify and address the concerns of vulnerable people in order to promote resilience and adaptation (Wisner et al. 2004). In this context, resettled refugees do not merely constitute a vulnerable group; the capabilities of former refugees should also be recognised when developing disaster response strategies at the neighbourhood and community levels. Importantly, this research indicates that developing relationships with neighbours may help to build community capacity and resilience. The importance of social networks in building community resilience has been documented for other natural disasters such as bush fires in Australia (Akama et al. 2013). Other community variables that can support adaptation among refugee communities in particular include effective leadership, stronger family relationships and community associations and enhancing linking social capital (Mason and Pulvirenti 2013).

There is increasing concern internationally about extreme environmental events (Black et al. 2012). At the same time, migration driven by economic, social, political, demographic and environmental factors constitutes a major force shaping global realities. While there is an extensive body of research examining the ways in which environmental change will shape migration pathways, this chapter provides an insight into the experiences of refugee migrants who experience environmental disaster in a site of settlement. In a world with substantial population mobility, people will migrate to areas that subsequently experience environmental vulnerability and disaster. This is an important and under-examined area for policy and programmatic focus.

Acknowledgements

This research has been funded by the National Health and Medical Research Council (NHMRC Project Grant 1027856) and the National Climate Change Adaptation Research Facility (NCCARF). We acknowledge the contribution of research participants and the research team, including Sabah Al Ansari, Elijah Buol, Saw Patrick Maw, Vivien Nsanabo, Gerald Onsando, Wilson Oyat, Suan Muan Thang and Moses Tongun.

References

Adger, W.N. (2003) Social capital, collective action, and adaptation to climate change. *Economic Geography* 79(4), 387–404.

Akama, Y., Chaplin, S. and Fairbrother, P. (2013) Role of social networks in community preparedness for bushfire. Paper presented at the *9th Annual International Conference of the International Institute for Infrastructure Renewal and Reconstruction*, Queensland University of Technology, Brisbane, 7–10 July.

Aldrich, D.P. (2012) *Building Resilience: Social Capital in Post-Disaster Recovery*. The University of Chicago Press, Chicago.

Bardsley, D.K. and Hugo, G. (2010) Migration and climate change: examining thresholds of change to guide effective adaptation decision-making. *Population and Environment* 32, 238–262.

Black, R., Adger, W.N., Arnell, N.W., Dercon, S., Geddes, A. and Thomas, D. (2011) The effect of environmental change on human migration. *Global Environmental Change* 21S, S3S11.

Black, R., Arnell, N.W., Adger, W.N., Thomas, D. and Geddes, A. (2012) Migration, immobility and displacement outcomes following extreme events. *Environmental Science and Policy* 27(S1), S32–S43.

Chandra, A., Acosta, J., Stern, S. et al. (2011) *Building Community Resilience to Disasters*. RAND Corporation, Santa Monica, CA.

Chaskin, R. (2008) Resilience, community and resilient communities: Conditioning context and collective action. *Child Care in Practice* 14, 65–74.

Colic-Peisker, V. (2009) Visibility, settlement success and life satisfaction in three refugee communities in Australia. *Ethnicities* 9(2), 175–199.

Correa-Velez, I. and Gifford, S. (2011) *SettleMEN: Health and Settlement Among Men from Refugee Backgrounds Living in South East Queensland*. La Trobe Refugee Research Centre, Melbourne.

Department of Immigration and Citizenship (2011) *Australia's Humanitarian Program 2012–13 and Beyond*. DIAC, Canberra.

Fielding, A. and Anderson, J. (2008) *Working with Refugee Communities to Build Collective Resilience*. Association for Services to Torture and Trauma Survivors, Perth.

Foresight (2011) *Migration and Global Environmental Change: Future Challenges and Opportunities. Final Project Report*. Government Office of Science, London.

Ginexi, E.M., Weihs, K., Simmens, S.J. and Hoyt, D.R. (2000) Natural disaster and depression: a prospective investigation of reactions to the 1993 midwest floods. *American Journal of Community Psychology* 28(4), 495–518.

Hutchinson, M. and Dorsett, P. (2012) What does the literature say about resilience in refugee people? Implications for practice. *Journal of Social Inclusion* 3(2), 55–78.

Kniveton, D., Schmidt-Verkerk, K., Smith, C. and Black, R. (2008) *Climate Change and Migration: Improving Methodologies to Estimate Flows, Migration Research Series No. 33*. International Organization for Migration, Geneva.

Leong, K.J., Airriess, C., Chia-Chen Chen, A. et al. (2007) From invisibility to hypervis-ibility: The complexity of race, survival, and resiliency for the Vietnamese-American community in Eastern New Orleans. In: Bates, K.A. and Swan, R.S. (eds) *Through The Eye of Katrina: Social Justice in the United States*. Carolina Academic Press, Durham, NC, pp. 169–185.

Loizos, P. (2000) Are refugees social capitalists? In: Baron, S., Field, J. and Schuller, T. (eds) *Social Capital: Critical Perspectives*. Oxford University Press, Oxford.

Mason, G. and Pulvirenti, M. (2013) Former refugees and community resilience: 'Papering Over' domestic violence. *British Journal of Criminology*, published online 29 January 2013, doi:10.1093/bjc/azs077.

McMichael, C., Barnett, J. and McMichael, A.J. (2012) An ill wind? Climate change, migration and health. *Environmental Health Perspectives* 120(5), 646–654.

Multicultural Development Association (2011) *Queensland Floods Commission of Inquiry 2011: Submission by the Multicultural Development Association*. Multicultural Development Association, Brisbane.

Nix-Stevenson, D. (2013) Human responses to natural disasters. *SAGE Open* July–September, 1–12.

O'Malley, P. (2011) Security after risk: Security strategies for governing extreme uncertainty. *Current Issues in Criminal Justice* 23, 5–15.

Osman, M., Hornblow, A., Macleod, S. and Coope, P. (2012) Christchurch earthquakes: How did former refugees cope? *New Zealand Medical Journal* 125(1357), 113–121.

Patton, M.Q. (2002) *Qualitative Research and Evaluation Methods*, 3rd edition. Sage Publications, Thousand Oaks, CA.

Porter, M. and Haslam, N. (2005) Predisplacement and postdisplacement factors associated with mental health of refugees and internally displaced persons. A meta-analysis. *JAMA* 294(5), 602–612.

Queensland Health (2011) *Self Reported Health Status 2011: Natural Disasters and Health, Queensland*. Queensland Health, Brisbane.

Schaefer, J.A. and Moos, R.H. (1998) The context for posttraumatic growth: life crises, individual and social resources, and coping. In: Tedeschi, R.G., Park, C. and Calhoun, L.G. (eds) *Posttraumatic Growth: Positive Changes in the Aftermath of Crises*. Laurence Erblaum Associates, London, pp. 99–125.

Scheffran, J., Marmer, E. and Sow, P. (2011) Migration as a contribution to resilience and innovation in climate adaptation: Social networks and co-development in Northwest Africa. *Applied Geography* 33(1), 119–127.

Tacoli, C. (2009) Crisis or adaptation? Migration and climate change in a context of high mobility. *Environment and Urbanization* 21, 513–525.

Tedeschi, R.G. and Calhoun, L.G. (2004) Posttraumatic growth: a new perspective on psychotraumatology. *Psychiatric Times* 21, 58–60.

United Nations (2013) *The Number of International Migrants Worlwide Reaches 232 Million*. Department of Economic and Social Affairs, Population Division, United Nations, New York.

United Nations High Commissioner for Refugees (2013) *Resettlement Fact Sheet*. United Nations High Commissioner for Refugees, Geneva. Available at http://www.unhcr.org/524c31a09.html (accessed 2 June 2014).

van den Honert, R.C. and McAneney, J. (2011) The 2011 Brisbane Floods: Causes, Impacts and Implications. *Water* 3(4), 1149–1173.

West, B.T., Welch, K.B. and Galecki, A.T. (2006) *Linear Mixed Models: A Practical Guide Using Statistical Software*. Taylor and Francis Ltd, Boca Raton, FL.

WHO (1996) *WHOQOL-BREF: Introduction, Administration, Scoring and Generic Version of the Assessment*. Programme on Mental Health, World Health Organization, Geneva.

Wisner, B., Blaikie, P., Cannon, T. and Davis, I. (2004) *At Risk: Natural Hazards, People's Vulnerability and Disasters*, 2nd edition. Routledge, London.

29 Vulnerability to climate change among disadvantaged groups: the role of social exclusion

ARUSYAK SEVOYAN AND GRAEME HUGO

Australian Population and Migration Research Centre, University of Adelaide, Australia

29.1 Background

In the last two decades there has been an increasing focus on the consequences of a changing climate for the social systems and how vulnerability to these effects varies between different social groups and areas (Adger 1999; Adger and Kelly 1999; Bohle et al. 1994; Nelson et al. 2010; Smit and Wandel 2006). This is a shift from what was known as the 'end-point' view of vulnerability, according to which the main problem was climate change itself and the reduction of greenhouse gas emissions and addressing the sensitivity of various socio-economic systems was the solution (O'Brien et al. 2007). However, there has recently been a shift to a 'start-point' view that recognises that vulnerability is a pre-existing state generated by multiple environmental and social processes, making it difficult to cope with changing climate conditions (Kelly and Adger 2000). Here, adaptive capacity is affected by vulnerability; the main problem is the socio-economic marginalisation and inequalities, and the solution lies in identifying and addressing these inequalities (O'Brien et al. 2007).

The 'start-point' view of vulnerability falls within the social constructivist framework, according to which vulnerability influences the socio-economic capacity of individuals to respond to different external stressors (Füssel 2005). Adger and Kelly (1999) propose that for a long-term response to climate change we need to understand the processes that shape current adaptive capacity and affect vulnerability to contemporary environmental stress. Vulnerability, according to them, is determined by availability of resources and entitlements of individuals and groups to call on these resources. Research on vulnerability of individuals and social groups should therefore start from an understanding of human use of resources.

Adger and Kelly (1999) argue that vulnerability is socially constructed and should be distinguished into individual and collective vulnerability. At the individual level, vulnerability is determined by access to resources, diversity of income sources and the social status of the individuals or households; at the community level meanwhile, vulnerability is determined by institutional and market structures and can be exacerbated by climate change impact (Adger and Kelly 1999).

Applied Studies in Climate Adaptation, First Edition. Edited by Jean P. Palutikof, Sarah L. Boulter, Jon Barnett and David Rissik.
© 2015 John Wiley & Sons, Ltd. Published 2015 by John Wiley & Sons, Ltd.

According to Wolf (2011) there are four dimensions of local context that shape vulnerability, including: perception of vulnerability and impacts; cognitive and behavioural aspects; social and institutional aspects; and values. The socio-demographic and economic background of individuals and groups shape their perceptions and attitudes towards climate change, which in its turn determines their behaviour and adaptive response (Wolf 2011). Values, taking the role of standards, guide the decisions and choices made at each level of climate change adaptation. However, individual actions of adaptation to climate change are affected by institutional and social context and social capital, which determine their ability to act collectively.

The importance of the social capital in the process of adaptation was also emphasised by Adger (2003) and Pelling and High (2005). Adger (2003) argued that, as the vertical social links become stronger between civil society and the state, the emerging co-operative social capital helps the process of adaptation to climate change. However, when the state fails to respond and provide support during environmental hazards, social capital takes over the supportive role of the state. The latter function of social capital is the one that, if left out from vulnerability assessments, will affect the predictions of future adaptation models and risk assessment, underestimating the ability of the social groups to cope with climate change impacts. Similarly, Pelling and High (2005) suggest focusing on informal social networks to increase our understanding of the social factors underlying the adaptive capacity of individuals and groups.

One framework for assessing climate vulnerability from a start-point approach is that of social exclusion which includes pre-existing socioeconomic disadvantage and social connectedness and also incorporates institutional structures, the community and the family contexts (Saunders et al. 2007). A recent project on the impact of climate change on disadvantaged groups in South Australia used the framework of social exclusion to study climate vulnerability at the community

and household levels (for the full project report see Sevoyan et al. 2013). The study used primary and secondary sources of information as well as quantitative and qualitative data and methods to explore the factors which make certain communities or disadvantaged groups more vulnerable to the negative impacts of climate change than others.

29.2 Vulnerability and social exclusion at the community level

To identify the most vulnerable local governmental areas in South Australia, dimensions of community disadvantage (such as the proportion of single parent or poor households, or the percentage of unemployed or low education individuals in the community, etc.) were derived using data from the 2011 Australian Census of Population and Housing (for the full list of indicators see Sevoyan et al. 2013, pp. 23–38). The analysis showed that there was a great deal of overlap in the spatial distribution of individual indicators. Combined, they presented the distribution of the level of social exclusion in the communities across South Australia (Fig. 29.1).

There is a distinct divide across metropolitan Adelaide: socially excluded groups are mostly concentrated in the north-western suburbs, while the southern and eastern suburbs showed relatively lower levels of exclusion. Outside of Adelaide, high levels of social exclusion were also observed in some regional cities and rural areas that had experienced severe economic setbacks due largely to drought, including Murray Bridge, Port Pirie, Berri and Barmera. However, it was observed that the factors affecting the level of social exclusion in metropolitan and non-metropolitan areas were different. If high levels of social exclusion in metropolitan areas were explained by higher proportions of unemployed and disabled individuals, as well as single parent and renting households, in non-metropolitan areas it was more likely explained by higher proportions of low-income households, individuals with poor education and an aged population. This

Social Exclusion Index

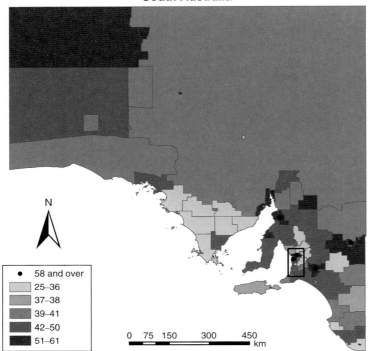
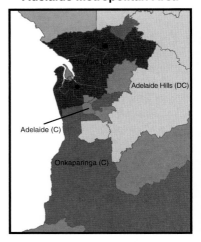

Figure 29.1 Spatial distribution of social exclusion at local government association level in South Australia in 2011. Source: Sevoyan et al. 2013. Based on data from ABS Census data from 2006 and 2011. For colour details please see Plate 21.

suggests that research and social policy aimed at reducing vulnerability and increasing resilience must consider metropolitan and non-metropolitan areas separately.

29.3 Vulnerability and social exclusion at the household level

Vulnerability at the household level was studied using quantitative and qualitative data from a survey on the impacts of climate change on disadvantaged groups, conducted in three South Australian local governmental areas with the highest levels of relative socio-economic disadvantage in 2011. They are shown on Figure 29.1 and include the local government area of Port Adelaide Enfield (2011 population 112,817) to represent the metropolitan area, Port Pirie (17,333) to represent a regional city and Berri/Barmera (10,582) to represent an essentially rural region. In each region a household survey was conducted of around 600 households and a number of in-depth interviews were carried out with key stakeholders. The survey involved stratification into six groups to represent major excluded communities: (1) Indigenous; (2) foreign born; (3) single parents; (4) aged or disabled; (5) unemployed; and (6) private renters or public housing tenants. In addition a control group category of the remaining population was included. Considering the high possibility of overlapping between the types of disadvantage, respondents were asked to identify where their households had more than one disadvantage.

The survey results were used to identify five different dimensions of vulnerability: (1) perception of vulnerability to climate impact; (2) perception of vulnerability to extreme weather; (3) perception of climate change as a household challenge; (4) perception of heat wave as a household challenge; and (5) perceived adaptive capacity.

The results of multivariate analysis showed that households with various types of disadvantage experienced higher levels of vulnerability and lower adaptive capacity than more advantaged households. Moreover, the increased vulnerability among different disadvantaged groups was explained by various dimensions of social exclusion. Indigenous, single parent and renting households therefore showed a marginally higher perception of vulnerability to climate change than the control group; this was mostly explained by levels of economic exclusion, however. The perception of vulnerability among migrant and unemployed households, on the other hand, was not significantly different from the control group. The aged/disabled households reported lower perceived vulnerability; this was mostly due to their lower awareness of climate change issues, however.

Similarly, Indigenous, single parent and renting households showed significantly higher vulnerability to extreme weather events than those with no disadvantages, but again this difference was explained by the level of economic exclusion. Meanwhile, the evidence suggested that the aged/disabled population had significantly higher vulnerability to extreme weather, controlling for all other factors. Having multiple disadvantages also made households significantly more vulnerable to extreme weather events compared to those with no disadvantages, net of other factors.

Heat waves were more likely to be recognised as household challenges among migrant and renting households, compared to the control group.

While this effect was explained by economic exclusion among renting households, the effect for migrant households remained after controlling for economic and social dimensions of social exclusion.

Compared to the control group, the level of perceived adaptive capacity was significantly lower among all disadvantaged groups except for the aged/disabled group. Interestingly, the lower perception of adaptive capacity was explained by economic exclusion among Indigenous, single parent, unemployed and renting households. Among migrant households, it was explained by low levels of social connectedness and climate awareness. When all the dimensions of social exclusion were controlled for however, the aged or disabled group was the only one that showed an increased perception of difficulty of adaptation compared to the control group.

To summarise, the findings from the quantitative analysis indicate that climate vulnerability among disadvantaged groups is mostly affected by the level of economic exclusion among them. However, it must be noted that social connectedness and climate awareness are also significant factors shaping their level of vulnerability.

For a deeper understanding of vulnerabilities, behavioural changes and barriers preventing disadvantaged households from adapting to the changing climate, in-depth interviews were conducted not only with disadvantaged households selected from among the survey participants, but also with a number of stakeholders from various governmental and non-governmental organisations working with disadvantaged population in the three South Australian sites included in the survey.

The reoccurring messages across households with different types of disadvantage showed that the hardships faced by them due to changing climate were not very different, and the solutions they found were also similar. The most important message that could be taken from this discussion is that households with multiple disadvantages were the most vulnerable to climate change impact.

Both disadvantaged household members and stakeholders working with them agreed that one of the biggest issues arising from extreme weather events was associated with health. This was especially concerning for those with pre-existing health problems. Vulnerability to the direct impact of extreme weather was also more prominent for those living in more disadvantaged conditions, such as locations prone to bushfires and floods and public/rental housing with poor insulation.

Along with vulnerabilities to the direct impact of extreme weather, many of the householders and stakeholders also recognised the indirect consequences of changing climate such as rising costs of living, underlying stress of dealing with the changing environment and social isolation. The Indigenous community, low-income families and those with health issues were especially vulnerable to the increasing costs of living, as they were least able to deal with the financial difficulties or make changes to their lifestyle and environment to adapt to the increasing costs.

Conversations with both the stakeholders and disadvantaged householders showed that certain characteristics make some of the disadvantaged groups more resilient to climate change impact than other more privileged population groups, however. Using cooler public spaces during hot weather and avoiding turning on air conditioning, living within their means and being more economical in their use of available resources are all small adjustments to the everyday life that make an economically less-privileged population more resilient and better equipped to adapt to changing environment. The interviews revealed that creativity and flexibility helped younger members of low-income population to be more resilient; life experience and better preparedness made older people more resilient; being better accustomed to the environment made Indigenous and country people more resilient; and better social connections made migrants and cultural groups more resilient. Meanwhile, those with better means might continue their lifestyle as before, willing to pay the increased costs for it while they can.

Although driven mostly by financial reasons rather than environmental awareness, many respondents had made significant changes to their lifestyle and living situation to decrease the direct and indirect impacts of climate change. Home improvements such as installing solar panels and water tanks and changing household appliances for more efficient models are some of adaptations that disadvantaged households were likely to make if they could afford them. However, along with financial and physical difficulties, a number of barriers exist that prevented them from taking more long-term adaptive actions. Households in the private rental market and public housing highlighted institutional barriers that needed to be addressed for better adaptation in the future. Restrictions on the number of solar panels that could be installed, public housing rules that restrict installation of solar panels and water tanks and regulations to water the lawn areas of rental properties were among the problems mentioned.

There were also more universal barriers such as lack of information, mentioned by other disadvantaged groups as well. While stakeholders complained about not being able to inform and send messages to these groups about better ways to deal with extreme weather and climate change, disadvantaged groups complained about the lack of sufficient, relevant and consistent information. Among the reasons for most of this miscommunication between the service providers and receivers, language and cultural barriers were mentioned by both sides as well as a lack of personalised information directed specifically at disadvantaged groups who were dealing with many other daily stressors and might not consider climate change to be relevant to them. The common problem of 'one size fits all' approach to public information dissemination seems to apply here. Members of disadvantaged households thought that a stronger message to the public, mandatory improvements to public housing and new structures for better energy efficiency were among the institutional changes that need to be made for better adaptation to climate change, and that individual behavioural changes were not sufficient.

It is apparent, especially from the qualitative study, that socially excluded groups in the study should not be depicted as 'passive victims' of the effects of social change and that there are many examples where they have shown agency, resilience and innovation in adapting to it. However, they are clearly highly constrained in this by the various dimensions of disadvantage that they experience.

29.4 From vulnerability to resilience: policy implications and recommendations for climate adaptation among disadvantaged groups

29.4.1 *Embedding climate change in the social inclusion agenda and vice versa*

In Australia in the last decade both climate change and social inclusion have emerged as important issues of political, community and media concern, and both have been the target of strong initiatives at all three levels of government, especially at the national level. However, these initiatives have almost entirely been undertaken in isolation from one another and not taken cognisance of the important linkages between them. Having accepted that there is an important social inclusion dimension to effective adaptation to climate change, it is essential that social inclusion elements be injected into climate change adaptation strategies not only at a national level but also at state and local levels. It needs to be recognised that the severity of the impact of negative climate change effects falls unequally across regions and upon different subgroups living in those areas.

However, it is equally important to include recognition of the potential effects of climate and environmental change in the national social inclusion agenda. Environmental impacts threaten to jeopardise the gains of wider policies and programs intended to reduce social exclusion in Australia; as has been demonstrated in the field study, environmental changes and climate change are already exacerbating and worsening the situation of the disadvantaged population. It is not just necessary to recognise that social exclusion increases vulnerability to the effects of climate change; in fact, environmental vulnerability can be seen as part of social disadvantage. It is one of the ways in which the lives of people are worsened and their ability to participate fully in society compromised.

29.4.2 *The importance of the local community and non-government organisations*

Effective adaptation to climate change is strongly influenced by local and community factors, and this is especially the case for disadvantaged groups. These include:
1. social connectedness, the extent to which people feel that there is a network of family, friends and community that they can rely on as a source of appropriate information and advice, as well as social and economic support, providing a sense of belonging and participation; and
2. local environmental circumstances; the disadvantaged are disproportionately located in less desirable locations which often have elevated risk of negative environmental effects.

There are a number of important policy messages here. It is important to empower and provide support for local bodies including local and regional government, local non-government organisations, community organisations and businesses in climate change adaptation. The engagement of these organisations to work effectively with federal, state and local governments to enhance the adaptive capacity of disadvantaged groups is an important priority. As Stanley et al. (2009) argue, there is a need for developing synergy between the state, community and civil society in order to develop social capital and strong communities.

It is also important to engage local communities and especially the disadvantaged and their advocates in service and support planning and delivery. The field study indicated that there is a widespread understanding of the realities of environmental change and pressure on scarce resources such as water in the population. Local communities are motivated, prepared and able to make changes in their lifestyle and behaviour towards more sustainable practices. However, they perceive that they lack the power to influence decision-making at any level. Their engagement can not only help in gaining wider involvement in these activities, but also lead to those activities being better suited to local needs and hence more cost effective. Clearly, governments at all levels need to recognise this and be more prepared to give communities complete information to enable them to make better decisions about lifestyle and behaviour. Governments also need to assist people in making those lifestyle and behaviour choices with appropriate assistance programs. This is especially the case with disadvantaged groups, who are much more limited in the resources available to them to facilitate making such choices.

29.4.3 The role of information

Information has an important role to play in adaptation to climate change in a number of ways. It provides people with an understanding of the nature and potential impacts of climate change so they can be motivated to take the actions needed to minimise its negative effects. Information provides people with the basis for taking initiatives or allowing behaviour in anticipation of future negative effects of climate change. It also provides knowledge on how to access subsidies and other resources that allow people to take actions to adapt to climate change.

Despite the widespread recognition of environmental change as an important issue influencing people's lives in Australia, the study found that information being accessed by disadvantaged groups is far from ideal in terms of its ability to help them adapt to climate change. In particular, the information available from official sources is often limited and not useful both in terms of its content and the way in which it is presented. Meanwhile, the information accessed from media and informal sources is often confusing, contradictory and of limited utility to the distinctive local situation. It is often 'middle

class' in orientation and doesn't resonate with excluded groups.

There is a clear role for better communication of information to support different groups, especially the disadvantaged, to adapt effectively to current and impending impacts of climate change. This is especially relevant to the culturally and linguistically diverse groups such as the Aboriginal population and immigrant groups. It is clear that 'one size fits all' approaches to communication are doomed to failure in the context of disadvantaged groups. Information messages need to be appropriate to the perceived needs of particular subgroups; a more segmented approach to information packaging and dissemination that takes this into account is needed.

29.5   Conclusion

Findings from a recent study in South Australia provide empirical evidence of the close associations between social exclusion, disadvantage and climate vulnerability suggested in the international literature. Higher levels of vulnerability among disadvantaged groups with respect to the effects of climate change compared to the rest of the community are part of a wider social vulnerability which makes them more at risk of experiencing negative effects of other shocks (e.g. economic downturns, personal tragedies, etc.). Part of the answer in a policy sense must be the support of wider initiatives to overcome disadvantage, to empower disadvantaged groups, to increase the economic resources available to them, to increase their social connectedness and sensitise all government and non-government activity to their needs. Building up this wider resilience must be a central part of any strategy to develop a more effective adaptive response to climate change.

References

Adger, N.W. (1999) Social vulnerability to climate change and extremes in coastal Vietnam. *World Development* 27(2), 249–69.

Adger, N.W. (2003) Social capital, collective action, and adaptation to climate change. *Economic Geography* 79(4), 387–404.

Adger, N.W. and Kelly, P.M. (1999) Social vulnerability to climate change and the architecture of entitlements. *Mitigation and Adaptation Strategies for Global Change* 4, 253–266.

Bohle, H.G., Downing, T.E. and Watts, M.J. (1994) Climate change and social vulnerability: toward a sociology and geography of food insecurity. *Global Environmental Change* 4(1), 37–48.

Füssel, H.M. (2005) *Vulnerability in Climate Change Research: A Comprehensive Conceptual Framework*, Breslauer Symposium, University of California International and Area Studies, UC Berkeley.

Kelly, P.M. and Adger, N.W. (2000) Theory and practice in assessing vulnerability to climate change and facilitating adaptation, *Climatic Change* 47, 325–352.

Nelson, R., Kokic, P., Crimp, S., Meinke, H., and Howden, S.M. (2010) The vulnerability of Australian rural communities to climate variability and change: Part I: conceptualising and measuring vulnerability. *Environmental Science and Policy* 13, 8–17.

O'Brien, K., Eriksen, S., Nygaard, L. and Schjolden, A. (2007) Why different interpretations of vulnerability matter in climate change discourses, *Climate Policy* 7(1), 73–88.

Pelling, M. and High, C. (2005) Understanding adaptation: what can social capital offer assessments of adaptive capacity? *Global Environmental Change* 15, 308–319.

Saunders, P., Naidoo, Y. and Griffiths, M. (2007) *Towards New Indicators of Disadvantage: Deprivation and Social Exclusion in Australia*, Social Policy Research Centre, Sydney, Australia.

Sevoyan, A., Hugo, G., Feist, H., Tan, G., McDougall, K., Tan, Y. and Spoehr, J. (2013) *Impact of Climate Change on Disadvantaged Groups: Issues and Interventions*. National Climate Change Adaptation Research Facility, Gold Coast.

Smit, B. and Wandel, J. (2006) Adaptation, adaptive capacity and vulnerability. *Global Environmental Change* 16, 282–292.

Stanley, J. (2009) *Promoting Social Inclusion in Adaptation to Climate Change: Discussion Paper, Monash Sustainability Institute Report 09/4*. Monash Sustainability Institute, Melbourne.

Wolf, J. (2011) Climate change adaptation as a social process. In: Ford, J.D. and Berrang-Ford, L. (eds) *Climate Change Adaptation in Developed Nations: From Theory to Practice*, Advances in Global Change Research no. 42. Springer, Netherlands.

30 Adapting the community sector for climate extremes

EMILY HAMILTON[1] AND KARL MALLON[2]

[1]Australian Council of Social Services, Australia
[2]Climate Risk Pty Ltd, Australia

30.1 Introduction

Even in the world's most economically developed countries, people experiencing poverty and social disadvantage (disadvantaged people) are among those at greatest risk from climate change and extreme weather impacts: those with the least resources have the least ability to cope, to adapt and to recover. As climate change accelerates, the stresses on already disadvantaged people will increase. At the same time, the number of people forced into situations of economic or social strain will also grow.

Community Service Organisations (CSOs) support people experiencing both short- and long-term disadvantage to manage everyday adversity, to respond in times of crisis and to develop both individual and structural solutions to entrenched disadvantage. Indeed, the critical and increasingly recognised role CSOs play in helping communities respond to and recover from natural disasters suggests they are an important part of the social infrastructure that communities will turn to for assistance to adapt to climate change.

As the impacts from climate change become more severe, CSOs will act as societies' shock absorbers for disadvantaged people facing the physical and social consequences of more frequent and severe extreme weather, as well as for those forced into disadvantage by such events.

Despite the risks to disadvantaged people from climate change and the role CSOs play in supporting communities' resilience to adversity, the community sector has been largely overlooked in climate adaptation research and policy settings across the developed world. Unfortunately, empirical evidence from many events indicates that such organisations are already failing during extreme weather events.

This chapter outlines the major findings from original research by the Australian Council of Social Service (ACOSS), the peak body for community services in Australia, and Climate Risk, Pty. Ltd., which was designed to address this research and policy gap (Mallon et al. 2013). The research was conducted in Australia, one of the world's wealthiest countries, with findings applicable to any economically developed country. Specifically, the *Adapting the Community Sector for Climate Extremes* project aimed to understand CSO vulnerability to extreme weather events and to explore their capacity to support individual and community adaptation through four interlinked research questions:

1. Will climate impacts to physical infrastructure negatively affect CSO service delivery capacity?

Applied Studies in Climate Adaptation, First Edition. Edited by Jean P. Palutikof, Sarah L. Boulter, Jon Barnett and David Rissik.
© 2015 John Wiley & Sons, Ltd. Published 2015 by John Wiley & Sons, Ltd.

2. Will CSO service delivery failure worsen the underlying vulnerability of disadvantaged people who access their services?

3. Do CSOs have inherent capacity to lessen the impacts of physical infrastructure failure on communities?

4. What are the barriers to CSO services delivery and are there ways to build adaptive capacity?

The project found that CSOs are highly vulnerable and poorly prepared to respond to climate change and extreme weather impacts and that, as a result, disadvantaged people will be at even greater risk of harm as climate change accelerates.

Importantly however, the project also found that well-adapted CSOs are a critical resource for building adaptive capacity; they have inherent skills, capacities and resources to make an unparalleled contribution to individual and community resilience to climate change.

The chapter concludes with a set of recommendations for adaptation policymakers and researchers which address the specific barriers to adaptation faced by CSOs. If fully implemented, these recommendations will ensure that CSOs across the developed world can continue to deliver effective services as climate change accelerates and also play a broader role in supporting adaptive capacity across communities, ensuring that disadvantaged people have equitable access to adaptation.

30.2 Poverty, inequality and climate change adaptation

30.2.1 Disadvantaged people will be first and worst affected

There is a wealth of literature on the differential impacts of climate change on rich and poor nations globally (e.g. UNFCCC 2006a, b, 2007a, b). This literature clearly demonstrates that *developing* countries are most vulnerable to climate change impacts because they have fewer resources to adapt socially, technologically and financially. There is also a growing body of literature about the particular ways that climate change will impact disadvantaged people living in *developed* countries (e.g. Hennessey et al.

2007; Altman and Jordan 2008; Macchi et al. 2008; Smith 2008; USGCRP 2008; Edwards et al. 2009; Ensor and Berger 2009; McMichael and Butler 2009; Stanley 2009; Victorian Government 2009; Horton et al. 2010; Porter and Abbott 2010; Reser et al. 2011; Green et al. 2012; Oven et al. 2012). This literature reveals that the following groups within developed countries will be first and worst affected by the full spectrum of climate change impacts including increasingly frequent and intense extreme weather events; reduced air and water quality; vector, food and water-borne diseases; and increasing costs for essential goods and services caused by impacts to food supplies, utilities and infrastructure:

- people on low incomes and people who are unemployed;
- homeless people and people living in poor-quality housing or in the private rental market;
- frail older people and those with chronic health conditions;
- Aboriginal and Torres Strait Islander people;
- single parents;
- newly arrived migrants and refugees;
- people with a disability and their carers;
- residents of places where extreme weather events such as cyclones, bushfires, storms and storm surges are likely to be more frequent or intense; and
- rural communities and farmers exposed to more frequent droughts and floods.

In addition, some research has demonstrated that climate change may also lead to an increase in interpersonal violence including domestic and family violence in some social, community and family contexts (Enarson 1999; Jenkins and Phillips 2008; Walker 2012).

30.2.2 Socio-economics: adaptation blind spot

Despite the growing recognition that disadvantaged people will be disproportionately impacted by climate change, to date adaptation research has given little consideration to the ways in which socio-economic factors affect adaptive capacity nor to the resource limitations of particular groups. This research project was conducted in Australia, one of the world's wealthiest countries, yet the evidence suggests that national wealth is not insulating the nation's most vulnerable (Stanley 2009). Given the already growing disparity in household wealth in Australia (ABS 2014), some authors suggest that a key risk from climate change is the exacerbation of relative poverty in Australia, as people on low incomes struggle to meet the rising costs associated with adaptation (Brotherhood of St Laurence 2007; ACOSS et al. 2008; Stanley 2009). For example, utility investment to increase the resilience of electricity and water supplies results in price increases for the associated commodities. In general the increased security of supply is good for society, but small increases in costs may make essential services unaffordable for those experiencing extreme poverty.

According to Stanley (2009), people who are likely to have the greatest difficulty adapting to climate change include those who are at risk of social exclusion, those who lack knowledge of climate change and possible sources of assistance, those who lack financial resources and those with a chronic physical or mental illness. Similarly, ACOSS et al. (2008) assert that low-income households will be adversely affected by climate change, largely because of the barriers they face to adaptation including that they:

• tend to live in areas more likely to be adversely affected by climate change and have far less ability to move or make other necessary adjustments to their living circumstances;
• spend a greater proportion of total weekly household budget on energy and water; and
• are often less able to introduce measures to reduce exposure to rising costs such as energy and water efficiency; and are less likely to have the financial capacity for purchasing disaster-related products and services such as insurance and preparedness kits.

For the same reasons, disadvantaged people are least likely to have capacity to meet rising insurance and recovery costs caused by worsening extreme weather events and rising sea levels (or to respond to the withdrawal of insurance altogether from highly exposed areas); or increasing food costs and food insecurity driven by changes in rainfall, less stable growing conditions and damage to crops due to extreme weather events (Garnaut 2008; QCOSS 2011; IPCC 2012).

30.2.3 Social disadvantage and adaptation: the evolving policy context

Based on the findings above it is clear that in order to ensure that climate change does not further entrench poverty and inequality within developed nations, it is critical that vulnerable groups are a central focus in any adaptation strategy (QCOSS 2011). However, policymakers around the world have been slow to address the social and economic barriers to adaptation they face in key national adaptation policy instruments. In Australia, the Commonwealth Government's 2010 position paper (Adapting to Climate Change in Australia) recognises that adaptation is the shared responsibility of governments and the community. However, it does not identify people or communities – particularly those most vulnerable within them or the organisations that deliver essential social and community services – as priority areas for adaptation. Similarly, the Productivity Commission's 2013 report (Barriers to Effective Climate Change Adaptation) overlooks structural social and economic barriers to adaptation which particularly affect disadvantaged people (Productivity Commission 2013).

However, some recent developments give cause for optimism. In its 2013 Climate Adaptation Outlook, which proposes a national assessment framework to assess progress in adapting to the impacts of climate change, the Australian Government identifies the capacity of disadvantaged groups to manage risks from climate change as a key outcome for adaptation (Commonwealth of Australia 2013). It also acknowledges both the role that CSOs play in achieving this outcome and the challenges they face in adapting to climate change. Likewise, the National Strategy for Disaster Resilience (NSDR) recognises that 'disasters do not impact everyone in the same way, and it is often our vulnerable community members who are the hardest hit' (COAG 2011, p. 2); and that 'non-government and community organisations are at the forefront of strengthening disaster resilience in Australia' (COAG 2011, p. ii). In addition, in July 2013 the Standing Council on Police and Emergency Management identified the building of resilience of vulnerable groups to disasters and enhancing engagement with the not-for-profit sector as two key priorities for implementing the NSDR.

Despite these policy developments, in practice the mechanisms required to resource the community services sector and its networks to build adaptive capacity and resilience to climate extremes, or to facilitate their active participation in adaptation planning or emergency management preparedness, response and recovery efforts, have yet to be identified. An extensive international literature review established a similar situation across the developed world. Given that social and economic disadvantage is increasingly recognised as a barrier to adaptation, this situation indicates that if developed country governments want to reduce the impacts of climate change on their most vulnerable people, priority must be given to identifying practical solutions to building resilience and adaptive capacity which recognise, resource and strengthen the inherent skills and capacity of CSOs.

30.2.4 The community services sector: an overlooked resource?

The Australian community services sector comprises thousands of large and small non-government, not-for profit organisations that provide a diverse range of essential social services in almost every community across Australia. They are particularly sensitive to the needs of disadvantaged people and communities. Services provided include: accommodation, care and respite services for older people, people with a disability and their carers; refuge, accommodation and casework services for people experiencing homelessness or domestic and family violence; drug and alcohol rehabilitation; emergency relief, financial counselling and microfinance; community legal and advocacy services; generalist and specialist counselling; child welfare, youth and family support; community health, allied health and mental health services; and education and employment services.

CSOs are embedded within their communities, deliver key services, have in-depth knowledge of local people, history, risks and vulnerabilities and are often best placed to understand and identify their support needs. The services they provide are a critical feature of Australian society, complementing government income support systems as well as the health and education systems. As such, CSOs comprise an essential component of the social infrastructure in the communities they serve, particularly in small and remote places where access to services is limited. Indeed, for many disadvantaged people these organisations can be the primary source of connection to the broader community and form the basis of their resilience to everyday adversity as well as in times of crisis.

Given its connection to communities and active role in supporting disaster resilience and recovery, the community services sector 'has a vital role to play in ensuring that actions adopted to prevent dangerous climate change are both effective in their outcomes and socially just in their implementation' (Edwards et al. 2009, p. 58). However, for all but a handful of organisations, inadequate recognition and resourcing of CSOs has severely limited their capacity to implement adaptation strategies and to participate in formal policy development and planning processes for both adaptation and emergency management. Because of this oversight, the preparedness of this critical sector has not occurred. Consequently, very little is understood about ways in which community-based social service delivery will be impacted by climate change or organisation capacity to mitigate and manage such impacts for vulnerable individuals and communities.

This knowledge gap is important because of the size and scope of the sector, its influence, the range of services it provides and the critical role it plays in increasing resilience to non-climate change-related adversity as well as in supporting individuals and communities to respond to and recover from the devastating impacts of extreme weather events and natural disasters. On the one hand, if highly vulnerable to climate change impacts themselves, CSOs may worsen climate impacts to social infrastructure and human settlements through the disruption or cessation of service provision. On the other hand, if well prepared and resilient they may lessen the risks from climate-driven physical infrastructure failure faced by communities and those most vulnerable within them, thereby shielding decision-makers from the true extent of these risks.

30.3 Filling the knowledge gap: key findings from the research

The *Adapting the Community Sector for Climate Extremes* project was the first attempt to undertake a comprehensive assessment of the resilience of the community services sector to climate change and extreme weather impacts. Key research methods employed included a series of 10 workshops with over 150 community sector professionals across Australia and a national survey of almost 500 CSOs. The workshops aimed to identify the particular risks of climate change and extreme weather events to CSOs, develop example adaptation strategies and consider opportunities that might arise from successful adaptation. The survey's key objectives were to identify: indicators of organisational vulnerability and resilience; the organisational features most vulnerable to climate change and extreme weather impacts; and key barriers to adaptation.

30.3.1 *Systems analysis of climate-driven CSO failure modes*

Data gathered through the workshop program were analysed using a 'rich pictures' systems approach. This enabled the identification of relationships and interdependencies between the key elements that contribute to a typical CSO operating system such as clients, staff and volunteers, essential infrastructure, financial services, suppliers, public and private sector institutions and peer organisations and the degree to which these relationships are reliant on or modified by disruptions in critical infrastructure or infrastructure commodities. It also allowed examination of relationships between CSOs and more abstract elements such as the economy and social capital. Once the key systems elements were identified, diagrammatic representations were created using non-hierarchical classification software which enabled the system to be viewed from the perspective of each element. This generalised systems analysis was then applied to organisational case studies presented during the workshops to explore the processes by which climate change impacts to infrastructure might lead to disruptions to an organisation's operational system and, eventually, to the failure of service delivery.

This approach was used to identify and analyse system failure modes. The failure mode exemplar presented in Figure 30.1 shows how power failure caused by a bushfire could lead to the total failure of service delivery by an outreach provider of Home and Community Care services to the elderly and people with a disability. In this case the failure mode is through the reliance on Voice-Over Internet Provided (VOIP) telecommunications for all incoming and outgoing communications with clients, other service providers and the emergency services. This exemplar also shows how the failure of CSO service delivery impacts on other elements of the social service system, including peer organisations and the state emergency services which in turn become strained by increased demand for their services.

30.3.2 *Quantifying CSO vulnerability*

Results from the national survey of CSOs paint a vivid picture of the sector's vulnerability to climate-driven physical infrastructure failure as well as the serious consequences of service failure

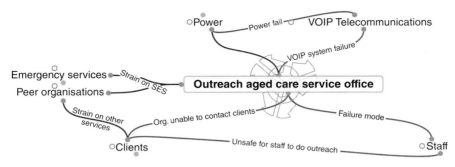

Figure 30.1 Failure mode exemplar. For colour details please see Plate 22.

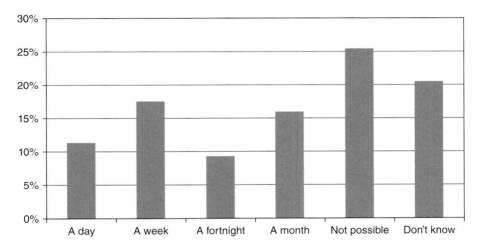

Figure 30.2 Building vulnerability and service capacity.

for vulnerable and disadvantaged sectors of the community. For example, one week after an extreme weather event which caused an inability to use their premises, 50% of organisations would be unable to operate. More seriously, 25% of organisations reported that damage caused by an extreme weather event might lead to its permanent closure (Fig. 30.2; $n = 399$). Organisations that primarily provide direct services to clients from an office or building are most at risk of permanent closure. This suggests that diversifying and mobilising service provision may be an effective way to reduce the vulnerability of CSO service delivery. Similarly, organisations in the lowest income quintile were almost twice as likely as those in the highest quintile to

report that significant damage to service centres and offices might cause the permanent closure of the affected services.

Respondents also reported high levels of CSO vulnerability to the failure of infrastructure utilities, particularly electricity (93%) and telecommunications (89%), with disruptions to these services having the potential to cause the complete cessation of service provision for some organisations.

Qualitative responses provide an insight into the specific ways in which the failure of these services could lead to service disruptions. For example, power and telecommunications failures could result in loss of access to client records and contact details, potentially causing a total loss of contact with clients during disasters. Loss of power could also disrupt organisations' ability to co-ordinate programs and appointments, to meet their contractual obligations and to pay their staff. Disruptions to water supplies could lead to serious workplace health and safety breaches, particularly for organisations providing accommodation and meal services, causing significant disruptions to service provision.

A number of other factors were also revealed to have an impact on organisations' capacity to provide effective services in the aftermath of an extreme event, including staff losses and increased demand for services. For example, 60% of respondents would not be able to provide services for more than a day if an extreme weather event seriously impacted or disrupted their staff. Further, 50% of organisations also predicted an extreme weather event would cause a short-term increase in demand for services, with 30% predicting that increase would be maintained over the long term. Organisations most likely to report long-term increased demand included those providing housing and homelessness, emergency relief and advocacy services. With CSOs typically struggling to meet demand for services with inadequate resources, such increases in demand would also likely lead to a considerable strain on organisational resources.

30.3.3   Little action across a range of risk controls

The project's findings also reveal that very few organisations within the sector have begun to take systematic action to manage, mitigate or transfer climate change and extreme weather risks. Respondents to the survey reported high levels of under-insurance against losses related to climate change and extreme weather impacts, particularly contract (over 40%) and income losses (45%) and losses caused by disruptions to business continuity (35%) and staff (over 40%). Respondents also reported undertaking low levels of risk management activity, including climate and extreme weather risk

assessments and developing appropriate plans to respond.

30.3.4   Experience matters

Significantly, the survey found that organisations that had experienced an extreme weather event in the previous 10-year period were more likely than those that had not to have taken concrete actions to build preparedness and response capacity. Indeed, the study identified three indicators of organisational resilience to climate change impacts: size, knowledge and experience. That is, larger organisational size, higher levels of self-reported knowledge about climate change and past experience of an extreme weather event were all positively correlated with robust responses to risk and therefore higher levels of resilience to climate change and extreme weather impacts.

30.3.5   Service failure heightens client vulnerability

The consequences of the total cessation of community-based social service delivery in response to an extreme event at a time when demand for services is increased are serious, for clients and for the community more broadly. Key themes that emerged include:
• that people would be at increased risk of homelessness, financial hardship, hunger, disease and ill-health;
• that those with pre-existing mental health conditions would likely suffer a deterioration in their condition (including an increased risk of suicide);
• that women and children already exposed would be at increased risk of experiencing domestic and family violence; and
• in a worst-case scenario, that people with high-level personal and health care needs and the homeless would be at increased risk of death if service provision failed.

One survey respondent working for a CSO that provides high-security accommodation for women and children at immediate risk of domestic and family violence reported that 'we are the only service of its type for 1500 km' and 'women and children's safety in the region would be extremely compromised' if the service was to fail in response to an extreme event. An aged care provider reported that its clients 'would die (no eating, no toileting, no showering, no getting out of bed)'.

30.4   Barriers to adaptation in CSOs

A clear understanding of the barriers to adaptation faced by different sectors in the community is critical for effective adaptation policy development and implementation. The *Adapting the Community Sector for Climate Extremes* project sought to understand barriers to adaptation for the community services sector as perceived by CSOs. CSOs identified a broad range of barriers through the workshop program and national survey including: lack of resources; rigid service contracts; and a lack of awareness about how climate change will affect the core business of organisations (Fig. 30.3).

30.4.1   Resourcing is critical

Participants identified lack of resources in the form of knowledge, skills and financial resources and capital as the key barrier to adaptation. However, lack of financial resources and capital emerged as a particularly significant barrier to investment in adapting service delivery and assisting clients to adapt to climate change and extreme weather impacts. This result is supported by findings from the annual ACOSS survey of CSOs nationally, which in 2011, 2012 and 2013 identified underfunding and lack of funding certainty as the key challenges for the sector into the future (ACOSS 2011, 2012, 2013).

The resourcing issue becomes particularly important in light of the project's findings about the impact of extreme weather events on demand for services. As discussed previously, CSOs perceive that extreme events will have significant impacts on demand for services in both the short and the long term. With organisations already struggling to meet demand with limited resources (ACOSS 2011, 2012), this increase in demand may itself become a source of strain or failure for CSO service delivery, at a time when their services will be needed more than ever.

30.4.2   Inflexible contracts for service delivery

The majority of Australian CSOs rely on government funding contracts to deliver services to the community. A clear theme that emerged through the workshops was that current funding arrangements create a significant barrier to adaptation. This is because service contracts fail to recognise the costs to CSOs of investing in adaptation or to compensate organisations for the financial and other impacts of extreme weather events to their operations, including as a result of the role they play in response and recovery. One example provided by a study participant clearly demonstrates the serious consequences for CSOs arising from rigid contracting processes. The organisation mobilised its entire workforce – on a voluntary basis – to participate in the response and recovery efforts to the 2009 Victorian Bushfire. It was not able to recover the costs expended in seconding its professionals to the response effort. More seriously, when the organisation resumed normal operations, it was informed by the government agency that provided its funding that it was to lose over one million dollars in funding because it had missed a contractual reporting deadline while participating in the disaster response.

30.4.3   Climate adaptation as core business

Lack of awareness of the risks and beliefs about the relevance of climate change adaptation to the sector's core mission of relieving poverty and advocating social justice (but not about the veracity of climate science itself) were also perceived by survey respondents to create a significant barrier to climate change adaptation. The workshop program highlighted that most participants had not previously thought about their organisation's exposure and vulnerability to

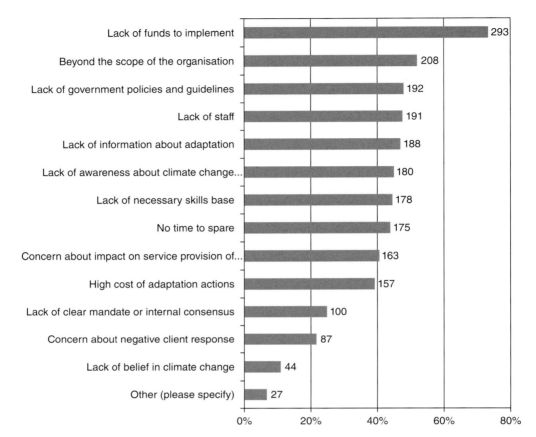

Figure 30.3 Barriers to adaptation.

climate change and extreme weather and their likely severe impacts on service delivery capacity or the flow on consequences for client wellbeing. Similarly, the survey revealed that a majority of organisations regard climate change adaptation as outside the sector's 'core business'.

These findings suggest that increasing risk awareness and supporting CSOs to classify climate change as a significant risk to 'core business' will be critical in driving adaptation across the community services sector. Findings from recent projects conducted in the UK to raise CSO awareness about how climate change will impact their clients and their service delivery capacity, and this project team's experience facilitating the Welfare Professional Climate Workshops, suggest

that when they are engaged directly on the issues CSOs are quickly able to identify ways in which climate change and extreme weather events will impact their clients and their service delivery; and to recognise the serious risks they face from these impacts; and the need for action to reduce vulnerability and increase resilience to those risks.

30.5   The role of adapted CSOs in community resilience

The project results have a positive story to tell with implications for domestic and international policymakers and for communities broadly: with adequate resources and the right support, CSOs have the desire and the inherent capacity to contribute to client and community resilience to climate change impacts.

30.5.1 *Unique access and unique means*

Through the project CSOs reported a strong desire to undertake a range of actions to adapt their services to climate change and extreme weather impacts to ensure they can continue to provide services to clients as climate change accelerates. They also reported having often unique and unparalleled localised specialist skills and assets to assist clients to better prepare for climate change impacts and extreme weather events, and to respond to them after they occur.

As Figure 30.4 indicates, the actions that organisations were most likely to have undertaken already include action to reduce utility bills

(42%) and develop a disaster management plan (27%). Fewer than 10% of organisations reported having developed a climate change adaptation plan, relocated their service centres or undertaken a climate change risk assessment. However, significant numbers expressed willingness to engage in a broad range of adaptation actions. For example, over 40% of organisations reported they would like to:

• assist clients to prepare for climate change and develop a climate change adaptation plan;
• work with other organisations to plan collaborative service provision during disasters (over 60%);
• take action to reduce their utility bills;
• upgrade organisational infrastructure to be resilient to climate change and extreme weather impacts;
• develop a disaster management plan; and
• undertake a climate change risk assessment.

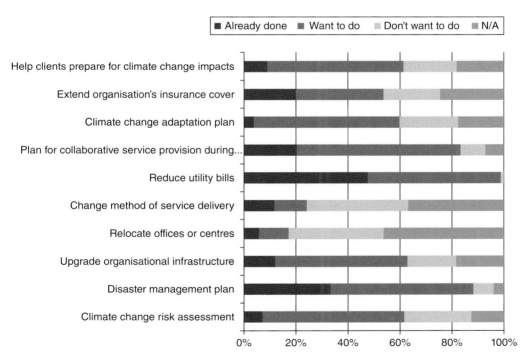

Figure 30.4 Actions organisations (*n* = 408) would like to take if resources were available. For colour details please see Plate 23.

30.5.2 Becoming part of the solution

An overwhelming majority of respondents reported that, with sufficient resources, they could also provide a range of services or assistance to their clients before and after extreme weather events to help them prepare for and respond to impacts. Before an event, over 70% reported that they could provide community education programs to teach clients about local extreme weather risks and to prepare for their potential impacts; over 60% also reported that they could warn clients about a predicted extreme weather event; and approximately 40% reported an ability to evacuate clients from high-risk areas before a predicted extreme weather event occurred.

After an event, almost 80% reported that with adequate resources they could provide specialist services related to their main area of service provision and over 70% reported an ability to contact and locate clients in the aftermath of an extreme weather event. Over 50% reported being able to provide volunteer management services, general or trauma counselling services and emergency relief to organisations and clients in need (Fig. 30.5).

Indeed, many organisations across the sector are already developing and implementing innovative programs to support climate adaptation for CSOs and their clients, particularly in the area of rising energy costs. For example, the West Australian Council of Social Service partnered with the West Australian Peaks Forum to deliver the *Climate Ready Communities Program*, which aimed to increase employment opportunities for disadvantaged people during the global economic crisis while at the same time increasing CSO preparedness for climate change. It did so by training people who were out of work to perform energy efficiency audits and then employing them to conduct free audits for participating CSOs and recommend cost-effective strategies to reduce energy and water use. Similarly, the Brotherhood of St Laurence is conducting a *Home Energy Efficiency Upgrade Project* in partnership with the Victorian Government, which will assist 1400 low-income households to upgrade to energy-efficient hot water systems, thereby reducing energy bills for consumers as well as greenhouse gas emissions. Similar programs could be designed to engage disadvantaged people in building CSO and household resilience to extreme weather events.

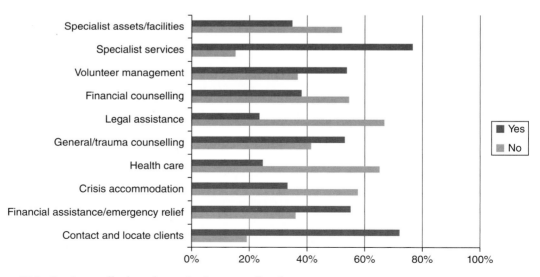

Figure 30.5 Services well-adapted organisations can offer after extreme events.

30.6 Recommendations

In order to continue developing and implementing innovative strategies to support adaptation for disadvantaged people, CSOs must first be supported to increase their own adaptive capacity. Recommendations presented by this project include: adequately resourcing the sector to invest in adaptation; raising awareness about the risks to the core business of CSOs as well as their capacity to support community adaptation; implementing practical programs to build CSO resilience and that of their clients; and ensuring CSOs are able to appropriately and effectively share risks. If fully implemented, they would increase CSOs adaptive capacity and ensure that the social impacts of climate change can be equitably managed in developed countries.

30.6.1 Resource investment in community sector adaptation

With many organisations already struggling to meet demand for services with limited financial resources, few have spare resources for the important task of adaptation. Effective adaptation will therefore require a significant investment of funds to support capacity and resilience building projects for CSOs and their clients, particularly for high-cost upgrades to infrastructure. Efficacy will require that funds target the thousands of small- and medium-sized organisations that are integral to the cohesion and resilience of their communities; and are tied to prompt, concrete action to reduce vulnerability through preparedness, resilience building and sharing risks.

In addition, amending service delivery contracts to recognise the increased costs to CSOs of adaptation and ensure they are not penalised for failing to meet contractual obligations due to participation in disaster response and recovery would provide greater flexibility to CSOs to support adaptive capacity across communities.

30.6.2 Raise awareness of risks and capabilities

The project identified the need for raising awareness at two levels: within the community services sector and across research and policy institutions. Within the sector, awareness must be built about the real and serious threat to CSOs core business from climate change: that is, the direct and indirect risks to disadvantaged people and the ways in which the climate-driven failure of social service provision will exacerbate those risks. Awareness must also be built across governments and the community of the critical role CSOs play in building community resilience to extreme events and the need for adequate resourcing to ensure they can continue to develop and implement innovative community-wide adaptation solutions which specifically address the needs of disadvantaged people.

30.6.3 Build resilience and adaptive capacity of CSOs and client groups

There are also a number of low-cost practical adaptation strategies that can be implemented immediately to build the adaptive capacity of CSOs and their client groups. These include developing and implementing robust, sector-specific risk assessment, service continuity, climate adaptation and disaster management plans as well as simple benchmarking systems to plot organisational and sector progress towards resilience. To ensure long-term efficacy, risk management knowledge and skills will need to be institutionalised within organisational governance structures and across the sector. Once developed, such tools could be easily adapted to assist those clients with increased exposure and vulnerability to plan, adapt and respond.

30.6.4 Support effective and appropriate risk sharing

Sharing risks through access to appropriate and affordable insurance and effective collaboration between organisations and sectors is also critical to supporting the adaptive capacity of CSOs. Improving CSO resilience through the implementation of strategies identified here will increase insurability and bring down the costs of insurance premiums, ensuring access to affordable and adequate insurance cover against risks.

Developing and strengthening collaborative networks between organisations within the community service sector as well as between it and the public and private sectors (including non-traditional partners such as utilities and emergency services) will enhance local-level and community-wide responses to climate change and extreme weather impacts; and ensure that the specific needs of people experiencing poverty and inequality are addressed appropriately in decision-making and in response to extreme events.

References

ABS (2014) *Household Wealth and Wealth Distribution, Australia, 2011–2012*. Canberra, Australian Bureau of Statistics. Available at http://www.abs.gov.au/ausstats/abs@.nsf/Latestproducts/6554.0Main%20Features22011%E2%80%9312?opendocument&tabname=Summary&prodno=6554.0&issue=2011%9612&num=&view= (accessed 13 June 2014).

ACOSS (2011) *Australian Community Sector Survey 2011 ACOSS Paper 173 Volume 1: National*. Australian Council of Social Service, Strawberry Hills.

ACOSS (2012) *Australian Community Sector Survey 2012 National Report ACOSS Paper 191*. Australian Council of Social Service, Strawberry Hills.

ACOSS (2013) *Australian Community Sector Survey 2013 National Report ACOSS Paper 202*. Australian Council of Social Service, Strawberry Hills.

ACOSS, ACF and CHOICE (2008) *Energy & Equity. Preparing households for climate change: efficiency, equity, immediacy*. Available at http://acoss.org.au/images/uploads/4204__EnergyEquity_low_res.pdf (accessed 13 June 2014).

Altman, J.C. and Jordan, K. (2008) *Impact of Climate Change on Indigenous Australians: Submission to the Garnaut Climate Change Review, CAEPR Topical Issue No. 3/2008*. Australian National University, Canberra.

Brotherhood of St Laurence (2007) *Equity in Response to Climate Change Roundtable 2007, Report of the Equity in Response to Climate Change Roundtable, Melbourne, 26 March 2007*. Brotherhood of St Laurence. Available at http://www.bsl.org.au/pdfs/Equity_in_Response_to_Climate_Change_Roundtable_report_2007.pdf (accessed 2 June 2014).

Commonwealth of Australia (2013) *Climate Adaptation Outlook: A Proposed National Adaptation Assessment Framework*. Department of Industry, Innovation, Climate Change, Science, Research and Tertiary Education, Canberra.

Council of Australian Governments (2011) *National Strategy for Disaster Resilience*. Attorney General's Department, Canberra.

Edwards, T., Fritze, J. and Wiseman, J. (2009) Community wellbeing in a changing climate: challenges and priorities for the Australian Community Services Sector. *Just Policy* 50, 80–86.

Enarson, E. (1999) Violence against women in disasters: a study of domestic violence programs in the United States and Canada. *Violence Against Women* 5(7), 747–768.

Ensor, J. and Berger, R. (2009) *Understanding Climate Change Adaptation: Lessons from Community-Based Approaches*. Practical Action Publishing, Rugby, UK.

Garnaut, R. (2008) *Garnaut Climate Change Review*. Cambridge University Press, Port Melbourne.

Green, D., Niall, S. and Morrison, J. (2012) Bridging the gap between theory and practices in climate change vulnerability assessments for remote Indigenous communities in northern Australia, *Local Environment*, 17(3), 295–315.

Hennessey, K., Fitzharris, B., Bates, B. et al. (2007) Australia and New Zealand Climate Change 2007. In: Parry, M., Canziani, O., Palutikof, J., van der Linden, P. and Hanson, C. (eds) *Impacts, Adaptation and Vulnerability. Contribution of Working Group II to the Fourth Assessment Report of the Intergovernmental Panel on Climate Change*. Cambridge University Press, Cambridge, UK, pp. 507–540.

Horton, G., Hanna, L. and Kelly, B. (2010) Drought, drying and climate change: emerging health issues for ageing Australians in rural areas. *Australian Journal of Ageing* 29(1), 2–7.

IPCC (2012) Summary for policymakers. In: Field, C.B., Barros, V., Stocker, V.F. et al. (eds) *Managing the Risks of Extreme Events and Disasters to Advance Climate Change Adaptation, A Special Report of Working Groups I and II of the Intergovernmental Panel on*

Climate Change. Cambridge University Press, Cambridge, UK and New York, USA, pp. 1–19.

Jenkins, P. and Phillips, B. (2008) Battered women, catastrophe and the context of safety after Hurricane Katrina. *Feminist Foundations* 20(3), 49–68.

Macchi, M., Oviedo, G., Gotheil, S. et al. (2008) *Indigenous and Traditional Peoples and Climate Change, Issues Paper*. IUCN, Switzerland.

Mallon, K., Hamilton, E., Black, M., Beem, B. and Abs, J. (2013) *Adapting the Community Sector for Climate Extremes*, National Climate Change Adaptation Research Facility, Gold Coast.

McMichael, A.J. and Butler, C.D. (2009) Climate change and human health: recognising the really inconvenient truth. *Medical Journal of Australia* 191, 595–596.

Oven, K.J., Curtis, S.E., Reaney, S. et al. (2012) Climate change and health and social care: defining future hazard, vulnerability and risk for infrastructure systems supporting older people's health care in England. *Applied Geography* 33, 16–24.

Porter, S. and Abbott, D. (2011) Disabled people and environmental hazard: a scoping study. *Nursing Times* 28, 839–852.

Productivity Commission (2013) *Barriers to Effective Climate Change Adaptation*. Australian Government, Canberra.

Queensland Council of Social Services (2011) *Climate Change: Adaptation for Queensland Issues Paper, QCOSS Submission, October 2011*. Queensland Council of Social Services, Brisbane.

Reser, J.P., Morrisesey, S.A. and Ellul, M. (2011) The threat of climate change: Psychological response, adaptations and impacts. In: Weisbecker, I. (ed.) *Climate Change and Human Wellbeing*. International and Cultural Psychology Series, Springer Publications, New York, pp. 19–42.

Smith, K.R. (2008) Introduction: mitigating, adapting, and suffering: how much of each? *Annual Review of Public Health*, 29(32), doi: 10.1146/annurev. pu.29.031708.100011.

Stanley, J. (2009) *Promoting Social Inclusion in Adaptation to Climate Change: Discussion Paper, Monash Sustainability Institute Report 09/4*. Sustainability Victoria, Melbourne.

UNFCCC (2006a) *Background Paper: Impacts, Vulnerability and Adaptation to Climate Change in Latin America*. UNFCCC Secretariat, Bonn, Germany.

UNFCCC (2006b) *Background Paper: Impacts, Vulnerability and Adaptation to Climate Change in Africa*. UNFCCC Secretariat, Bonn, Germany.

UNFCCC (2007a) *Vulnerability and Adaptation to Climate Change in Small Island Developing States: Background Paper for the Expert Meeting on Adaptation for Small Island Developing States*. UNFCCC Secretariat, Bonn, Germany.

UNFCCC (2007b) *Background Paper: Impacts, Vulnerability and Adaptation to Climate Change in Asia*. UNFCCC Secretariat, Bonn, Germany.

USGCRP (2008) *Analysis of the Effects of Global Change on Human Health and Welfare and Human Systems (SAP 4.6)*. US Environmental Protection Agency, Washington DC.

Victorian Government Department of Social Service (2009) *January 2009 Heatwave in Victoria: an Assessment of Health Impacts*. Victorian Government Department of Human Services, Melbourne, Victoria.

Walker, R. (2012) *The Relationship Between Climate Change and Violence: A Literature Review*. South East Community Health Partnerships, Dandenong, Victoria.

Section 6

Indigenous experience of climate change

31 Continuity and change: Indigenous Australia and the imperative of adaptation

MEG PARSONS

School of Environment, University of Auckland, New Zealand

31.1 Introduction

There is mounting evidence that climate change will disproportionately affect poor and marginalised populations, including Indigenous peoples, and reinforce current social inequity (IPCC 2007; Ford et al. 2010). Within Australia, inequalities exist between ethnicities, classes, males and females, age groups and places (Matthew et al. 2010). Inequalities between Indigenous (Aboriginal and Torres Strait Islander) and non-Indigenous Australians are among the worst in the developed world (Anderson et al. 2006; Sutton 2009). For many Aboriginal and Torres Strait Islanders access to and opportunities for employment and education is limited, the provision of social services, community infrastructure and healthcare remain deficient and social problems are endemic (ABS 2006; Biddle 2009; Langton et al. 2012). The impacts of climate change threaten to exacerbate the existing social, economic, institutional and health problems faced by Indigenous communities and create new challenges and opportunities for Indigenous groups. For both ethical and practical reasons, it is critical that we understand the nature of current inequalities, and the ways in which climate change adaptation can allow for more equitable and just outcomes for Indigenous Australians.

In the present day there are two main ethnic groups of Indigenous Australians: Aboriginal people from the Australian continent and nearby offshore islands, and Torres Strait Islanders. In 2006, these two groups made up 2.5% of the total Australian population (517,000 people; ABS 2006). From this total, 90% identified themselves as Aboriginal, 6% as Torres Strait Islander and 4% as both Aboriginal and Torres Strait Islander (Anderson et al. 2006; ABS 2010). The Australian Indigenous population is comparatively young, with a median age of 21 years compared to 37 years for the non-Indigenous population. The majority of Aboriginal people and Torres Strait Islanders live in non-remote areas, either in regional areas (43%) or major cities (32%). Just 25% of the Indigenous population (combined Aboriginal and Torres Strait Islander populations) live in remote areas (Biddle 2009; ABS 2010).

Indigenous peoples are often confined to the margins of international climate change discussions, and comparatively little research on Indigenous climate change adaptation has been undertaken outside of the North American Arctic. Furthermore, existing research into climate change vulnerability and adaptation for Indigenous peoples has been widely criticised for being limited in scope and focus, rearticulating colonial

Applied Studies in Climate Adaptation, First Edition. Edited by Jean P. Palutikof, Sarah L. Boulter, Jon Barnett and David Rissik.

imaginings and making various assumptions based on climate-related risks without an appreciation of local histories, institutions or cosmologies (Cameron 2012; Veland et al. 2012). The following four chapters by Horne and Mantel (Chapter 32), Haynes et al. (Chapter 33), Tran et al. (Chapter 34) and Nursey-Bray et al. (Chapter 35) represent just a small portion of the emergent research on Australian Indigenous experiences of and responses to climate change and extremes.

In this essay I propose two key challenges for the task of adapting to climate change for Indigenous communities. The first is to move beyond current conceptualisation of Indigenous as 'traditional' and consider the diversity of Indigenous communities. The second key challenge relates to future planning and the need to consider how adaptation relates to social justice and Indigenous rights.

31.2   Embracing the diversity of Indigenous Australia

Recently geographer Emilie Cameron asked the important question: how does critical research into the spatialisation of Indigeneity and colonial governmentality intersect with the climate change adaptation literature? Cameron argues, in the context of the Canadian Arctic, that the vast majority of Indigenous-focused climate change adaptation research overlooks how historic and contemporary experiences of colonisation continue to shape Indigenous societies (Cameron 2012). The central premise of much of the work is that to be 'Indigenous' is to be 'traditional' and 'local', whereby Indigenous people are tied to their local environment, historic practices and beliefs and pre-modern modes of living. In the context of Inuit experiences of climate change, Cameron observes, scholars typically describe the 'Indigenous' dimensions of climate change adaptation in terms of Inuit 'traditional' activities (hunting, gathering, cultural ceremonies) and 'traditional' knowledge (about weather, water, land, plants and animals) (Cameron 2012). This discursive tie (traditional = Indigenous) contributes to

a series of closures around who and what are legitimate areas of scholarly endeavour. For instance, Indigenous people's use and management of biodiversity is often discussed, but not people's use of diesel generators to power their homes and communities, the petrol that runs their vehicles and never the mineral deposits that are extracted from their lands. The most important form of Indigenous knowledge and practice is therefore the traditional and local. Proceeding from this premise, Cameron argues, 'the vector through which locals are understood to be vulnerable to climate risk is through their Indigeneity: it is through engaging in "traditional" cultural and economic practices such as hunting, camping, or travelling' (Cameron 2012, p. 108).

Similarly in the Australian context, Indigenous interests are often compartmentalised into cultural heritage values within natural resource management planning as well as more broadly in government decision-making (Jackson 2006). Finn and Jackson's work with Indigenous groups in the Daly River in the Northern Territory demonstrates how water management continues to emphasise 'cultural' elements of Aboriginal groups' relationships with river systems, and perpetuates a view that only certain social practices and interests (chiefly non-market subsistence activities) were legitimate and worthy of protection at the expense of others (primarily economic) (Finn and Jackson 2011). While the establishment of a separate category associated with Indigenous communities within climate change adaptation may serve to improve the status of Indigenous peoples and to strengthen their voices in decision-making about adaptation, there remains a durable logic that relegates Indigenous interests into the realm of symbolic, cultural and traditional (beliefs, values and traditional ecological knowledge) while the practical and material (the political economy) sphere remains the domain of non-Indigenous (white) Australians. This is not to say that consideration of Indigenous cultural values, sites of cultural heritage, knowledge (including traditional ecological knowledge) and wider Indigenous cosmologies are not critical for climate change adaptation, but rather that by compartmentalising

Indigenous issues into 'traditional' and 'cultural' scholars limit what questions are asked, what problems are identified and what solutions are found. Furthermore, from this intellectual standpoint, white Australian culture (or cultures) remains invisible, the taken for granted norm, as civilisation, as progress, as modernity, while Aboriginal and Torres Strait Islander cultures are seen as unchanging traditions, referencing only back to the past rather than ahead into the future. The imperative of adaptation requires us not only to examine how the past conditions the present, but also to consider how the future is conditioned by the adaptation (and mitigation) choices made in the present.

If adaptation is intended to maintain or enhance the provision of things that Indigenous groups value, then vulnerability assessments and adaptation planning needs to consider the diversity of Australian Indigenous communities; this includes an appreciation of the importance of historical and contemporary processes of change. For instance, the present-day location of many Indigenous communities is a consequence of colonial land confiscations, programs of forced relocation and discriminatory government policies. Indeed the geographical shift in where Indigenous people live is undoubtedly one of the most certain and visible measures of the ways in which colonisation has impacted on Australia's Indigenous peoples over the last 200 years or so (Taylor 2011). In 1788 the Indigenous population was spread throughout the continent and islands, with traditional clan estates varying in population densities (generally higher in coastal, riverine and lower inland areas). Today, the Indigenous population is clustered into residential arrangements that are less determined by traditional connections to clan estates (homeland or country) and more focused on regional towns and specific suburbs in the major cities (Biddle 2009; Taylor 2011).

However, scholarly attention remains largely focused on the impacts of climate change on remote (and therefore supposedly 'traditional') Indigenous people, and little is known about the ways in which climate change is (or will) affect urban Indigenous communities. Indeed, as the work of Biddle (2009) and Taylor (2011) demonstrates, Indigenous Australians are highly mobile and are far more likely to migrate between remote, regional and metropolitan areas on an ongoing basis than non-Indigenous Australians. For instance, a great degree of research has been conducted on the impacts of climate change on the islands of the Torres Strait and the vulnerability of and capacity of Torres Strait Island communities to adapt to changing climate conditions, yet limited attention has been given to Torres Strait Islanders living outside their ancestral homelands. The metropolitan centres of Townsville, Cairns and Brisbane, for example, each have larger Torres Strait Islander populations than all the home islands combined (Beckett 1987). This settlement pattern, a result of more than a century of migration (both forced and voluntary), has created diverse and multiple identities. Torres Strait Islanders living in metropolitan centres continue to affiliate with their ancestral islands, but have established new ways of expressing their cultural identities including the promotion of a shared pan-island Torres Strait Islander cultural identity (Davis 2004). Understanding how urban Indigenous communities experience and respond to climate-related risks, as well as how climate change may influence or interact with Indigenous migration patterns, is essential for equitable and inclusive climate change adaptation plans and policies.

31.3 Future planning and social justice

The second key challenge for adaptation for Indigenous people is the need to consider adaptation not simply as a series of short-term actions designed to reduce vulnerability, but rather as an ongoing process which engenders just outcomes for Indigenous communities. Questions of justice lie at the centre of discussions of how Indigenous communities adapt to changing environmental conditions (Schlosberg 2012). There are unavoidable justice dimensions associated with adaptation, which determines the winners and losers of decisions about how, when and where to adapt. For instance, who decides and benefits from the relocation and resettlement of low-lying island and coastal Indigenous populations, whose interests are represented in adaptation policymaking and implementation and who ultimately decides what constitutes 'good' and 'sustainable' adaptation as opposed to 'bad' and 'maladaptive' adaptation (Eriksen et al. 2011)?

To answer these questions we might look to a number of international treaties that provide formal definitions of climate justice based on the concept of universal human rights. For instance, the United Nations Covenant on Civic and Political Rights states that 'in no case may a people be deprived of its own means of subsistence'. Similarly the United Nations Declaration on the Rights of Indigenous Peoples maintains that:

> Indigenous peoples have the right to maintain and develop their political, economic and social systems or institutions, to be secure in the enjoyment of their own means of subsistence and development, and to engage freely in all their traditional and other economic activities (Article 20, UNGA 2007, p. 7).

Accordingly governments are obliged to take 'appropriate actions' to mitigate any adverse environmental, social, economic or cultural impacts on Indigenous groups (UNGA 2007, p. 12). However, the question of what constitutes 'appropriate' adaptation actions for Indigenous peoples remains largely undefined in international treaties. It is this area of climate change adaptation where more progress could be made through reframing the focus of adaptation science and policy away from the vulnerability of 'traditional' Indigenous people, towards a focus on Indigenous individuals' and communities' abilities to lead the type of lives they (both as individuals and members of broader community, clan or language groups) value and their ability to choose to live different kinds of lives (their opportunities).

The extent to which governments can or should make decisions about adaptation for Indigenous communities is problematic because of the multiplicity of ways governments have intervened in the lives of Indigenous people in the name of civilisation, science and protection. Scientific knowledge, in particular, has often been used to justify government actions as being urgent necessities (Bashford 2004; Parsons 2010; Cameron 2012). In early twentieth-century Australia, for instance, the apparent crisis of Aboriginal population decline and then-current scientific knowledge was used to justify extensive interventions in Aboriginal families (including the forcible removal of Aboriginal children from their mothers) as a means to 'protect' the Australian nation (Haebich 2000; Parsons 2009). Indeed, some scholars warn that current depictions of Indigenous groups as being highly vulnerable to the impacts of climate change rearticulates colonial imaginings of Indigenous people as being passive victims (so-called 'doomed races') of external stimuli (in this instance global warming) who require outside (non-Indigenous) intervention to save them (Howitt et al. 2011; Cameron 2012; Veland et al. 2012). This is not to say that climate change does not present risks to Indigenous peoples nor that responses to climate change should be delayed indefinitely, but rather that adaptive actions need to address the concerns and issues identified by Indigenous communities themselves rather than simply another top-down intervention into Indigenous lives. Adaptation policies, plans and projects, as the following chapters attest, need to draw on the present-day knowledge, historic experiences and aspirations of Indigenous communities, and seek to harness and build on local knowledge and skills to enable successful Indigenous-driven adaptation.

31.4 Conclusion

If adaptation options are to be sustainable, effective and equitable in Australia (and elsewhere), they must be built on the social values people hold and the cultural settings that shape and influence those value systems. Since adaptation to climate change is multifaceted and embedded in local contexts of risk aversion, decision-making and social value systems, it is crucial that the diversity of contemporary Aboriginal and Torres Strait Islander communities, which range from very remote outstations to large urban centres, is acknowledged and, as far as possible, incorporated into adaptation thinking, planning and practices. Furthermore, thought needs to be given not only to how adaptation can be facilitated among Indigenous communities, but also to the justice issues inherent in adaptation.

References

Australian Bureau of Statistics (2006) *Population Distribution: Aboriginal and Torres Strait Islander Australians.* Australian Bureau of Statistics, Canberra.

Australian Bureau of Statistics (2010) *The Health and Welfare of Aboriginal and Torres Strait Islander Peoples* (cat. no. 47074.0). Available at http://www.abs.gov.au/AUSSTATS/abs@.nsf/ProductsbyCatalogue/2AFBAD91D361725ACA2577D80012373E?OpenDocument (accessed 12 June 2014).

Anderson, I., Crengle, S., Kamaka, M. L., Chen, T., Palafax, N. and Jackson-Pulver, L. (2006) Indigenous Health in Australia, New Zealand and the Pacific. *The Lancet* 367, 1775–1785.

Bashford, A. (2004) *Imperial Hygiene: A Critical History of Colonialism, Nationalism and Public Health.* New York: Palgrave Macmillan.

Beckett, J. (1987) *Torres Strait Islanders: Customs and Colonialism.* Cambridge University Press, New York.

Biddle, N. (2009) *The Geography and Demography of Indigenous Migration. CAEPR Working Paper No. 58/2009.* Centre for Aboriginal Economic Policy Research, Canberra.

Cameron, E.S. (2012) Securing Indigenous politics: A critique of the vulnerability and adaptation approaches to the human dimensions of climate change in the Canadian Arctic. *Global Environmental Change* 12, 103–114.

Davis, R. (ed.) (2004) *Woven Histories, Dancing Lives: Torres Strait Islander Identity, Culture and History.* Aboriginal Studies Press, Canberra.

Eriksen, S., Aldunce, P., Bahinipati, C.S. et al. (2011) When not every response to climate change is a good one: Identifying principles for sustainable adaptation. *Climate and Development* 3, 7–20.

Finn, M. and Jackson, S. (2011) Protecting Indigenous values in water management: a challenge to conventional environmental flow assessments. *Ecosystems* 14(8), 1232–1248.

Ford, J.D., Berrang-Ford, L., King, M. and Furgal, C. (2010) Vulnerability of Aboriginal health systems in Canada to climate change. *Global Environmental Change* 20, 668–680.

Haebich, A. (2000) *Broken Circles: Fragmenting Indigenous Families, 1880–2000.* Fremantle Arts Centre Press, Fremantle.

Howitt, R., Havene, O. and Veland, S. (2011) Natural and unnatural disasters: Responding with respect to Indigenous rights and knowledges. *Geographical Research* 20(1), 47–59.

IPCC (2007) *Climate Change 2007: Impacts, Adaptation and Vulnerability. Contribution of Working Group II to the Fourth Assessment Report of the Intergovernmental Panel on Climate Change.* Parry, M.L., Canziani, O.F., Palutikof, J.P., van der Linden, P.J. and Hanson, C.E. (eds), Cambridge University Press, Cambridge.

Jackson, S. (2006) Compartmentalising culture: the articulation and consideration of Indigenous values in water resource management. *Australian Geographer* 37(1), 19–31.

Langton, M., Parsons, M., Leonard, S. et al. (2012) *National Climate Change Adaptation Research Plan for Indigenous Communities.* National Climate Change Adaptation Research Facility, Gold Coast.

Matthew, R., Barnett, J., McDonald, B. and O'Brien, K. (eds) (2010) *Global Environmental Change and Human Security.* MIT Press, Cambridge MA.

Parsons, M. (2009) *Spaces of Disease: The Creation and Management of Aboriginal Health and Disease in Queensland 1900–1970.* PhD Thesis, University of Sydney.

Parsons, M. (2010) Segregating Race: Sir Raphael Cilento, Aboriginal health and leprosy management in twentieth century Queensland. In: Konishi, S. and Nugent, M. (eds) *Aboriginal History.* Volume 34. ANU E Press, Canberra.

Schlosberg, D. (2012) *Justice, Ecological Integrity, and Climate Change.* In: Thompson, A. and Bendik-Keymer, J. (eds) *Ethical Adaptation to Climate Change: Human Virtues of the Future.* MIT Press, Cambridge MA, pp. 165–183.

Sutton, P. (2009) *The Politics of Suffering: Indigenous Australia and the End of Liberal Consensus.* Melbourne University Press, Melbourne.

Taylor, J. (2011) Postcolonial transformation of the Australian Indigenous population. *Geographical Research* 49(3), 286–300.

United Nations General Assembly (2007) *61/295 United Nations Declaration on the Rights of Indigenous Peoples*, resolution/adopted by the General Assembly, Sixty-first Session, 13 September 2007. United Nations, Geneva.

Veland, S., Howitt, R., Dominey-Howes, D., Thomalla, F. and Houston, D. (2012) Procedural vulnerability: Understanding environmental change in a remote indigenous community. *Global Environmental Change* 23 (1), 314–326.

32 Housing, households and climate change adaptation in the town camps of Alice Springs

RALPH HORNE[1] AND ANDREW MARTEL[2]

[1]*College of Design and Social Context, RMIT University, Australia*
[2]*Faculty of Architecture, Building and Planning, University of Melbourne, Australia*

32.1 Introduction

Housing for Indigenous people living in rural and remote Australia has been long characterised by an acute shortage of dwellings, poor-quality construction and a building stock ill-suited to Indigenous lifestyles and preferences. Progress in addressing this issue has relied upon a range of research into the study of traditional Indigenous dwellings, traditional socio-spatial properties of Indigenous settlements, the composition of Indigenous 'households', housing and health and the use of inside and outside domestic space (Heppell 1979; Heppell and Wigley 1981; Ross 1987; Memmott 1988; Pholeros et al. 1993). As a result, Aboriginal housing design in Australia is a specialised field within housing studies, combining anthropology and an understanding of cultural differences in Aboriginal domiciliary behaviour along with conventional design disciplines (Memmott 1989, p. 115).

All housing imposes conditions on households that enable the performance of some practices, while hindering others. When housing constrains traditional relationships and practices, then severe stress can result (Reser 1979; Memmott 1988). As Ross notes, 'Inappropriate housing and town planning have the capacity to disrupt social organisation, the mechanisms for maintaining smooth social relations, and support networks' (Ross 1987, p. 6). Memmott (1988, p. 34) lists the housing stress factors for Indigenous people as including: lack of protection from the weather; living in squalor; overcrowding; alcoholism; domestic violence; widespread ill-health; insecurity due to temporary tenure; and the threat of forced eviction. Many of these factors persist today, with the notable addition of climate change.

In this chapter, we explore the adaptive capacity of Indigenous households using the case study area of the town camps of Alice Springs in central Australia. Jurisdictionally, Alice Springs is in the Northern Territory where, over the last 50 years, the frequency of extremely warm days and nights has increased and the average annual maximum temperature has increased by 0.12°C per decade and the minimum temperature by 0.17°C per decade (Hennessy et al. 2004). Alice Springs currently averages 90 days over 35°C and 17 days over 40°C. By 2030 this is expected to

Applied Studies in Climate Adaptation, First Edition. Edited by Jean P. Palutikof, Sarah L. Boulter, Jon Barnett and David Rissik.
© 2015 John Wiley & Sons, Ltd. Published 2015 by John Wiley & Sons, Ltd.

increase to between 96 and 125 days over 35°C and to 21–43 days over 40°C (Hennessy et al. 2004, p. 34). Alice Springs currently has on average 23 'hot spells' (3–5 days over 35°C in a row) per year. By 2030, this is expected to grow to 25–33 over 35°C hot spells, and up to 10 very hot (over 40°C for 3–5 days) spells (Hennessy et al. 2004).

Reports by the CSIRO (Wang et al. 2010, 2011) examine the impact of such predicted climate change effects on the performance of residential houses across Australia, indicating that a 2°C rise in global temperature may increase emissions by 49% in Alice Springs (Wang et al. 2011), reflecting higher cooling requirements of between 61% and 101% to 2050 (Wang et al. 2010). Predicted changes to energy consumption are of course averages, and every individual household will demonstrate a different energy-use profile that reflects social practices, occupancy levels, age profiles, types of household appliances and so on. However, the predictions clearly indicate that maintaining indoor domestic comfort in Alice Springs is likely to become increasingly difficult or expensive in terms of energy and water consumption.

32.2   The town camps of Alice Springs

The town camps of Alice Springs comprise 18 communities of Aboriginal people located around this principal town in central Australia. They consist of permanent residents and also have influxes of visitors (generally extended family members) at various times. There are more visitors than in the community of westernised cultures, and these visitors may also stay longer. The town camps have existed in some form or other since the establishment of Alice Springs itself (then called Stuart) in the late nineteenth century (Heppell and Wigley 1981). However, considerable efforts were made by the local (white) population, supported by various levels of government, to ensure that the camps were not made permanent (Rowley 1970; Heppell and Wigley 1981). This ensured that, for over a century, the camps were denied land tenure,

water, electricity and sanitation services. Despite this, Alice Springs developed as an important centre for Aboriginal people from the outlying areas of the central desert region of Australia, with people coming to town to visit relatives, for employment, to receive medical treatment, attend to government or police business and for shopping and recreation. Visitor numbers can swell the permanent town camp population by up to 50% (Heppell and Wigley 1981; Foster et al. 2005).

Greater certainty of tenure began to emerge in the 1960s and following a change in Commonwealth policies in the early 1970s. By the late 1970s, 12 town camps had applied for leaseholds and there was a considerable expansion of house construction and social services in the camps (Heppell and Wigley 1981; Coughlin 1991). Tangentyere Council was established to assist Aboriginal communities establish legal entitlement to land, shelter and essential services, building maintenance and management, education, training, employment and income security (Coughlin 1991). It has retained this role and is not therefore a 'standard' Local Authority, rather, it is a non-government organisation focussed upon securing and supplying services to the town camp communities.

Currently, there are 284 houses in the 18 official town camps, with a population estimated to be between 2000 and 3000 permanent residents and visitors (Tangentyere Council *pers. comms.*; Foster et al. 2005). Each camp is a small community, substantially made up of residents connected by ties of friendship, kinship and common language. Traditional ties, customs and laws are strong, as is the association with 'traditional' land or country. Residents of a town camp identify as members of a group with shared mutual interests and responsibilities (Heppell and Wigley 1981). Camps are mostly geographically orientated towards traditional lands, with camps on the west, east, north and south of central Alice Springs reflecting the direction of the traditional lands of the residents. As the building of permanent houses and infrastructure has developed since the 1980s, the structure of the town camps has modified and become more conventionally planned. Most recently, the town camp leases have been 'leased back' to the Commonwealth Government through the Office of Township Leasing in return for new houses, renovations and refurbishments to existing houses, and a shift from a community housing management program to a 'mainstreamed' public housing program (Commonwealth Ombudsman 2012).

32.3   Method and approach

Previous studies have examined aspects of power and water usage, energy efficiency and tenant satisfaction with housing performance (post-occupancy evaluations or POE) in discrete Indigenous communities. These include energy audit studies, tenancy management and POE-type studies. For example, the Tangentyere *Energy Efficiency Study* (Tangentyere Research Hub 2011) reported a POE of 100 house upgrades/refurbishments, including the fitting of evaporative air conditioning (known as 'swampies'), radiant panel heating, one-shot booster solar hot water switches (where solar systems can be boosted by electric backup when solar hot water is not available or has run out) and timers, external shading, window and door upgrades and compact fluorescent lighting. Information guides were distributed and, generally, the upgrades were welcomed and appeared to lead to lower spending on energy. Other studies include a CoolMob (Darwin) post-audit survey of 70 homes following an energy efficiency campaign, and a Western Australia Home Energy Efficiency Engagement Refit Program where 100 houses in five remote Aboriginal communities received roof insulation, solar hot water systems, low flow water devices, external shading, energy efficient appliances and educational materials. Repairs and maintenance have also been an area of project focus, for example with the *Maintaining Houses for Better Health* program (Mansell and Sowerbutts 2011) developing a systems approach to sustainable repair and maintenance and seeking to increase capacity development of Indigenous housing service providers. The program aimed to develop a tenancy management database that integrates the survey/audit process (data collection), the practical day-to-day operations of housing repair together with maintenance, quality assurance and analysis and reporting functions.

The Centre for Appropriate Technology POE of Alice Springs town camp housing (CAT 2013) reports on newly constructed houses, refurbished and rebuilt houses in the town camps, finding good levels of satisfaction and also some dissatisfaction; the latter concerning the yards surrounding houses and the relative lack of storage space inside dwellings (most particularly in the kitchen). The survey did not specifically inquire about comfort levels related to climate conditions.

Building on this earlier work, the study reported in this chapter specifically focused on adaptation to changing climate from the perspective of the householder experience. In order to investigate responses to hot and cold weather, the project starts with an understanding of current social practices for keeping cool or warm. The ways in which some practices are maintained over time, some are modified, new practices emerge and still others fade away all have implications for community and household adaptation in the face of changing environmental and climatic conditions.

What are social practices? They have been defined variously as 'a routinized type of behaviour' (Reckwitz 2002, p. 249) and a 'temporally and spatially dispersed nexus of doings and sayings' (Schatzki 1996, p. 89). In simplified terms a practice is both an entity in itself, a specific thing that exists in a society such as the practice of driving a car and something that is performed by an individual (a 'carrier' of a practice), where a person performs the act of driving a car. A practice can exist for a time without being performed by a carrier, whether in the minds of potential carriers or stored as instructions in books or online. To be active and influential, a practice must 'capture' carriers willing to repeatedly perform the practice; carriers of practices are often unreliable, however. That is, people will modify how practices are performed through changed circumstances, personal preference or caprice and, if enough carriers modify how a practice is performed, then that practice is changed. What happens in a society is not fixed but is continually reproduced by doing, and in that reproduction may change if circumstances change (Shove et al. 2012).

Social practice theorists posit at least three primary elements that shape social practice form and take-up: material settings of the practice; skills and knowledge required to 'do' the practice; and rules and common understandings about how a practice should be done. Recent studies of household resource use that adopt a social practices perspective further divide the latter into two categories (see Sofoulis 2005; Maller 2011; Strengers 2011).

In order to understand social practices in a changing climate, following various changes in housing and tenancy arrangements in the town camps, we characterise social practices as consisting of four elements:

1. Materials: the physical things around us; in housing terms, these would include the form of the dwellings, services, appliances, furniture and so on.

2. Rules: the things that must be done (or that must not be done).

3. Practical knowledge: the technical know-how and skill required to perform a particular practice.

4. Common understandings: these are the set of understandings about cultural practices that people consider to be 'normal' or common to a particular social group. An often unspoken assumption about 'what people like me, do' in different situations.

Changes to any of these elements may modify how a practice is performed, and hence change the profile of that practice. Moreover, as practices (particularly those performed within a household) are often connected to other practices (or 'bundled'), changes in a particular practice may cascade into multiple changes across a series of practices (Shove et al. 2012).

To investigate practices, interviews were conducted in town camp houses with households by researchers from RMIT University and the Tangentyere Research Hub. The first set of interviews was aimed at recruiting a broad cross-section of participants from different town camps, house types, language groups, age ranges and genders (Maxwell 2005, pp. 88–89; Weiss 1994, p. 24). Interviews were conducted with 'house bosses' (the head of the household, usually these householders keep the house and do not have paid employment), supplemented in a second phase with people in paid employment. Separate interviews were conducted with housing service provider stakeholders. At the time of the interviews, the residents were living in either new houses or houses that had had improvements to their general condition (including new safety screens for doors and windows, for example) and new appliances installed (such as air conditioners). In addition, the residents are now formally public housing tenants of the Northern Territory Government, and are subject to a changed set of tenancy management rules and guidelines.

The research focused on engaging people to tell their stories about what they did to achieve comfort during periods of hot or cold weather. The interviewees were encouraged to use their own words in describing the various practices performed in their households, and to reflect on the reasons behind the preference for those practices. The interviewees were also asked to reflect on the effect that past extreme weather events had had on them, and on their feelings about the current tenancy arrangements (including options for reporting faults) and the state of their house.

32.4 Managing comfort and climate change in the town camps

The results of the study show that the practice of keeping cool in the town camps is evolving. A variety of ways of keeping cool ('practice variants') were used by residents. We observed both active (that is, requiring the use of metered power) and passive practices, and both 'stable' and changing or emerging practices. Many residents chose to hose down their front yard rather than turn on air conditioning or, where they had adapted to using the latter, were still able to return to the hose technique.

The material elements of cooling practice were therefore important; developing or emerging cooling practice variants overwhelmingly involved new housing hardware and active energy use, with the widespread uptake of the newly installed evaporative air conditioners (the 'swampies'). Importantly, the shift was often associated with concerns for others or with shifts in socialising.

> Air conditioner more comfortable for daughter (2 years old), she gets uncomfortable if outside in the heat. (Respondent 15)
>
> Used to all sit outside out back. Big space and verandah with trees. All fruit trees planted by herself. Now, prefer the luxury of the air conditioner. (Respondent 17)

The most significant emergent practical knowledge element was associated with costs of power use of the newly installed air conditioners. Town camp households are already adept at managing household energy use, and air conditioner usage was generally viewed through the prism of overall energy use. All town camps are on power cards, which are debit-based cards, purchased and topped up in selected retail outlets and inserted into the meter to enable power activation. Once the power card runs out of credit, the meter switches power off. There is therefore an immediate relationship between power cost and use, and 13 of 31 interviewees mentioned cost directly.

> You have to be a millionaire to keep it running. It chews a lot of power yeah. (Respondent 8)
>
> Turn all of the power off (at the fusebox) to save power for the air conditioner. (Respondent 7)

The rules imposed by systems of provision associated with cooling practice are generally associated with reporting faults, etc. and the system of tenancy management and repairs and maintenance had changed in the year leading up to the study. Of perhaps more significance, common understandings of how practices of cooling 'should' be done were also highly influential. Notably, access to the technologies was regulated by house bosses. Visitors' ability to gain permission to independently use house hardware such as the swampies varied, and they would often contribute power cards or cash to compensate for power consumption.

> They need to ask me permission to use it, but they can turn it on. If they want to then I let them. (Respondent 32)
>
> It depends on if my wife has put it on, they don't put it on if she is not here. (Respondent 34)
>
> I don't let them touch the switches, I don't have the air conditioner on during the day, so they go outside and sit down. That's why I don't let them, they might break it. (Respondent 36)

The equivalent material, knowledge, rules and common understandings of keeping warm in cold conditions were also of interest, and a similar analysis was conducted of these comfort practices. As with dealing with the heat, practice variants were spread over a variety of passive and active methods. The preference for sitting outside wrapped in blankets and by a fire (either next to the house or elsewhere in the yard) was noted by 28 of 31 respondents. Some confusion over the status of lighting fires in the tenancy contract existed, with many believing that it was not allowed (but that they continued to do it anyway). The ruling that the provision of air conditioners is standard, whereas heating equipment must be provided by tenants, was also clearly significant in shaping how these technologies are governed and maintained.

Nevertheless, inside dwellings, electric heaters (brought from stores) remain the most popular method of internal heating, although most residents noted their relative expense (in power use) (28 of 31 respondents). As with cooling, flexibility and multiple overlapping practices featured although there was a generational shift away from open fires, with the young often associating them negatively with smoke and smells. There was also a significant material shift following the removal of pot-bellied stoves ('fire buckets') during the refurbishments:

> We make fire when no power. But we use heaters inside. Only in the morning to heat up the house, and only in the lounge and kitchen. (Respondent 16)

Don't have fire buckets, that how we used to
keep warm but now I have to go by housing rules.
But still I want that fire bucket because it's good.
(Respondent 3)

32.5   Discussion and conclusion

Using social practices for comfort as a way of
understanding how households adapt to changing
climate and other dynamic factors around
housing provision, a range of adaptive practices
and vulnerabilities to climate change are revealed.
Diversity in practices exists and this is a sign of
adaptive capacity. Town camp residents retain
variants of previous practice and embrace new
practice variants, which have emerged since the
refurbishments and provision of new housing
over the couple of years prior to the study.
Moreover, town camp residents have a clear
understanding and many experiences of dealing
with extreme weather events and are (at least)
bilingual, bi-cultural and have strong cultural
identities in Indigenous practice while participating in 'mainstream' economic and social life
in Alice Springs and throughout Australia.
As such, the town campers are well placed to
adapt to changing circumstances, including
changing climate conditions. However, there
exists a capacity imbalance arising from poverty
and both chronic and periodic overcrowding,
which remains an entrenched problem and cause
of community stress.

While heating and cooling practices vary, they
are replicated across camps suggesting a degree of
social and cultural cohesion around common
understandings of appropriate comfort practices.
Current 'hands-on' tenancy management practices by the Central Australian Affordable
Housing Company also encourage cross-camp
transmission of knowledge to residents about
climate-related practices.

The uptake in air conditioner use has changed
the balance of cooling practices, shifting them
towards both indoor and power-based practices.
This has brought cooling practices into potential
competition for resources with other power-based

activities practiced in the home such as cooking,
showering and watching TV, further complicating
power management by the household.

We also find that air conditioner cooling
practice has become embedded into practices of
'raising healthy kids' as well as practices of
dealing with heat. This should be further investigated as strategies aimed at providing passive
'hot-weather-friendly' community infrastructure (including shading and water features) for
pre-school children may act to lower the power
use of households (with young children) over the
hot summer months.

Current constraints in adaptive capacity are
'capacity' based (money, access to technology
and services), rather than being 'education or
knowledge-gap' related, so the current tenancy
and property management focus on practical,
technical, education platforms regarding housing
hardware may not be addressing the most pressing
adaptation threats. Moreover, 'educational campaigns' characterise the town camper population as
'lacking' necessary information and skills. In
addition, they foster a teacher–student (or manager–
client) relationship which limits the possibility of
information exchange both ways or the development
of, in the terms of Elvin et al. (2010), a 'recognition space' of meaningful engagement between
housing management and residents.

The results also have profound implications
for housing design. Since social practices essential
to climate adaptation are conducted outside
dwellings as well as within them, there are implications for climate adaptive housing design both
internally and externally to dwellings. Across
these two domains, the current research indicates that there has been relative neglect of design
considerations for the spaces outside dwellings,
leading to loss of tree cover, outside shading,
verandahs, open fires areas and related material
infrastructure.

The current town camp management strategies impose limitations on the four social practice elements, which influence their uptake.
• The housing asset is the primary focus of
housing management strategies; this could be
expanded to include broader design considerations

including the environment immediately surrounding the dwelling.

• Rules are in place primarily to protect and preserve the housing asset; these could take more account of the implications of changing practices.

• Practical knowledge is disseminated in education programs to support 'proper' house use; these could take more account of how social practices are constituted and the constraints associated with other elements of practice.

• Common understandings are often ignored, or when considered are viewed as stemming from poor practical knowledge and so can be 'fixed' by targeted education programs; the obduracy of common understandings and the linkages between practices could be accounted for more explicitly in provision of housing and related services.

In light of this research, the existing focus of housing providers on the material aspects and governance of the housing stock should be broadened. Moreover, while education programs are important in building community capacity through expanding practical knowledge, such programs in the future should move explicitly beyond behaviour change and take appropriate account of (1) practice elements (material, rules, common understandings and practical knowledge) and (2) dynamic relations between practices. Future social housing design guidelines should take due account of the importance of broader design considerations including the curtilage of the property in promoting and enabling adaptive climate practices in town camps. Specific responsibility for climate adaptation planning and resourcing should be assigned and plans and actions instituted to equip town campers with ongoing climate adaptive capacity.

Given predictions that climate change will bring pressure for increased cooling capacity in households (with a reduction in heating requirements), continued changes in associated practices are likely and, moreover, changes in any elements of practices for dealing with hot or cold weather are likely to prompt shifts in the balance of practice variants across different households. There is therefore a need to track changing practices as they relate to climate change stressors in order to monitor adaptive capacity and detect any potential declines in this critical resource in the face of increasing climate change.

Acknowledgements

The authors wish to thank Paula Arcari from RMIT University and Denise Foster and Audrey McCormack from Tangentyere Research for assistance with the research data collection.

References

Centre for Appropriate Technology (2013) *Housing Experience: Post Occupancy Evaluation of Alice Springs Town Camp Housing 2008–2011.* Centre for Appropriate Technology, Alice Springs, Northern Territory.

Commonwealth Ombudsman (2012) *Remote Housing Reforms in the Northern Territory*, Report No. 03/12. Commonwealth of Australia, Canberra.

Couglin, F. (1991) *Aboriginal Town Camps and Tangentyere Council: The Battle for Self-Determination in Alice Springs.* Masters thesis, School of Humanities, La Trobe University, Melbourne.

Elvin, R., Peter, S., Porter, R. and Young, M. (2010) *Discordance, Mobility and Agency: Dilemmas for Housing Reform in Northern Territory Aboriginal Settlements,* Desert Knowledge Cooperative Research Centre, Report 76, Alice Springs.

Foster, D., Mitchell, J., Ulrik, J. and Williams, R. (2005) *Population and Mobility in the Town Camps of Alice Springs: A Report Prepared by the Tangentyere Council Research Hub.* Desert Knowledge Cooperative Research Centre, Report 9, Alice Springs.

Hennessy, K., Page, C., McInnes, K., Walsh, K., Pittock, B., Bathols, J. and Suppiah, R. (2004) *Climate Change in the Northern Territory,* Consultancy report for NT Department of Infrastructure, Planning and Environment. CSIRO, Melbourne University, Melbourne.

Heppell, M. (ed.) (1979) *A Black Reality: Aboriginal Camps and Housing in Remote Australia.* Australian Institute of Aboriginal Studies, Canberra.

Heppell, M. and Wigley, J. (1981) *Black Out in Alice: A History of the Establishment and Development of Town Camps in Alice Springs.* Development Studies

Centre, Monograph No.26. Australian National University, Canberra.

Maller, C. (2011) Practices involving energy and water consumption in migrant households. In: Newton, P. (ed.) *Urban Consumption*. CSIRO, Melbourne, pp. 237–250.

Mansell, B. and Sowerbutts, T. (2011) Maintaining Houses for Better Health (MHBH). Affordable Housing Company. Available at http://www. affordablehousingcompany.com.au/wp-content/ uploads/2011/08/Hollows-MHBH-Report-July-2011. pdf (accessed 3 June 2014).

Maxwell, J.A. (2005) *Qualitative Research Design: An Interactive Approach*, 2nd edition. Sage Publications, Thousand Oaks.

Memmott, P. (1988) Aboriginal housing: The state of the art (or non-state of the art), *Architecture Australia* June, 34–47.

Memmott, P. (1989) The development of Aboriginal housing standards in central Australia: The case study of Tangentyere Council. In: Judd, B. and Bycroft, P. (eds) *Housing Issues No.4: Evaluating Housing Standards and Performance*. RAIA Education Division, Canberra.

Pholeros, P., Rainow, S. and Torzillo, P. (1993) *Housing for Health: Towards a Healthy Living Environment for Aboriginal Australia*. Healthhabitat, Newport Beach, NSW.

Reckwitz, A. (2002) Toward a theory of social practices: A development in cultural theorizing. *European Journal of Social Theory* 5(2) 243–263.

Reser, J. (1979) A matter of control: Aboriginal housing circumstances in remote communities and settlements.

In: Heppell, M. (ed.) *A Black Reality: Aboriginal Camps and Housing in Remote Australia*. Australian Institute of Aboriginal Studies, Canberra.

Ross, H. (1987) *Just for Living: Aboriginal Perceptions of Housing in Northwest Australia*. Aboriginal Studies Press, Canberra.

Rowley, C. (1970) *The Destruction of Aboriginal Society*. Australian National University Press, Canberra.

Schatzki, T. (1996) *Social Practices: A Wittgensteinian Approach to Human Activity and the Social*. Cambridge University Press, Cambridge.

Shove, E., Pantzar, M. and Watson, M. (2012) *The Dynamics of Social Practice: Everyday Life and How It Changes*. Sage Books, London.

Sofoulis, Z. (2005) Big water, everyday water: A socio-technical perspective. *Continuum* 19(4), 445–463.

Strengers, Y. (2011) Beyond demand management: Co-managing energy and water practices in Australian households. *Policy Studies* 32(1), 35–58.

Tangentyere Research Hub (2011) *Energy Efficiency Study*. Tangentyere Council, Alice Springs.

Wang, X., Chen, D. and Ren, Z. (2010) Assessment of climate change impact on residential building heating and cooling energy requirement in Australia. *Building and Environment* 45, 1663–1682.

Wang, X., Chen, D. and Ren, Z. (2011) Global warming and its implication to emission reduction strategies for residential buildings. *Building and Environment* 46, 871–883.

Weiss, R. (1994) *Learning from Strangers: The Art and Method of Qualitative Interview Studies*, The Free Press, New York.

33   Indigenous experiences and responses to Cyclone Tracy

KATHARINE HAYNES[1], DEANNE K. BIRD[1]
AND DEAN B. CARSON[2,3]

[1]Risk Frontiers, Macquarie University, Australia
[2]School of Medicine, Flinders University
[3]The Northern Institute, Charles Darwin University, Australia

33.1   Introduction

Early on Christmas morning 1974 Tropical Cyclone Tracy (herein referred to as 'Tracy'), a Category 4 (Australian) storm, devastated the Northern Territory city of Darwin killing 71 people and leaving only 6% of the city's housing habitable (see Fig. 33.1 for the location of Darwin in relation to Australia). Due to the extreme level of devastation and the fear of another cyclone remobilising fallen debris, it was decided that the majority of Darwin's c. 46,000 population be evacuated. Almost 25,000 people were evacuated through the Darwin airport and approximately 12,500 left by vehicle (Stretton 1976). The systematic failure of Darwin's building stock and the related humanitarian disaster provided the impetus for an overhaul of building regulations and construction practices throughout Australia (Mason and Haynes 2010; Mason et al. 2013). The published and grey literature documents well the response and recovery process, and the subsequent organisational adaptations (Haynes et al. 2011). However, while the societal impacts on the non-Indigenous population are well documented within this literature little attention has been paid to the specific experiences of the Indigenous population (Haynes et al. 2011).

As Tracy occurred nearly 40 years prior to this research, it was considered essential to document now the stories of those still alive for future generations. In particular, the research aimed to explore if the impact and recovery from Cyclone Tracy differed for Indigenous and non-Indigenous groups due to inherent sociocultural or political factors or pressures. Marked differences exist today in relation to socio-economic wellbeing and Indigenous people are significantly disadvantaged when compared to the non-Indigenous Australian population (Howitt 2012). This makes them potentially more vulnerable to environmental changes; however, little research has investigated the differential impacts of past events on Indigenous and non-Indigenous groups. This research aimed to document and examine any trends and themes important to current and future risk reduction and adaptation practices and what had changed since 1974, if anything.

33.1.1   Hazard context

On average a cyclone passes within 200 km of Darwin every 1–2 years. The expected recurrence interval of an event similar to or stronger than

---

Applied Studies in Climate Adaptation, First Edition. Edited by Jean P. Palutikof, Sarah L. Boulter, Jon Barnett and David Rissik.

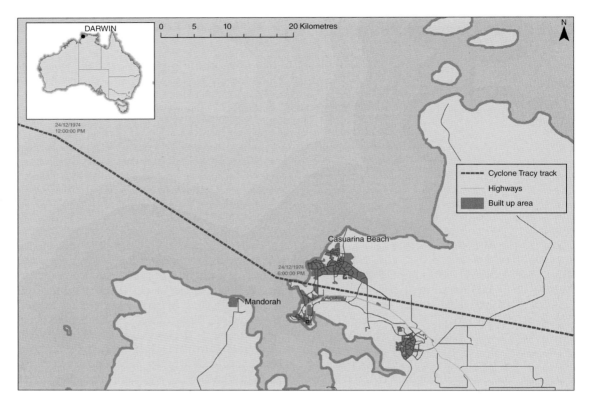

Figure 33.1 Darwin and surrounds, showing the track of Cyclone Tracy. Source: Reproduced with permission of James O'Brien, Risk Frontiers. For colour details please see Plate 24.

Tracy (itself a Category 4 event) impacting Darwin is longer than 100 years (Mason and Haynes 2010).

Cyclones are principally viewed by non-Indigenous people as negative events, dangerous to lives and property. In comparison, cyclones are culturally and economically important to Indigenous people. Doohan (2004) notes that Australian Indigenous people view cyclones as creative entities that bring renewal but also as a punishment when inappropriate engagement with the natural or supernatural world has taken place. Other extreme climatic events such as flooding and lightning-induced bushfires are similarly viewed as elements of a living landscape with which Indigenous people are intertwined (Rose 1996; Doohan 2004; Rose 2005).

33.1.2 Demographic and socio-political context

The local Indigenous people in Darwin are the Larrakia people, many of whom lived in public housing at the time of Tracy. In addition, there were members of the 'Stolen Generation'— Indigenous children who, through a series of laws, practices and policies, were forcibly removed from their families between 1910 and 1970 and re-settled in missions and homes, including Darwin (Human Rights and Equal Opportunity Commission 1997). There were also Indigenous people from remote communities in the Northern Territory (and perhaps beyond) living in Darwin, the majority of whom were housed in hostels (Haynes et al. 2011). Those from remote communities had moved temporarily or permanently into Darwin as part of the urbanisation process that continues to this day (Taylor and Carson 2009).

Experience and cultural knowledge of cyclones was likely to be tied to 'country' (a term Indigenous people use to describe the lands with which they have a traditional or cultural attachment; Weir et al. (2011)). For example, it may be expected that Larrakia people have knowledge of cyclones while desert people do not. While there is no information readily available about socio-economic status, it can be assumed that at least some of the Indigenous populations (particularly those from remote communities and those living in hostels and town camps) were substantially economically disadvantaged when compared to non-Indigenous populations. Much of the non-Indigenous population of Darwin was transitory, living in Darwin for relatively short periods of time for work. Consequently, much of the non-Indigenous population did not regard Darwin as 'home' in any long-term sense.

33.1.3 Political context

The national political context at the time promoted the inclusion of Indigenous people as part of Australian society to a far greater extent than had previously been considered. The 27 May 1967 constitutional referendum was still relatively fresh in people's minds as a statement about the status of Indigenous people. One question sought to remove two references within the Australian Constitution that discriminated against Aboriginal people. One was in reference to Aboriginal people not being counted as part of the Australian population. The other was in reference to the Australian Government having power to make special laws for Aboriginal people, different to those developed for people of any other race (visit http://www.naa.gov.au/collection/fact-sheets/fs150.aspx for further details). While the language of inclusion had emerged strongly, there was not yet the language of 'disadvantage' and separate treatment (in health, education, economic engagement, etc.) that has since emerged (e.g. Lea 2008; Johns 2011).

33.1.4 Demographic context

Darwin had a small population with a relatively small (in contemporary terms) proportion of Indigenous people (Haynes et al. 2011). At the 1971 Census, 2275 Indigenous people were counted in Darwin. This represented 6.5% of the Darwin population and 9.7% of the total Indigenous population of the Northern Territory. The Indigenous population was much younger than the non-Indigenous population, with 45% aged less than 15 years of age compared with 33% of non-Indigenous people. The small total population size meant that there was closer contact between all populations. There was also a very mixed non-Indigenous population as a legacy of in-migrations of people of Chinese, Greek, Italian and other heritages (Carson et al. 2010). It should be noted that as Tracy made landfall on Christmas Eve, many non-Indigenous people might have been absent from their homes, spending Christmas with their families interstate. In addition, many of the itinerant Aboriginals would also have returned to their communities for Christmas.

33.1.5 Social context

Consistent with the demographic context, the small size of Darwin meant people mixed a great deal (Haynes et al. 2011). Sports teams were generally mixed Indigenous/non-Indigenous (there were no specific 'Indigenous' sports clubs, for example, as are found today). Schools and workplaces were also mixed environments, with far fewer programs separating Indigenous and non-Indigenous workers or students than exist in contemporary Darwin society (Haynes et al. 2011).

33.1.6 Cultural context

Present-day Indigenous people often distinguish between Larrakia Indigenous people and other Indigenous people who may be from relatively near (e.g. Tiwi Islands, Katherine, Borroloola) or distant places (Taylor 2009). The acknowledgement of specific Indigenous cultures was not so strong in 1974. Land rights legislation introduced in 1975, for example, began a much more explicit process of separate identification of different Indigenous peoples in Australia and particularly in northern Australia (Haynes et al. 2011).

33.2   Methodology

33.2.1   In-depth face-to-face interviews

Qualitative interviews were conducted face-to-face between April and May 2011. Nineteen Indigenous and six non-Indigenous respondents were interviewed (for a full demographic breakdown of respondents, see Haynes et al. 2011). Interviews were arranged through a variety of Indigenous organisations, including the Emotional and Social Well Being Division of Danila Dilba, the Larrakia Nation Aboriginal Corporation, Warddeken Land Management Limited, the Stolen Generation and the Northern Land Council. These organisations were targeted because they have strong links with Indigenous groups in the Darwin area.

Indigenous people were asked to tell their story of Tracy and were prompted to focus on:
1. personal preparation undertaken by them or their family;
2. how Tracy impacted them or their family;
3. their or their family's recovery process and how, if at all, their lives changed after Tracy;
4. if they received assistance;
5. how they think the Indigenous community changed after Tracy;
6. how Tracy impacted on the distribution of food, goods, services and livelihood opportunities; and
7. how they view their current cyclone risk.

Key government and NGO personnel and health care workers who were in these positions during Tracy were asked to discuss the same topics as above but in relation to how Tracy impacted their organisation and the wider community. Interviews were recorded and transcribed. A thematic analysis was then undertaken

where the data were coded in relation to the themes and issues of interest to the study and also in terms of new themes emerging from the data.

33.2.2   Oral history transcripts

Data were collected from the Oral Histories Unit at the Northern Territory Archives from February to May 2011. The Northern Territory Archives hold a collection of oral histories from people who have played a prominent role in Darwin life, and who lived or continue to live in Darwin and the surrounding regional areas. The collection was examined via computer-based searches for accounts from Indigenous people who experienced Tracy and from key government/NGO officials/health care workers who were involved in the response and recovery process.

Twelve transcripts were identified as containing information useful for the project. These were treated as raw data and coded and analysed in the same way as the interview data.

33.3   Results

The results are presented around the main themes identified and discussed by respondents: warnings and preparedness; impacts; aftermath and evacuation; returning to Darwin; longer-term recovery; and resilience. These themes are also important for disaster risk reduction and adaptation policy. For a full presentation of the results see Haynes et al. (2011).

33.3.1   Warnings and preparedness

Respondents noted that formal warnings concerning the approach of Tracy went mostly unheeded as people were complacent or busy with Christmas preparations and celebrations.

> ... we heard about the cyclone coming on the radio but I didn't think much of it as I had five kids to look after ... Every year you hear cyclones coming and we don't take any notice because Darwin is a cyclone city... (Indigenous mother aged 35 at the time of Tracy)

I always took notice, to say get things ready with the torches and the food and water. Particularly that year there was many warnings … well there was but no one was listening 'cos they was all busy with Christmas time. Nobody was listening to radios. (Indigenous mother in her early 30s at the time of Tracy)

Both Indigenous and non-Indigenous respondents discussed natural signs in the environment that warned of an approaching cyclone. It was stated that some Indigenous people picked up on these signs and tried to warn others. However, many of the Indigenous respondents said that because of their cultural heritage being mixed with an Australian education, they no longer took the natural environmental signs seriously. This was something that changed following Tracy and respondents discussed becoming more in-tune with their traditional knowledge alongside the official emergency management procedures.

A few days before the cyclone, [an Aboriginal man] just walked in out of nowhere … he said to our boss 'Old lady' you know 'you'd better go away. Big wind and rain… it's going to all come'… Well Tracy hit us alright … I really believed him you know but being of mixed education I thought 'Oh the white man knows better'. (Indigenous female in her early 30s working as a hospital ward maid at the time of Tracy)

Indigenous people also associate spiritual beings with cyclones.

[Prior to Tracy] in September or something … there was an unknown creature spotted around Mandorah that was described as a rainbow serpent by non-Aboriginal fisherman and there were several sightings, so when I … understood that connection … I was a bit older … I realised that to me that was a precursor warning … And where it was seen is a part of a rainbow serpent dreaming track that goes from Casuarina Beach to Mandorah. So the fact that it was seen along the dreaming track, just before the cyclone, to me is quite significant. (Larrakia female aged 4 years at the time of Tracy; Fig. 33.1 for the location of Mandorah and Casuarina Beach with respect to Darwin)

Dreaming tracks or songlines refer to invisible pathways stretching across various parts of Australia. They represent journeys of ancestral beings, marking locations of significance in time and place and linking cultural groups through language, stories and song; Bird et al. 2013.

33.3.2 Impacts

Many Indigenous and non-Indigenous respondents told similar stories of how they sheltered from the cyclone at home. Some sheltered in bathrooms, some in bedrooms and others moved from upstairs to downstairs as the eye passed over or as the house disintegrated around them.

I said 'come on. We're all going to the bathroom'… and we was all panicking … we felt that the bathroom was safer. And we got into the tub and put a mattress over our head … we all waited there til the next morning, all cramped up in the bathtub. (Indigenous female in her early 30s working as hospital ward maid at the time of Tracy)

Respondents' response to the destruction was one of complete shock. Many respondents noted how they thought they were the only ones who were going through the disaster, and that the storm was very localised in their area or street and in some cases home. Many also discussed the eerie silence that ensued, as well as the loss of leaves and wildlife. Those who had lived in Darwin their whole lives talked about feeling lost as the natural and built environment had completely changed.

The street was … just about gone. There were two houses left … and we were one of them. And we just could not believe it … I even get emotional now, talking about it … we just stood there and looked at this utter devastation and thought my God, you know. How did we live through that? It was extraordinary. (Indigenous male aged 23 years at the time of Tracy)

It was suggested the homeless Indigenous population within Darwin would have been most vulnerable to the initial impacts due to their lack of shelter.

I mean there would have been a couple of hundred people living in the long grass ... they had nowhere to go. Where do they go? They were on foot. They didn't have cars or anything. These were the people like they are now, they are itinerants ... they're not listening to the radio. (Indigenous male aged 23 years at the time of Tracy)

The term 'long grass' refers to homeless people, predominantly Aboriginal, living in the long grass on the outskirts of Darwin.

33.3.3   Aftermath and evacuation

Indigenous respondents felt that in terms of the emergency management response and evacuation, all Darwin residents were treated fairly and equally. The evacuation, although strongly encouraged, was voluntary. Priority for air evacuation was given to those in need of medical attention, women and children, and the elderly. This meant that families were often divided with many men remaining in Darwin or having to evacuate themselves by car. A number of people had suggested to Major General Alan Stretton (Australian Army Officer in charge of Darwin's recovery from Tracy) that Indigenous people should be forcibly moved out of the Darwin area, due to health risks. But Stretton refused to follow that advice.

I informed the minister that as far as I was concerned, in Darwin, everyone was a human being. I didn't care what nationality they were, coloured or anything else, they would all be treated the same. (Stretton 1975, p. 7)

Nevertheless, Stretton reported that he did make one exception to the evacuation policy, which often divided family groups. Considering that those Indigenous people who opted to evacuate to a southern state had most probably never left the Darwin area, Stretton 'modified the regulations in their favour by saying that any Aborigines going south could move in a family group, and they could fly too' (Stretton 1975, p. 7). Additionally, Stretton gave the local airline authorisation to fly others back to their communities around the Darwin area, if that was their choice.

This humane gesture was quite surprising since Stretton was adamant on all other aspects of the evacuation. Stretton reports that he had pressure from Greek, Italian and UK officials wanting to arrange specific evacuations for their citizens, yet he refused. As far as Stretton was concerned, he wasn't going to give any group priority over another. For further details of the evacuation see Haynes et al. (2011).

Indigenous respondents discussed the cold, loneliness and the culture shock of evacuating south. However, many recounted the generosity and kindness of people across Australia who hosted and assisted them.

... poor mum, I don't know how she coped ... we didn't see him [dad] for 12 months ... we had to do schooling down there and that was like a culture shock for us because it was like ... uniforms and a totally new school and it was horrible for me because I didn't want to be there. (Indigenous female aged 9 years at the time of Tracy)

33.3.4   Returning to Darwin

Indigenous respondents all noted that once they had been evacuated they quickly became homesick, referring to their strong connection to country and community. Many returned within a few months and all but one of the respondents had returned within a year. In comparison, many non-Indigenous people never returned to Darwin.

The whole of Australia is beautiful but there is no place like home ... this is what we do we care for our country, we protect our country. My mother's brother used to protect this country and we took over. We have been looking after it for centuries, the Larrakia people have always looked after this country. (Indigenous female aged 13 years at the time of Tracy)

Respondents discussed the need to prove you had accommodation before you could officially return. This worked well for those evacuees who had family with habitable accommodation in Darwin to stay with. Others lived in caravans and demountables while their home was rebuilt, or before they were moved to new social housing. When asked about assistance, many mentioned help from friends and family.

> As long as you had the accommodation here ... and we did ... because Mum and Dad still were here and my sister's house had been patched up and you know, so we had somewhere to stay ... Within a year we were back and life started all over again ... went back to the same job. (Indigenous male aged 23 years at the time of Tracy)

33.3.5 Longer-term recovery

Although many people in Darwin would not have been insured, the non-Indigenous people spoken to, who were all home owners, were insured to some degree and discussed receiving money for their homes and contents.

> Insurance just paid out because there was no argument about it and then we'd only just bought the house and it was enough money to pay the house off. (Non-Indigenous male aged 46 years and working with Aboriginal Affairs at the time of Tracy)
>
> We'd been insured for six thousand dollars ... which was all our contents ... so we got our money ... [I'd] never seen so much money in one hit before ... (Non-Indigenous male aged 33 years and a police officer at the time of Tracy)

In comparison, all our Indigenous respondents suggested that they did not have contents insurance, although questions about insurance were not directly asked. The majority of Indigenous respondents discussed the struggle of saving up to replace their possessions. This would suggest that Indigenous people suffered more financially following Tracy. However, all people, whether they were insured or not, were eligible to receive money from the disaster fund.

> You just had to wait until you earned enough money to ... get back on your feet in the way of clothing and household goods. (Indigenous female in her early 30s working as hospital ward maid at the time of Tracy)

> When we moved to Katherine we went to nothing ... we were actually living in tents for about a year until we could get accommodation ... finally housing commission give us a home ... So yeah, we struggled a lot. (Indigenous female aged 9 years at the time of Tracy)

33.3.6 Resilience

When asked about coping emotionally or getting assistance for contents and lost belongings, respondents considered that Indigenous people are more resilient as they are not as reliant on material possessions. It was also discussed how Indigenous people could rely on the land for food and had many family connections and people from whom they could get help. They were therefore more self-sufficient than non-Indigenous people who did not have the same level of connection with the land and community, particularly those who had recently arrived.

> My view is ... because my people had been through a lot before the cyclone ... we were people that didn't have much at all ... we could go for fishing and hunting, so fortunately we connected to the land compared to others ... so when you mention about the urgency of getting this and that, we sort of went along with it because ... we wasn't the type to worry about other stuff. (Indigenous female aged about 8 years at the time of Tracy)

However, some Indigenous respondents discussed their detachment from their Indigenous heritage and culture. They felt that they had integrated into Darwin 'city life' and were no longer as attached to their environment. Some respondents noted that Tracy had awoken an interest in the environment and their cultural heritage.

One respondent discussed an increasing risk from future cyclone impacts due to coastal development and the loss of the mangroves which create a natural buffer along the coast. This respondent also argued that Darwin is due for another large cyclone as a form of punishment as Indigenous people are not upholding their cultural obligations in caring for country and are allowing the environment to be degraded. She believed that the lack of care for country is worse now than in 1974. If a severe cyclone were to occur today, many Indigenous respondents indicated that they would go to an official shelter or travel inland.

33.4 Discussion and conclusion

A significant amount of research has demonstrated that the impacts of disasters are felt disproportionately worldwide due to inherent underlying socio-cultural, economic and political vulnerabilities (Wisner et al. 2004, 2012). These include a range of interrelated factors (such as race and ethnicity, gender, age, disabilities, economic status, political representation, health, education, housing security, etc.) that shape how people can prepare and reduce their risks from, are impacted by, cope with and recover from disasters. For example, Hurricane Katrina exposed significant vulnerabilities and inequities among the impacted populations in Louisiana, Mississippi and Alabama, especially within the city of New Orleans. In particular, a number of studies have demonstrated that the underlying factors of race and socio-economic capital heightened exposure and impacts, the ability to evacuate and long-term patterns of migration and recovery (Cutter and Emrich 2006; Branshaw and Trainor 2007; Myers et al. 2008).

In contrast, this study has identified that the Indigenous experience of Tracy was very similar to that of the broader Darwin population. Respondents did not have the sense that Indigenous people were treated substantially differently to non-Indigenous people in terms of evacuation procedures, health care or resettlement. However, the general perception was that Indigenous people were more resilient than non-Indigenous residents as they were not so reliant on material possessions and could rely on the land for food and their direct and extended family for help. Other differences were noted in relation to the strong cultural connection with country both before and after Tracy. This connection was highlighted in discussions of traditional

knowledge, or lack thereof, of early warning signs and in relation to Indigenous people being more eager to return to Darwin after the evacuation than non-Indigenous people.

The extent to which the current generation of Indigenous people in Darwin would be impacted by a cyclone in similar or different ways to 1974 is determined not only by the lived and inherited experience of those who were in Darwin for Tracy, but by the many ways in which 'Indigenous Darwin' (and 'non-Indigenous Darwin') has changed in the past 37 years. The political, demographic, social, cultural and physical contexts are very different now to what they were then, particularly in the way that there has been an increasing separation of Indigenous and non-Indigenous people (and even between different Indigenous peoples) in many of these contexts. The creation of a separation has allowed Indigenous people in Darwin to connect more deeply with their cultural identity, and some participants in this research saw that as a positive aspect in terms of their preparedness for future events. At the same time, the 'equality' of treatment of Indigenous and non-Indigenous people that occurred with the response to Tracy was also seen as a positive aspect in helping the Darwin community as a whole cope with the initial shock and generate the collective energy needed for rebuilding.

The advancement of forecasting procedures has meant that the provision of warnings is now much improved since Tracy. Furthermore, a significant amount of general preparedness material is provided throughout the year. Other research (Li 2009b) has noted that long-term Darwin residents have similar cyclone risk perceptions to experts in the field. As would be expected, however, short-term residents (particularly those who did not experience Tracy) are less aware.

Nevertheless, Li (2009a) notes that exposure to the surge risk within Darwin is increasing. Similarly, one respondent discussed an increasing risk due to coastal development and the loss of mangroves. The number of people and assets located within the surge zone has increased despite the recognition of the storm surge risk by the government, which identifies land use planning as an adaptive approach.

It is unlikely that a cyclone like Tracy would have the same impact today due to the building codes now in place, and only people within the surge zone would need evacuating. Despite recognising the ability of their homes to withstand cyclonic winds, many Indigenous respondents noted that in the event of a severe cyclone they would prefer to go to an official shelter or alternatively travel inland. This research has demonstrated that much of the Indigenous population living within Darwin is fairly urbanised and, although many retain their Indigenous cultural heritage, they are able to work with the standard Australian emergency management warnings and procedures currently in place.

However, this may not to be the case for the transient itinerants or Indigenous people living in communities in regional areas outside of Darwin. Moreover, these Indigenous Australians are likely to feel the impacts of climate change the hardest. This is due to a continuing cycle of political, social and economic marginalisation leading to disadvantages in health, education and employment (Howitt 2012). These inequalities, which have led to an 11-year gap in life expectancy between Indigenous and non-Indigenous groups in Australia (COAG 2009), constrain the ability of many Indigenous people to cope with and adapt to external shocks and climate stresses. However, Indigenous people are also likely to adopt coping strategies, such as migration, and may be resilient due to a culture that has learnt to exist and adapt to a harsh environment (Bird et al. 2013). Apart from a few studies (see Doohan 2004; Veland et al. 2010; Bird et al. 2013), little research has examined the vulnerability and adaptive capacity of Indigenous communities in Northern Australia (Green et al. 2009). Furthermore, despite increasing recognition of their vulnerability and resilience there has, to date, been little allowance for meaningful policy input from Indigenous people (Petheram et al. 2010).

Acknowledgements

We would like to thank all the participants who gave up their time to share their stories of Cyclone Tracy. We are grateful to the following organisations for their assistance: the Emotional and Social Well Being Division of Danila Dilba, the Northern Territory Archives Service, the Larrakia Nation Aboriginal Corporation, Warddeken Land Management Limited, the Stolen Generation and the Northern Land Council. We also acknowledge the assistance of Professor Steve Larkin who acted both as a research participant and co-author on the original NCCARF report. Thanks are also due to Jean Palutikof and Sarah Boulter from NCCARF for all their assistance.

References

Bird, D., Govan, J., Murphy, H. et al. (2013) *Future Change in Ancient Worlds: Indigenous Adaptation in Northern Australia.* National Climate Change Adaptation Research Facility, Gold Coast, Australia.

Branshaw, J. and Trainor, J. (2007) Race, class, and capital amidst the hurricane Katrina diaspora. In: Brunsma, D. L., Overfelt, D. and Picou, J. S. (eds) *The Sociology of Katrina: Perspectives on a Modern Catastrophe.* Rowman and Littlefield, Lanham, MD.

Carson, D., Schmallegger, D., Campbell, S. and Martel, C. (2010) *Whose City is it? A Thinking Guide to Darwin.* Charles Darwin University Press, Darwin.

Council of Australian Governments (2009) *National Indigenous Reform Agreement (Closing the Gap).* Council of Australian Governments, Australia.

Cutter, S.L. and Emrich, C.T. (2006) Moral hazard, social catastrophe: The changing face of vulnerability along the hurricane coasts. *Annals of the Academy of Political and Social Science* 604, 102–112.

Doohan, K. (2004) *Helping Whitefellas to See: Community Engagement and Cultural Practices in Emergency Situations in Remote Regions (the Kimberley). Processes for Cross-Cultural Engagement.* Paper presented to the Western Regional Science Association meeting. Wailea Marriott Resort, Maui.

Green, D., Jackson, S. and Morrison, J. (2009) *Risks from Climate Change to Indigenous Communities in the Tropical North of Australia*. Department of Climate Change and Energy Efficiency, Canberra.

Haynes, K., Bird, D.K., Carson, D., Larkin, S. and Mason, M. (2011) *Cyclone Tracy Part 2: Institutional response and Indigenous experiences of Cyclone Tracy*. National Climate Change Adaptation Research Facility, Gold Coast.

Howitt, R. (2012) Sustainable indigenous futures in remote indigenous areas: relationships, processes and failed state approaches. *GeoJournal* 77, 817–828.

Human Rights and Equal Opportunity Commission (1997) *Bringing them home: National inquiry into the separation of Aboriginal and Torres Strait Islander children from their families*. Commonwealth of Australia, Canberra.

Johns, G. (2011) *Aboriginal Self-Determination: The White Man's Dream*. Connor Court Publishing, Melbourne.

Lea, T. (2008) *Bureaucrats and Bleeding Hearts: Indigenous Health in Northern Australia*. University of New South Wales Press, Sydney.

Li, G.M. (2009a) (Mal)adapting to tropical cyclone risk: the case of 'Tempestuous Tracy'. *The Australian Journal of Emergency Management* 24, 44–51.

Li, G.M. (2009b) Tropical cyclone risk perceptions in Darwin, Australia: a comparison of different residential groups. *Natural Hazards* 48, 365–382.

Mason, M. and Haynes, K. (2010) *Cyclone Tracy Part 1: The Engineering Response*. National Climate Change Adaptation Research Facility, Gold Coast.

Mason, M., Haynes, K. and Walker, G. (2013) Cyclone Tracy and the road to improved wind resistant design. In: Boulter, S., Palutikof, J., Karoly, D. and Guitart, D. (eds) *Natural Disasters and Adaptation to Climate Change*. Cambridge University Press, Cambridge, pp. 87–94.

Myers, C.A., Slack, T. and Singelmann, J. (2008) Social vulnerability and migration in the wake of disaster: the case of Hurricanes Katrina and Rita. *Population and Environment* 29, 271–291.

Petheram, L., Zander, K.K., Campbell, B.M., High, C. and Stacey, N. (2010) 'Strange changes': Indigenous perspectives of climate change and adaptation in NE Arnhem Land (Australia). *Global Environmental Change* 20, 681–692.

Rose, D.B. (1996) *Nourishing Terrains: Australian Aboriginal Views of Landscape and Wilderness*. Australian Heritage Commission, Canberra.

Rose, D.B. (2005) Indigenous water philosophy in an uncertain land. In: Botterill., L.C. and Wilhite., D.A. (eds) *From Disaster Response to Risk Management. Advances in Natural and Technological Hazards Research*. Springer, Dordrecht, The Netherlands, pp. 37–50.

Stretton, A. (1975) TS7909. Northern Territory Archives Service, NTRS 226, Typed transcripts of oral history interviews with 'TS' prefix, 1979-ct. Northern Territory Archive, Darwin.

Stretton, A.B. (1976) *The Furious Days: The Relief of Darwin*. William Collins Publishers Pty Ltd, Sydney.

Taylor, A. and Carson, D. (2009) Indigenous mobility and the Northern Territory emergency response. *People and Place* 17, 29–38.

Taylor, J. (2009) Indigenous demography and public policy in Australia: population or peoples? *Journal of Population Research* 26, 115130.

Veland, S., Howitt, R. and Dominey-Howes, D. (2010) Invisible institutions in emergencies: Evacuating the remote Indigenous community of Warruwi, Northern Territory Australia, from Cyclone Monica. *Environmental Hazards* 9, 197–214.

Weir, J.K., Stacey, C. and Youngetob, K. (2011) *The Benefits Associated with Caring for Country: Literature Review*. Institute of Aboriginal and Torres Strait Islander Studies (AIATSIS) for the Department of Sustainability, Environment, Water, Population and Communities, Canberra.

Wisner, B., Blaikie, P., Cannon, T. and Davis, I. (eds) (2004) *At Risk: Natural Hazards, People's Vulnerability and Disasters*. Routledge, London.

Wisner, B., Gaillard, J. and Kelman, I. (eds) (2012) *The Handbook of Hazards and Disaster Risk Reduction*. Routledge, Oxon.

34 Indigenous governance and climate change adaptation: two native title case studies from Australia

TRAN TRAN[1], JESSICA K. WEIR[2], LISA M. STRELEIN[1]
AND CLAIRE STACEY[1]

[1]Australian Institute of Aboriginal and Torres Strait Islander Studies, Australia
[2]Institute for Culture and Society, University of Western Sydney, Australia

You know, in our Dreamtime [*jilas*] they're all connected … in our Dreaming stories which is the foundation of our native title … Our connection to those water holes and the Dreaming stories attached to them are very important. (Joseph Edgar, Chair, Karajarri Traditional Lands Association, 26 November 2012)

34.1 Introduction

Climate change has generated interest in the roles and responsibility of Indigenous peoples in adapting to and mitigating change, as their culture and social organisation is deeply embedded in land and water. At the same time, the remoteness and socio-economic disadvantage that many Indigenous peoples experience has meant that Indigenous communities are presented in a position of vulnerability in climate change adaptation discourses (Cameron 2012; Veland et al. 2013). This framing of Indigenous peoples as vulnerable, combined with the often-held presumption that ecological, physical, economic and technological perspectives are culturally neutral (Adger et al. 2009; Dovers 2009), disables productive collaborations between Indigenous and non-Indigenous institutions for climate change adaptation.

In Australia, there is an opportunity for broader perspectives on the governance of climate change adaptation to be embraced as part of the belated recognition of Indigenous peoples' prior ownership of the continent through the legal concept of 'native title', as recognised in the historic High Court *Mabo* decision in 1992 (*Mabo and Others v Queensland* (No. 2) [1992] HCA 23). Native title while imperfect, is recognition of prior and ongoing Indigenous connections to land and waters and is one means through which Indigenous peoples in Australia have sought to assert their authority under Australia's property law system. The way in which native title legislation is devised and applied has an impact on the ability of an Indigenous group to be recognised, regardless of whether or not they are traditional owners (a term widely used in Australia to identify the first peoples or first nations of an area). The authors note that not all Indigenous people can successfully apply to have native title recognised on their lands, nor do all Indigenous people wish to.

Applied Studies in Climate Adaptation, First Edition. Edited by Jean P. Palutikof, Sarah L. Boulter, Jon Barnett and David Rissik.
© 2015 John Wiley & Sons, Ltd. Published 2015 by John Wiley & Sons, Ltd.

Already, more than 100 Aboriginal and Torres Strait Islander groups have successfully applied to the Federal Court for legal recognition of the relationships they hold with their lands and waters, known as 'country'. While this exercise is greatly circumscribed by laws that govern what rights and interests can be recognised, where these applications can be made, Indigenous peoples' determined native title lands and waters now comprise over 22% of Australia's total land mass, covering large tracts of the coast and interior including urban areas, regional towns and Indigenous communities. It is likely that this figure will double once the backlog of native title applications is addressed.

Native title law recognises Indigenous peoples' rights to continue to make decisions about the use and management of their land and waters, according to their own systems of laws and customs. This ongoing decision-making role implicitly relies on the enduring governance systems of Indigenous peoples, which now have a new corporate form as institutionalised through the native title process. The Commonwealth's *Native Title Act 1993* (*Native Title Act*) requires the establishment of Registered Native Title Bodies Corporate (RNTBCs). RNTBCs are established by native title holders to hold or manage their native title rights and interests, and are required to facilitate the interests of other parties on native title lands such as government and private industry (Martin 2003). This decision-making relates to activities central to climate change adaptation such as the provision of housing and other community infrastructure, land and water management, environmental conservation, mining and resource extraction, agricultural development and so on. The diverse parties applying to RNTBCs for decision-making outcomes include different levels of government and government departments, private companies and Aboriginal and Torres Strait Islander councils.

There have been alternative analyses of climate change adaptation from broader social and ethical perspectives recognising a diversity of knowledge, and how they form consequent rules and institutions (Adger et al. 2009). This same literature points to how decision-making processes by institutions based on the knowledge practices of the cultural majority can occur without recourse to other cultural perspectives. More inclusive and diverse governance is critical to addressing climate change adaptation and ensuring that power relations embedded in and between institutions do not marginalise Indigenous peoples. This chapter considers the opportunity offered by the formal recognition of Indigenous peoples' property rights in Australia under native title for stronger climate change institutions, and the challenges that persist despite this recognition.

34.2   Native title, native title corporations and climate change adaptation

RNTBCs reflect structures of community governance that have been articulated and recognised under the *Native Title Act*; indeed, they form the evidence required for proving native title. These community governance structures are based on connections to land that form the basis of identity for Aboriginal and Torres Strait Islander peoples. Prior to *Mabo*, a range of different entities existed to represent Indigenous governance, including the councils that govern remote Indigenous communities and formed on the basis of local government elections. RNTBCs, and the many organisations with which they hold immediate dealings, must now work out their roles and responsibilities in relation to each other. RNTBCs are however caught in a 20-year policy impasse between the Commonwealth and State and Territory governments as to who is responsible for funding the establishment and operation of these organisations (Bauman et al. 2013; McGrath et al. 2013; Strelein and Tran 2013). In the absence of any predictable settlement or compensation framework and with no sustainable funding regime, RNTBCs often operate without staff, an office, office equipment or even meeting space to perform their legal obligations under the *Native Title Act*, as well as meet their own community's expectations which can include the delivery of cultural, social, environmental and economic outcomes (Bauman et al. 2013).

Despite the challenges of capacity, RNTBCs are intended to exist in perpetuity and draw on long-standing governance structures that have survived the colonisation of Australia: the decision-making authority of traditional owners. Native title recognition of Indigenous peoples' rights to make decisions and be involved in decision-making processes about their country means that planning over native title lands need to occur with RNTBCs. To facilitate decisions, the directors of the RNTBC (who are drawn from members of the native title group) must communicate and often consult on the matter with the group, a process usually involving recourse to Elders and the reaching of consensus, rather than majority votes (e.g. Glaskin 2007; Bauman et al. 2013). As part of this process, and as traditional owners, the native title group has to consider consequences over very long time frames that encompass responsibilities to ancestors, country and future generations. Significant for the broad relevance of RNTBCs, this process of reaching agreement is critical for the business of the many parties with interests in native title lands (Bauman et al. 2013). Indeed, the agendas of RNTBCs are often in concordance with long-term land-use planning, management and policy objectives of governments (Bauman et al. 2013). While *Mabo* and the *Native Title Act* provide legal impetus for greater engagement with the decision-making authority of traditional owners, the slow response to providing RNTBC support is hampering the operation of all parties working with the native title sector as well as government policy more broadly. In the interim, the capacity for RNTBCs to deliver upon sustainable social, cultural, environmental and economic outcomes for their communities goes widely unexplored.

The overtly technical, scientific and managerial response to climate change (Hulme 2011) has the potential to further entrench the marginalisation of Indigenous peoples' contributions to climate change adaptation. Part of the policy failure to support RNTBCs includes a failure to value the symbolic and cultural values of Indigenous relationships with country, which are not included in the 'calculus' of climate adaptation policymaking

(Adger et al. 2011, p. 2). Indigenous connections to country, and the holistic knowledge arising out of these connections, form the basis of their laws and customs and are therefore central to RNTBC responsibilities to hold and manage their native title. It is our argument that the failure to invest in Indigenous forms of land and water governance via RNTBCs, including greater cultural literacy and recognition of the place of Indigenous laws and customs in contemporary society, is a form of serious maladaptation.

34.3 Research context: Bidyadanga and Kowanyama

Bidyadanga is a small community established on the western edge of the Great Sandy Desert, Western Australia and supported by the fresh groundwaters of the Canning Basin. The small community of Kowanyama is on the tropical banks of Magnificent Creek in Cape York Queensland, which is part of a river system larger than the Nile delta. Both communities are located very close to the coast and experience cyclonic and flooding activity. Both communities have a recent history as mission lands (Bidyadanga from the 1950s to 1970s and Kowanyama from the early 1900s until 1967), and then becoming Aboriginal communities managed under community governance legislation (*Aboriginal Communities Act 1979* (WA); *Community Services (Aborigines) Act 1984* (Qld)). Kowanyama has a population of around 1,150 people of whom around three-quarters are the traditional owners of the area. The 800 residents of Bidyadanga are primarily Aboriginal people, but comprising diverse groups who either voluntarily or forcibly moved in from surrounding desert areas. As a result, the Karajarri traditional owners are now minority residents within the community. In both places, most of the residents have low incomes, are usually welfare dependent and experience poor health and poverty (Tran et al. 2013).

The role of RNTBCs in Bidyadanga and Kowanyama in climate change adaptation was the subject of our 18-month action research project (Tran et al. 2013). This interdisciplinary research was undertaken using qualitative methods, including convening workshops, fieldwork stays, semi-structured interviews and participant observation (Tran et al. 2013). Our two RNTBC partners for this research were the Karajarri Traditional Lands Association Aboriginal Corporation RNTBC (KTLA) and the Abm Elgoring Ambung Aboriginal Corporation RNTBC (Abm Elgoring Ambung). Our research questions were focused on socio-institutional challenges and opportunities facing RNTBCs in climate change adaptation. We compared the experiences of the two RNTBCs to identify blockages and opportunities in the laws, policies and relationships that determined their interactions with land and water management institutions.

34.3.1 Karajarri Traditional Lands Association

Our first case study partner, the KTLA, have native title rights and interests over large parts of Karajarri country including desert and saltwater country, with the coastal strip including numerous *jilas* (fresh water springs), mangroves, beaches, burial and midden sites and other key cultural places. Karajarri *Pukarrikarra* (dreaming and creation) stories revolve around the different freshwaters that they describe as 'living' (Yu 1999), connecting water to life in the desert, and are a formal part of Karajarri culture, law and economy. The underground waters support localised ecosystems, Karajarri hunting and camping, permanent outstations, the community of Bidyadanga, pastoral stations and limited horticultural and pastoral activities. Along this coastal strip, cyclones and storm surges are contributing to coastal erosion and saltwater infiltration of the fresh groundwater, with lasting effects on the meaning of these places as well as their capacity to support life. As desert country, temperature increases are also a very real concern and changing rainfall patterns may affect how *jilas* are recharged by underground water flows. With urban, pastoral and industrial growth in the Kimberley, including coal seam gas development and mining, the pressures of development on groundwater are increasing.

Karajarri concerns about decision-making in their country, including accessing *jilas*, led to their successful pursuit of three native title determinations encompassing more than 33,000 km² of land and water (*Nangkiriny v Western Australia* [2002] FCA 660; *Nangkiriny v Western Australia* [2004] FCA 1156; and shared country in *Hunter v State of Western Australia* [2012] FCA 690). Karajarri native title is recognised as either exclusive possession or non-exclusive possession native title. Exclusive possession native title is usually described as 'to possess, occupy, use and enjoy [the determination area] to the exclusion of all others', including the right to make decisions about access; non-exclusive native title possession rights and interests only prevail as far as they do not conflict with the rights and interests of certain other parties such as pastoralists. Exclusive possession native title ensures that Karajarri have decision-making responsibilities over their native title including rights of access. Karajarri have exclusive possession native title over areas that were formerly radical Crown title (unallocated Crown lands) and the community of Bidyadanga because of its land tenure history. Their non-exclusive possession native title is recognised over the pastoral leases that line the coast.

Native title recognition has propelled Karajarri, through their RNTBC, into formal land-use planning and development on their country, as well as meeting corporate administrative responsibilities and community expectations. These responsibilities include negotiating matters with the Bidyadanga Aboriginal Community Council who are responsible for, and funded by the state government to, administer the Bidyadanga (La Grange) Aboriginal community. Importantly, recognition has enabled Karajarri to assert their position as traditional owners where they were previously a minority in the community (Edgar 2011). However, in the absence of funding for the KTLA, their work has been carried out sporadically by volunteers without an office or administrative support, draining the leadership capacities of key people who have paid and unpaid commitments elsewhere (Weir 2011). Karajarri have more recently been successful at sourcing Federal environmental monies which have cross-subsidised the KTLA (Tran et al. 2013) and negotiated an RNTBC office with the Western Australian state government, although they remain without staff.

34.3.2 *Abm Elgoring Ambung*

Our second case study partner, Abm Elgoring Ambung, represents the interests of four main language groups – the Yir Yoront (sometimes called Kokomenjen), the Koko Bera, Kunjen and Koko Berrin – whose close relationships are based upon a continuing connection to lands located in the Gulf of Carpentaria spanning over 37,000 years. The native title lands of the Kowanyama people includes a mix of forest country which has woodland cover, perennial rivers and permanent freshwater lagoons in the river beds (where there are also key cultural sites), as well as saltwater country consisting of tidal rivers, saltmarshes and grass plains that are flooded in the wet season (Monaghan 2005). The proximity of the community to the coast and large Mitchell delta means that the area is prone to flooding and cyclones with road access cut off for significant periods throughout the year. Further, storm surges can lead to the loss of heritage on coastal areas and the salination of wetlands, threatening important ecological and cultural areas (Sinnamon and Frank 2010).

The Yir Yoront, Koko Bera, Kunjen and Koko Berrin peoples established Abm Elgoring Ambung as their RNTBC to manage their joint native title determinations in 2009 and 2012 (*Kowanyama People v State of Queensland* [2009] FCA 1192 Part A; *Greenwool for and on behalf of the Kowanyama People v State of Queensland* [2012] FCA 1377 Part B and C). These determinations settle two of four separate claims with a small area south of the Staaten River to be determined later. The native title claim process is only partway through, with the entire claim area covering 19,800 km² and overlapping Deed of Grant in Trust land tenures (Trust lands) granted to Aboriginal peoples under state legislation. Significantly, the Kowanyama people have exclusive native title possession over the Trust land area, including the Kowanyama township as well as 16,396 km² of pastoral property and non-exclusive rights and interests over sea and tidal areas. Among the rights and interests recognised is the right to live and erect structures on pastoral lease areas, which traditional owners had previously been excluded from accessing.

The Trust land titles are vested in the Kowanyama Aboriginal Shire Council (Shire Council) that also supports the Kowanyama Aboriginal Land and Natural Resource Management Office (KALNRMO 1994). With native title recognition, Abm Elgoring Ambung is in the process of negotiating how land management functions will be transferred to the RNTBC from these two organisations. There is limited administrative funding and support for the RNTBC, which previously had one temporary full-time manager (funded by the Kowanyama Shire Council). Abm Elgoring Ambung struggles, like the KTLA, to receive financial support to carry out its functions.

In both case studies it is perhaps unsurprising that law and custom and day-to-day cultural practice is embedded in and regenerates a deep understanding of country. This capacity to understand country and the impact of changes to climate, combined with the responsibilities of community development and managing country, place RNTBCs in a unique position to contribute to climate change adaptation planning, decision-making, monitoring and implementation.

34.4 Research findings: adapting to native title

The experiences of Abm Elgoring Ambung and the KTLA in negotiating their relationship with state government land managers, local community structures and federal funding bodies reflects the history of conflict and dispossession and the ongoing failure to provide infrastructure and services for growing communities, all circumscribing the health and wellbeing of the Karajarri and Yir Yoront, Koko Bera, Kunjen and Koko Berrin peoples and their RNTBCs.

Despite native title constituting the recognition of ancient jurisdictions with an affinity to country, and establishing a legal right and responsibility for decision-making, our fieldwork research demonstrated that RNTBCs are persistently viewed as a 'latecomer' to the planning process and are often only included with respect to a few specific provisions in the *Native Title Act*. In particular, ensuring successful climate change adaptation within these two communities is challenged by former approaches that did not recognise the identity of Aboriginal peoples. Through the community or shire councils, the governance of many aspects of both Kowanyama and Bidyadanaga has been based on state legislation that does not recognise traditional community dynamics and governance systems. The community or shire councils have been responsible for community services and recreation, local infrastructure, education and training, land management and other services based on municipal arrangements. These inflexible arrangements are now transforming with native title recognition and the formation and presence of RNTBCs within the two communities.

One of the stumbling blocks to the greater involvement of RNTBCs in planning is the lack of clarity on the interaction between native title and planning. Elsewhere this question has been examined in detail (Wensing 2011; Tran et al. 2013, pp. 59–63). It is clear that native title holders have legal rights to be involved in decisions that affect their rights and interests, and that compensation procedures can be triggered if their native title is adversely affected. For example, planning treatments such as zoning clearly seek to make long-term decisions about land use that conflict with the decision-making rights of exclusive possession native title holders. Zoning can also conflict with many other native title rights and interests, including the right to take resources whether recognised as exclusive or non-exclusive possession native title. The *Native Title Act* provides some relief to native title holders under section 211, by exempting personal, communal and cultural use from any licensing regime unless for public health and safety or conservation concerns.

It has been acknowledged that the lack of development of Aboriginal land-use and planning schemes are a significant barrier to Indigenous participation in environmental and natural resource management (Winer et al. 2012), which can be applied more broadly to priorities such as infrastructure and housing. However, ongoing uncertainties about native title forces traditional owners to 'live with' the limitations of legal forms of engagement under the *Native Title Act*. As a consequence, despite being the 'holder' of recognised native title rights and interests, the KTLA and Abm Elgoring Ambung can only engage with land and water planning and management regimes based on the limited legal processes established under the *Native Title Act*. The complexity of the legal interactions between recognised native title and the myriad of land and water planning and management regimes reflects the unsettled nature of the scope of native title rights and interests and ideally should not be resolved through costly litigation.

There is a legal mechanism for working out these matters under the *Native Title Act*, known as an Indigenous Land Use Agreement (ILUA). It is a valuable, but very formal, process; it involves hiring lawyers to help negotiate and determine the agreement, and is not necessarily suited to the cohesive and daily management of ongoing governance roles and relationships. In Bidyadanga for example, the KTLA entered into an ILUA with the Bidyadanga Community Council and the State of Western Australia concerning the provision of 16 houses, a tip, a cyclone shelter and basketball court, an arts and culture centre and additional sewerage infrastructure, clearly developments that largely fall under the compass of climate change adaptation. With exclusive possession native title rights to Bidyadanga, the KTLA are a necessary part of decision-making for these developments. However, progress was hampered by significant community tension around native title, a situation exacerbated by the absence of KTLA funding and the lack of state policy and support for RNTBCs. While many important matters were addressed in the ILUA – including communication on what native title meant for everyone in Bidyadanga and negotiating an office for the KTLA – the process took four years (Edgar 2011; Weir 2013).

Another ILUA is now needed in Bidyadanga to address more comprehensive planning issues including inconsistencies in tenure and development, the identification of the future land requirements of the Bidyadanga community and the divestment of Aboriginal Lands Trust land to the KTLA. As these native title issues are facing, or will be facing, many Aboriginal communities across the state, the Western Australian government has proposed a format for ILUAs 'resolving' the negotiation of development activities over native title lands (DPCWA 2012). It is implicit that the aim is to set native title processes aside to facilitate development, thereby eliminating the leveraging power that RNTBCs sometimes exercise to gain access to planning processes. Expediency is important but should not be achieved through consistently disadvantaging native title holders and their corporate entities, and thereby entrenching disregard and opposition to the governance changes wrought by native title in local communities and more broadly. Instead, a funded operational RNTBC is a good governance investment to ensure that the RNTBC is able to participate effectively in planning, provide guidance to other institutions involved in planning and to ensure decisions made are culturally, socially and environmentally sustainable.

In our Kowanyama case study we found that government funding for the Shire Council supported local government decision-making processes over the existing cultural governance institutions of the traditional owners, inadvertently generating conflict within the community over who has the authority to make decisions over land and water management. For example, Trust lands introduced in Queensland as a part of 'self-management' policies in the 1980s compete with recognised native title rights and interests. Trust lands can be transferred to Aboriginal shire councils whose membership can include those with a 'historical association' with the area (under the *Aboriginal Land Act 1991* (Qld); under this Act, previously granted

lands such as Trust lands are transferable as inalienable freehold land). Significantly, changes to the governance structures of Aboriginal shire councils in Queensland mean that the ability to maintain community control of shire council functions is increasingly at risk of being transferred to the state government (*Local Government (Community Government Areas) Act 2004* (Qld)), with no guarantee of direct Indigenous representation. At the same time, RNTBCs are not viewed as legitimate land management institutions and are also excluded from regional planning in Queensland, which seeks to engage with Aboriginal shire councils rather than RNTBCs (DSDIP 2012).

Drawing on its limited capacity, Abm Elgoring Ambung is keen to work with the Kowanyama Shire Council to examine tenure reform and the transfer of Trust lands in the context of regional planning, yet lacks formal mechanisms to negotiate how decision-making over land management occurs within the community. There is also a significant opportunity for Abm Elgoring Ambung to partner with the KALNRMO and benefit from its historical experience and engagement with processes for asserting the self-governance and sovereignty of the community. In order to do this however, both Abm Elgoring Ambung and the Shire Council must be in a position to clarify their roles and functions, discuss whether the existing representation structures are appropriate and agree on opportunities for greater collaboration.

From a climate change adaptation perspective, it makes good sense to engage traditional owners in land-use planning and decision-making. They have established positions of authority and knowledge in country, which can contribute a much more nuanced understanding of the role and long-term impacts of planning decisions. Indigenous peoples' responsibilities for country are intergenerational; they are tied to their cultural survival. This guides the prioritisation of issues in ways that non-Indigenous institutions are only just learning to appreciate in the face of climate change. By drawing on these strengths RNTBCs are achieving many outcomes (Bauman et al. 2013; Tran et al. 2013), largely through the volunteer hours of RNTBC directors, contradicting the depiction of Indigenous communities as 'vulnerable' and instead informing an understanding of RNTBCs as representing committed and engaged communities.

34.5   Conclusion

There is no doubt Karajarri, Yir Yoront, Koko Bera Kunjen and Koko Berrin country is particularly vulnerable to climate change, given where their country is located and the socio-economic circumstances of their people. There are however, many opportunities to reduce the risks and strengthen the existing resilience and agency of the people who live here, including the recognised traditional owners. Clearly, clarifying and supporting the roles and functions of RNTBCs and other community organisations in a way that creates space for Indigenous governance structures to influence long-term land-use planning is crucial to supporting community resilience and adaptation.

The *Mabo* decision was based on the principle of non-discrimination before the law, that is, that the property rights of Australia's Indigenous peoples should be recognised as part of a fair and just society. The legal and symbolic force of *Mabo* has very practical consequences for how we live in Australia under climate change. As noted by Koko Bera traditional owner Leslie Gilbert, one of the key components of sustainable adaptation is the recognition of Indigenous decision-making authority.

> When people come into Kowanyama they should come and see not only the council but also the PBC which is the board of the traditional owners. Other people who want to do business in Kowanyama should also talk to the PBC and not only the Council itself. The true organisation is not in a way that one is trying to control the other. The thing is that elected councillors they have to work under the local government structures and the act and from where we are [the PBC] we are trying to work with the structure/governance from the land. (Leslie Gilbert, Director, Abm Elgoring Ambung RNTBC, 12 December 2013)

Acknowledgements

The authors acknowledge and thank Karajarri woman Anna Dwyer from the Nulungu Centre for Indigenous Studies, who assisted with fieldwork and interviews. The authors also acknowledge and thank their research partners the Karajarri Traditional Lands Association RNTBC and Abm Elgoring Ambung RNTBC. We also acknowledge the funding support of the National Climate Change Adaptation Research Facility.

References

Adger, W.N., Dessai, S., Goulden, M. et al. (2009) Are there social limits to adaptation to climate change? *Climatic Change* 93(3–4), 335–354.

Adger, W.N., Barnett, J., Chapin, I.F.S. and Ellemor, H. (2011) This must be the place: Under representation of identity and meaning in climate change decision-making. *Global Environmental Politics* 11(2), 1–25.

Bauman, T., Strelein, L.M. and Weir, J.K. (2013) Navigating complexity: living with native title. In: Bauman, T., Strelein, L.M. and Weir, J.K. (eds) *Living With Native Title: The Experiences of Registered Native Title Corporations*. AIATSIS Research Publications, AIATSIS, Canberra, pp. 1–26.

Cameron, E.S. (2012) Securing Indigenous politics: A critique of the vulnerability and adaptation approach to the human dimensions of climate change in the Canadian Arctic. *Global Environmental Change* 22(1), 103–14.

Department of Premier and Cabinet Government of Western Australia (2012) *Guide to the Government Indigenous Land Use Agreement and Standard Heritage Agreements, Version 1.2*. Government of Western Australia, Perth.

Department of State Development Infrastructure and Planning (2012) *Regional Planning in Cape York*. Department of State Development Infrastructure and Planning, Queensland. Available at http://www.dsdip.qld.gov.au/resources/factsheet/regional/cape-york-regional-planning-factsheet.pdf (accessed 3 June 2014).

Dovers, S. (2009) Nomalizing adaptation. *Global Environmental Change* 19, 4–6.

Edgar, J. (2011) Indigenous land use agreement: building relationships between Karajarri traditional owners, the Bidyadanga Aboriginal Community La Grange Inc.

and the Government of Western Australia. *Australian Aboriginal Studies* 2, 50–62.

Glaskin, K. (2007) Outstation incorporation as precursor to a Prescribed Body Corporate. In: Weiner, J.F. and Glaskin, K. (eds) *Customary Land Tenure and Registration in Australia and Papua New Guinea: Anthropological Perspectives*. ANU E-press, Canberra.

Hulme, M. (2011) Reducing the future to climate: a story of climate determinism and reductionism. *Osiris* 26(1), 245–266.

Kowanyama Aboriginal Land and Natural Resource Management Office (1994) *Strategic Directions for Natural Resource Management on Aboriginal Lands of Kowanyama*. Kowanyama Aboriginal Land and Natural Resource Management Office, Kowanyama.

Martin, D. (2003) *Rethinking the Design of Indigenous Organisations: The Need for Strategic Engagement*. Centre for Aboriginal Economic Policy Research, CAEPR Discussion Paper 248, Canberra.

McGrath, P.F., Stacey, C. and Wiseman, L. (2013) An overview of the Registered Native Title Bodies Corporate regime. In: Bauman, T., Strelein, L.M. and Weir, J.K. (eds) *Living with Native Title: The Experiences of Registered Native Title Corporations*. AIATSIS Research Publications, AIATSIS, Canberra, pp. 27–64.

Monaghan, J. (2005) *Natural Disaster Risk Management: Kowanyama Stage 1 Report*. Unpublished report. Kowanyama Aboriginal Land and Natural Resource Management Office, Queensland.

Sinnamon, V. and Frank, A. (2010) *Climate Change on the Northern Gulf of Carpentaria Plains*. Paper presented at the North Australian Indigenous Land and Sea Management Alliance Climate Change Adaptation Workshop, Darwin, 21 April 2010.

Strelein, L. and Tran, T. (2013) Building Indigenous governance from Native Title: Moving away from 'fitting in' to creating a decolonised space. *Review of Constitutional Studies* 18(1), 19–47.

Tran, T., Strelein, L., Weir, J., Stacey, C. and Dwyer, A. (2013) *Changes to Country and Culture, Changes to Climate: Strengthening Institutions for Indigenous Resilience and Adaptation*. National Climate Change Adaptation Research Facility, Gold Coast.

Veland, S., Howitt, R., Dominey-Howes, D., Thomalla, F. and Houton, D. (2013) Procedural vulnerability: understanding environmental change in a remote Indigenous community. *Global Environmental Change* 23, 314–326.

Weir, J.K. (2011) *Karajarri: A West Kimberley Experience in Managing Native Title*. Australian Institute of Aboriginal and Torres Strait Islander Studies, Research Discussion Paper 30, Canberra.

Wensing, E. (2011) Improving planner's understanding of Aboriginal and Torres Strait Islander Australians and reforming planning education in Australia. *Proceedings of the 3rd World Planning Schools Congress*, Perth (WA), 4–8 July 2011, paper no. 112.

Winer, M., Robertson, H. and Murphy, H. (2012) Environmental justice for Indigenous people in the Cape York Peninsula: Enabling potential and navigating constraints. In: Pearson, N., Fridriksson, G., Schuele, M. et al. (eds) *2009–2012 Policy and Research: Key Policy Papers and Submissions from 2009–2012*. Cape York Institute for Policy and Leadership, Cairns, pp. 89–100.

Yu, S. (1999) *Ngapa Kunangkul (Living Water): Report on the Aboriginal Cultural Values of Groundwater in the La Grange Sub-Basin*, for The Water and Rivers Commission of Western Australia, December 2009. Government of Western Australia, Perth.

35 Indigenous adaptation to climate change: the Arabana

MELISSA NURSEY-BRAY[1], DEANE FERGIE[2],
VERONICA ARBON[3], LESTER-IRABINNA RIGNEY[4],
ROB PALMER[5], JOHN TIBBY[1], NICK HARVEY[1],
LUCY HACKWORTH[2] AND AARON STUART[6]

[1]*Geography, Environment and Population, University of Adelaide, Australia*
[2]*Anthropology of Native Title Societies, University of Adelaide, Australia*
[3]*Wilto Yerlo, University of Adelaide, Australia*
[4]*Division of the Deputy Vice Chancellor Vice President, University of Adelaide, Australia*
[5]*AuConsulting, Plympton, South Australia, Australia*
[6]*Arabana Association, Centacare Catholic Family Services Country SA, Australia*

35.1 Introduction

Indigenous peoples are projected to be disproportionately affected by the impacts of climate change (Macchi 2008). In Australia, many Indigenous peoples and communities are particularly vulnerable to climate change (Green et al. 2009; Beer et al. 2013). Some of the issues that may be of particular concern for Indigenous peoples include sea-level rise, the impact of more intense extreme weather events, flooding, drought and heat (Macchi 2008). The capacity of Indigenous peoples to respond to these changes will depend upon a number of factors including their level of exposure to extreme events, availability of natural resources, the location and quality of housing, the ability to live on or visit ancestral lands, the status of property rights, poverty, the extent and level of social networks, health impacts and livelihood diversification (Adger 1999; Kelly and Adger 2000; Ford et al. 2010; Adger et al. 2011; Veland et al. 2013). This chapter presents some insights into how

the Arabana people, traditional owners of the Kati Thanda–Lake Eyre region (see Fig. 35.1), Australia plan to respond to these factors and the journey they took to build their own adaptation strategies.

As a region, Kati Thanda–Lake Eyre is Australia's largest salt lake, located 647 km northeast of Adelaide in the state of South Australia. Its catchment is located in the states of Queensland, South Australia, New South Wales and the Northern Territory. Its drainage basin is over 1.2 million km^2 and, at 15.2 m below sea level in its eastern perimeter, is Australia's lowest point. The region is the traditional land for the Arabana, the Anangu Pitjantjatjara Yankunytjatjata (APY) and other Indigenous peoples who have inhabited the area for thousands of years. Today, there are about 57,000 people living in the basin working in pastoralism, tourism, mining, petroleum exploration and township-based work such as retail, education, medical and other services. The Arabana people want to

Applied Studies in Climate Adaptation, First Edition. Edited by Jean P. Palutikof, Sarah L. Boulter, Jon Barnett and David Rissik.
© 2015 John Wiley & Sons, Ltd. Published 2015 by John Wiley & Sons, Ltd.

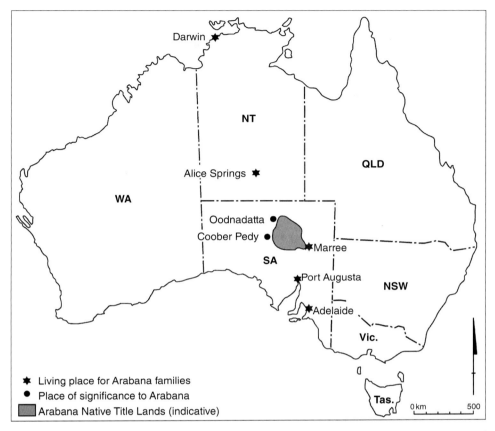

Figure 35.1 The lands of the Arabana people. Source: Map reproduced with permission of Christine Crothers, University of Adelaide.

develop options for adapting to predicted change in their country in this context. (Note that the term 'country' has a very specific meaning to Australian Aboriginal and Torres Strait Islander peoples, and refers to the land and seas to which a group is spiritually and physically affiliated. 'Country' is a term that expresses holistic connection to land and sea which is understood as an integrated entity).

35.2   Method

Drawing on the work of Ford et al. (2006), Green et al. (2010), Cameron (2012), Howitt et al. (2012) and Veland et al. (2013), in conjunction with a team of researchers from the University of Adelaide, the Arabana conducted a community-based interdisciplinary vulnerability assessment in three phases: (1) identification of specific impacts; (2) a values assessment and risk perception study; and (3) development of adaptation strategies at local scales. The project specifically incorporated the social and cultural dimensions of vulnerability via these steps. Case studies have demonstrated the value of vulnerability assessments and the importance of incorporating the social and cultural dimensions (Naess et al. 2006) and of determining resilience (Berkes and Jolly 2001; Folke 2006; Ford et al. 2006; Gallopin 2006; Smit and Wandel 2006).

A central principle of this project was its application throughout of participatory action research principles in ways that focused respectfully on the experiences, interests and aspirations of the Arabana (McTaggart 1997; Kovach 2010). Secondly, it explored how climate change will impact Arabana lives by building relationships and using a research design that gave legitimacy to the voices of the Arabana people (Rigney 1999; Arbon 2008). Engagement and relationship building within and external to the team ensured that this project embarked on a process to draw together, as a community of interest, diverse Aboriginal and non-Aboriginal positions informed by Indigenous knowledge, science and social science.

The project also met the research protocols of Indigenous Australia at a very high level (AIATSIS 2011) while also recognising the diversity and dispersal of Arabana people. The project was conducted between November 2011 and February 2013. We started by commissioning a study that reviewed the scientific estimates of the likely impacts of climate change on Arabana country (Tibby 2012). Using this as a starting point, the research team then undertook fieldwork which included the conduct of extensive semi-structured interviews and workshops in all the places across Australia where Arabana people live. Over 100 Arabana people were interviewed and participated in the project directly while other Arabana people were notified of the project (and invited to participate) through Facebook, minutes of meetings from the Arabana Association and via a national mail out, which included introductory DVDs and many descriptive flyers. Overall, the team aimed to understand the inter-disciplinary context of risk, adaptation and social sustainability and how these dimensions affected the way in which the Arabana perceived and would respond to climate change impacts. A workshop was then convened where all researchers presented the research results to Arabana to receive face-to-face feedback, thus authenticating the validity and interpretation of data collected. On the second day of the workshop, and based on the results, the

Arabana people and the research team worked together to develop the climate change adaptation strategy, which they then launched at the conclusion of the project.

This was therefore a journey that began as a research project but ended in a community-owned product, now in the process of being implemented. What key lessons can be drawn from this experience? We argue that the key insight from this project is that, to build effective climate change adaptation strategies in Indigenous contexts, it is necessary to 'join the dots' between historical experience, Indigenous values about country, and the available portfolio of adaptation actions in order to arrive at culturally appropriate outcomes that build and enhance adaptive capacity.

35.3   Impacts of climate change on Arabana country

Our scientific study found that climate change will impact both Arabana people and their country. Given the dispersal of the Arabana people from Darwin in the north to Adelaide in the south, major population centres of the Arabana span very large temperature and rainfall gradients across the country. Summer season monsoon rainfall dominates in the north, while in the south the majority of rain falls in winter. Rainfall variability is very high in the region (van Etten 2009), but significant historical trends have been observed. For example, there were substantial increases in rainfall observed in the second half of the twentieth century in the north, with declines experienced in the southern coastal region (CSIRO and BoM 2007).

Looking into the future, there is a consistent pattern of projected temperature rise in the study region in the twenty-first century, with increases that may exceed 4 °C by the end of the century for central Australia (CSIRO and BoM 2007). Even if there are significant reductions in the rate of greenhouse gas emissions, the region is still projected to experience warming through the twenty-first century (CSIRO and BoM 2007). In general terms, temperature increases will be greater in inland regions.

Consistency of rainfall projections varies through the region with greater coherence between different computer-based climate models in the south of the region where declining rainfall is projected in most, but by no means all, scenarios (CSIRO and BoM 2007; Gibbs et al. 2013). As latitude decreases, so does consistency. In the northern part of the study region (particularly where influenced by summer-dominated monsoonal rainfall), increases in rainfall are a distinct possibility (CSIRO and BoM 2007).

Detailed assessments of climate change effects on hydrology have been undertaken for the arid South Australian section of the study area using downscaling of the most accurate climate models (Gibbs et al. 2013). This analysis highlights uncertainties associated with estimates of future water availability; although the 'most likely' scenario (based on model agreement) is that there will be declines in rainfall and declines in large recharge events (defined as months with >100 mm rainfall). Some models predicted increased water availability including large recharge events (Gibbs et al. 2013).

To summarise, in a report commissioned by this project, Tibby (2012) showed that while projections of temperature increases display a consistent pattern across the region and reductions in rainfall are likely in the southern part of the region, it is also possible that increases in the average rainfall and high magnitude rainfall events will occur.

Interestingly, these results are complemented by first-hand observations of the Arabana people about changes they have experienced over time. Firstly, Arabana people identified key risks to their country including concerns about water access and availability. For generations, Arabana people relied upon hidden wells and artesian springs for their water; these are now threatened by drying, pollution by salt intrusion or by reduced water flow. They have witnessed a reduction of and changes to various bush tucker (wild harvested or hunted food) species, particularly reptiles. Places where Arabana people camped just 20 years ago are degrading and in some places the vegetation is dying out completely. Plants are becoming scarce and flowering for much shorter periods of time. Arabana people are worried about the impacts of flooding and erosion on their sacred and cultural sites: they perceive that sacred sites are in danger of being washed away as flood levels from nearby creeks get closer each year. Arabana are also concerned about how cultural knowledge is going to be transmitted over time, especially if physical sites are changing; they argued strongly that for climate change impacts to be dealt with Arabana need to be supported in trying to develop livelihood opportunities that would enable them to return to, work on and manage their country (Nursey-Bray et al. 2013).

35.4 Historical experience (including colonisation) matters

Arabana people have a long and continuous history of experiencing change and responding through adaptation. One dimension of this experience is that we found Arabana people have a significant heritage of adaptation indicated by their social viability in a region that already has many of the characteristics that are predicted to be the broader challenges of this coming phase of climate change (that is, inhospitable extremes and dramatic unpredictability). For example, a source of wellbeing was connection to country (Ganesharajah 2009). Arabana people identified that country remains important even though they may not have resided within country for more than 10, 20 or 30 years. Arabana demonstrated adaptation and resilience to more recent ongoing and intense challenges, such as the invasion of their country and colonisation of their lives.

Colonisation has and continues to have multiple impacts on Indigenous societies in Australia. Western diseases killed whole groups of Aboriginal peoples (Campbell 2002). Aboriginal people were dislocated and forcibly removed from their land and seas to make way for the colonists (Reynolds 1987; Broome 1994). Many religious denominations built missions across Australia where, for the 'benefit' of Indigenous peoples, they were punished for practicing their culture and forbidden to speak their own languages. Within the last half of the twentieth century, whole generations of children were taken from their parents and sent to foster care (HREOC 1997; Haebich 2000). As Dudgeon et al. (2010, p. 30) describe, 'the effects were a form of cultural genocide of Indigenous Australians, through the loss of language, family dispersion and the cessation of cultural practices'. It is easy to make the mistake that this is simply a thing of the past rather than a lived reality today (Green et al. 2009). In this context, when working with Australian Aboriginal peoples it is crucial to acknowledge that climate change adaptation planning will be occurring in the context of colonisation.

Over and again Arabana people linked the present climate change issue to discussion of past colonial impacts such as working on the railway or being moved around by missionaries. These experiences are not divorced from one another at any point. As the findings show, Arabana people still have within their living memory the experience of being members of or affected by the stolen generation (a policy in Australia where Aboriginal children were taken from their parents), of living in a mission, of receiving rations and being beaten for speaking the Arabana language (Nursey-Bray et al. 2013). The fact that the Arabana people now live in many other states and regions in Australia (far from their original lands) is a function of the dislocation, removal and dispossession caused by colonisation and its associated impacts.

As such, the development of effective adaptation needs to be underpinned by understanding that the history of colonisation has an ongoing legacy and that this fact, in combination with the ongoing issues that climate change and other environmental changes are having on Indigenous communities (Baldwin 2009), creates ongoing tension between the Arabana people and other stakeholders in the region. As Rigney (2011) put it, the settler state needs to settle with whom it colonises.

35.5 Adaptation is not just about climate change: correlating values and adaptation actions

All Arabana people, no matter where they live, want adaptation for Arabana country to happen:

> You definitely need adaptation for the country. To keep it to be the beautiful place that it is. The place is changing so need to keep up with the change. That place is my heritage, is my family; I want to take my kids out one day to show it to them. Much harder to do in towns and cities than communities and lands. (Alice Springs respondent 3, 2012)

Given the current geographical dispersal of the Arabana people, this is an important finding; it gives some direction to developing adaptation for many Indigenous peoples across Australia who have affiliation with country but live elsewhere. Further, we found that discussions of key concerns and ways forward in adapting to climate change were couched within a discourse around Arabana values. For example, identity, its importance and value in promoting connection to country and affirming Arabana culture wherever they lived was reiterated. The importance of history, as already discussed, was an underpinning refrain that set the context for discussions around the value of culture and country. Water and its importance to life and the maintenance of livelihoods was strongly asserted as a cultural and livelihood value.

In developing the adaptation strategy, Arabana people were therefore clear that the strategies identified were strongly influenced by and deeply embedded in their culture. The options identified as priorities in the adaptation strategy (see Table 35.1) also provide multiple other and added benefits to the Arabana as a whole. Pursuing them will therefore address many other issues. This suggests that, while identifying impacts is an important dimension of obtaining the baselines needed for developing adaptation options, it is in fact the social, cultural and economic values underpinning responses (perceived or otherwise) that will hinder or help adaptation now and into

Table 35.1 Priorities in the Arabana adaptation strategy.

Value	Adaptation options suggested
History	Cultural keeping centres and enhanced use of information and communications technology, getting youth involved
Place	Going back to country; re-stocking native vegetation and wildlife
Water	Access, developing partnerships with government, mining
Livelihoods	Pastoralism on Aboriginal lands such as at Finniss Springs (their old pastoral property which is also on country), cultural tourism, employment as rangers
Culture	Cultural revitalisation programs and centres
Country	Setting up ranger stations; re-vegetation programs

the future. If these values change, it is likely that priorities for adaptation will also. It also highlights an ability by the Arabana to conceptualise change in a very holistic manner.

What emerges from these priorities is that adaptation for Arabana people is not just about adapting to climate or environmental change but ensuring culture is maintained and that Arabana identity is reinforced and built up over time as well. Adapting to change therefore also involves cultural maintenance as a form of adaptation. This finding reinforces the wider literature that highlights the importance of thinking about the cultural dimensions of climate change adaptation (Petheram et al. 2010; Adger et al. 2011; Cameron 2012; Veland et al. 2013). It also goes beyond the individual to address overall societal practice and is a means to redress the damage caused by colonisation. It provides a pathway to reconnect to culture and rebuild livelihoods so that Arabana can return to their country whether to visit or to live.

35.6 'Joining the dots' between adaptation and values: communication is key to process and practice

Finally, we argue that understanding Indigenous values and aspirations for adaptation is not enough. A key component of our project was to employ a communications expert to help build the connections needed between academic researchers and Indigenous peoples' observations and understanding about change and country. This bridge was critical to the success of the project as it formed a link between the research team and the Arabana project participants, enabling Arabana people from all the places they live to genuinely contribute to and obtain ownership of the adaptation strategy.

Communicating climate change adaptation is recognised as a difficult domain which often results in confusion (Somerville and Hassal 2011, p. 48) and presents a 'major marketing challenge' (Ungar 2007, p. 85) to communicators. This was the case with communicating climate change adaptation to the Arabana people and generating the interest required for true community engagement. The communications in the project were important because without 'buy-in' from the Arabana communities, the project was destined to have limited success.

Contemporary communication methodologies argue that presenting climate change as an insurmountable negative challenge facing society results in disengagement (Moser and Dilling, 2007; Futerra Sustainability Communications 2010). Doherty and Clayton (2011) have argued that those most reactive to climate change messaging are often the most educated and that the level of reaction is related more to political orientation, liberal vs. conservative, rather than to levels of engagement with the scientific majority view on the likely impacts of the changes and the consequential requirements for adaptation actions.

We therefore initially framed the project as a positive exercise that called on Arabana people to get involved. We produced a four-minute video and a colourful brochure that was distributed to over 100 individuals from a database provided to the team by the Arabana Board of Directors, encouraging a positive engagement with the project.

The response was negligible. In trying to understand why this was the case, we found that Indigenous Australians have effectively 'compartmentalised their experiences into distinct 'whitefella' and 'blackfella' domains' (Greer and Patel 2000, p. 312). The deployment of a mainstream communications approach meant we were effectively using 'whitefella' tools to talk with 'blackfellas'. It was never going to work. Giannelli and Rabkin (2007, p. 116) argue the success of a climate change communications initiative requires that you 'not only speak, but listen to' those you are trying to communicate with. We subsequently incorporated different worldviews, such as the notion of kinship and how that system impacts upon the way Arabana peoples communicate with each other. An understanding of the kinship structures in an Arabana context enabled the research team to gain a firmer understanding of the different family groups and thus how to communicate with them in a culturally appropriate way. Knowing how cultural links were integrated across the different family structures meant that our team was in a position to communicate with the right people who then dispersed information throughout their kinship groups, enabling a genuine two-way flow of information and ideas.

The strategic understanding of culture has been used in other research contexts to encourage Indigenous participation in community-based projects, including Leonard et al. (2013) who found that: 'The integration of such traditional knowledge into local and regional planning is necessary for Indigenous peoples to develop, initiate and collaborate with climate change adaptation measures' (Leonard et al. 2013, p. 3).

We were therefore able to exchange information between the project team and Arabana peoples.

Given the significant decline in support around the world for taking urgent action to deal with this issue (Hanson 2012), this approach has the potential to make a significant contribution for reframing the way governments, environmental non-government organisations and others communicate climate change adaptation to society.

In our case study we also measured our communication success by the level of participation by Arabana peoples in the project and then the extent to which the Native Title Body Corporate (an authorised statutory body under Australian law, constituted to represent the Native Title holders and their interests) adopted the recommendations made in the draft adaptation plan presented to the committee. Wolf and Moser (2011) have written on the importance of obtaining individual participation and engagement with the issue of climate change to encourage action to 'deal with the impacts and identify, develop, support, and implement climate solutions' and how 'involving [individuals] is not an option but an imperative' (Wolfe and Moser 2011, p. 551). We have found consistency with this theory. Our demonstrable achievements provide clear evidence to support theoretical frameworks that contend that individual participation in climate change responses will enhance the likelihood of success in such an endeavour.

35.7 Summary

Cumulatively, our results show that climate change is affecting and will continue to affect Arabana country and other places where they live. History, both the long-term history of adaptation over millennia (for survival) as well the ongoing legacy of colonisation, matters. It still informs and drives Indigenous reaction to the climate change conversation. We find that Arabana people married their values with adaptation priorities and did not divorce the impact of climate change from other risks. As such, adaptation is more than about climate change, it encompasses change for the better across the board, including building the capacity to obtain and maintain livelihoods.

This is consistent with work by Petheram et al. (2010), Leonard et al. (2013) and Veland et al. (2013). Suggested adaptations were holistic rather than individual, and acknowledged the variable contexts of Arabana across time and scale. Finally, we argue that for adaptation to work in an Indigenous context, communications can be used for more than the preparation of flyers and brochures, but as a driving force to help enact change, transmit knowledge and help implement adaptation. Trying to develop an adaptation process that is inclusive along the way, as much as seeking inclusivity at an endpoint, is invaluable and our communications focus helped us to do this. For example, ensuring that there is room to discuss what may seem to be 'non-climate' related issues is essential and facilitates a much wider and inclusive result. As Veland et al. (2013) note, without embedding a (communication) process that accommodates multiple fields and frames of inquiry, the research can inadvertently create 'procedural vulnerability'. For policymakers there are many lessons here, as summarised in Box 35.1.

Box 35.1　Lessons for policymakers

- Acknowledge that adaptation in Indigenous contexts cannot be extricated from the history of colonisation.
- When conducting vulnerability assessments, find ways of assessing both social and biophysical vulnerability.
- Embed justice and equity in all adaptation options and policy.
- Recognise that adaptation needs to incorporate the lived reality of Indigenous peoples and the fact that they live in urban as well as remote places: adaptation policy needs to integrate place and scale.
- Design adaptation in ways that will assist building community capacity and ownership.
- Develop communications about climate change not only at the local scale.
- Employ communications expertise into community-based risk-assessment processes.

Arabana people are not alone in asserting their right to undertake and manage the impacts of climate change. For example, Indigenous peoples have already begun the process of documenting their climate histories and traditional knowledge about climate change (e.g. Riedlinger and Berkes 2001; Ford and Smit 2004; Ford et al. 2006; Laidler 2006; IPCC 2007; Leduc 2007; Laidler et al. 2009; Pearce et al. 2009; Turner and Clifton 2009; Cameron 2012). Other studies have focused on how Indigenous peoples can develop adaptation strategies and how the vulnerability of Indigenous peoples to climate change can be assessed (IPCC 2007).

Arabana people are now working on the implementation phase of their adaptation strategy. However, as with all the other challenges they face, they cannot do this alone. Government, researchers and private industry have an opportunity to support and learn from the Arabana. Ultimately, the challenge facing the Arabana – that of reconciling economy and the environment, making distinctions between urban and remote experience and dealing with the legacy of a collective past – is a universal one. It is one we need to work on together in order to adapt to the difficulties the future will bring.

As Arabana Elder Ken Buzzacott concludes (pers. comm. June 2012):

> We got to talk together, stick together and try to do something there … work together, and everybody … can get success.

References

Adger, N. (1999) Social vulnerability to climate change and extremes in coastal Vietnam. *World Development* 27(2), 249–269.

Adger, W. N., Brown, K., Nelson, D. R. et al. (2011) Resilience implications of policy responses to climate change. *WIREs Climate Change* 2, 757–766.

AIATSIS (2011) *Guide for Ethical Research in Indigenous Studies*. Australian Institute of Aboriginal and Torres Strait Islander Studies, Canberra, ACT.

Arbon, V. (2008) *Arlathirnda Ngurkarnda Ityirnda: Being-knowing-doing: De- Colonising Indigenous*

Tertiary Education. Post Pressed, Tenneriffe, Queensland.

Baldwin, A. (2009) Carbon nullius and racial rule: race, nature and the cultural politics of forest carbon in Canada. *Antipode* 41(2), 231–255.

Beer, A., Tually, S., Kroehn, M. et al. (2013) *Australia's Country Towns 2050: What Will a Climate Adapted Settlement Pattern Look Like?* National Climate Change Adaptation Research Facility, Gold Coast.

Berkes, F. and Jolly, D. (2001) Adapting to climate change: social-ecological resilience in a Canadian Western Arctic community. *Conservation Ecology* 5(2), 18.

Broome, R. (1994). *Aboriginal Australians*, 2nd edition. Allen and Unwin, Sydney.

Cameron, E. (2012) Securing Indigenous politics: a critique of the vulnerability and adaptation approach to the human dimensions of climate change in the Canadian Arctic. *Global Environmental Change* 22, 103–114.

Campbell, J. (2002) *Invisible Invaders: Smallpox and Other Diseases in Aboriginal Australia, 1790–1880*. Melbourne University Press, Melbourne.

CSIRO and Bureau of Meteorology (BOM) (2007) *Climate Change in Australia. Technical Report*. CSIRO, Melbourne.

Doherty, T. and Clayton, S. (2011) The psychological impacts of global climate change. *American Psychologist* 66, 265–276.

Dudgeon, P., Wright, M., Paradies, Y. et al. (2010) The social, cultural and historical context of Aboriginal and Torres Strait Islander Australians. In: Purdie, N., Dudgeon, P. and Walker, R. (eds) *Working Together: Aboriginal and Torres Strait Islander Mental Health and Wellbeing Principles and Practice*. Commonwealth of Australia, Canberra, ACT, pp. 25–43.

Folke, C. (2006) Resilience: the emergence of a perspective for social-ecological systems analyses. *Global Environmental Change* 16, 253–267.

Ford, J.D. and Smit, B. (2004) A framework for assessing the vulnerability of communities in the Canadian Arctic to risks associated with climate change. *Arctic* 57, 389–400.

Ford, J., Smit, B. and Wandel, J. (2006) Vulnerability to climate change in the Arctic: a case study from Arctic Bay, Canada. *Global Environmental Change* 16, 282–292.

Ford, J., Pearce, T., Duerden, F., Furgal, C. and Smit, B. (2010) Climate change policy responses for Canada's Inuit population: the importance of and opportunities for adaptation. *Global Environmental Change* 20, 177–191.

Futerra Sustainability Communications (2010) SIZZLE: The New Climate Message. Futerra, UK.

Gallopin, G. (2006) Linkages between vulnerability, resilience, and adaptive capacity. *Global Environmental Change* 16, 293–303.

Ganesharajah, C. (2009) *Indigenous Health and Wellbeing: The Importance of Country. Native Title Research Report 1/2009*. Australian Institute of Aboriginal and Torres Strait Islander Studies, Canberra, ACT.

Giannelli, L. and Rabkin, S. (2007) Listening to the audience: San Diego hones its communications strategy by soliciting residents' views. In: Moser, S. and Dilling, L. (eds) *Creating a Climate for Change*. Cambridge University Press, Cambridge, pp. 105–118.

Gibbs, M., Alcoe, D. and Green, G. (2013) *Impacts Of Climate Change On Water Resources. Phase 3 Volume 4 South Australian Arid Lands Natural Resources Management Region. DEWNR Technical Report 2013/06*. Government of South Australia, Adelaide.

Green, D., Jackson, S. and Morrison, J. (eds) (2009) *Risks from Climate Change to Indigenous Communities in the Tropical North of Australia*. Department of Climate Change and Energy Efficiency, Canberra, ACT.

Green, D., Billy, J. and Tapim, A. (2010) Indigenous Australians' knowledge of weather and climate. *Climatic Change* 100, 357–354.

Greer, S. and Patel, C. (2000) The issue of Indigenous world views and accounting. *Accounting, Auditing and Accountability Journal* 13, 307–329.

Haebich, A. (2000) *Broken Circles: Fragmenting Indigenous Families 1800–2000*. Fremantle Arts Centre Press, Fremantle, Western Australia.

Hanson, F. (2012) *The Lowy Institute Poll 2012: Australia and New Zealand in the World, Public Opinion and Foreign Policy*. Lowy Institute for International Policy, Sydney, New South Wales.

Howitt R., Havnen O. and Veland S. (2012) Natural and unnatural disasters: responding with respect for Indigenous rights and knowledges. *Geographical Research* 50, 47–59.

Human Rights and Equal Opportunity Commission (1997) *Bringing Them Home: Report of the National Inquiry into the Separation of Aboriginal and Torres Strait Islander Children from their Families*. Commonwealth of Australia, Canberra, ACT.

IPCC (2007) Summary for Policymakers. In: Solomon, S., Qin, D., Manning, M. et al. (eds) *Climate Change 2007: The Physical Science Basis. Contribution of Working Group I to the Fourth Assessment of the Intergovernmental Panel on Climate Change.* Cambridge University Press, Cambridge.

Kelly, P. and Adger, N. (2000) Theory and practise in assessing vulnerability to climate change and facilitating adaptation. *Climatic Change* 47, 325–352.

Kovach, M. (2010) *Indigenous Methodologies: Characteristics, Conversations, and Contexts.* University of Toronto Press, Toronto.

Laidler, G. (2006) Inuit and scientific perspectives on the relationship between sea ice and climate change: the ideal complement? *Climatic Change* 78, 407–444.

Laidler, G., Ford, J.D., Gough, W.A. et al. (2009) Travelling and hunting in a changing Arctic: assessing Inuit vulnerability to sea ice change in Igloolik, Nunavut. *Climatic Change* 94, 363–397.

Leduc, T. (2007) Sila dialogues on climate change: Inuit wisdom for a cross-cultural interdisciplinarity. *Climatic Change* 85, 237–250.

Leonard, S., Mackenzie, J., Kofod, F. et al. (2013) *Indigenous Climate Change Adaptation in the Kimberley Region of Northwestern Australia: Learning from the Past, Adapting in the Future: Identifying Pathways to Successful Adaptation in Indigenous Communities.* National Climate Change Adaptation Research Facility, Gold Coast.

Macchi, M. (2008) *Indigenous and Traditional Peoples and Climate Change.* IUCN Issues Paper, IUCN, Gland.

McTaggart, R. (ed.) (1997) *Participatory Action Research: International Contexts and Consequences.* State University of New York Press, Albany.

Moser, S. and Dilling, L. (eds) (2007) *Creating a Climate for Change.* Cambridge University Press, Cambridge.

Naess, L., Norland, I., Lafferty, W. and Aall, C. (2006) Data and processes linking vulnerability assessment to adaptation decision making on climate change in Norway. *Global Environmental Change,* 16, 221–233.

Nursey-Bray, M., Fergie, D., Arbon, V. et al. (2013) *Community based adaptation to climate change: The Arabana, South Australia.* National Climate Change Adaptation Research Facility, Gold Coast.

Pearce, T., Ford, J.D., Laidler, G. et al. (2009) Community collaboration and climate change research in the Canadian Arctic. *Polar Research* 60, 10–27.

Petheram, L., Zander, K., Cambell, B., High, C. and Stacey, N. (2010) 'Strange changes': Indigenous perspectives of climate change and adaptation in NE Arnhem Land (Australia). *Global Environmental Change* 20, 681–692.

Reynolds, H. (1987) *Frontier: Aborigines, Settlers and Land.* 1st edition, Allen and Unwin, Sydney, New South Wales.

Riedlinger, D. and Berkes, F. (2001) Contributions of traditional knowledge to understanding climate change in the Canadian Arctic. *Polar Record* 37, 315–328.

Rigney, L. (1999) Internationalization of an Indigenous anticolonial cultural critique of research methodologies: a guide to Indigenist research methodology and its principles. *Wicazo Sa Review* 14(2), 109–122.

Rigney, L. (2011) Can the settler state settle with whom it colonises? Reasons for hope and priorities for action. In: Maddison S. and Brigg M. (eds) *Unsettling the Settler State, Creativity and Resistance in Indigenous Settler-State Governance.* The Federation Press, Sydney, New South Wales, pp. 206–212.

Smit, B. and Wandel, J. (2006) Adaptation, adaptive capacity and vulnerability. *Global Environmental Change* 16, 282–292.

Somerville, R.C.J. and Hassal, S.J. (2011) Communicating the science of climate change. *Physics Today* 64(10), 48–53.

Tibby, J. (2012) *Science Report: Arabana NCCARF Project.* Peer-reviewed report for Arabana Climate Change Project. University of Adelaide, South Australia.

Turner, N.J. and Clifton, H. (2009) "It's so different today": Climate change and indigenous lifeways in British Columbia, Canada. *Global Environmental Change* 19, 180–190.

Ungar, S. (2007) Public scares: changing the issue culture. In: Moser, S. and Dilling, L. (eds) *Creating a Climate for Change.* Cambridge University Press, Cambridge, pp. 81–89.

van Etten, E.J.B. (2009) Inter-annual rainfall variability of arid Australia: Greater than elsewhere? *Australian Geographer* 40, 109–120.

Veland, S., Howitt, R., Dominey-Howes, D., Thomalla, F. and Houston, D. (2013) Procedural vulnerability: understanding environmental change in a remote Indigenous community. *Global Environmental Change* 23, 314–326.

Wolf, J. and Moser, S. (2011) Individual understandings, perceptions, and engagement with climate change: insights from in-depth studies across the world. *WIREs Climate Change* 2, 547–569.

Section 7
Settlements and housing

36 Contextualising the challenge of adapting human settlements

WILLIAM D. SOLECKI

Institute for Sustainable Cities, CUNY Hunter College, USA

Human settlements find themselves on the front lines of climate change. They demand the attention of policymakers because several factors, such as high population density and complex interdependent urban systems, make them extremely vulnerable to climate change. However, they also hold rich opportunities for successful adaptation. With respect to vulnerability, it is now widely realised that most people in the world live in cities, and that urban growth is projected to continue well into the twenty-first century, nearly doubling to 6.3 billion people by mid-century. Urban agglomerations include concentrations of interconnected systems, to which extreme events and other climate-related events can result in resource shortages and infrastructure shocks associated with cascading failures and system collapses. At the same time, settlements are often located in places that are exposed to climate extremes. Densely populated settlements are found near or on coastlines where increases in sea level and large storm surges will more often threaten livability and critical infrastructure. Furthermore, increasingly frequent and intense floods and droughts will put even greater demands on water supplies that are often already scarce. In other contexts, communities along the urban–rural interface must respond to the potential growing risk of wildfire. Finally, cities have effects on their own environment, the impacts of which are exacerbated by climate change. Among their other environmental impacts, cities and urbanisation processes create urban heat islands (UHIs) and pollute air and water, alter local weather patterns and fill or degrade valuable urban area wetlands.

As concern for climate change has grown within human settlements, the demand for adaptation action also has increased. Meaningful adaptation action presents a variety of challenges for cities and towns, as well as for individual homeowners. Human settlements are hubs of global economic activity. As centres of wealth and innovation, such communities often have the best tools and greatest human and capital resources for tackling the challenges of climate change. More cities throughout the world are moving towards the leading edge of climate change action. While high-profile cities such as Durban, London, Melbourne, Mexico City, New York and Seattle have been associated with climate change planning and activities, many other cities, large and small, in developing and developed country contexts have been involved in climate change adaptation and mitigation work. This has happened individually and via global-scale networks such as the C40, the World Mayor's Summit, ICLEI, the Urban Climate Change Research Network and the Rockefeller Foundation's 100 Resilient Cities program. These efforts have included formal governance structures as well as a variety of informal civil-society-based activities.

With settlements moving forwards with more aggressive climate action several significant challenges remain, such as limited financial resources and the need for updated risk and vulnerability data. As these challenges and barriers to progress are recognised and in some cases met, forward movement of successful adaptation in these communities can be enhanced. Some challenges specifically relate to conditions of climate change, others more specifically to cities and urbanisation while others connect to the interaction between the two (urban climate change). With respect to specifically climate change, not enough information is available regarding the benefits and costs of adaptation and how to ensure that adaptation strategies best provide co-benefits for addressing other urban environmental challenges including water shortages, air pollution and solid waste management. In all settlement contexts, funding for adaptation is scarce. Within developing country contexts, such funding can often be virtually non-existent. In many developed countries, municipal governments are facing stagnant or declining revenues at a time of increasing financial responsibility and aging infrastructure. Together these hamper climate adaptation activities.

The question of how to most effectively connect climate adaptation and development of these settlements is a pressing challenge. Critical questions remain as to how to simultaneously promote climate adaptation and urban economic development that will overcome social inequities and the demands of enhanced quality of life for urban residents. These conditions are especially pressing in: developed country cities facing decline and fiscal stress; and in rapidly growing cities in developing countries where rapid population growth, lack of comprehensive planning and governance capacity and limited infrastructure create

large concentrations of vulnerable urban residents with limited resiliency.

Another challenge relates to the issue of scale and integration present within settlements. Much contemporary settlement growth and development takes place in the context of individual property rights and relatively independent jurisdictions, where regulations, taxation and public policy largely focus on lessening conflicts and the spread of negative impacts from one property to another. Zoning and building codes and standards by definition are often designed to limit the spread of negative externalities from one site to another. Successful adaptation will partly hinge on the extent to which adaptation strategies can be developed and implemented to promote 'positive externalities', where actions taken by individual home-owners or municipalities can benefit themselves as well as neighbours. At the property level, policies to encourage individual homeowners to implement urban heat island mitigation strategies (i.e. tree planting, cool roof activities) will, if done at scale, provide neighbourhood-scale positive impacts. At the metropolitan scale, region-wide adaptations that provide the greatest level of benefits to the largest number of municipalities are a critical public policy goal. This type of integrative adaptation can be particularly challenging when attempting to connect the large number of formal and informal stakeholders and stakeholder organisations present within metropolitan regions.

It also can be argued that the current striving for urban adaptation can be better understood by evaluating past urban environmental challenges and opportunities. While the development of practical adaptation practices in settlements seems daunting, it is important to recognise that environmental challenges similar to those defined by current climate change have been met before in these places as part of the ongoing urbanisation process, and that cities have been quite dynamic and active in their response to challenges in the past. In many ways cities always remain – as urban historian Thomas Bender notes – 'unfinished', and are being constantly built and rebuilt both literally and figuratively. It is during this constant process of renewal and future possibilities that the opportunities to respond to environmental challenges have been seized in the past. As example, the rise of the 'sanitary city' in more developed countries during the late nineteenth century was evident as both (1) economic development and the application of new science and technology focused on hygiene and sanitation engineering; as well as (2) a re-conceptualisation of urban life as 'modern' and free of the congestion, filth and disease which had characterised most cities until then.

Today, the dynamism of cities continues. In New York City for example, even infrastructure as fundamental as the city's water supply and sewage treatment facilities are continuing to be built and rebuilt. This includes the construction of a massive third water main through the hard rock deep below the streets of Manhattan, a multi-billion dollar water treatment facility under a public park in the Bronx, the implementation of extensive watershed protection in the exurban green belt 150 km from midtown and the dramatic integration of green infrastructure technology into storm water management planning in heavily degraded urban wetlands. In this context, the challenge of urban sustainability can be conceived as the next in a series of major transitions and transformations that cities have responded to since the dawn of the urban-industrial era.

These potential transitions and transformations can take several forms or appear in several contexts and scales, such as: discrete and spatially specific gradual population and development retreat from highly exposed locations along urban shorelines; the broad-scale strengthening of local social capacity via social media and a variety of informal networks; and the (re)imaging of urban life as a moderate-density public-transit-oriented lifestyle and not the more automobile suburban ideal which dominated popular conscientious in many locales throughout the twentieth century. Crucial for understanding and managing these transitions and transformations in cities are fundamental questions about their character and quality. We need to better understand the connection between vulnerability and resilience and how dynamic climate change risks – either chronic or acute (i.e. extreme events) – accelerate or direct ongoing transition and transformation, and what the significant tipping points are that promote opportunities for orchestrated regime shifts. In the case of urban settlements and housing, we must answer the question of how, why and when do residents decide to change or not change their living situation in response to climate cues and under what conditions to situate climate adaptation strategies and benefits, especially in an ever more dynamic climate context.

37 Climate change and the future of Australia's country towns

ANDREW BEER[1], SELINA TUALLY[1], MICHAEL KROEHN[1], JOHN MARTIN[2], ROLF GERRITSEN[3], MIKE TAYLOR[4], MICHELLE GRAYMORE[5] AND JULIA LAW[1]

[1]Centre for Housing, Urban and Regional Planning, University of Adelaide, Australia
[2]LaTrobe University, Australia
[3]Central Australian Research Group (CARG), Charles Darwin University, Australia
[4]University of South Australia, Australia
[5]Water in Drylands Collaborative Research Program, University of Ballarat, Australia

37.1 Introduction

This chapter considers the future of Australia's country towns in the context of anticipated climate change, where it is assumed Australia faces moderate climate change as predicted by the CSIRO's MK3.5 model (2°C rise by 2050). We argue that the future of Australia's country towns is not simply a product of climate change and its manifestation in terms of altered rainfall patterns, increased average temperatures and more frequent extreme events. Instead, the future of Australia's country towns will be determined by their capacity to adapt, which in turn will be affected by their stock of community assets including social, economic, human and natural capital (Cocklin and Dibden 2005). Some locations will be more sensitive than others when exposed to climate change, some settlements will adapt well and others are likely to maladapt or simply disappear. The implications of climate change for Australia's country towns are complex and successful adaptation will be a product of rural and regional

Australia's ability to sustain current and new economic activities in the face of climate change, the willingness of government to invest in infrastructure, the level of resilience demonstrated by local populations and the ability of local communities to mobilise resources (Beer et al. 2013).

Developing an understanding of the challenges facing Australia's country towns under climate change is critical as: rural and regional Australia is the source of food production for the nation; regional communities are responsible for the majority of Australia's export earnings, with the mining sector alone accounting for 45% of export income; significant numbers of Australians are at risk from adverse climate change impacts, simply because of their place of residence; economic dislocation associated with failed regional communities could impose a significant burden on national fiscal capacity; and economic opportunities could be lost if we fail to recognise embedded opportunities to better adapt to a changing environment. However, the impact of climate change on country towns has been

Applied Studies in Climate Adaptation, First Edition. Edited by Jean P. Palutikof, Sarah L. Boulter, Jon Barnett and David Rissik.

relatively unexplored in the literature and policy. This absence stands in contrast to the volume of material on the impact of climate change on Australia's metropolitan areas and coastal settlements. This chapter briefly considers Australia's country towns within the context of contemporary and anticipated social, economic and climatic trends, before moving on to discuss the estimation of vulnerability within these communities. The chapter then considers both the implications for rural communities and the paths potentially available to governments across Australia.

37.2   Australia's country towns in context

Rural and regional Australia is often characterised in the media and in some academic accounts by its adaptability and innovation (Tonts et al. 2012), and these communities have demonstrated over many decades that they have the capability to respond – albeit variably – to economic, social and environmental challenges. These include natural disasters, drought and the broad economic consequences of changing productivity and commodity markets, globalisation and deregulation. Climate change presents rural and regional Australia with an additional hurdle. Along with related and unrelated economic, social and environmental pressures, climate change provides a new and more substantial hurdle that carries with it intergenerational impacts that could fundamentally change rural and regional Australia.

Economic conditions have a central impact on the functioning, social structure and wellbeing of all settlements, and especially country towns. Australia's system of country towns evolved from the late eighteenth century to the twentieth century as European colonisers appropriated Indigenous lands for agriculture and pastoralism. While some country towns were established without forward planning, others were the product of deliberate attempts to 'open up' the country and establish an efficient system of agricultural production. This history is significant because it is the country towns established through these processes that are now confronted by a climate-

change-affected future. In most instances these are places whose economic foundation was firmly rooted on agriculture and, in many instances, broad-scale dry-land farming. While some country towns have grown over time and developed a mix of industries, others have not prospered; over the past 50 years there has been a decline in the number and economic vibrancy of some places, especially smaller centres (Baum et al. 1999).

Agriculture lies at the heart of much economic activity in rural Australia. It remains an important source of export earnings and plays a fundamental role in ensuring food security for Australia and other nations. Agriculture, fisheries and forestry industries contribute approximately 3% of Australian Gross Domestic Product (GDP) and 14% (AU$ 31 billion) of Australia's total industry exports (CSIRO 2012). A considerable volume of research has considered the likely impacts of climate change on agricultural productivity in Australia (Preston and Jones 2006; Steffen et al. 2011). Steffen et al. (2011) summed up the complexities associated with attempting to draw conclusions around the impact of climate change on agriculture. They noted that:

> … for most types of Australian agriculture, water availability is a more important climate-related parameter than temperature, making temperature-based estimates of 'dangerous climate change' … less useful … Second, because of Australia's high natural climate variability, many producers and agricultural industries already have high adaptive capacity with regard to climate pressures. Third, given that Australia is a large continent spanning tropical, sub-tropical and temperate climatic zones, there is enormous variability in the types of agriculture undertaken in the country, in the sensitivity of the various industries to a changing climate and in their adaptive capacity. (Steffen et al. 2011, p. 205)

Climate change is therefore likely to have highly variable impacts on Australian agriculture as some places will experience increased rainfall while others are confronted by deficits relative to their expectations; tropical, sub-tropical and temperate regions will be affected by climate change in different ways; and, it is argued, Australian farmers appear to be 'pre-adapted' for climate change as many have already developed systems for dealing with a highly variable climate.

The impact of climate change on river flows will have far-reaching consequences for Australian agriculture. Some 80% of all profits in Australian agriculture are generated through irrigation and a reduction in the right – or opportunity – to draw water may hamper some districts, especially those with substantial investment in infrastructure. The scale of this potential impact is considerable, with Preston and Jones (2006) estimating that an increase in average temperatures of less than 1°C will reduce flows in the Macquarie River basin by up to 15%, while a 3°C change would cut Murray Darling flows by 16% to 48%.

Importantly, not all climate change impacts are likely to have a negative influence on the economies of country towns. Steffen et al. (2011) noted that the Margaret River wine region may come under increased growth pressure as competing locations for agriculture, viticulture and tourism activities are adversely affected. Ward (2009) examined the potential impact of climate change on Western Australia's wine industries. She concluded that while climate change will have an impact, 'the wine regions of South West Western Australia (SWWA) will remain ideally suited to further viticulture development for the production of high quality grapes and wine' (p. iii). Blackmore and Goodwin (2009) came to a comparable conclusion from their analysis of the Hunter wine region. Other research has however highlighted that different wine regions will be affected to variable degrees by climate change. Webb et al. (2005, 2006) found there were significant varietal differences (with pinot noir most sensitive and chardonnay and shiraz relatively unaffected) and differences in both quality and quantity of production by region. They estimated that returns on grape production in the Riverina (New South Wales) would decline by 16% to 2030, but only fall 3.5% in the Yarra Valley (Victoria) and 1.6 % in the Coonawarra (South Australia).

Harle et al. (2007) noted that while climate change will have an impact on Australian wool production between now and 2030, it would not represent a major shock to the industry. Ludwig et al. (2009) found that climate change is unlikely to affect wheat yields in Western Australia under moderate climate change, because both current and projected rainfall reductions take place in June and July when rainfall often exceeds crop demand. Clearly, there are a number of pivotal agricultural industries unlikely to experience reduced production levels in the near future, or at least not on a uniform basis.

The mining industry is a second important industry for Australia's country towns and, while it is not normally considered vulnerable to climate change, it remains at risk. Cleugh et al. (2011) suggest the mining industry is likely to be confronted by hotter, drier and more challenging weather (including storms and flooding). The most likely impact will be increased operational and maintenance costs. The CSIRO (2012) also notes that while every stage of mining is potentially influenced by climate change, the direct production stage is most at risk. The 2010–11 floods in Queensland highlighted this vulnerability, with many mines flooded and some inundated for months. Other research has suggested that all parts of the supply chain are at risk with ports, rail lines, bridges and roads subject to the vagaries of extreme weather conditions (Moser and Boykoff 2013). Some risks may be dealt with by simply strengthening existing infrastructure or building new infrastructure to a higher standard, while others may be an inescapable part of a climate-change-affected world.

Tourism in country towns is potentially very vulnerable to the impacts of climate change as much non-metropolitan tourism relies heavily on both quality of life and natural amenity, which may be at risk as a consequence of climate change. Moreover, country towns may lack the financial and other resources to deal effectively with climate change via infrastructure investment or other adaptation strategies.

Climate change clearly represents a substantial challenge for the industry base of many country towns, but this challenge should be thought of as one of adaptation rather than survival. Even agriculture, which at face value would appear to be an exposed sector, is confronted by both opportunities and risks. These risks are not evenly distributed and some regions face potentially greater challenges than others, with those forecast to see the greatest declines in rainfall – such as the southwest of Western Australia – confronted by greater hurdles. That said, the available evidence suggests that most established industries will continue via the application of new technologies and management strategies. New agricultural and other industries will also emerge and some will be linked to climate change adaptation, while others will simply reflect shifting opportunities.

37.3   Adaptation and Australia's country towns

Climate change will result in a range of impacts for Australia's country towns including: shifts in agricultural productivity; more frequent extreme weather events; and changing local environments and the diminution of resources, including major river systems. These ecologically driven changes will interact with long-term demographic, economic and social shifts to produce complex outcomes. Many rural and regional settlements and communities will be especially vulnerable to climate change because of an already marginal position within the settlement system, their dependence on agriculture, their economic reliance on natural resources, the impact of extreme weather events and the need for infrastructure investment in excess of the capacity of governments or the community.

Adaptation has been defined by the IPCC and United Nations Development Program respectively as:

> Adjustment in natural or human systems in response to actual or expected climatic stimuli or their effects, which moderates harm or exploits beneficial opportunities. (IPCC 2001)

> A process by which strategies to moderate, cope with and take advantage of the consequences of climatic events are enhanced, developed, and implemented. (UNDP 2005)

Moreover, Fünfgeld and McEvoy (2011) note:

> … climate change adaptation can be considered a process of continuous social and institutional learning, adjustment and transformation. Understanding adaptation as an ongoing process of learning is particularly relevant for local and regional scale decision making. (Fünfgeld and McEvoy 2011, p. 6)

This perspective on adaptation highlights the importance of human factors in responding to environmental challenges. Places and individuals better able to adapt and adjust will have enhanced prospects compared to those trapped in past production systems or ways of thinking. Critically, factors such as educational attainment, access to information and vibrant social capital will contribute to the success of some communities and towns.

McEvoy et al. (2010) have recently argued that successful adaptation is affected by local institutional capacity, the 'inconsistency of regulations' and local economic difficulties, which may adversely affect implementation (see also Moser and Boykoff 2013). Additionally:

> Achieving a better understanding of future risk, and therefore how best to adapt, requires an integrated assessment of both climate and non-climate scenarios … adaptation will be further influenced by a complex array of interacting factors including the perception of potential risk by decision-makers, political and institutional context, issues such as cultural heritage, the availability of financial resources. (McEvoy et al. 2010, p. 792)

There is a substantial body of work concerning the physical infrastructure impacts of climate change and this research considers the impact of the predicted increased frequency and intensity of extreme weather events including heat waves, storms (and storm surges), floods and, in southern Australia, bushfires and prolonged heat waves. This literature emphasises the need to recognise weather-related impacts on new and existing transport infrastructure (roads, rail, ports, bridges), housing, commercial and industrial buildings and energy infrastructure. Some infrastructure will be affected by path dependency, with suppliers and consumers alike potentially unable to move away from current models of provision to more effective and climate appropriate technologies.

In seeking to understand how country towns are likely to change over time, analyses need to incorporate both climate-induced and socially driven change. Rosenzweig and Wilbanks (2010) summarised this:

> For human and human-managed systems, projecting longer-term climate change impacts and costs is complicated by the fact that systems will be changing for other reasons as well (demographic, economic, technological, institutional), and climate change impacts will depend on interaction with these other changes. (Rosenzweig and Wilbanks 2010, p. 105)

The adaptation and future of Australia's country towns is not simply a story of a changing climate and its impacts. Instead, a range of demographic, social and economic changes will continue to reshape Australia's urban settlement system, with country towns likely to experience some of the sharpest change. It is important to acknowledge that many country towns are unlikely to survive in the medium- to long-term, even in the absence of climate change. Those that continue will have new functions, new methods of operating as a community and as an economic system and be making use of infrastructure that differs from that currently on the ground.

37.4   Modelling the vulnerability of rural settlements to climate change

The previous sections in this chapter have shown that Australia's country towns are potentially vulnerable to the impacts of climate change and that

the level of vulnerability varies considerably. A key question is therefore: which settlements are most vulnerable to the impacts of climate change and what are the implications for individual communities and public policy? This section sets out the processes used to construct a vulnerability index for Urban Centres and Localities (UCLs) in rural and regional Australia. This index was developed for all non-metropolitan UCLs in Australia more than 50 km from the coast. UCLs are a standard unit of data collection and reporting for the Australian Bureau of Statistics (ABS). Importantly, UCLs are the best approximation currently available for urban places. They are contiguous urban areas that have been identified by the ABS since the 1960s and they provide insights into how settlements change both independently of, and in conjunction with, their hinterlands.

Vulnerability indices have been used extensively in analysing the adverse impacts of climate change and emanate from an array of academic disciplines (Adger et al. 2004; Brooks et al. 2005; Nelson et al. 2005; Pearson and Langridge 2008). Indicator-based assessments are a widely used tool in climate change analyses, although they have been criticised for their apparent blindness to complex relationships, their inability to capture the circumstances of individual cases or observations and their failure to shed light on the underlying processes that generate change or risk (Armitage and Plummer 2010). Preston et al. (2011) note there are a number of different frameworks for assessing vulnerability including the use of risk-hazard models, social vulnerability models, pressure and release models and expanded vulnerability models. Nelson et al. (2005) however highlight that while there is a diverse array of frameworks, there is a convergence in the generic attributes of vulnerability applied to climate change. It is the convergence of exposure, sensitivity and adaptive capacity that guides the model of vulnerability developed in this study. The Allen Group (Allen Consulting 2005) argued that an assessment of vulnerability requires identifying and understanding both exposure and sensitivity to climate change in order to assess potential impacts, and these are then evaluated

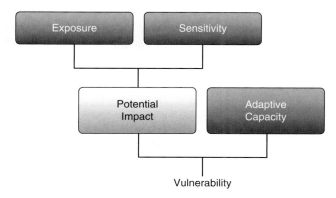

Figure 37.1 Assessing vulnerability to climate-change-related impacts. Source: Allen Consulting 2005
© Commonwealth of Australia 2013/CC-BY-3.0-AU.

against local adaptive capacity. These relationships are summarised in Figure 37.1.

For the purposes of this research: data for each UCL were collected for the exposure variable using forecast change in total rainfall and change in mean surface temperature in the moderate change scenario; sensitivity was assessed using the percentage of the workforce employed in agriculture and a remoteness variable; and adaptive capacity was constructed as a product of population, percentage of households with an internet connection, level of educational attainment and highest year of schooling. The latter two variables were calculated using data from the Australian Bureau of Statistics. Data for climate change scenarios were extracted from the OzClim Climate Change Scenario Generator, for which there are a number of global climate models available for projections of future climates for Australia. The degree to which there will be change in temperature and precipitation varies substantially by scenario or model, particularly when projected over long time frames. This variability guided the selection of the global climate model and we selected the CSIRO Mk3.5 model, which has an m-skill score (a measure of climate model performance; CSIRO 2007) that shows that it is a suitable and reliable predictor of the Australian climate (CSIRO 2007). The CSIRO Mk3.5 model is one of moderate – as opposed to substantial or

muted – climate change and this middle road was selected in order to present a more likely set of outcomes. The CSIRO Mk 3.5 model predicts annual average precipitation decreases across all of Australia, except for the east coast. The model also predicts that the annual pattern of temperature change per degree of global warming (PDGW) will increase across all of Australia.

It is important to acknowledge that the vulnerability assessment developed through this model reflects the processes of mainstream Australian society. It does not adequately capture the potential adaptability of Indigenous communities, but instead assumes that their pathway to adaptation will follow that of the formal economy.

The results of this model are presented in Figure 37.2 and Tables 37.1 and 37.2, and they clearly show that remote inland settlements are most at risk in a climate-change-affected Australia. Indeed, Table 37.1 shows that many of the most at-risk communities are Indigenous communities in remote locations. Nine of the twenty most at-risk settlements are to be found in the Northern Territory (NT), a jurisdiction with relatively limited resources for climate change adaptation. No Tasmanian or Victorian community appears in the list of the 20 most vulnerable communities. Additionally, it is clear that many parts of the established cropping lands in the southeast of Australia appear to face a

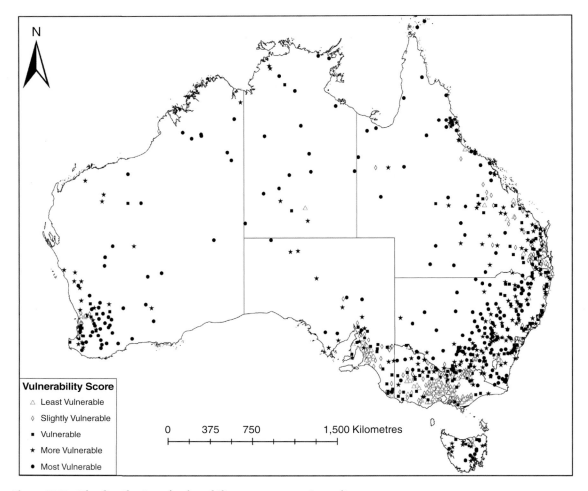

Figure 37.2   The distribution of vulnerability scores across Australia.

relatively muted risk, while settlements in Western Australia's agricultural lands appear to face a greater threat than those in South Australia or Victoria. The level of vulnerability appears high throughout NSW also, and this may be partly a function of the distance of many of these centres from Sydney or one of the other capitals. High levels of dependence on potentially marginal farming lands may also have affected the results. Table 37.2 shows that the least vulnerable places are located close to the capitals (Crafers-Bridgewater and Summertown in SA) or larger

settlements with diverse economies, such as Bendigo. No WA or NT settlement appears on the list of the 20 least vulnerable communities, reflecting the apparently greater risk in these two jurisdictions when compared with southern and eastern Australia. Critically, geography and location appear to matter when it comes to both the risks for country towns associated with climate change and their capacity to adapt.

Importantly, at both the national and state levels, country towns already at risk because of economic restructuring, demographic change or

Table 37.1 Most vulnerable non-coastal UCLs.

State	Location	Vulnerability scores	Rank
WA	Marble Bar	0.763	1
NT	Ampilatwatja	0.755	2
NT	Willowra	0.750	3
NT	Ali Curung	0.748	4
NT	Kaltukatjara	0.748	5
WA	Warburton	0.738	6
Qld	Boulia	0.737	7
NSW	Tottenham	0.734	8
NSW	White Cliffs	0.731	9
WA	Jigalong	0.729	10
Qld	Camooweal	0.726	11
NT	Alpurrurulam	0.726	12
NT	Kintore	0.722	13
NT	Nyirripi	0.721	14
NT	Titjikala	0.720	15
SA	Ernabella	0.719	16
WA	Balgo	0.714	17
Qld	Quilpie	0.711	18
NSW	Goodooga	0.708	19
NT	Yuendumu	0.707	20

Table 37.2 Least vulnerable non-coastal UCLs.

State	Location	Vulnerability scores	Rank
SA	Crafers-Bridgewater	0.229	1
SA	Summertown	0.254	2
SA	Oakbank	0.300	3
Vic	Beaconsfield Upper	0.300	4
Vic	Ballarat	0.304	5
SA	Hahndorf	0.305	6
Vic	Menzies Creek	0.305	7
Vic	St Andrews	0.307	8
SA	Mount Barker	0.310	9
SA	Uraidla	0.312	10
Qld	Mount Nebo	0.313	11
SA	Echunga	0.314	12
Vic	Emerald	0.314	13
Qld	Mount Glorious	0.317	14
Vic	Mount Macedon	0.318	15
SA	Balhannah	0.324	16
NSW	Gundaroo	0.325	17
Vic	Macedon	0.329	18
Vic	Greendale	0.330	19
Vic	Warburton	0.331	20

other processes appear to be most vulnerable to climate change. Climate change may therefore compound the difficulties facing such places and serve to tip some settlements onto a downward path.

It is important that the results of this index are read as a starting point for thinking about the adaptability of Australia's country towns. They do not constitute a definitive statement. This index inevitably reflects the values and metrics embedded in the aggregate measure and may provide a false sense of the true picture in any one place, or group of places, because of the failure to incorporate locality-specific factors. The index is also predicated on a European-centric understanding of adaptability, and most likely underestimates the capacity of Indigenous communities to transform themselves in the face of climate change. That said, the results also suggest that relying upon mainstream approaches to facilitate the adaptation of Indigenous communities is unlikely to be successful. The index was

constructed as an indicator of long-term social and economic sustainability or resilience, and does not include data on a range of factors we might consider relevant to debates around Australia's pattern of settlement and climate change. There is not, for example, a measure associated with bushfire risk or flood because Australians, to date, have largely rebuilt in the aftermath of such events. However, managing bushfire risk is one of the key adaptations some towns have embarked upon over the past decade.

There is a human dimension to this analysis that cannot and should not be ignored. Over the course of the conduct of this research, many communities ranked as among the most vulnerable to climate change specifically, and with some force, rejected the findings. Many highlighted their strong local bonding social capital as a key indicator of their resilience but also noted sustainability crises of the recent past, including severe threats to water supply and declining economic circumstances.

37.5 Conclusion

This chapter has considered the impact of climate change on Australia's country towns. It has argued on the basis of other published work that: country towns are likely to be exposed to variable levels of risk; climate change will have complex impacts on all rural and regional industries, with some faring relatively well; and infrastructure and community attitudes/skills will be important in the process of adaptation. The chapter presented an index of vulnerability for Australia's country towns which showed that – contrary to expectations – country towns in Australia's rural heartlands had a modest exposure to the negative impacts of climate change while remote and very remote settlements faced considerable risks. These communities were also assessed as having little capacity to bring about positive change if we assume they are to rely upon mainstream processes to adapt to a changing climate. This finding is concerning as these settlements are already some of the most vulnerable places in the nation. There is no doubt that forward planning and action by governments, industries and communities alike is needed to adapt Australia's country towns to their future climate and would appear to carry considerable benefits relative to the potential costs. Some of the actions governments and communities should consider include: measures to diversify the economic base of country towns; the assessment and, where necessary, reconfiguration of infrastructure; measures to enhance human capital; and steps to further both bridging and bonding social capital at the local level.

References

Adger, W., Brooks, N., Kelly, M., Bentham, G. and Eriksen, S. (2004) *New Indicators of Vulnerability and Adaptive Capacity, Technical Report 7*. Tyndall Centre for Climate Change Resources.

Allen Consulting (2005) *Climate Change Risk and Vulnerability: Promoting an Efficient Adaptation Response in Australia, Report to AGO*. Department of Environment and Heritage, Australia.

Armitage, D. and Plummer, R. (2010) *Adaptive Capacity and Environmental Governance*. Springer-Verlag, Berlin.

Baum, S., O'Connor, K., Mullins, P. and Davis, R. (1999) *Community Opportunity and Vulnerability in Australia's Cities and Towns: Characteristics, Patterns and Implications*. UQ Press, St Lucia.

Beer, A., Tually, S., Kroehn, M. et al. (2013) *Australia's Country Towns 2050: What Will A Climate Adapted Settlement Pattern Look Like?* National Climate Change Adaptation Research Facility, Gold Coast.

Blackmore, K. and Goodwin, I. (2009) *Analysis of the Past Trends and Future Projections of Climate Change and their Impacts on the Hunter Wine Industry*. HCCREMS and the University of Newcastle, New South Wales.

Brooks, N., Adger, W. and Kelly, P. (2005) The determinants of vulnerability and adaptive capacity at the national level and the implications for adaptation. *Global Environmental Change* 15, 151–163.

Cleugh, H., Stafford-Smith, M., Battaglia, M. and Graham, P. (2011) *Climate Change, Science and Solutions for Australia*. CSIRO Publishing, CSIRO, Collingwood, Victoria.

Cocklin, C. and Dibden, J. (2005) *Sustainability and Change in Rural Australia*. UNSW Press, Kensington.

CSIRO (2007) *Climate Change in Australia*. CSIRO and Australian Bureau of Meteorology, Melbourne.

CSIRO (2012) *Primary Industries, Enterprises and Communities Adapting to Climate Change*. SIRO, Victoria. Available at http://www.csiro.au/en/Organisation-Structure/Flagships/Climate-Adaptation-Flagship/Adaptive-Primary-Industries.aspx (accessed 4 June 2014).

Fünfgeld, H. and McEvoy, D. (2011) Framing Climate Change Adaptation in Policy and Practice, VCCCAR Project: Framing Adaptation in the Victorian Context, Working Paper 1, RMIT University, Melbourne.

Harle, K., Howden, S., Hunt, L. and Dunlop, M. (2007) The potential impact of climate change on the Australian wool industry by 2030. *Agricultural Systems* 93, 61–89.

IPCC (2001) *Climate Change 2001: Synthesis Report: A Contribution of Working Groups I, II, and III to the Third Assessment Report of the Intergovernmental Panel on Climate Change*. Cambridge University Press, Cambridge.

Ludwig, F., Milroy, S. and Assenya, S. (2009) Impacts of recent climate change on wheat production systems in Western Australia. *Climatic Change* 92, 495–517.

McEvoy, D., Matczak, P., Banaszak, H. and Chorynski, A. (2010) Framing adaptation to climate-related extreme events. *Mitigation and Adaptation Strategies for Global Change* 15, 779–795.

Moser, S. and Boykoff, M. (2013) *Successful Adaptation to Climate Change*. Routledge, London.

Nelson, R., Kokic, P., Elliston, L. and King J. (2005) Structural adjustment: a vulnerability index for Australian broadacre agriculture. *Australian Commodities* 12(1), 171–179.

Pearson, L. and Langridge, J. (2008) Climate change vulnerability assessment: Review of agricultural productivity, CSIRO Climate Adaptation Flagship Working Paper No.1. CSIRO, Melbourne.

Preston, B. and Jones, R. (2006) *Climate Change Impacts on Australia and the Benefits of Early Actions to Reduce Greenhouse Gas Emissions*, Australian Business Roundtable on Climate Change. CSIRO, Victoria.

Preston, B., Yeun, E. and Westaway, R. (2011) Putting vulnerability to climate change on the map: a review of approaches, benefits and risks. *Sustainability Science* 6, 177–202.

Rosenzweig, C. and Wilbanks, T.J. (2010) The state of climate change vulnerability, impacts and adaptation research. *Climatic Change* 100, 103–106.

Steffen, W., Sims, J., Walcott, J. and Laughlin, G. (2011) Australian agriculture: Coping with dangerous climate change. *Regional Environmental Change* 11, 205–214.

Tonts, M., Argent, N. and Plummer, P. (2012) Evolutionary perspectives on rural Australia. *Geographical Research* 50(3), 291–303.

UNDP (2005) *Adaptation Policy Frameworks for Climate Change: Developing Strategies, Policies and Measures*. Cambridge University Press, Cambridge.

Ward, G. (2009) *Preparing for a Changing and Variable Climate, Final Report to the GWRDC*. Department of Agriculture and Food, Western Australia.

Webb, L., Whetton, P. and Barlow, S. (2005) Impact on Australian viticulture from greenhouse induced temperature change. In: Zerger, A. and Argent, R.M. (eds) *MODSIM 2005 International Congress on Modelling and Simulation*. Modelling and Simulation Society of Australia and New Zealand, December 2005, pp. 1504–1510.

Webb, L., Whetton, P. and Barlow, S. (2006) Potential impacts of projected greenhouse gas induced climate change on Australian viticulture. *Wine Industry Journal* 21, 16–20.

38   Robust optimisation of urban drought security for an uncertain climate

MOHAMMAD MORTAZAVI-NAEINI[1],
GEORGE KUCZERA[1], ANTHONY S. KIEM[2], LIJIE CUI[1],
BENJAMIN HENLEY[1,4], BRENDAN BERGHOUT[3]
AND EMMA TURNER[3]

[1]School of Engineering, University of Newcastle, Australia
[2]School of Environmental and Life Sciences, University of Newcastle, Australia
[3]Hunter Water Corporation, Australia
[4]School of Earth Sciences, University of Melbourne, Australia

38.1   Introduction

The majority of the world's population, including most Australians, live in large urban centres that cannot function without adequate water supply. Failure of bulk water systems to supply minimum water requirements for an extended period would most likely result in disastrous social and economic losses. Recent experience in Australia with drought and a shifting climate has highlighted the vulnerability of urban water supplies to 'running out of water' and has triggered major investment in water source infrastructure that ultimately will run into tens of billions of dollars. With the prospect of rapid population growth in major cities the provision of acceptable drought security will become more pressing, particularly in the face of considerable uncertainty about future climate change (e.g. Dessai et al. 2009).

Scenarios describe plausible future states of the system (Mahmoud et al. 2009). They may be model parameters or forcing time series that are exogenous to the system, meaning they affect system performance but cannot be changed by the system. In the case of urban water supply, decisions need to be made in the face of significant uncertainty about future scenarios. Much of this uncertainty (such as natural climate variability) can be described using probability distributions that have been inferred from past data on system behaviour. In some cases however, while the future scenarios may be identifiable there is insufficient data and prior knowledge to meaningfully assign probabilities to such scenarios. In such cases, there is almost complete reliance on the assignment of subjective probabilities. Such uncertainty is referred to as 'deep uncertainty' to emphasise that traditional probability approaches are not likely to be meaningful.

One such deep uncertainty is future climate change. Dessai et al. (2009) argue that the accuracy of model predictions of future climate is limited by 'fundamental irreducible uncertainties' such as knowledge limitations (e.g. cloud physics and sub-grid variability), the inherent chaotic nature of climate and uncertainty about human behaviour to reduce greenhouse gas emissions. They argued that such uncertainties may be extremely difficult to quantify and recommend against over-reliance on projections made by

Applied Studies in Climate Adaptation, First Edition. Edited by Jean P. Palutikof, Sarah L. Boulter, Jon Barnett and David Rissik.
© 2015 John Wiley & Sons, Ltd. Published 2015 by John Wiley & Sons, Ltd.

climate models and favour solutions that are robust across a range of possible futures.

This chapter considers the problem of finding good and robust solutions for urban bulk water systems in the presence of deep uncertainty about future climate change. It briefly describes the concept of robust multi-objective optimisation and then illustrates its application in a case study. The reader is referred to Mortazavi-Naeini et al. (2013) for a fuller treatment of this topic.

38.2 Robust multi-objective optimisation

Urban water agencies are responsible for planning and operating bulk water systems in a way that minimises economic, social and environmental costs while providing an acceptable level of drought security. A two-pronged strategy is used to manage drought security.

1. *Risk mitigation* seeks to develop medium- to long-term strategies that affect water use and long-lead-time infrastructure associated with surface and subsurface water storage, harvesting and recycling to manage the probability of initiating the drought contingency response.

2. *Drought contingency* represents the immediate response to severe drought to avoid running out of water. As storage declines during a drought, trigger events initiate short-term responses such as restrictions/rationing and short-lead-time (and usually very expensive) source augmentation.

The number of feasible solutions to efficiently manage drought security may be very large. Trial-and-error search, even by experienced engineers and planners, cannot adequately explore this space. Multi-objective optimisation offers a quantum improvement over trial-and-error search and more closely reflects actual decision-making. Here we are looking for solutions that maximise drought security, minimise operating and investment costs and minimise social impacts. There is no one best solution. Multi-objective optimisation identifies the Pareto set of solutions, which expresses the optimal trade-offs between competing objectives (Deb 2001). This Pareto set provides the best set of solutions for

negotiation, something trial-and-error search cannot guarantee.

This in itself is not sufficient, however. Risk-averse decision-makers will shun solutions that are optimal in an average sense but produce unacceptable outcomes if certain scenarios actually occur. Indeed, Mahmoud et al. (2009) observe that 'scenario results are of limited value if the involved uncertainty is not properly considered'. Decision-makers are interested in solutions that are not only efficient in an average sense but also robust in the presence of significant uncertainty about future scenarios.

A natural way to characterise robustness is to minimise the spread or difference in performance between the worst and best scenario. The core idea is to introduce the minimisation of spread into the multi-objective optimisation problem. This is illustrated by an example involving minimisation of expected cost. Figure 38.1 shows how the introduction of a second objective, namely the minimisation of cost spread between scenarios, leads to an optimal trade-off set of solutions. In this example, solution A is the most efficient (smallest expected present worth cost) but least robust (greatest cost spread), while solution B is the most robust but least efficient. Decision-makers will be interested in exploring the trade-offs between robustness and efficiency. For example, solution C may interest decision-makers as it is slightly more costly than solution A but substantially less sensitive to differences between the worst and best scenarios.

38.3 Case study: problem formulation

We illustrate the robust multi-objective optimisation approach using a case study based on the Lower Hunter bulk water system. This system provides drinking water to over 200,000 properties in the Newcastle region of New South Wales, Australia. Water is harvested from a combination of surface storages and groundwater resources as shown in Figure 38.2. Major components of the bulk water system include Chichester Dam, Seaham Weir on the Williams River from which water is pumped to the off-river storage

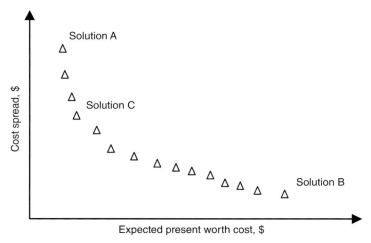

Figure 38.1   Trade-off between expected cost and spread of costs across scenarios.

Grahamstown Dam, and the Tomago and Tomaree groundwater sources.

We formulate the robust optimisation problem using the approach described by Cui and Kuczera (2010) and Mortazavi et al. (2012) that combines an adaptation of the Watkins and McKinney (1997) formulation with an evolutionary multi-objective algorithm. This is implemented in a software tool called WATHNET5 (Kuczera et al. 2009) that uses network flow programming and scripts to simulate and optimise complex water resource systems.

38.4   Construction of uncertain future climate change scenarios

The construction of future climate scenarios has to incorporate the very considerable influence of natural climate variability, the identification problems associated with short hydro-climatological instrumental records and the concern about how human-induced climate change may increase the frequency and severity of future extreme events, including droughts and floods. There have been attempts (see review in Kiem and Verdon-Kidd 2011) to utilise climate model outputs to determine how anthropogenic climate change may affect water resources and, on the basis of

this information, to develop water resource management strategies to deal with the projected risks. However, the uncertainty associated with future climate projections is known to be significant (e.g. Kiem and Verdon-Kidd 2011) and is magnified further when attempting to make inferences at the regional scale. This is especially the case for precipitation (e.g. Lim and Roderick 2009) and hydro-climatic extremes (see IPCC 2012). The uncertainty is so high that projections of future drought risk, over either the short (seasonal up to 5 years) or long (more than 10 years into the future) term, currently have limited practical usefulness for water resource managers and/or government policymakers (Bates et al. 2011).

However, methods do exist to quantify risks associated with climate variability and/or change that do not rely solely on climate model outputs (e.g. McMahon et al. 2008; Verdon-Kidd and Kiem 2010). These methods involve utilising stochastic modelling approaches that replicate the important statistics associated with historical data (e.g. interannual to multidecadal natural climate cycles, dry spells, wet epochs, extreme high and low rainfall, etc.) and then utilising the best available climate modelling information to provide a range of plausible future scenarios.

This study focuses on the future climate around 2070 (i.e. the average climate during

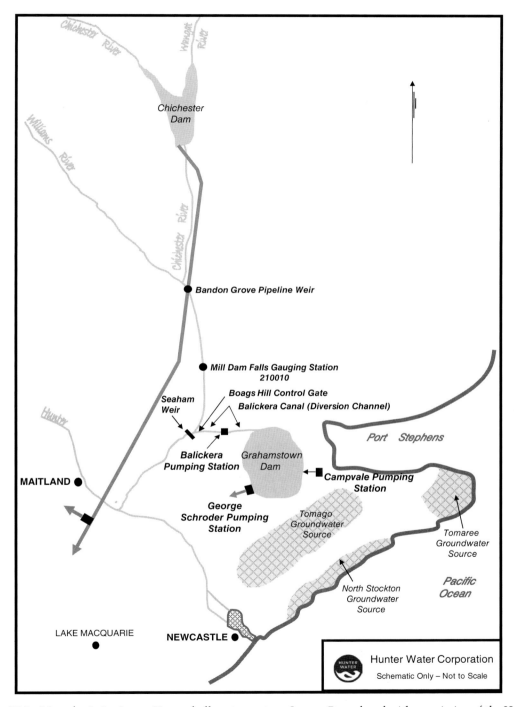

Figure 38.2 Map of existing Lower Hunter bulk water system. Source: Reproduced with permission of the Hunter Water Corporation. For colour details please see Plate 25.

2060–2079). To simulate the performance of the Lower Hunter system, streamflow replicates representative of the 2070 climate were developed using the following procedure.

1. Two climate scenarios were selected to represent the range in likely 2070 climate conditions: these scenarios will be referred to as 'dry' 2070 and 'wet' 2070 scenarios.

2. For each future scenario, observed historic rainfall and potential evapotranspiration (PET) time series were perturbed using the selected 2070 climate change factors.

3. A stochastic multi-site model of rainfall and PET was calibrated to the perturbed historic data.

4. The stochastic model generated 10,000 50-year replicates of daily rainfall and PET representative of the perturbed climate.

5. The stochastic rainfall and PET time series were input to calibrated rainfall–runoff models to produce 10,000 50-year replicates of monthly streamflow at multiple sites.

The climate change factors used in step (2) were obtained from CSIRO-BoM (2007), which is based on the Coupled Model Intercomparison Project 3 (CMIP3) model results as used in IPCC (2007), reported as a range in Table 38.1. This range was derived as follows.

• For the A1F1 emission scenario, the projected range for global warming based on all global climate models (GCMs), run several times, is 1.74–4.64°C in 2070.

• Using the 50th percentile estimate of local change in rainfall and PET per degree of global warming, a range of climate change factors for the Lower Hunter was obtained. The upper bound for rainfall and lower bound for PET of this range was used to develop the 'wet' 2070 scenario and the lower bound for rainfall and upper bound for PET used to produce the 'dry' 2070 scenario.

The 50th percentile was used in recognition of the numerous studies (e.g. Kiem and Verdon-Kidd 2011) and other more recent work produced as part of the South East Australian Climate Initiative (SEACI, http://www.seaci.org/) that have demonstrated that of the 23 GCMs used to develop future climate projections by Climate Change in Australia (CSIRO and BoM 2007), fewer than 8 can be considered close to realistic for Australian hydro-climatology and that current best practice is to use the projections where there is most agreement across the models rather than the extremes.

38.5   Decisions

Decisions or options represent the 'levers' that control the performance of the bulk water system. Table 38.2 summarises the decisions that can be changed by the optimiser along with permissible ranges. A particular set of decisions represents a solution or portfolio of options. The decisions are grouped into three categories:

1. *Infrastructure*, representing new assets such as surface water reservoirs, desalination plants and rainwater tanks;

2. *Operational*, representing variables that control system operation such as reservoir balancing and pumping triggers; and

3. *Drought contingency*, representing variables that determine the response to severe drought to avoid running out of water.

Figure 38.3 provides more detail on the drought contingency plan used in this case

Table 38.1 Climate change factors (% change to 1990 baseline mean) (CSIRO-BoM, 2007) using the 2070 'best estimate' (50th percentile) high emissions scenario.

Variable	Scenario	Winter	Spring	Summer	Autumn
Rainfall	2070 wet	−10	−10	+2	−2
	2070 dry	−20	−20	−2	−5
PET	2070 wet	+12	+4	+8	+12
	2070 dry	+16	+8	+12	+16

Table 38.2 List of decision variables.

Decision	Description	Lower limit	Upper limit	Category
1	Level 1 restriction storage trigger	0.00	0.90	Drought
2	Restriction trigger increment	0.05	0.25	contingency
3	Emergency restriction trigger	0.0	1.0	
4	Severe rationing demand scaling factor	0.4	0.7	
5	Storage trigger to initiate desalination plant construction	0.0	1.0	
6	Desalination plant capacity (ML/day)	0	500	
7	Desalination plant construction lead time (months)	0, 12, 24, 36 or 60		
8	Desalination plant trigger	0.0	1.0	Operational
9	New surface water storage capacity (ML)	Either 0 or in range 100,000–230,000		Infrastructure
10	Chichester to Newcastle pipeline capacity expansion (ML/day)	0	50	Infrastructure
11	Grahamstown base gain	8000	12000	Operational
12	Grahamstown incremental gain	20	200	Operational
13	Grahamstown pumping trigger	0.0	1.0	Operational
14	Number of domestic rainwater tanks	0	100,000	Infrastructure

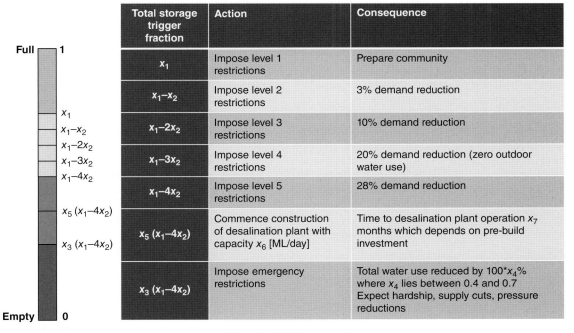

	Total storage trigger fraction	Action	Consequence
Full — 1	x_1	Impose level 1 restrictions	Prepare community
x_1	$x_1 - x_2$	Impose level 2 restrictions	3% demand reduction
$x_1 - x_2$	$x_1 - 2x_2$	Impose level 3 restrictions	10% demand reduction
$x_1 - 2x_2$	$x_1 - 3x_2$	Impose level 4 restrictions	20% demand reduction (zero outdoor water use)
$x_1 - 3x_2$	$x_1 - 4x_2$	Impose level 5 restrictions	28% demand reduction
$x_1 - 4x_2$	$x_5 (x_1 - 4x_2)$	Commence construction of desalination plant with capacity x_6 [ML/day]	Time to desalination plant operation x_7 months which depends on pre-build investment
$x_5 (x_1 - 4x_2)$ $x_3 (x_1 - 4x_2)$ Empty — 0	$x_3 (x_1 - 4x_2)$	Impose emergency restrictions	Total water use reduced by 100*x_4% where x_4 lies between 0.4 and 0.7 Expect hardship, supply cuts, pressure reductions

Figure 38.3 Drought contingency plan. For colour details please see Plate 26.

study. As total storage drops, the first responses (yellow zone) impose 'normal' restrictions with the principal focus on outdoor water use. As storage continues to decline, construction of a desalination plant (orange zone) and imposition of emergency restrictions (red zone) are triggered.

It is important to note that the case study does not consider the full suite of possible decisions available to manage the Lower Hunter bulk water system; information on plausible decision variables has been generalised for demonstration purposes.

38.6   Objectives and constraints

The optimiser searches through the feasible decision space. It uses objectives to guide its search towards 'good' or so-called optimal solutions. In this case study, two objectives were optimised.
1. Minimise the total expected present worth (EPW) cost consisting of capital, operating, normal and emergency restriction costs.
2. Minimise the cost spread or the difference between the total EPW cost for the dry and wet climate scenarios.

The quantification of the social costs of restrictions in monetary terms is subject to considerable uncertainty. However, by monetising all costs, the visualisation of the trade-off between efficiency and robustness is simple to communicate. In reality water planning is likely to involve more than two objectives reflecting social, environmental, financial, equity and risk values. The multi-objective optimisation methodology is quite general and can be applied to more than two objectives. Beyond three objectives however, it becomes difficult to visualise and understand the trade-offs.

The constraint was imposed that, in the 10,000 50-year replicates of dry and wet climate, the system is not permitted to run out of water. Without this constraint the least-cost solution would be to supply no water.

38.7   Case study: results and discussion

Results are presented for two demand scenarios, the 1.28 × current demand scenario which represents projected demand in 2060, and the 2 × current demand scenario which represents a highly stressed system.

38.7.1   Scenario 1: 2070 uncertain climate, 2060 demand

Figure 38.4 presents two sets of approximate Pareto-optimal solutions for the 2060-demand scenario. The first set is produced when all the decisions in Table 38.2, including the surface water storage options, are active. In this case, Figure 38.4 shows that there is a limited trade-off between total EPW cost and robustness. The total EPW cost ranges from AU$ 360 million to 485 million, while the cost spread across the dry and wet 2070 climate scenarios ranges from AU$ 80,000 to 13.4 million. The proposition of increasing robustness by reducing the spread from AU$ 13.4 million to 80,000 for an increased community investment of AU$ 125 million is not compelling. A closer inspection of the Pareto-optimal solutions reveals that the chance of restrictions is always less than 1.7% regardless of climate scenario and the chance of emergency restrictions is at worst 0.08%. There is little difference between the solutions other than an increase in the new surface water source capacity; all solutions excluded rainwater tanks and desalination from their portfolios. Overall, there is little sensitivity to future climate change uncertainty.

However, the situation changes radically if the new surface water option is removed from the decision space; this corresponds to setting decision 9 to 0. The total EPW cost increases significantly and falls within the range AU$ 812–3460 million, while the cost spread across the dry and wet 2070 climate scenarios falls within the range AU$ 6–102 million. Without the option of a surface water reservoir, desalination is adopted in all solutions. The least robust solution opts for a 50 ML/day desalination plant operating

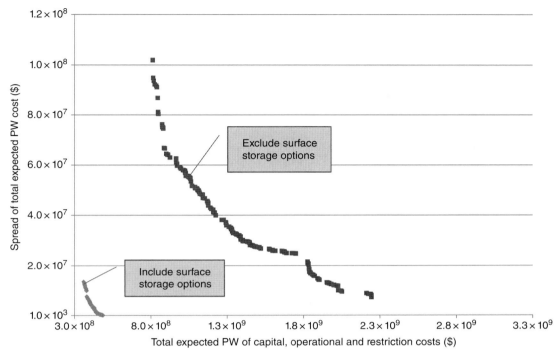

Figure 38.4   Pareto optimal fronts for the 2060 demand scenario. For colour details please see Plate 27.

whenever storage drops below 43%, while the most robust solution opts for an almost maximum rollout of rainwater tanks together with a much larger 350 ML/day desalination plant which is operated whenever storage drops below 5%.

38.7.2   Scenario 2: 2070 uncertain climate, 2 × current demand

Figure 38.5 presents the approximate Pareto-optimal solutions for the 2 × current demand scenario with all the decisions in Table 38.2 activated. Table 38.3 presents a summary of the six solutions labelled on Figure 38.5. Unlike the 2060 demand scenario, there is a strong trade-off between efficiency and robustness. The total EPW cost falls within the range AU$ 1092–2943 million, while the spread falls within the range AU$ 1–455 million. For solution 1 the spread represents 42% of the total EPW cost. Unfortunately,

reducing the spread requires a disproportionately large expected investment. The reduction of spread by AU$ 53 million when moving from solution 1 to 2 is offset by an increase in total EPW cost of AU$ 73 million, while moving from 4 to 5 reduces the spread a further AU$ 174 million in return for an increase in total expected cost of AU$ 596 million.

Unlike the Pareto solutions for the 2060 demand scenario that do not favour the desalination option, the Pareto solution set for the 2 × current demand scenario opts for desalination in all cases. The following features of the Pareto solution set have significant implications for managing the system.

1. All solutions opt for the maximum capacity of the new surface water source. This suggests that if it were technically feasible to increase the upper bound on the storage capacity, the optimiser would exploit the additional capacity.

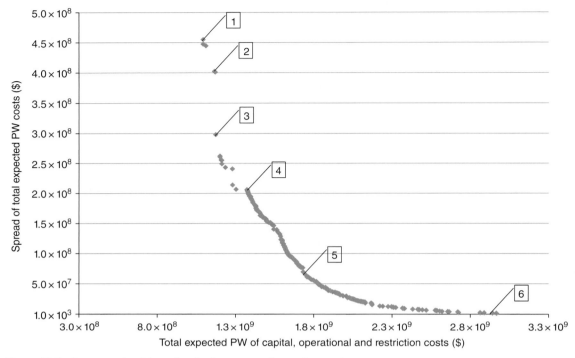

Figure 38.5 Pareto optimal front for the 2 × current demand scenario.

2. All solutions adopt desalination with zero lead time, meaning the plant is commissioned at the start of the simulation and will supply water whenever total storage drops below the trigger. The pressing need for an additional source of water forces the optimiser to embrace desalination as a permanent infrastructure rather than a drought contingency measure. Solutions 1 to 3 (more efficient, less robust) opt for the small and cheap desalination option, while solutions 4 to 6 (less efficient, more robust) opt for the larger and more costly option. Stakeholders would need to trade-off objectives to decide on a preferred strategy.

3. The increasing investment in desalination is accompanied by a decreasing restriction trigger, resulting in a lower frequency of restrictions and emergency restrictions. The increased financial cost of desalination was more than offset by the reduction in the social cost of normal and emergency restrictions.

4. Solution 1 exhibits the greatest difference in restriction frequency between the dry and wet climates. Only solutions 5 and 6 (less efficient, more robust) produce restriction frequencies less than 5% for both the dry and wet climate scenarios, which is a common level of service criterion for Australian water utilities.

38.8   Conclusions

The Lower Hunter case study has demonstrated the capability of robust optimisation to assist water managers to adapt to uncertain future climate change. There was considerable variability in the sensitivity of solutions to uncertainty about future climate change. For the 2060 demand scenario, there was little difference in expected present worth cost between the dry and wet future climate scenarios, provided the new surface water source was included in the options.

Table 38.3 Summary of labelled solutions on Pareto front in Figure 38.5.

	Solution	1	2	3	4	5	6
Objectives	Total expected PW of capital, operational and restriction costs ($m)	1092	1165	1173	1372	1968	2943
	Spread of total expected PW cost ($m)	455	402	298	205	31	1
Frequency	Chance of restrictions in any month (%) for dry and wet scenario	16.5 (dry) 6.3 (wet)	14.9 (dry) 5.6 (wet)	10.4 (dry) 3.4 (wet)	6.5 (dry) 2.0 (wet)	2.4 (dry) 0.82 (wet)	0.31 (dry) 0.13 (wet)
	Chance of emergency restrictions in any month (%) for dry and wet scenario	1.0 (dry) 0.32 (wet)	0.91 (dry) 0.28 (wet)	0.71 (dry) 0.21 (wet)	0.32 (dry) 0.11 (wet)	0.055 (dry) 0.032 (wet)	0.0 (dry) 0.0 (wet)
Decisions	Level 1 restriction storage trigger	0.394	0.394	0.331	0.281	0.253	0.199
	Restriction trigger increment	0.057	0.054	0.051	0.051	0.051	0.051
	Emergency restriction trigger (expressed as % storage)	0.759 (12.6%)	0.751 (13.4%)	0.768 (9.7%)	0.681 (5.2%)	0.511 (2.5%)	0.641 (0%)
	Severe rationing demand scaling factor	0.678	0.677	0.696	0.686	0.697	0.546
	Storage trigger to initiate desalination plant construction	–	–	–	–	–	–
	Desalination plant capacity (ML/day)	17	17	17	50	74	125
	Desalination plant construction lead time(month)	0	0	0	0	0	0
	Desalination plant trigger	0.497	0.493	0.493	0.591	1.000	0.997
	New surface water source capacity (ML)	229918	229953	229934	229929	229918	222930
	Chichester to Newcastle pipeline capacity expansion (ML/day)	8	12	6	2	7	8
	Grahamstown base gain	8623	8101	8590	8719	8117	9623
	Grahamstown incremental gain	32	59	47	56	69	34
	Grahamstown pumping trigger	0.989	0.997	0.998	0.992	0.808	0.735
	Number of domestic rainwater tanks	36	25050	50232	91	245	50223

In contrast, for the 2 × current demand scenario, there was a considerable trade-off between economic efficiency and robustness.

This variability in outcomes highlights the fact that (possibly large) uncertainty about future climate may not necessarily produce significantly different performance trajectories. The key consideration is that of the exogenous stress imposed on the system. This not only represents the system drivers, namely population (and consequent demand) and future climate change, but also the scope of the decision space that determines the possibilities explored by the optimiser. For example, if the new surface water source option were not available to add more storage capacity to the system or if the desalination option were not available to provide a climate-independent source of water, the stress level would be considerably higher, producing very different performance outcomes and solution sets. Indeed, many Pareto optimal solutions did hit the upper limit on the new surface water source storage capacity, suggesting that making more opportunities available for storage could produce better outcomes.

Robust optimisation does not produce a single answer. Its goal is to identify a range of 'good' solutions (from the available solution space) that forms the basis for negotiation between relevant parties towards a preferred solution.

The technology to implement robust optimisation in the context of urban bulk water systems has advanced to the point where decision-makers and stakeholders have considerable latitude in specifying objectives, constraints and decisions. The task of meaningfully formulating the robust optimisation problem is not trivial; the difficulty in codifying the values of decision-makers and stakeholders should not be underestimated.

Finally, it needs to be stressed that the robust optimisation results are conditional on the credibility of the estimated range of potential climate change impacts on rainfall and potential evapotranspiration, and also on the credibility of stochastic modelling of natural climate variability. While the climate change factor approach used in this study was regarded as best practice at the time of writing, alternative methods were available which may have yielded different ranges. Resolving this issue is considered an important goal in the quest for identifying robust adaptation strategies in water resource planning.

References

Bates, B., Bunn, S., Baker, P. et al. (2011) *National Climate Change Adaptation Research Plan for Freshwater Biodiversity*. National Climate Change Adaptation Research Facility, Gold Coast.

CSIRO and Bureau of Meteorology (2007) *Climate Change in Australia: Technical Report*. CSIRO Publishing, Melbourne.

Cui, L. and Kuczera, G. (2010) Coping with climate change uncertainty using robust multi-objective optimization: Application to urban water supply systems. In: Engineers Australia (eds) *Climate Change 2010: Practical Responses to Climate Change*. Engineers Australia, Barton, A.C.T, pp. 283–292.

Deb, K. (2001) *Multi-objective Optimization using Evolutionary Algorithms*. John Wiley and Sons, Inc.

Dessai, S., Hulme, M., Lempert, R. and Piekle, Jr., R. (2009) Do we need better predictions to adapt to a changing climate. *Eos, Transactions of American Geophysical Union* 90(13), 111–112.

IPCC (2007) Summary for policymakers. In: Solomon, S., Qin, D., Manning, M. et al. (eds) *Climate Change 2007: The Physical Science Basis. Contribution of Working Group I to the Fourth Assessment Report of the Intergovernmental Panel on Climate Change*. Cambridge University Press, Cambridge, United Kingdom and New York.

IPCC (2012) *Managing the Risks of Extreme Events and Disasters to Advance Climate Change Adaptation. A Special Report of Working Groups I and II of the Intergovernmental Panel on Climate Change*. Field, C.B., Barros, V., Stocker, T.F. et al. (eds) Cambridge University Press, Cambridge, UK, and New York, NY, USA.

Kiem, A.S. and Verdon-Kidd, D.C. (2011) Steps toward "useful" hydroclimatic scenarios for water resource management in the Murray-Darling Basin. *Water Resource Research* 47, W00G06, doi: 10.1029/2010WR009803.

Kuczera, G., Cui, L., Gilmore, R. and Graddon, A. (2009) Addressing the shortcomings of water resource simulation models based on network linear programming. *Proceedings of the 31st Hydrology* and *Water Resources Symposium*. Engineers Australia, Newcastle.

Lim, W.H. and Roderick, M.L. (2009) *An Atlas of the Global Water Cycle Based on the IPCC AR4 Climate Models*. ANU E Press, Canberra.

Mahmoud, M., Liu, Y., Hartmann, H. et al. (2009) A formal framework for scenario development in support of environmental decision-making. *Environmental Modelling and Software* 24, 798–808.

McMahon, T.A., Kiem, A.S., Peel, M.C. et al. (2008) A new approach to stochastically generating six-monthly rainfall sequences based on Empirical Model Decomposition. *Journal of Hydrometeorology* 9, 1377–1389.

Mortazavi, M., Kuczera, G. and Cui, L. (2012) Multiobjective optimization of urban water resources: Moving toward more practical solutions. *Water Resource Research* 48, W03514, doi: 10.1029/2011WR010866.

Mortazavi-Naeini, M., Kuczera G., Kiem A.S., Henley, B., Berghout, B. and Turner, E. (2013) *Robust Optimisation of Urban Drought Security for an Uncertain Climate*. National Climate Change Adaptation Research Facility, Gold Coast.

Verdon-Kidd, D.C. and Kiem, A.S. (2010) Quantifying drought risk in a non-stationary climate, *Journal of Hydrometeorology* 11(4), 1019–1031.

Watkins Jr, D.W. and McKinney, D.C (1997) Finding robust solutions to water resources problems. *Journal of Water Resources Planning and Management* 123(1), 49–58.

39 How to cope with heat waves in the home

WASIM SAMAN[1], STEPHEN PULLEN[2]
AND JOHN BOLAND[3]

[1] *School of Engineering, University of South Australia, Australia*
[2] *School of Natural and Built Environments, University of South Australia, Australia*
[3] *School of Information Technology and Mathematical Sciences, University of South Australia, Australia*

39.1 Introduction

Climate change is leading to an increased frequency and severity of heat waves in many parts of the world. Spells of several consecutive days of unusually high temperatures have led to increased mortality rates for the more vulnerable in the community as well as increased levels of thermal discomfort. The problem is compounded by escalating energy costs and increasing peak electrical demand as people are becoming more reliant on air conditioning. Domestic air conditioning during heat waves is the primary determinant of peak power demand that has been a major driver of higher electricity costs.

Researchers have only recently started investigating the means to reduce the serious consequences of heat waves in dwellings. Recent examples include the work by Porritt et al. (2011, 2012) who investigated the potential for overheating of a Victorian terraced house in the UK during future heat waves. It determined the effectiveness of measures aimed at lessening heat wave effects and found that overheating could be avoided by passive means. Studying overheating in 3456 dwellings in London, Mavrogianni et al. (2012) found dwelling type, age and insulation were the main determinates of overheating and that all were more important determinates of overheating potential than orientation.

This chapter presents some of the findings of a multidisciplinary research project that set out to develop a framework to evaluate the potential impacts of heat waves in Australia. In particular it presents a technical, social and economic approach to adapt Australian residential buildings to ameliorate the impact of heat waves in the community and reduce the risk of its adverse outcomes in various climatic regions.

39.2 How will climate change affect weather data?

In order to enable the selection of air conditioning systems to provide thermal comfort in particular locations, the American Society for Heating, Refrigeration and Air Conditioning Engineers (ASHRAE 2005) has provided design value guidelines for the external dry bulb (DB) air temperature with coincident wet bulb (WB) temperature (and vice versa). The method for determining the summer design conditions is based on historical maximum temperatures and coincident humidity measurements. For this approach, the 0.4%, 1% and 2% levels of exeedance of seasonal temperatures (dry bulb temperature with coincident wet bulb temperature, and wet bulb temperatures with the coincident value of the dry bulb temperature) are calculated from the cumulative distribution function for the whole season.

Table 39.1 presents the ASHRAE data for all Australian capital cities. The second and third columns list the values of the dry bulb temperature and corresponding wet bulb temperature. These values are normally used to represent the outside summer conditions in the selection and sizing of comfort air conditioning equipment for buildings. The last two columns are the design conditions used when the humidity level is more significant than the temperature in equipment selection such as cooling towers. The choice of exceedance level in building design is dependent on how critical it is to maintain indoor tempera-

ture when outside temperatures exceeds design conditions.

The summer design conditions provided by the Australian Institute for Refrigeration, Air Conditioning and Heating (AIRAH 2007), on the other hand, is set by investigating the dry and wet bulb temperature values recorded at 15:00 hours each day.

39.3   Future design temperatures using historical data

It is clear that design guidelines based on past climate conditions are unlikely to be adequate for future climate change conditions. A systematic method to determine future design conditions could be undertaken by altering historical data (i.e. dry bulb and wet bulb temperatures) consistent with climate change, from which to construct a frequency distribution of temperatures in the same way as present-day design guidelines.

To demonstrate this type of approach, an analysis was conducted on the current and future typical meteorological year (TMY) data. The full methodology for modifying TMY to reflect the prediction of climate change is available in Saman et al. (2013). The modifications were based on estimates of the expected increases in dry bulb (DB) developed by Watterson et al. (2007). These estimates were produced using 23 global climate models from the CMIP3 database and a statistical downscaling method. The corresponding wet bulb (WB) temperature was found using the predicted DB design temperature, assuming a constant relative humidity. Watterson et al. (2007) also suggest a single value for the change in relative humidity for each location. The predicted change is very small, equating to on average +0.2% for 2030 and +0.6% for 2070 across all cities presented. This change has an almost negligible impact on the wet bulb temperature.

Table 39.2 shows the current and future design data, based on the TMY with the estimated increases in maximum DB temperature for 2030 and 2070 applied. By comparing the current DB temperatures to the ASHRAE values it can be

Table 39.1   ASHRAE summer design temperatures for each of three exceedance levels for Australia's capital cities (DB: dry bulb; CWB: corresponding wet bulb; WB: wet bulb; CDB: corresponding dry bulb).

City	Exceedance (%)	DB	CWB	WB	CDB
Adelaide	0.4	39.8	17.4	23	27.1
	1	38.2	21.5	22.3	30.6
	2	36.6	19.4	21.5	42.6
Perth	0.4	38.9	18.8	24.2	35.2
	1	37.3	22.5	23.3	37.3
	2	35.7	23.2	22.6	26
Darwin	0.4	33.9	23.8	27.6	33.2
	1	33.4	25	27.3	33.1
	2	32.9	27.5	27.1	31.6
Brisbane	0.4	33.7	20.4	25.9	31
	1	32.4	22.2	25.3	32.2
	2	31.4	24.4	24.8	30
Melbourne	0.4	37.8	18.6	22.7	36.5
	1	35.6	19.4	21.8	23.9
	2	33.8	19.1	21	27.6
Sydney	0.4	33.2	23.3	24	28.9
	1	31	18.4	23.4	28.3
	2	29.1	24.8	22.9	23.9
Hobart	0.4	32.7	18.3	20.1	32.3
	1	29.8	17.8	19.2	25.3
	2	27.2	15.8	18.4	22.6

Table 39.2 Examples of current and future external dry bulb design data determined from the TMY.

City	Exceedance level	Current	2030	2070
Adelaide	0.4	39.3	40.1	40.9
	1	37.8	38.7	39.4
	2	36.4	37.3	38
Perth	0.4	37.5	38.7	39.6
	1	36.4	37.6	38.5
	2	35.3	36.5	37.4
Darwin	0.4	34.1	35.6	36.4
	1	33.6	35.1	36
	2	33.2	34.7	35.5
Brisbane	0.4	31.2	32.8	33.5
	1	30.7	32.2	33
	2	30.1	31.7	32.4
Melbourne	0.4	33.9	34.6	35.5
	1	32.6	33.4	34.3
	2	31.4	32.2	33
Sydney	0.4	32.7	34	34.4
	1	31.5	32.8	33.2
	2	30.4	31.8	32.2
Hobart	0.4	27.6	28.6	29
	1	26.1	27.1	27.6
	2	24.8	25.8	26.3

Table 39.3 Indicative 2030 and 2070 summer design temperatures at different levels of exceedance for Australian capital cities.

City	Percentage exceedance level	2030		2070	
		DB	CWB	DB	CWB
Adelaide	0.4	40.6	17.8	41.3	18.1
	1	39	22	39.7	22.5
	2	37.4	19.9	38.1	20.3
Perth	0.4	40.1	19.5	41	20
	1	38.5	23.3	39.4	24
	2	36.9	24.1	37.8	24.8
Darwin	0.4	35.4	25	36.2	25.6
	1	34.9	26.2	35.7	26.9
	2	34.4	28.8	35.2	29.6
Brisbane	0.4	35.3	21.5	36	22
	1	34	23.4	34.7	24
	2	33	25.8	33.7	26.4
Melbourne	0.4	38.5	19	39.4	19.5
	1	36.3	19.8	37.2	20.4
	2	34.5	19.6	35.4	20.2
Sydney	0.4	34.8	24.6	34.9	24.7
	1	32.6	19.5	32.7	19.6
	2	30.7	26.3	30.8	26.4
Hobart	0.4	33.9	19.1	34.3	19.4
	1	31	18.6	31.4	18.9
	2	28.4	16.6	28.8	16.9

seen how the current TMY fails to encompass extreme periods in the future, in that the extreme values for all probability levels are exceeded in the 2030 and 2070 projections.

Alternatively, indicative future design temperatures can be determined by simply adding the predicted maximum temperature increases to all DB conditions determined by using the ASHRAE method only. Table 39.3 presents these approximate values for comparison.

39.4   Impact on air conditioning energy use

By using the methodology for estimating the impact of global warming on key weather parameters in 2030 and beyond, we estimated the selection and anticipated energy consumption of air conditioners in future years to model the air conditioning requirements within a building (Saman et al. 2013). We determined that, by 2030, it is

likely that all mainland cities in Australia will use more electricity for cooling than for heating. Adelaide, Melbourne, Perth and Darwin are likely to experience a small increase in heating and cooling costs, while dramatic increases are predicted in Sydney and Brisbane. On the other hand, Hobart can expect an overall reduction in heating and cooling costs in view of dramatic reductions in heating demand.

We also undertook monitoring of indoor temperature and humidity in groups of households in South Australia, New South Wales and Queensland as part of a broader investigation to better understand household behaviour and responses during hot weather (Saman et al. 2013). The monitoring confirmed the important role played by air conditioning in increasing the total peak electricity demand. We estimated that, on average, 38% of total peak demand in Australia is due to residential air conditioning. With climate change it is anticipated that peak demand in Adelaide, Melbourne, Darwin and Perth will experience small increases; Sydney and Brisbane are likely to experience dramatic increases however, well beyond current peak power demand growth, and it is likely that this will further impact on electricity costs above the projected increases outlined above.

39.5 Case studies: dwelling design options to cope with heat waves

Adaptation to extremes of heat has been ongoing for centuries in many parts of the world (Palmer et al. 2012, 2013). The lessons from the past regarding dwelling design in hot climates were reviewed in preparing our research and have led to the proposal of considering a 'cool retreat' in future housing design or refurbishment. This is a room or zone within a dwelling that is located and designed to be little affected by external weather conditions and provide comfortable conditions during a heat wave while using minimal energy. The benefit to the dwelling occupants is that, by restricting their movement and most activities to the cool retreat for the relatively short period of time of extreme heat, they can be assured of a thermally comfortable environment at minimal running cost. A cool retreat can be either part of the design of a new dwelling or retrofitted to existing housing.

Although applicable to dwellings occupied by all sections of the community, the focus of the case study research was on households that are more vulnerable (both in terms of health and financial means) to heat waves such as those occupied by the elderly, infirm, infants and those with a lower ability to pay high electricity bills. Given that the life cycle of many types of dwellings can be measured in decades, we also considered both existing and new dwellings for possible design solutions in determining how householders might maintain comfort conditions in their homes during future heat waves without consuming larger quantities of energy (Bennetts et al. 2012).

While the popularity of semi-detached houses and flats is increasing (Saman et al. 2013), it is clear that detached houses will still dominate the dwelling mix in Australia in the next two decades. We therefore considered five house design case studies in this investigation, representing different types of dwellings typical of the Australian housing stock both now and anticipated in the future.

1. A single-storey house (78 m²) with two bedrooms similar to new public housing for members of the community on lower incomes, retirement homes and assisted accommodation for senior citizens.

2. A three-bedroom family brick veneer dwelling (150 m²) typical of project homes built on new housing estates and with an internal open plan design.

3. A two-storey three-bedroom house (189 m²) designed for a narrow site such as may arise from the subdivision of a large lot in an established residential area. This case study features a large open volume to be cooled during heat waves resulting from open plan design and an open staircase to the upper level.

4. A three bedroom, two-storey apartment (159 m²) on the upper stories of a four-level development featuring an open plan design. This case study reflects medium density apartment living, predicted to grow in cities in future decades.

5. A two-bedroom single-level apartment (82 m²) on the third floor of a seven-storey block with other apartments on each side as well as above and below.

Each case study dwelling type was analysed for its thermal performance: first in its existing form ('base case'); second with improvements to the existing design as might be carried out by retrofitting ('retrofit measures'); and third as an improved design consistent with new-build dwellings ('modified design'). The thermal analysis was carried out using the AccuRate software (Delsante 2004) which is widely used for residential building energy research (Wang at al. 2010; Morrissey et al. 2011) and is the basic tool used by the Australian Nationwide House Energy Rating Scheme (NatHERS). The software estimates heating and cooling energy demand to maintain comfort temperatures in a dwelling in a selected climate zone as well as a star rating summarising thermal performance. In its free-running mode, AccuRate can be used for determining resultant temperatures within dwellings (i.e. with no heating and cooling).

The retrofit measures applied to each of the five base cases included improved glazing, external blinds, increased insulation, roof-space foil, a light coloured roof, roof-space vent, light-coloured external walls and ceiling fans. The resultant internal temperatures of the main living area and bedroom were determined using AccuRate for each retrofit measure. This was followed by the application of a combination of these measures, that is, as a retrofit package, and the internal temperatures and cooling energy use were evaluated.

The modified designs of each base case utilised a range of features including different but commonly available materials, use of earth-coupled designs such as basements, integrating 'cool retreats' which could also be used for normal daily living, incorporation of courtyards or atriums, separation of living spaces normally included in open plan designs, locating living spaces away from areas susceptible to excessive solar gains and using unconditioned spaces to separate living areas and external heating loads.

We adopted the upper limits of residential thermal comfort suggested by Peeters et al. (2009) of bedrooms at 26 °C, or 29 °C if there is increased air movement (a ceiling fan), and 30 °C for the living/kitchen area.

We analysed the building performance in three case study locations: Adelaide (South Australia), Amberley near Brisbane (Queensland) and Richmond near Sydney (NSW). Four-day hot-period data were extracted for the three locations in order to test thermal performance in an extreme heat event. Table 39.4 summarises the features of the four-day hot periods in the three locations and Figure 39.1 shows the outdoor temperatures during the four-day hot period extracted from the Adelaide weather file as an example of the temperature patterns considered.

The following section summarises the case study analysis in brief for the base case, retrofit measures and modified designs for all five housing designs and the three locations. The full set of analyses and results are available in Saman et al (2013).

39.6   Thermal performance of the case study locations

When analysed in free-running mode in Adelaide during the four-day hot period, house design 1 (single-storey house) in its base case form is within the thermal comfort limit for just 33% of the time in the living/kitchen area and 9% of the time in the first bedroom. The thermal analyses for Amberley and Richmond were similar, although internal resultant temperatures were slightly lower than those in Adelaide.

The retrofit measures individually had little effect on the free-running temperatures in the dwelling. The combination of all the retrofit measures resulted in a moderate lowering of the maximum temperatures, for example from 38.8 °C to 35.6 °C in the living/kitchen area and an increase from 33% to 53% for the proportion of time in the thermal comfort range. The analysis showed that continuous comfort conditions

Table 39.4 Features of four-day hot periods for three Australian locations.

Location	Maximum temperature (°C)	Period >35 °C (hours)	Other features
Adelaide (South Australia)	44	25	Three hot nights
Amberley (Queensland)	43	22	Three nights in high 20s
Richmond (New South Wales)	40	14	Minimum temperature of 21 °C

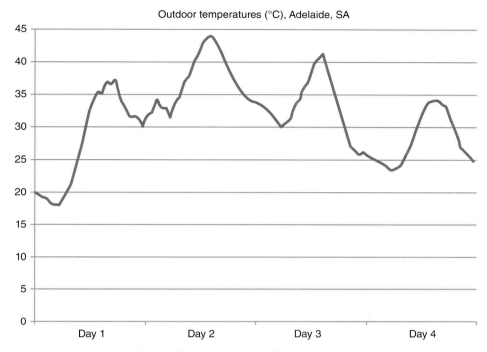

Figure 39.1 External temperature during a four-day hot period from Adelaide.

could not be maintained using passive retrofit measures alone, and this was the case for all three locations. When analysed using active cooling (e.g. air-conditioning) however, the AccuRate simulations showed that the combination of retrofit measures had a significant effect. In Adelaide, the cooling energy load for the living/kitchen and first bedroom reduced from 483 MJ to 259 MJ and the peak demand from 5.6 kW to 3.2 kW. Significant energy reductions were also found for the Amberley and Richmond locations.

The modified designs for the case studies incorporate cool retreats where residents are expected to restrict their occupancy to particular rooms or zones within the dwelling during a hot period. A comparison of the modified designs with the base cases shows substantial reductions in cooling energy load and peak demand. For example, the utilisation of a modified second bedroom as a cool retreat in house design 1, as illustrated in Figure 39.2, results in a cooling energy load of 13.3% of that used in the base case and a peak demand of 17.9% of the base case during the hot period. For the locations of Amberley and Richmond, reductions in cooling energy load and peak demand are also substantial but not quite as large.

House designs 2–5 showed broadly similar trends to the analysis of house design 1 with regard to thermal performance with the inclusion of retrofit measures. This is also true of the performance of the modified designs although, with house designs 2 (three-bedroom house) and 3 (two-storey, three-bedroom house), the reductions in cooling energy load and peak load were even more substantial than in house design 1. This is because these modified designs incorporated basements and utilised the effects of earth coupling to reduce heat loads. The modified design of house design 2 incorporated a basement under the garage area whereas for the modified house design 3, the basement was essentially a

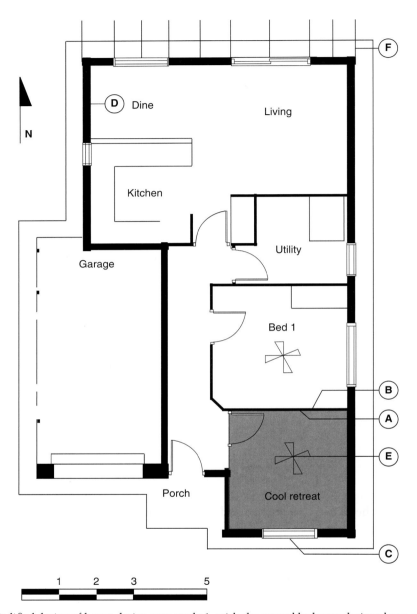

Figure 39.2 Modified design of house design case study 1, with the second bedroom designed as a cool retreat.

floor of the two-storey house with the upper-storey at ground level.

The five house designs combined with retrofit measures showed that significant reductions in cooling energy load and peak demand during a four-day hot period could be achieved. This result is important for planning upgrades of the existing dwelling stock and creating more comfortable conditions within homes. When the dwelling designs were further modified to create cool

Table 39.5 Comparison of cooling energy, cooling area and peak demand during four-day hot period between the base case (unmodified) dwelling design and modified design that includes a cool retreat for five house types. Results shown are for simulations for Adelaide.

	Cooling during heat wave (MJ)	Conditioned area (m²)	Cooling/area (MJ m⁻²)	Peak demand (kW)
House design case study 1 (single-storey house)				
Base case	483	44.8	10.8	5.6
Cool retreat	64	10.2	6.3	1.0
House design case study 2 (three-bedroom, brick veneer)				
Base case	645	87.3	7.4	10.2
Cool retreat	15	17.6	0.9	1.6
House design case study 3 (two-storey, three-bedroom)				
Base case	963	132.5	7.3	18.8
Cool retreat	58	35.1	1.7	2.9
House design case study 4 (two-storey apartment)				
Base case	1051	124.3	8.5	16.3
Cool retreat	129	25	5.2	1.9
House design case study 5 (single-storey, two-bedroom apartment)				
Base case	700	75.4	9.3	10.7
Cool retreat	176	30.2	5.8	2.8

retreats, the cooling energy load required for comfort conditions was a small proportion of that required in the base case version as illustrated in Table 39.5 for Adelaide, but also reflected in the results for the Amberley and Richmond. Basement cool retreats appear to be especially suited to maintaining comfort conditions during heat waves. The research described needs to be followed up with a detailed cost–benefit analysis of the retrofit measures and modified designs. There may also be other benefits in adopting these design options, including the reduction of greenhouse gas emissions from the residential sector although life-cycle analyses will be required to quantify such benefits. The question of changes in the building regulations to mandate the use of cool retreats and possible maladaptations would also need to be addressed. For example, the inclusion of energy efficiency measures to create comfortable conditions in winter by maximising glazing may create an overheating risk in summer, particularly during heat waves.

39.7 Cost implications and householder attitudes

Our research demonstrated that significant savings in energy costs and peak demand can be made through housing design. Successful implementation of this relies on householder affordability and behaviour changes, however. Social research (key informant interviews, online surveys; Saman et al. 2013) conducted as part of our project demonstrated a willingness among householders to change behaviour rather than spend money. This included the appropriate use of external shades and curtains or moving to a cooler room during heat waves. Around half of the respondents were also willing to spend up to $2000 to achieve better thermal comfort, although 30% of respondents considered they were not in a position to spend any money at all. Some retrofit options had negligible additional costs. For example, one of the most effective methods for reducing cooling demand in existing dwellings is to modify their roofs by increasing their total solar reflectance, adding reflective foil and increasing thermal insulation. The cost of these modifications can be negligible when building a new home or upgrading an existing roof. Other options (e.g. adding a basement) represent a potentially significant cost and may be unaffordable and therefore impractical.

39.8   Outcomes and recommendations

Our research has demonstrated that a combination of responses is necessary in order to adapt to heat waves. This includes behavioural change during heat waves, reconfiguration of house design and appropriate use of air conditioning. Reflection of some of the proposed house design measures to enable a better coping ability during heat waves in future building regulations is likely to be necessary. In Australia, for example, we recommend the inclusion of climate data and air conditioning design calculations that are adjusted to reflect a changing climate in the Nationwide House Energy Rating Scheme as an important step toward adapting housing stock. It is essential that these guidelines incorporate the whole of air conditioning systems in regulations, ensuring all regulations apply to all new systems rather than just those installed in new buildings.

In addition to considering reducing annual energy and power demand for existing housing, special attention must be paid to minimise peak cooling demand in new buildings. The inclusion and use of cool retreats has been demonstrated to provide thermal comfort at dramatically reduced power consumption, and we expect it to be an important tool in adapting to extreme heat.

The proposed integrated approach and measures highlighted in this study can collectively reverse the current compounding health risks associated with climate change. In themselves, each measure would achieve limited success due to the potential negative impact of other factors. The complementary nature of each component will however deliver a framework for adapting households and diminish the risks associated with heat waves to individuals, as well as reducing the need for augmenting the electricity infrastructure.

An important element of this holistic approach will involve considerable investment in increased community awareness; the findings of our study indicate that existing information and awareness campaigns has a limited effect on informing and influencing Australian responses to heat waves. This information should include detailed advice on how to plan for heat waves. Advice on how to make a dwelling more comfortable for heat wave conditions and the types, suitability and energy consumption of air conditioners should be included. Those with low incomes and elderly individuals who are the least able to afford energy use for air conditioning should be a priority for intervention and assistance. One potential mechanism for this is the provision of government grants and financial incentives to assist these groups to adapt their homes to extreme heat.

Acknowledgements

This chapter is a summary of research carried out in the research project 'A framework for adaptation of Australian households to heat waves' and the support of NCCARF is gratefully acknowledged. Other members of the research team contributed indirectly to this chapter by conducting the work and obtaining the findings that underpin this chapter and their efforts are gratefully acknowledged. They include Wendy Miller, Barbara Pocock, Martin Belusko, David Whaley, Helen Bennetts, Barbara Ridley, Jasmine Palmer, Jian Zuo, Tony Ma, Natalie Skinner, Janine Chapman, and Nicholas Chileshe.

References

American Society of Heating, Refrigerating and Air Conditioning Engineers, Inc. (2005) *ASHRAE Handbook: Fundamentals*. Atlanta, USA.
Australian Institute for Refrigeration, Air Conditioning and Heating (2007) *AIRAH Technical Handbook*, Edition 4. Australian Institute for Refrigeration, Air Conditioning and Heating, Australia.

Bennetts, H., Pullen, S. and Zillante, G. (2012) Design strategies for houses subject to heat waves. *Open House International Journal* 37, 29–38.

Delsante, A. (2004) *A Validation of the 'AccuRate' Simulation Engine Using BESTEST*. Report for The Australian Greenhouse Office. CSIRO, Melbourne.

Mavrogianni, A., Wilkinson, P., Davies, M., Biddulph, P. and Oikonomou, E. (2012) Building characteristics as determinants of propensity to high indoor summer temperatures in London dwellings. *Building and Environment* 55, 117–130.

Morrissey, J., Moore, T. and Horne, R.E. (2011) Affordable passive solar design in a temperate climate: An experiment in residential building orientation. *Renewable Energy* 36(2), 568–577.

Palmer, J., Bennetts, H., Chileshe, N., Pullen, S.F., Zuo, J. and Ma, T. (2012) Heat wave risks and residential buildings. In: Skates, H. (ed.) *Proceedings of the 46thAnnual Conference of the Australian and New Zealand Architectural Science Association (ANZAScA). Building on Knowledge: Theory and Practice*. Griffith University, Queensland.

Palmer, J., Bennetts, H., Pullen, S., Zuo, J., Ma, T. and Chileshe, N. (2013) The effect of dwelling occupants on energy consumption: The case of heat waves in Australia. *Architectural Engineering and Design Management, Special Issue: The Impact of the Building Occupant on Energy Consumption* 10, 40-59.

Peeters L., de Dear R., Hensen J. and D'haeseleer, W. (2009) Thermal comfort in residential buildings: Comfort values and scales for building energy simulation. *Applied Energy* 86, 772–780.

Porritt, S., Shao, L., Cropper, P. and Goodier, C. (2011) Adapting dwellings for heat waves. *Sustainable Cities and Society* 1(2), 81–90.

Porritt, S. M., Cropper, P.C., Shao, L. and Goodier, C. (2012) Ranking of interventions to reduce dwelling overheating during heat waves. *Energy and Buildings* 55, 16–27.

Saman, W., Boland, J., Pullen, S. et al. (2013) *A Framework for Adaptation of Australian Households to Heat Waves*. National Climate Change Adaptation Research Facility, Gold Coast, 242 pp.

Wang, X., Chen, D. and Ren, Z. (2010) Assessment of climate change impact on residential building heating and cooling energy requirement in Australia. *Building and Environment* 45(7), 1663–1682.

Watterson, I., Whetton, P., Moise, A. et al. (2007) Chapter 5. Regional climate change projections. In: CSIRO and Bureau of Meteorology (eds.) *Climate Change in Australia*. CSIRO, Melbourne.

40 Pathways for adaptation of low-income housing to extreme heat

GUY BARNETT, R. MATTHEW BEATY, JACQUI MEYERS, DONG CHEN AND STEPHEN MCFALLAN

CSIRO Land and Water Flagship, Australia

40.1 Introduction

There is growing global awareness of the effects of excess heat on human health following unprecedented death tolls from heat waves in Europe (Katsouyanni et al. 1988; Hemon and Jougla 2004), the United States (Klinenberg 2003; Luber and McGeehin 2008) and Australia (Nitschke et al. 2007; Schaffer et al. 2012). This coincides with predictions of more frequent and severe heat waves as a result of climate change (Meehl and Tebaldi 2004). At the same time, the sensitivity of populations to excess heat is also increasing, due to ageing and the growing prevalence of chronic health conditions (Bambrick et al. 2011). Together, this is expected to result in a significant rise in future public health burden (McMichael et al. 2006).

Nonetheless, significant potential remains to address increased heat-related health risk and population sensitivity through climate adaptation (Hajat et al. 2010). So far, much of the emphasis has been on broad population-level measures such as heat alert systems and public awareness campaigns to inform vulnerable individuals and communities on how to protect themselves during hot weather (Lowe et al. 2011). This is appropriate, as we know there are various socio-economic and individual risk factors for health impacts during hot weather that characterise the most vulnerable

(Conti et al. 2005; Loughnan et al. 2010; Ma et al. 2012). What we know far less about is the effectiveness of various biophysical adaptation actions.

Most of the work on the thermal performance of housing has been undertaken in Europe, the USA and New Zealand, generally focusing on the health impacts of cold weather (Howden-Chapman and Chapman 2012). Of the studies that looked at hot weather, a study in France found that people living in homes without insulation and with bedrooms directly below the roof were at greatest risk of dying during a heat wave (Vandentorren et al. 2006). A similar study in the USA identified those living above the ground floor within multi-storey dwellings were at increased risk of non-fatal heat stroke (Klinenberg 2003). There has been very little research on the heat-related health risks associated with characteristics of Australian housing.

One of the major factors limiting the ability to undertake good quantitative research on both climate change impacts and adaptation options in the housing sector has been the difficulty in securing high-quality data on housing assets and their occupants. Here we summarise the findings of a study where this problem was overcome by partnering with two social housing authorities collectively responsible for 35% of Australia's social housing (Barnett et al. 2013).

Applied Studies in Climate Adaptation, First Edition. Edited by Jean P. Palutikof, Sarah L. Boulter, Jon Barnett and David Rissik.

The study was undertaken using a multi-level methodology, which combined building-level assessments of housing thermal performance with broader assessment of the socio-ecological neighbourhood context in which these houses are located to determine the influence of 'place' on climate vulnerability and adaptation options. We argue, just as the provision of social housing has been a societal response to social disadvantage, identifying climate adaptation pathways for low-income households ensures fairness in Australia's climate change response.

40.2   Assessing housing performance

A social housing dataset comprising 142,410 housing assets was prepared using information provided by project partners, covering all major climate zones identified in the Building Code of Australia (ABCB 2009) with the exception of alpine areas (Fig. 40.1). Using this social housing dataset, a simple housing typology was developed to classify the 142,410 housing assets into common housing types. The main parameters used in the classification were the type and period of construction as well as the materials used for wall cladding and roofing. Ten housing types were identified, ranging from single detached houses to units in high-rise apartment buildings. The final typology is presented in Table 40.1 and was used to classify assets in the social housing dataset for which there were sufficient data (108,878 assets, or 76% of the portfolio).

The thermal performance and indoor environment of each housing type was assessed using the dynamic building simulation software *AccuRate*, commonly used within Australia to assess the energy efficiency of residential buildings (Wang et al. 2010). Simulations were performed using baseline and future weather input for a reference city in each climate zone (Fig. 40.1). Weather data were compiled in the form of a typical meteorological year (TMY), a format regularly used in building simulation (Chen et al. 2012). The baseline TMY weather files were prepared using historical weather data around 1990. Projections of

future climate were derived from the *OZClim* climate change projection tool for Australia (Ricketts and Page 2007) for the years 2030, 2050 and 2070. These projections were then converted into future TMY weather files using a morphing approach developed by Belcher et al. (2005).

AccuRate simulates the thermal environment within a building based on detailed input on the building location, orientation, materials and construction, as well as occupant behaviour such as the operation of windows and blinds. Various assumptions were necessary. While building occupant behaviour was kept constant, four housing scenarios were developed for each house type to reflect major differences in building quality and construction. These were a 'worst case', 'base case', 'cheap retrofit' and 'expensive retrofit' scenario, covering the range of performances that might be expected given the housing variability in each housing type. For instance, the 'worst case' scenario combines poor building orientation, a dark-coloured roof and no ceiling insulation to represent the worst possible thermal performance. The 'base case' assumes good building orientation and a light-coloured roof, but no ceiling insulation. The 'cheap retrofit' builds on the 'base case' through weatherproofing to reduce air leaks and infiltration and the addition of ceiling insulation and window shading. This was expanded in the 'expensive retrofit' to include double glazing, more insulation and better window frames.

Housing performance was measured using the discomfort index (DI; Thom 1959), a widely used indicator of heat-related health risk (Epstein and Moran 2006). It was calculated in *AccuRate* using estimates of wet-bulb and dry-bulb temperatures for zones within each house type. DI is expressed as an hourly value over the period of a year, represented by the baseline and future TMY weather files. A threshold value of DI > 28 was identified by Epstein and Moran (2006) as indicating severe environmental heat load and increased risk of heat-related illness. It is based on physiological research and has been adopted in this study.

The results are presented in Figure 40.2. Comparing differences across the climate zones, those represented by the reference cities of

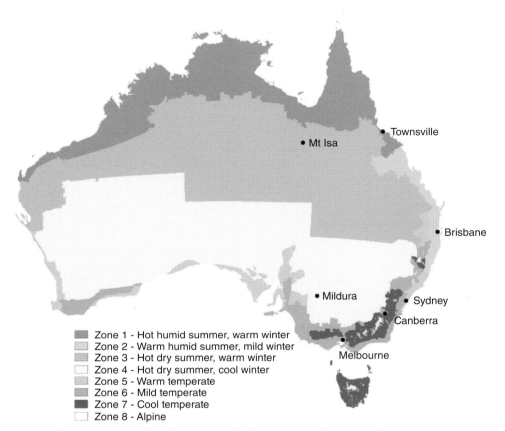

Figure 40.1 The climate zones used in this project and reference cities that were selected as representative of each climate zone. Source: Australian Building Codes Board 2009. Image modified with permission from the Australian Building Codes Board (ABCB) www.abcb.gov.au. For colour details please see Plate 28.

Table 40.1 Description of the classification used to represent low-income housing types.

Code	Type	Subfloor	External Wall	Year Built	Bedrooms	Count	Percent
A1	House	Slab on ground	Brick/block veneer	Pre-2005	3 or 4	18,648	17.1
B1	House	Low-set raised	Brick/block veneer	Pre-2005	3 or 4	6589	6.1
B2	House	Low-set raised	Fibro/weatherboard	Pre-2005	3 or 4	2393	2.2
C1	House	High-set raised	Brick/block veneer	Pre-2005	3 or 4	2417	2.2
C2	House	High-set raised	Fibro/weatherboard	Pre-2005	3 or 4	10,101	9.3
D1	House	Slab on ground	Brick/block veneer	Post-2005	3 or 4	506	0.5
E1	Flat	Low-rise	Concrete/brick/block	Anytime	1	29,200	26.8
F1	Flat	Low-rise	Concrete/brick/block	Anytime	2 or 3	32,002	29.4
G1	Flat	High-rise	Concrete/brick/block	Anytime	1	1946	1.8
H1	Flat	High-rise	Concrete/brick/block	Anytime	2 or 3	5076	4.7
					Total	108 878	100.0

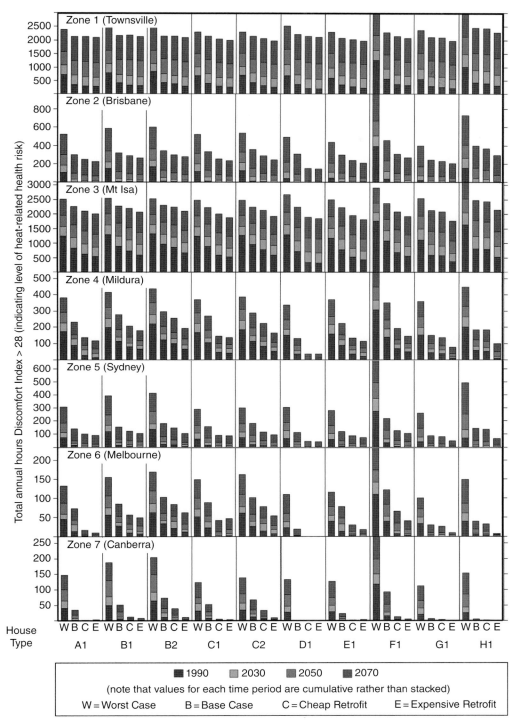

Figure 40.2 Summary of the total annual hours the discomfort index (DI) exceeds the threshold value of 28 for each house type, within each reference city, for reference and future climates. For colour details please see Plate 29.

Mt Isa, Townsville and Mildura were subject to periods of severe heat-related health risk (DI > 28) for all house types under all housing scenarios. This is most evident in the 'worst case' scenario of each house type. Cheap and expensive retrofits reduce the level of severe heat-related health risk, but cannot completely mitigate the impact.

In the other reference cities of Brisbane, Sydney, Melbourne and Canberra, there were large differences between 'worst case' and 'base case' scenarios for all house types. It appears that severe heat-related health risk (DI > 28) can be mitigated in most situations in these locations through the cheap or expensive retrofit options. This holds true particularly in the short term (2030) and, to a lesser extent, the medium term (2050). As such, building retrofits may have a role in 'buying time' by improving the thermal performance of older assets sufficiently so that they either reach the end of their expected asset life or enable the staggering of asset disposal.

40.3   Incorporating people and place

The interactions between people, housing and neighbourhood have been viewed as a complex socio-ecological system, with the vulnerability of this system defined by the widely accepted component measures of exposure, sensitivity and adaptive capacity (Smit and Wandel 2006). In this study, exposure was regarded as the combined thermal performance of the housing and neighbourhood in which people live, sensitivity as the prevalence of heat-related health risk factors in the resident population and adaptive capacity as the biophysical opportunities to ameliorate the potential impact (exposure + sensitivity). As these components are difficult to measure directly (Preston et al. 2011), indicators have been used and are described as follows.

40.3.1   Heat exposure

The thermal performance of housing assets was determined using the *AccuRate* software to calculate the total annual hours of DI > 28, providing a building-scale measure of heat exposure. At the neighbourhood scale, a 100 m buffer was placed around each housing asset using ArcGIS 10.1 (ESRI 2012) to quantify the amount of impervious surface in the buffer. Impervious surface cover was determined by linear spectral un-mixing of Landsat Mosaic satellite imagery sourced from Geoscience Australia. Impervious surface cover is correlated with land surface temperature (Yuan and Bauer 2007), providing a proxy measure of neighbourhood exposure. Both indicators were then summed, with weights of 0.6 (building) and 0.4 (neighbourhood) used to derive a single, aggregate measure of exposure.

40.3.2   Resident sensitivity

The demographic and socio-economic characteristics of residents living in each of the housing assets were not available due to reasons of confidentiality and privacy. As a proxy, data from the 2011 Australian Census of Population and Housing were used to characterise the social housing residents based on the neighbourhood in which the housing assets were located. Four indicators of resident sensitivity to heat-related health impacts were selected. These included the proportion of low income households with people aged 65 years and older, children aged 0–4 years, those living alone and those requiring assistance. The indicators were then summed with equal weight (0.25) providing a measure of resident sensitivity, assigned to the individual housing assets.

40.3.3   Adaptive capacity

At the scale of the housing asset, 'adaptive capacity' was defined as the potential to reduce the total annual hours of DI > 28. It was calculated by subtracting the 'base case' performance of a housing asset from the performance achieved through 'expensive retrofit'. It therefore represents the maximum potential to improve indoor thermal comfort via building retrofit. The greening potential of the neighbourhood was calculated as the amount of 'plantable space' in a 100 m buffer around each house, quantified using a grass fraction layer derived from the Landsat Mosaic satellite imagery described earlier. The final measure was proximity to 'cool places' (defined by libraries, museums, shopping centres, etc.) located within a 400 m buffer of the housing asset, representing air conditioning that may be available for respite from heat. The indicators were summed, with weights of 0.6 (building) and 0.2 for each neighbourhood measure.

40.4 Vulnerability and adaptation pathways

Based on the methods outlined so far, a vulnerability score was developed for each asset in the social housing dataset providing a standardised score to enable the identification of the most vulnerable housing assets and the efficacy of climate adaptation. The results are shown in Figure 40.3, which is a plot of the component scores for potential impact (exposure + sensitivity) and adaptive capacity for each individual housing asset for which we had vulnerability data (103,809).

Four broad climate adaptation pathways for our social housing portfolio are identified in Figure 40.3, which include 'early disposal', 'priority upgrades', 'learning analogues' and 'buying time'. These pathways apply to the housing assets that were characterised by the top and bottom quintiles of distribution in potential impact and adaptive capacity scores. About 5% (5031) of assets mapped to the 'early disposal' climate adaptation pathway (Fig. 40.3, top left). These were mostly located in the climate zones with hot and humid summers and pose the greatest heat-related health risk to occupants who are likely to be increasingly reliant on air conditioning to maintain a safe indoor thermal environment. While air conditioning is often considered a potentially maladaptive technology for low-income households (Farbotko and Waitt 2011), there is evidence it is protective against heat-related mortality (Bouchama et al. 2007) and hospitalisation (Ostro et al. 2010). Most social housing in Australia is generally not provided with air conditioning, so the relocation of

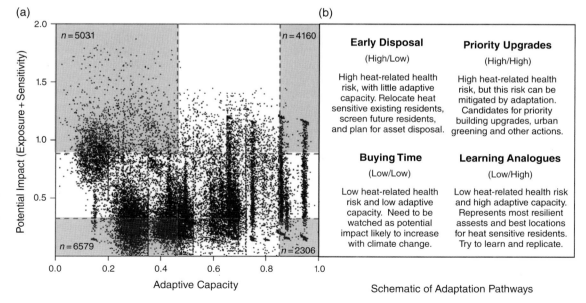

Figure 40.3 (a) Scatter plot of 103,809 housing assets by their potential impact and adaptive capacity scores with red dashed lines indicating the bottom and top quintiles of the data distribution and (b) corresponding adaptation pathways identified. For colour details please see Plate 30.

heat-sensitive occupants to safer accommodation will be important while plans are made to dispose of these housing assets.

A further 4% (4160) of housing assets also had high potential impact but with high adaptive capacity (Fig. 40.3b, top right), meaning there is scope to reduce heat-related health risk through adaptation. These assets should be candidates for 'priority upgrades' to improve building thermal performance. There is also a key role for urban greening, with a number of modelling studies in other parts of the world highlighting the urban cooling benefits of vegetation (Gill et al. 2007; Jenerette et al. 2011). Of interest in this regard is the closer examination of the 'learning analogues' pathway comprising 2% (2306) of housing assets (Fig. 40.3, bottom right). These are the most climate resilient assets in the portfolio and thus the best location for heat-sensitive residents. There are opportunities to learn from these assets to determine why they perform so well and what role the location, building quality and neighbourhood context has played. The final adaptation pathway is 'buying time' which represents 6% (6579) of the housing assets (Fig. 40.3, bottom left). These assets have low adaptive capacity, but currently present little heat-related health risk. They potentially 'buy time' while the portfolio is being transformed.

The focus of this study has been social housing, yet the housing performance modelling and findings around heat-related health risk factors are equally applicable to other forms of low-income housing such as private rental and owner/purchasing. The main difference will be in the adaptation pathways available. For instance, those in private rental are impacted by the 'split incentive' whereby landlords have little incentive to invest in climate adaptation on behalf of their tenants unless they can recoup their costs through increased rent within a reasonable time period. Barriers to adaptation are also evident in the social housing sector, particularly public housing, where housing assets are typically older and often with a significant maintenance backlog that would require attention before any adaptation can be considered.

40.5   Conclusions

Low-income households and the quality of the housing and neighbourhoods they live in can play a significant role in either exacerbating or ameliorating heat-related health risk. Based on our vulnerability analysis of 103,809 social housing assets distributed across most major climate zones in Australia, we found only 5% currently fall into the category of high potential impact with low adaptive capacity, that is, housing that presents immediate risk to occupants with very little capacity for improvement through building retrofit. These assets should be targets for disposal at the earliest opportunity, with heat-sensitive occupants moved to safer accommodation or provided with air conditioning. Overall however, the findings suggest there is significant opportunity for climate adaptation to reduce the heat-related health risk facing low-income households through asset disposal, building upgrades, urban greening and the use of 'cool places' for respite. Action on all of these fronts will be most effective.

References

Australian Building Codes Board (2009) *Australia Wide Climate Zone Map*. Australian Building Codes Board, Canberra.

Bambrick, H.J., Capon, A.G., Barnett, G.B., Beaty, R.M. and Burton, A.J. (2011) Climate change and health in the urban environment: Adaptation opportunities in Australian cities. *Asia-Pacific Journal of Public Health* 23(2), 67s–79s.

Barnett, G., Beaty, R.M., Chen, D. et al. (2013) *Pathways to Climate Adapted and Healthy Low Income Housing*. National Climate Change Adaptation Research Facility, Gold Coast.

Belcher, S., Hacker, J. and Powell, D. (2005) Constructing design weather data for future climates. *Building Services Engineering Research and Technology* 26(1), 49–61.

Bouchama, A., Dehbi, M., Mohamed, G., Matthies, F., Shoukri, M. and Menne, B. (2007) Prognostic factors in heat wave-related deaths: A meta-analysis. *Archives of Internal Medicine* 167(20), 2170–2176.

Chen, D., Wang, X.M. and Ren, Z.E. (2012). Selection of climatic variables and time scales for future weather preparation in building heating and cooling energy predictions. *Energy and Buildings* 51, 223–233.

Conti, S., Meli, P., Minelli, G. et al. (2005) Epidemiologic study of mortality during the summer 2003 heat wave in Italy. *Environmental Research* 98(3), 390–399.

Epstein, Y. and Moran, D.S. (2006). Thermal comfort and the heat stress indices. *Industrial Health* 44(3), 388–398.

ESRI (2012). *ArcGIS 10.1*. Redlands, California.

Farbotko, C. and Waitt, G. (2011). Residential air-conditioning and climate change: Voices of the vulnerable. *Health Promotion Journal of Australia* 22, S13–S16.

Gill, S., Handley, J., Ennos, A. and Pauleit, S. (2007) Adapting cities for climate change: The role of the green infrastructure. *Built Environment* 33(1), 115–133.

Hajat, S., O'Connor, M. and Kosatsky, T. (2010) Health effects of hot weather: From awareness of risk factors to effective health protection. *Lancet* 375(9717), 856–863.

Hemon, D. and Jougla, E. (2004). The heat wave in France in August 2003. *Revue D Epidemiologie Et De Sante Publique* 52(1), 3–5.

Howden-Chapman, P. and Chapman, R. (2012). Health co-benefits from housing-related policies. *Current Opinion in Environmental Sustainability* 4(4), 414–419.

Jenerette, G.D., Harlan, S.L., Stefanov, W.L. and Martin, C.A. (2011) Ecosystem services and urban heat riskscape moderation: Water, green spaces, and social inequality in Phoenix, USA. *Ecological Applications* 21(7), 2637–2651.

Katsouyanni, K., Trichopoulos, D., Zavitsanos, X. and Touloumi, G. (1988) The 1987 Athens heatwave. *Lancet* 2(8610), 573–573.

Klinenberg, E. (2003) Review of heat wave: Social autopsy of disaster in Chicago. *New England Journal of Medicine* 348(7), 666–667.

Loughnan, M.E., Nicholls, N. and Tapper, N.J. (2010) The effects of summer temperature, age and socioeconomic circumstance on Acute Myocardial Infarction admissions in Melbourne, Australia. *International Journal of Health Geographics* 9(41), doi: 10.1186/1476-072X-9-41.

Lowe, D., Ebi, K.L. and Forsberg, B. (2011) Heatwave early warning systems and adaptation advice to reduce human health consequences of heatwaves. *International Journal of Environmental Research and Public Health* 8(12), 4623–4648.

Luber, G. and McGeehin, M. (2008). Climate change and extreme heat events. *American Journal of Preventive Medicine* 35(5), 429–435.

Ma, W.J., Yang, C.X., Tan, J.G., Song, W., Chen, B. and Kan, H. (2012) Modifiers of the temperature-mortality association in Shanghai, China. *International Journal of Biometeorology* 56(1), 205–207.

McMichael, A.J., Woodruff, R.E. and Hales, S. (2006) Climate change and human health: Present and future risks. *Lancet* 367(9513), 859–869.

Meehl, G.A. and Tebaldi, C. (2004) More intense, more frequent, and longer lasting heat waves in the 21st century. *Science* 305(5686), 994–997.

Nitschke, M., Tucker, G.R. and Bi, P. (2007) Morbidity and mortality during heatwaves in metropolitan Adelaide. *Medical Journal of Australia* 187(11–12), 662–665.

Ostro, B., Rauch, S., Green, R., Malig, B. and Basu, R. (2010) The effects of temperature and use of air conditioning on hospitalizations. *American Journal of Epidemiology* 172(9), 1053–1061.

Preston, B.L., Yuen, E.J. and Westaway, R.M. (2011) Putting vulnerability to climate change on the map: A review of approaches, benefits, and risks. *Sustainability Science* 6(2), 177–202.

Ricketts, J.H. and Page, C.M. (2007) A web based version of OzClim for exploring climate change impacts and risks in the Australian region. *Modsim 2007: International Congress on Modelling and Simulation*, pp. 560–566.

Schaffer, A., Muscatello, D., Broome, R., Corbett, S. and Smith, W. (2012) Emergency department visits, ambulance calls, and mortality associated with an exceptional heat wave in Sydney, Australia, 2011: A time-series analysis. *Environmental Health* 11(3), doi: 10.1186/1476-069X-11-3.

Smit, B. and Wandel, J. (2006) Adaptation, adaptive capacity and vulnerability. *Global Environmental Change: Human and Policy Dimensions* 16(3), 282–292.

Thom, E.C. (1959). The discomfort index. *Weatherwise* 12(2), 57–61.

Vandentorren, S., Bretin, P., Zeghnoun, A. et al. (2006) August 2003 heat wave in France: Risk factors for death of elderly people living at home. *European Journal of Public Health* 16(6), 583–591.

Wang, X.M., Chen, D. and Ren, Z.G. (2010) Assessment of climate change impact on residential building heating and cooling energy requirement in Australia. *Building and Environment* 45(7), 1663–1682.

Yuan, F. and Bauer, M.E. (2007) Comparison of impervious surface area and normalized difference vegetation index as indicators of surface urban heat island effects in Landsat imagery. *Remote Sensing of Environment* 106(3), 375–386.

41 Climate change adaptation in the rental sector

LESLEY INSTONE[1], KATHLEEN J. MEE[1], JANE PALMER[2], MIRIAM WILLIAMS[1] AND NICOLA VAUGHAN[3]

[1]*Centre for Urban and Regional Studies, University of Newcastle, Australia*
[2]*Institute for Sustainable Futures, University of Technology Sydney, Australia*
[3]*Policy and Participation, Housing and Community Services, ACT Government, Australia*

41.1 Introduction

This chapter examines both assets and barriers for climate adaptation in the rental housing sector through a case study of renters and housing managers in regional Australia (Instone et al. 2013). Housing is recognised as both a significant site of greenhouse gas emissions that contribute to climate change, and a site where adaptations to climate change will need to be undertaken (Gabriel et al. 2010a). Rental housing represents an important and under-researched sector in climate change adaptation. Worldwide, the rental sector comprises a significant proportion of housing stock (Hammar 2004; Eurostat 2013); however the social, economic and legislative context of rental housing varies between countries and frames diverse possibilities for climate change adaptation.

In Australia the rental sector comprises 27% of housing (Gabriel et al. 2010b) and research has indicated that there is a poor standard of rental accommodation and a low take-up of retrofitting in the sector (Toohey and Fritze 2009). Low-income households predominate in rental housing in Australia (Gabriel et al. 2010b) and the housing of poorer tenants is likely to be of lower quality (UN-HABITAT 2008; Hulse et al. 2011b)

and more vulnerable to environmental threat (Gabriel et al. 2010a). Most research on adaptive capacity in relation to housing has focused on home owners and purchasers (Sullivan 2007; Stanley 2009; Gabriel et al. 2010a); housing tenants, especially low-income tenants, are often forgotten in considering the impact of climate change in the housing sector.

While research has identified landlord–tenant relationships as a barrier to climate change adaptation in Australia (Dillon et al. 2009; Tenants Union of Victoria 2010), the mediating role of housing managers has been neglected and tenants have generally been positioned as passive, bound by the regulations of their tenancies and their meagre financial capacity (Gurran et al. 2008; Toohey and Fritze 2009). Our project tackled this research gap by focusing attention on the assets of tenants and housing managers as key actors in current adaptive practices, their potential for building capacity and strategies for overcoming barriers. Assets in this context were defined as the 'natural, physical, social, financial and human capital' (Moser and Satterthwaite 2010, p. 237) that enable people and communities to adapt and meet their needs. Barriers are those factors – social, cultural, economic, structural and technological – which limit adaptive capacity

Applied Studies in Climate Adaptation, First Edition. Edited by Jean P. Palutikof, Sarah L. Boulter, Jon Barnett and David Rissik.
© 2015 John Wiley & Sons, Ltd. Published 2015 by John Wiley & Sons, Ltd.

and adaptive action (Barnett et al. 2009). As well as identifying assets and barriers, the objectives of the research were to link the adaptive capacities of key actors – tenants, property managers and landlords – and to identify productive entrypoints for interventions which would support and enhance existing assets in the rental sector.

Our research was undertaken in Newcastle, New South Wales which has a population of nearly 340,000 people and over 36,000 public and private rental properties (based on the combined Local Government Areas of Newcastle and Lake Macquarie; ABS 2012a, b). Our study included in-depth semi-structured interviews with 22 tenants and 17 housing managers in the public and private housing sectors. Tenants self-selected in response to posters and postcards advertising the study, so respondents tended to have an interest in environmental issues and climate change. Private sector housing managers were directly recruited from local real estate agencies and public housing managers from the Northern Region Office of Housing NSW, the major provider of public rental accommodation in the state. As well as interviews three focus groups were conducted, one of tenants and two with a mixture of housing managers and tenants. We also analysed a wide range of secondary sources including media articles, sustainable renting guides and legislative and policy documents. While the research was designed to ensure a focus on adaptation issues for low-income tenants, the findings and recommendations arising from the project can inform further research on rental sector adaptation in other jurisdictions.

41.2 Theoretical underpinnings and methodological approach

The research was undertaken within the framework of priorities established by the National Climate Change Adaptation Research Facility to advance understanding of:
1. vulnerability and adaptive capacity of individuals, communities, businesses and industries;
2. barriers and limits to adaptation;

3. governance and institutional arrangements necessary to ensure that adaptation is as effective, efficient and equitable as possible (Barnett et al. 2011).

In this context, potential barriers to adaptation have been defined as not only the extent of climate change itself, but 'the context in which climate change impacts are experienced: demographic, economic, social, psychological, technological and political factors' (Barnett et al. 2009, p. 24). Rather than focusing solely on barriers however, we aimed to highlight the active role of both tenants and housing managers in the rental sector, including low-income tenants. We therefore employed a novel methodology for Australian climate change research in adopting a 'pro-poor asset-based' approach, described below.

Asset-based community development (ABCD) is a community development theory, model and practice developed by Kretzmann and McKnight (1993) in the United States. It helps a community to identify its existing assets – social, physical, financial, natural and human – and uses dialogue between key actors to build new coalitions and expand the possibilities for change (Moser 2011; Haines and Green 2012). This approach enabled us to shift the focus of our project from adaptation barriers towards the skills and capacities that all actors, especially tenants and housing managers, bring to adaptation practices in the rental sector.

Pro-poor asset-based adaptation to climate change (PACCA) (Prowse and Scott 2008; Moser and Satterthwaite 2010; Moser 2011) is an approach developed for any context where marginalised groups encounter the challenges posed by climate change. It is therefore applicable to the Australian rental housing sector that is characterised by low-income households and a low standard of housing (Toohey and Fritze 2009; Gabriel et al. 2010a; Hulse et al. 2011b). Pro-poor adaptation advocates for improving, not just protecting, the conditions of disadvantaged groups. In our research we developed an approach that incorporated the insights of both PACCA and ABCD to identify and address assets and barriers across the rental sector as a whole, and in the everyday practices of tenants and property managers.

Rental housing has received scant attention in the climate change literature. There are significant issues for adaptation, however: a mal-adapted tenant–landlord culture; the 'split incentive' where landlords have no incentive to install energy- and water-saving technologies which will benefit tenants; and constraints on tenants' rights to adapt their homes (Dillon et al. 2009; Gabriel et al. 2010a). Moreover, key actors in the housing sector – real estate agents and property managers who are the interface between the tenant and the property owner – have been largely ignored in research thus far. Most research on relationships in the private rental sector focuses directly on the landlord and tenant, or conflates landlords and property managers (e.g. Lister 2004, 2005; Bierre et al. 2010; Grineski and Hernández 2010). There is also limited literature on the ways in which property managers engage with tenant and landlord practices to enhance the sustainability of dwellings, despite property managers' potential role as 'knowledge brokers' who could improve the adaptive capacity of the rental sector through their role in mediating relationships between landlords and tenants.

In light of our asset-based approach and aim to make renters more visible in discussions on climate change adaptation, we focused particularly on the home as a site of 'resource management' (Mee and Vaughan 2012, p. 149) and the everyday home-making practices that tenants adopt to make their homes more sustainable. There is a dearth of research on tenants' everyday practices, experiences of sustainability and climate change adaptation, but there is a significant (and growing) body of literature on material practices of household sustainability in Australia and elsewhere (Lane and Gorman-Murray 2011) as well as literature concerned with environmental behaviour change at the scale of the household (Reid et al. 2010; Evans 2011). This framed our study of what tenants were already doing, what practices could be built upon and the challenges that hindered adaptation practices. By adopting an asset-based approach with a pro-poor orientation, we aimed for a multi-stranded 'more-than-adaptation'

focus that went beyond the protection of 'material assets' in the form of dwellings.

41.3 Results: tenants and housing managers talk about climate change adaptation

> … particularly in an old house like this and particularly in rentals where … people aren't that inclined to spend a lot of money, it comes down to … the individual practices of the people in the house. (Tenant 9, house/private landlord)

We found that the tenants who participated in our research were motivated by concerns about the impact of human activity on the environment, and exercised this concern through everyday sustainable household practices such as buying green power, reducing energy use, buying second-hand goods, growing their own food and saving water, as well as through engagement with community or political organisations. Tenants believed however that their capacity to act in the home was inhibited by a lack of care from some landlords and property managers about the sustainability and liveability of rental housing, and felt that property managers could be unhelpful in negotiating tenant requests for modifications to their homes. Private sector tenants believed that negative social and political attitudes to renters in Australia were an obstacle to changing tenancy conditions and improving housing. Short leases, insecure tenancy and the requirement to return the property to the original condition at the end of the tenancy provided a major disincentive for tenant investment in home modifications such as water tanks, insulation and solar power.

> If I knew I was going to be living here for a while, then I might talk to the landlord about paying for a system myself because I guess us as renters would be the ones saving on electricity and things. (Tenant 18, townhouse/private manager)

Tenants were also concerned that improvements to properties were likely to result in rent increases, and property managers noted that mandatory improvements were likely to force some 'marginal' landlords to sell their investment property and thus put greater pressure on the rental market.

> If you put more barriers in place, there's already a shortage of housing and housing is costing a lot of money. I just think you got to be careful that if you put more barriers up, either landlords that exist are just going to go it's all too hard, I'm selling, and they'll get out of the market, which again, there's not too many new investors coming in… (Private Property Manager 5)

There was a wide range of views among property managers about the importance or reality of climate change, from belief to some scepticism. Housing managers influence sustainability and climate change adaptation through property maintenance, regulatory-led change, incentive-led change, systematic programs to upgrade properties, retrofitting and new housing in the case of public providers. For property managers in the private sector, busyness and lack of resources were seen as constraints on their capacity to advocate or arrange for sustainability modifications to the properties they managed. Mediation between landlord and tenant was seen as a major part of the property manager's role, which left little time for promoting the sustainability of rental housing.

Private housing managers expressed frustration with changes to tenancy laws, such as recently introduced requirements for water-saving devices and smoke alarms which were cumbersome to implement and an example of the difficulties of engaging landlords with sustainability investments in their properties. Nevertheless they saw legislation as necessary to motivate landlords to make improvements to their properties, while noting a range of attitudes of landlords in this regard. On the other hand, some pointed out the significance of this mediating role in improving sustainability.

> You know, if something's broken [the landlord will] go ahead and fix it but they're not really looking at taking the next step to insulation and water compliance and those sorts of things on

their own without a push from me. So I guess the more I get to know this portfolio the more I get to know the owners, the more I can get them to do. (Private Property Manager 4)

> [Policymakers] should actually speak to people that manage rental property and find out what the story is and what they need to be doing. (Private Property Manager 2)

Public sector housing managers who were interviewed saw the public housing sector as a policy leader in sustainability and adaptation, and strongly believed that the organisation's recent housing projects were exemplary in their sustainability standards (Housing NSW 2011).

> So we don't provide massive homes but we provide smaller homes with a higher standard of environmental focus. (Public Housing Practitioner 6)

However they noted the complexity of the task and felt constrained by a lack of resources (human and financial), the busy reactive nature of their work and the size and bureaucracy of the organisation that made change difficult. Public housing managers recognised that providing a home for vulnerable people is critical in responding to climate change. They regarded the security of tenure enjoyed by many public housing tenants as enhancing their capacity to adapt their dwellings to be more sustainable or to respond to climate change.

Both public and private sector managers suggested that the provision of advice to tenants on energy or water efficiency needed to be done thoughtfully, noting the potential for advice to be patronising or intrusive. This aspect of the housing manager's role, along with their role in providing advice to landlords on maintenance, improvements and responses to tenants' requests, positions housing managers as crucial 'knowledge brokers' in improving the adaptive capacity of the rental sector. Some understanding of this role could be seen in private housing managers who expressed a strong need for more authoritative information and training in order to be able to advocate convincingly to landlords for sustainability improvements in rental properties.

So therefore if they had a course or something property managers could go – and again it would have to be right across the board – to learn how to convince landlords to be more environmentally friendly even if it cost them a few dollars. That might make a bit of training, how would you have those skills? Because at the moment it's not a priority for them. (Private Property Manager 3)

The shortage of rental housing in the case study area was acknowledged by both tenants and housing managers as a critical constraint on tenants' ability to influence the market through exercising a preference for more sustainable dwellings. The lack of housing options was perceived by tenants as disempowering, and the possibility of a lease being terminated was a reason for not seeking improvements to properties to make them more adaptive or sustainable.

I was starting to get desperate. I was starting to get really scared. Like I moved in on the Friday and started work on the Monday. That was how close it was. So environmental stuff was the least of my worries at that point. (Tenant 10: flat/private manager)

Both property managers and tenants acknowledged the 'split incentive', where investment by landlords (e.g. solar power or water-saving devices) resulted in savings for tenants. However both also acknowledged that incentive schemes such as subsidies or rebates and short tenancy contracts discouraged a tenant's own investment in a rental property because the tenant was generally not eligible for incentives, nor would they occupy the house for long enough to receive a return on their investment; instead it was the landlord who would benefit in the long term.

41.4 Assets and barriers to climate adaptation

I basically came to the point where I gathered I only have so much power as an individual and the best I can do is to set a good example, educate others and just talk to as many people as possible and be as passionate as I can about it. So that's the go. (Tenant 19, house/private manager)

Our analysis of documents, focus group discussions and interview transcripts revealed a number of assets in the rental sector for adapting to climate change. We found that tenants brought a range of strong assets to adaptation, including concern about sustainability and climate change, an array of adaptive household practices already embedded in their everyday lives, a desire to modify their homes to support these practices, engagement with the wider community on sustainability issues and a range of information networks. Barriers which limited the capacity of individuals and organisations to exercise these assets related to the quality, availability and affordability of housing stock, attitudes of stakeholders to climate change adaptation and sustainability, lack of resources, public perceptions of tenancy and economic and regulatory constraints – particularly restrictive tenancy laws – which acted as disincentives or obstacles to climate change adaptation.

Property managers also brought strong assets to adaptation in the sector including organisational systems for maintenance and improvement of physical housing stock, a critical 'knowledge broker' role for both tenants and landlords, disaster preparedness systems (in some cases) and experience in implementing legislated changes. For property managers, barriers to adaptation included a lack of leadership (particularly in the private sector) on climate change issues, limited time and resources, some scepticism about the need for adaptation and some negative experiences with implementing legislative changes, including the implications of these processes on their relationship with landlords and/or tenants.

41.5 Recommendations and outcomes

From our research we have developed a number of recommendations (Instone et al. 2013) and resources for stakeholders (e.g. http://www.youtube.com/user/SustainingRentalLife, https://www.facebook.com/SustainingRentalLife, brochure and best-practice information guides for landlords, property managers and tenants) which, while specific to a particular Australian context, offer an overview of potential areas for best-practice interventions in the rental sector elsewhere. Key arenas for action include changing cultures of landlord–housing-manager–tenant relations; enhancing rental housing provision and quality; and fashioning tenancy provisions that facilitate rather than hinder adaptation.

We see the need for an enhanced role for property managers as advocates and knowledge brokers for sustainability and adaptation in the rental sector. Such a role could be supported by training, professional development and information resources. This could be strengthened by a new landlord culture of 'ethical investment' in rental housing that provides both incentives for landlords to upgrade properties and opportunities for tenants to practice sustainability and adapt to climate change. In this regard, the public housing sector in Australia should build on its potential to be a leader in climate change adaptation and direct resources to retrofitting existing stock to increase the environmental performance of dwellings.

Security of tenure and length of leases are crucial attributes shaping adaptation potential in the rental sector. Diverse tenancy models are in place in other countries that are more flexible and more supportive of tenants in 'making a home'. In Germany for example, tenants are able to make use of gardens and tenancy agreements can specify a tenant's responsibilities for maintaining the garden (Hulse et al. 2011a). In the Netherlands, tenants on long leases can make alterations that do not need to be reversed when the tenant vacates the property (Hulse et al. 2011a).

Amending tenancy conditions to enhance the active contribution of tenants to climate change adaptation may require innovative legislative frameworks that reflect climate change and equity imperatives as well as protection for landlords and tenants. This would need to be supported by the development of information resources for both tenants and housing managers.

Governments have a responsibility to take action on affordable rental housing supply (in both the public and private housing sectors) to increase tenants' options for adaptive housing, especially when rental markets are unusually tight. Governments at all levels need to make explicit in policy and legislation any potential climate adaptation contributions and impacts in the rental sector; this includes setting regulatory sustainability standards for residential rental buildings and providing incentives for adaptive retro-fitting and environmental upgrades of older rental housing. Equity, and the role of governments in activating the adaptive capacity of marginalised and disadvantaged citizens, are important elements here.

41.6 Conclusion

Our asset-based and pro-poor approach and the methods employed – literature and document reviews, interviews, focus groups and transcript analysis – revealed capacities and barriers that required specific, local responses. However, the resources and recommendations arising from our study developed out of a theoretical and methodological approach with wider application. Policy settings and structural factors in the rental sector will vary between countries offering diverse assets and constraints. However, the conceptual frame of 'more-than-adaptation' – within which socio-economic concerns such as equity and cultural aspects, for example landlord–tenant relationships, are embedded alongside material practices – offers a multi-stranded approach with potential value in enhancing the adaptive capacity of rental sectors elsewhere in Australia and overseas. The methodology utilised in this case study offers a framework for researchers, policymakers and stakeholder organisations in rental housing sectors elsewhere who wish to find out more about the relationships between stakeholders and their assets (capacities, knowledge and practices) and the barriers which can marginalise and reduce the sector's adaptive capacity. The strength of our conclusions and recommendations lies in the adaptive capacities waiting to be discovered.

References

ABS (2012a) *2011 Census of Population and Housing: Basic Community Profile for Lake Macquarie LGA14650.* Australian Bureau of Statistics, Canberra. Available at http://www.censusdata.abs.gov.au/census_services/getproduct/census/2011/communityprofile/LGA14650 (accessed 6 June 2013).

ABS (2012b) *2011 Census of Population and Housing: Basic Community Profile for Newcastle LGA15900.* Australian Bureau of Statistics, Canberra. Available at http://www.censusdata.abs.gov.au/census_services/getproduct/census/2011/communityprofile/LGA15900 (accessed 6 June 2013).

Barnett, J., Dovers, S., Hatfield-Dodds, S. et al. (2009) *National Climate Change Adaptation Research Plan: Social, Economic and Institutional Dimensions (Consultation Draft November 2009).* National Climate Change Adaptation Research Facility, Gold Coast.

Barnett, J., Dovers, S., Hatfield-Dodds, S. et al. (2011) *National Climate Change Adaptation Research Plan: Social, Economic and Institutional Dimensions.* National Climate Change Adaptation Research Facility, Gold Coast.

Bierre, S., Howden-Chapman, P. and Signal, L. (2010) 'Ma and Pa' landlords and the 'risky' tenant: discourses in the New Zealand private rental sector. *Housing Studies* 25(1), 21–38.

Dillon, R., Learmonth, B., Lang, M., et al. (2009) *Energy Efficiency for Low-Income Renters in Victoria.* Just Change Australia, Melbourne. Available at http://www.justchangeaustralia.org/images/JustChange.pdf (accessed 12 June 2014).

Eurostat (2013) *Housing Statistics.* European Commission. Available at http://epp.eurostat.ec.europa.eu/statistics_explained/index.php/Housing_statistics#Tenure_status (accessed 6 June 2014).

Evans, D. (2011) Consuming conventions: sustainable consumption, ecological citizenship and the worlds of worth. *Journal of Rural Studies* 27, 109–115.

Gabriel, M., Watson, P., Ong, R., Wood, G. and Wulff, M. (2010a) *The Environmental Sustainability of Australia's Private Rental Housing Stock. AHURI Final Report No.159.* Australian Housing and Urban Research Institute, Melbourne.

Gabriel, M., Watson, P., Ong, R., Wood, G. and Wulff, M. (2010b) *The Environmental Sustainability of Australia's Private Rental Housing Stock. AHURI Positioning Paper No.125.* Australian Housing and Urban Research Institute, Melbourne.

Grineski, S.E. and Hernández, A.A. (2010) Landlords, fear, and children's respiratory health: An untold story of environmental injustice in the central city. *Local Environment: The International Journal of Justice and Sustainability* 15 (3), 199–216.

Gurran, N., Hamin, E. and Norman, B. (2008) *Planning for Climate Change: Leading Practice Principles and Models for Sea Change Communities in Coastal Australia.* University of Sydney, Sydney.

Haines, A. and Green, G.P. (2012) *Asset Building and Community Development.* Sage Publications, Thousand Oaks.

Hammar, M. (2004) A multitude of tenure forms in South East Asia. *Global Tenant,* April 2014, 3.

Housing NSW (2011) *Environmental Sustainability in Housing NSW 2011/12- 2013/14.* NSW Government, Ashfield.

Hulse, K., Milligan, V. and Easthope, H. (2011a) *Secure Occupancy in Rental Housing: Conceptual Foundations and Comparative Perspectives.* Australian Housing and Urban Research Institute, Melbourne.

Hulse, K., Milligan, V. and Easthope, H. (2011b) *Secure Occupancy in Rental Housing: Conceptual Foundations and Comparative Perspectives. AHURI Final Report No. 170.* Australian Housing and Urban Research Institute, Melbourne.

Instone, L., Mee, K., Palmer, J., Williams, M. and Vaughan, N. (2013) *Climate Change Adaptation and the Rental Sector.* National Climate Change Adaptation Research Facility, Gold Coast.

Kretzmann, J.P. and McKnight, J.L. (1993) *Building Communities from the Inside Out: A Path Toward Finding and Mobilizing a Community's Assets.* Center for Urban Affairs and Policy Research, Northwestern University, Evanston.

Lane, R. and Gorman-Murray, A. (2011) *Material Geographies of Household Sustainability.* Ashgate Publishing, Farnham.

Lister, D. (2004) Young people's strategies for managing tenancy relationships in the private rented sector. *Journal of Youth Studies* 7(3), 315–330.

Lister, D. (2005) Controlling letting arrangements? Landlords and surveillance in the private rented sector. *Surveillance and Society* 2(5), 513–528.

Mee, K. and Vaughan, N. (2012) Experiencing home. In: Smith, S.J., Elsinga, M., O'Mahony, L.F. et al. (eds) *International Encyclopedia of Housing and Home,* vol. 2. Elsevier, Oxford, pp. 146–151.

Moser, C. (2011) A conceptual and operational framework for pro-poor asset adaptation to urban climate change. In: Hoornweg, D., Freire, M., Lee, M.J. et al.

(eds) *Cities and Climate Change: Responding to an Urgent Agenda, Volume 1*. The World Bank, Washington, pp. 225–253.

Moser, C. and Satterthwaite, D. (2010) Towards pro-poor adaptation to climate change in the urban centres of low-and middle-income countries. In: Mearns, R. and Norton, A. (eds) *The Social Dimensions of Climate Change: Equity and Vulnerability in a Warming World*. The World Bank, Washington, pp. 231–258.

Prowse, M. and Scott, L. (2008) Assets and adaptation: an emerging debate. *IDS Bulletin* 39 (4), 42–52.

Reid, L., Sutton, P. and Hunter, C. (2010) Theorizing the meso level: the household as a crucible of pro-environmental behaviour. *Progress in Human Geography* 34(3), 309–327.

Stanley, J. (2009) *Promoting Social Inclusion in Adaptation to Climate Change. Report 09/4 For Sustainability Victoria*. Monash Sustainability Institute, Melbourne.

Sullivan, D. (2007) *Climate Change: Addressing the Needs of Low-Income Households in the Private Rental Market: Background Paper*. Brotherhood of St Laurence, Melbourne.

Tenants Union of Victoria (2010) *Submission to Prime Minister's Task Group on Energy Efficiency*. Tenants Union of Victoria. Available at http://www.tuv.org.au/articles/files/bulletins/100517_PM_Energy_Efficiency_Task_Force.pdf (accessed 12 June 2014).

Toohey, S. and Fritze, J. (2009) *A Future Focussed Housing Standard: The Case for Rental Housing Standards to Help Vulnerable Households Adapt to Climate Change*. VCOSS, Melbourne.

UN-HABITAT (2008) *Housing the Poor in Asian Cities, Quick Guide 7*. UNESCAP, Bangkok.

Section 8

Adaptation and disaster management

42 Practical adaptation: past, present and future

Adaptation and Impacts Research Group, Meteorological Services Canada, Canada

In the climate change community it is common practice to speak of 'adaptation' as though it were something entirely new. Certainly, adaptation to *anthropogenic* climate change is new, since the impacts and speed of our current changing climate are unprecedented. However, adaptation to different climates has been a success story for millennia. Humans have spread around the globe, adapting to diverse weather variabilities and extremes. Adaptation seems to have been highly successful overall, as exemplified by the proven capacity of humans to exist and thrive in tropical and temperate forest zones, in prairies and savannah grasslands, in mountain regions and in semi-arid and sub-arctic conditions. Part of this success may be attributed to an experiential learning process through which people have learned wherever possible to stay away from and avoid the most hazardous locations, and otherwise to adapt their building designs and livelihoods to all but the most extreme events. There are lessons to be learned from historical adaptation to climate and weather that are relevant to the current search for practical adaptation to anthropogenic climate change.

In more recent times this traditional wisdom seems to have been lost or largely ignored. The growth of population and the need for more living space has led to the increasing occupation of lands and sites known to be hazardous. Despite a great increase in scientific knowledge about the spatial and temporal distribution of hazardous areas and events, and their magnitudes and probabilities, exposure continues to grow. More important than simple exposure is vulnerability. Despite the substantial improvements in building designs and standards, and the availability of higher-quality and more resistant materials, much of the expansion of human settlements in infrastructure, housing and public facilities such as hospitals and schools is of inferior quality. In short, people worldwide are becoming less well adapted than before. Weather-related losses and damage have risen from an annual average of approximately US$ 50 billion in the 1980s to US$ 165 billion over the last decade to 2010 (MunichRe 2013).

In other words, we have a current and growing adaptation deficit. Is there a chance that practical adaptation to anthropogenic climate change could halt and then reverse this trend? For that to happen there would have to be a much greater understanding and recognition of the underlying causes of the adaptation deficit. It has been proposed that a new generation of integrated forensic disaster investigations be developed to seek out and reveal such causes (Burton 2010;

Applied Studies in Climate Adaptation, First Edition. Edited by Jean P. Palutikof, Sarah L. Boulter, Jon Barnett and David Rissik.
© 2015 John Wiley & Sons, Ltd. Published 2015 by John Wiley & Sons, Ltd.

IRDR 2011). Such work may eventually lead to the creation of an Intergovernmental Panel for Disasters and/or a National or International Disaster Safety Board.

Much can be done to improve practical adaptation at the local level. If current trends continue without a sea change in the pattern of unsustainable development however, the adaptation deficit will continue to grow. There is a current tendency to think of adaptation as a limited, place-based activity, only responding to the unique combination of circumstances in each locality. A broader view of adaptation is required both temporally and spatially. There is a need to increase the sense of responsibility on the part of decision-makers, practitioners and policymakers from local to global levels.

An illustration of the growing temporal requirements can be seen in the recently announced Asian Development Bank funding for the Coastal Climate-Resilient Infrastructure Project in Bangladesh (ADB 2012, 2013). The ADB has provided a loan of US$ 20 million with co-financing of US$ 68 million from the International Fund of Agriculture and Development and the KfW Development Bank. There is an additional US$ 30 million from the Strategic Climate Fund and US$ 31 million from the Government of Bangladesh (ADB 2012). This US$ 160 million project aims to increase climate resilience and reduce vulnerability by 'upgrades to around 540 kilometers of roads, bridges and culverts, as well as improvements to rural markets'. The project will also 'build and improve cyclone shelters and animal shelters' (ADB 2013). This project has a good short-term rationale and may well serve to improve the local economy and reduce current risks from tropical cyclones and sea-level rise. To the extent that the project is successful it may be expected to encourage the population to remain in this hazardous area and may also attract more if the economy strengthens as forecast. However, tropical cyclones and floods are predicted to increase in frequency and magnitude with climate change, and there is a potential for the dislocation of some 35 million people by 2050 (ADB 2012, 2013; IPCC 2012). Furthermore, the global temperature increases that have already occurred and the additional increase to which the planet is already committed make it almost certain that large parts of the project area will be permanently under sea level. This is likely to be an example of successful and practical adaptation in the short term, but ultimately it will fail to bring about sustainable development. Such projects have been described as palliative adaptation (Tompkins et al. 2008; Dickinson and Burton 2014).

A second caveat about practical adaptation at the local and place-based level is that the consequences of climate-related disaster are less confined in spatial terms. The integration of the global economy, coupled with the speed of communication and large expansions in the movement of goods and people, means that the consequences of climate and weather events are no longer restricted to one locality, region or country. Practical adaptation everywhere must begin to take account of the transmission of impacts from distant places. Climate change adaptation will have to take account of the cascading epidemic or pandemic nature of impacts.

To be successful, practical adaptation has to learn lessons from the past, find ways of improvement to address the current adaptation deficit and, in the future, avoid the pitfalls of palliative adaptation and pandemic impacts.

Acknowledgements

The author acknowledges the assistance of Thea Dickinson in the preparation of this essay.

References

ADB (2012) *Press Release: ADB, Partners to Help Protect Rural Bangladesh from Climate Impacts.* Asian Development Bank. Available at http://www.adb.org/news/bangladesh/adb-partners-help-protect-rural-bangladesh-climate-impacts (accessed 7 June 2014).

ADB (2013) Project Data Sheet: Coastal Climate-Resilient Infrastructure Project. Asian Development Bank. Available at http://www.adb.org/printpdf/projects/45084-002/main (accessed 12 February 2014).

Burton, I. (2010) Forensic disaster investigations in depth: A new case study model. *Environment Magazine* 52, 5.

Dickinson, T. and Burton, I. (2014) Palliative climate change planning and its consequences for youth. In: Elder, J. (ed.) *The Challenges of Climate Change: Children in the Front Line, Innocenti Insight*. UNICEF Office of Research, Florence.

IPCC (2012) *Managing the Risks of Extreme Events and Disasters to Advance Climate Change Adaptation. A Special Report of Working Groups I and II of the Intergovernmental Panel on Climate Change*. Field, C.B., Barros, V., Stocker, T.F. et al. (eds) Cambridge University Press, Cambridge, UK, and New York, NY, USA.

IRDR (2011) *Forensic Investigations of Disasters: The FORIN Project*. IRDR FORIN Publication No. 1. Integrated Research on Disaster Risk, Beijing.

MunichRe (2013) *Press release: Natural Catastrophe Statistics for 2012 Dominated by Weather Extremes in the USA*. Reinsurance Press, Munich, Germany. Available at http://www.munichre.com/en/media_relations/press_releases/2013/2013_01_03_press_release.aspx (accessed 14 February 2014).

Tompkins, E.L., Lemos, M.C. and Boyd, E. (2008) A less disastrous disaster: Managing response to climate-driven hazards in the Cayman Islands and NE Brazil. *Global Environmental Change-Human and Policy Dimensions* 18, 736–745.

43 Community resilience to disaster in four regional Australian towns

College of Arts, Society and Education, James Cook University, Australia

... I don't think anyone else can understand the desperation you feel when it's your place. You'd do anything ... The wind changed and it (the fire) came back at us. We'd already lost two hundred and fifty acres of our farm and you're trying to save every last bit ... people say it's only grass. They don't understand! It's not only grass... that grass is our life! (A farmer quoted in Boon et al. 2012, p. 319)

43.1   Introduction

Disasters and climate change impacts have cross-scale effects, disrupting functioning across multiple levels of socio-ecosystems. In a global context of increasing incidence of natural disasters there is widespread interest in understanding patterns of resilience exhibited by communities and individuals that have experienced natural disaster in order for governments and organisations to instigate appropriate adaptation practices. For Australia, climate change risk scenarios for the future (2030) show a high probability of increased average temperatures, sea-level rises and water cycle implications, including higher intensity and frequency of floods, storm surges and droughts (Bureau of Meteorology and CSIRO 2012).

The aim of this research was to identify significant generic factors that supported community resilience to four disaster-impacted communities: Beechworth and Bendigo in Victoria and Ingham and Innisfail in Queensland, Australian regional towns recovering from bushfire (Black Saturday fires, 2009), prolonged drought (c. 1997–2010), multiple successive floods (2009) and cyclone (Cyclone Larry, 2006) respectively. The study sites were located in northern, tropical and southern, temperate areas of Australia, respectively (Fig. 43.1).

The category 5 Cyclone Larry resulted in the evacuation of over 300 people from Innisfail and surrounds; 30 people were injured and around 19,000 houses, sheds and garages required repair. Insurance claims came to over AU$ 369 million from 25 different insurance companies. In Ingham the two severe riverine floods that impacted the town reached the third-highest level recorded in Ingham's history. Forty people were evacuated from their homes, while many residents were isolated by floodwaters causing them to be trapped in their homes for a week. Emergency Management Queensland estimated that 65% of the Hinchinbrook Shire, about 2900 residences and businesses, were affected by the floodwaters, with initial estimates of infrastructure damage of AU$ 120 million. The Beechworth bushfires, part

Applied Studies in Climate Adaptation, First Edition. Edited by Jean P. Palutikof, Sarah L. Boulter, Jon Barnett and David Rissik.
© 2015 John Wiley & Sons, Ltd. Published 2015 by John Wiley & Sons, Ltd.

Figure 43.1 Location of the four study sites: Innisfail, Ingham, Bendigo and Beechworth.

of the more extensive Black Saturday fires which affected 78 towns and left about 7500 people homeless, claimed two lives, burnt through 30,000 hectares and destroyed several homes. The Bushfires Royal Commission estimated the cost of the Black Saturday fire disaster at AU\$ 4.4 billion (Boon et al. 2012). A creeping, slow onset drought affected the people of the City of Greater Bendigo, with impacts ranging from deteriorating mental health to economic and infrastructure ramifications: farmers repeatedly faced crop failures, the death of trees and starving stock due to low rainfall and changes to water allocation rules and policies (Kiem et al. 2010); heat stress on humans led to increased rates of hospital admissions and deaths (particularly among the elderly, people under physical stress and those with cardiovascular disease); and public amenities deteriorated as recreational facilities, parks and gardens declined due to water restrictions, impacting community connectedness and mental health (Boon et al. 2012).

43.2 Community and individual resilience links and definitions

Community resilience can be difficult to assess. At the community level resilience is understood as the ability of communities to withstand external shocks to their social infrastructure (Adger 2000). Therefore in assessing community resilience the economic, institutional, social and ecological dimensions of a community must be considered (Adger 2000). Community resilience is clearly related to the stability of its population and therefore it also depends on the resilience of individuals; it can only be fostered if relevant stakeholders who operate within a community's economic and social systems are also resilient.

Arbon et al. (2012) defined a disaster resilient community as one '... where members of its population are connected to one another and work together, so that they are able to function and sustain critical systems, even under stress; adapt to changes in the physical, social or economic environment...' (Arbon et al. 2012, p. 7). Community resilience is then an amalgam of the resilience of individuals, families and organisations of the community. Population movement might therefore be evidence of community instability. The IPCC (2007) view migration in response to climate-change-induced disasters as a failure to adapt. The IPCC (2007) urged for adaptation to a changing environment *in situ* rather than the abandonment of a location. In the wake of significant external stress such as a natural disaster, population displacement is often an indicator of the breakdown of a community's social resilience. In sum, resilient communities are believed to prevent community erosion post-disaster by having a set of adaptive capacities and strategies for promoting effective disaster readiness and responses (Norris et al. 2008b).

Integral to community resilience is the personal resilience of individuals. It has been defined as the 'dynamic process wherein individuals display positive adaptation despite experiences of significant adversity or trauma' (Luthar and Cicchetti 2000, p. 858). Personal resilience, thought to be the result of both a personality trait and developmental effects (Boon et al. 2012), arises from interactive processes across multiple levels of functioning including social interactions with family, peers and community systems. Risk or adversity can be the result of psychological, environmental or socio-economic factors that are

associated with an increased probability of negative outcome.

To examine and assess community resilience it was therefore important to look at both macro- (community level) and micro- (individual level) factors that promoted resilience. Bronfenbrenner's (1979) bioecological systems theory offered the most appropriate lens through which to analyse community resilience across the four disaster impacted towns. While Bronfenbrenner's theory has been used extensively, for example in the context of child development, this was the first time it was used to model disaster resilience. Bronfenbrenner (1979) structured an individual's social context into five areas:

1. microsystem: where the individual participates directly;

2. mesosystem: microsystem members' interactions, such as communications between members of a microsystem;

3. exosystem: entities and organisations that might be accessed by the individual or their family;

4. macrosystem: the politics, views and customs that represent the cultural fabric of the individuals' society; and

5. chronosystem: the elements of time as they relate to events in the individual's environment (Fig. 43.2).

Using this theoretical lens it was possible to structure the research in a way that allowed for the extraction of both the most salient influences on resilience and their location within the ecosystem so that policy recommendations could be made to enhance future adaptation practices. Moreover, the observed macro factors of each community could be accounted for in context to help decide whether the disaster experienced had eroded the community as a whole, suggesting signs of lack of resilience.

43.3   Methods

Given the considerations about community resilience described above, a key hypothesis underpinning the research was that individuals

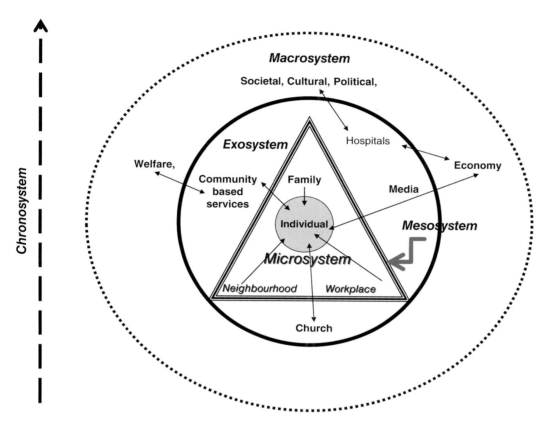

Figure 43.2 Conceptual scheme of Bronfenbrenner's (1979) systems and their interactions.

who remained in the disaster-impacted communities were likely to be resilient to disaster and therefore their views and attitudes could be used to obtain generic variables that support resilience, as well as disaster-specific factors that helped communities. Three of the disasters under study were fast-onset (wildfire, flood and cyclone) and the other (prolonged drought) was slow-onset, mirroring a likely future climate change scenario.

A stepwise mixed-methods research design was adopted over the three-year period of the research. Demographic data were used to profile communities for comparisons, to determine representativeness of samples and to compare community functioning, pre- and post-disaster, to assess disaster impacts and community level resilience. To determine what factors helped individuals'

disaster resilience, individual and group interviews were conducted with 186 people from the 4 communities. Results from the interviews were used to construct surveys in order to generalise results across the four sites and across other rural communities. The surveys were completed by a sample of 1008 people from the four sites approximately six months after the completion of the interview phase. Completed surveys were returned from:

1. Ingham, Queensland (multiple floods), representing 287 households;
2. Innisfail, Queensland (Cyclone Larry), representing 231 households;
3. Beechworth, Victoria (Black Saturday wildfires), representing 249 households; and
4. Bendigo, Victoria (Millennium Drought), representing 241 households.

It was therefore possible to triangulate and generalise the participants' views. Rasch analysis was used to quantify the factors identified from survey analyses and to compare the resilience of individuals across the four disaster-impacted communities. These factors were then modelled by structural equation modelling (SEM) to detect the underlying interactions and influences upon disaster resilience, using Bronfenbrenner's theory (1979) as a guiding framework.

43.4   Results and discussion

The first step in assessing community resilience was to examine key demographic characteristics of the four communities to ascertain whether each community's socio-economic fabric had been eroded by the disaster and whether the population in each place was stable. Extensive examination of 2001, 2006 and 2011 census data for each community indicated a stable population and economic picture as measured by employment rates, population numbers, school enrolments, business entities and other socio-demographic indicators. These data indicated that the four communities were not significantly changed pre- and post- disaster, suggesting overall macro community resilience (Boon et al. 2012). Results for the Bendigo drought show different trends to previous studies of drought-affected rural communities (e.g. Kiem et al. 2010). This could be explained by the large population of Bendigo and its economy, which is not just based on agriculture. The main economic drivers of Bendigo are health care, education, manufacturing and social assistance (Roggema et al. 2014).

Survey samples used to generalise interview findings were broadly reflective of each community's demographics. Sample characteristics which were important in interpreting results showed community samples were well matched by numbers and gender, although Queenslanders were younger than Victorians and had completed fewer years of formal education. Of the four study towns, the Ingham community had the highest disaster resilience as reported through interviews

and confirmed by the survey and census data. Compared to the other three samples, the Ingham sample had the highest level of employment and respondents had been living in their community longest. Most respondents from Ingham and Innisfail received financial assistance from the state or federal government as a result of the disaster, unlike the Victorian samples. Both Queensland samples had more experience of weather-related disasters and economic disadvantage than their Victorian counterparts possibly because, in the case of Bendigo, there had not been a major drought in the area since the World War II drought in 1937–1945 (Verdon-Kidd and Kiem 2009). The Innisfail sample, whose town had sustained most damage as a result of the cyclone, reported the least preparedness for future disaster events by way of building or contents insurance cover. On the other hand, Bendigo residents reported the highest level of insurance for future disaster events. These differences likely reflect the higher rates of home ownership and higher levels of education of the Victorian samples compared to the Queensland samples.

Disaster resilience for fast-onset events such as flood, cyclone and wildfire, and to some extent for slow-onset events, is strongly predicted by adequate preparedness (e.g. Cutter et al. 2008; Gissing et al. 2010), which is in turn dependent on clear and timely communications (Cutter et al. 2008). Preparedness has the potential to confer community and individual resilience (Kumagai et al. 2004). The majority of survey respondents stated that they were warned well in advance of the acute-onset events (flood, cyclone and fire) and, even in the case of the drought, 44% of the Bendigo sample stated they received the first warning in time to prepare for the drought. Queenslanders were significantly more emotionally prepared for the flood and cyclone than the Victorians were for wildfire and drought, and they received more help from friends, neighbours and family during those events than the Victorians. This help took the form of emotional and/or material support by way of communications, emergency equipment and/or viands or shelter. It is noteworthy that Bendigo residents overall stated they were not prepared for drought and were the least supported by friends/family, community services, communications and local government; this is a probable reflection of both the nature of the event (slow onset, drought) and the much bigger size of Bendigo which has a population of 10–11 times that of the other three towns, likely leading to less community cohesion.

Accurate and trusted communications have been constantly cited as important in assisting the process of preparation (Colten et al. 2008). Results here showed that neighbours, family or local community members were important sources of communication for those in Ingham, while the Country Fire Authority/State Emergency Services were more important for Innisfail and Beechworth.

Prior experience was consistently noted during interviews as being important for preparedness and a precursor to disaster resilience. Despite some Queensland interview participants' conten-tions that prior experiences sometimes led to complacency and a 'wait and see' attitude, SEM analyses showed household preparedness was sig-nificantly predicted by prior experiences of disas-ters and other traumatic events.

A number of general patterns were consis-tently found across the four communities. SEM analyses (Fig. 43.3) detected links between the various factors that supported resilience, which showed previously suspected but unreported associations between resilience and adaptability and the influences of other community-based factors upon individuals' resilience. Moreover, SEM analyses confirmed the study's hypothesis that individuals remaining in a community were resilient, since no links between resilience and a desire to leave the community were identified. In all communities a desire to leave the community, accompanied by very low resilience and adapt-ability, was predicted by poor health in immediate family or friends, their microsystem, and lack of

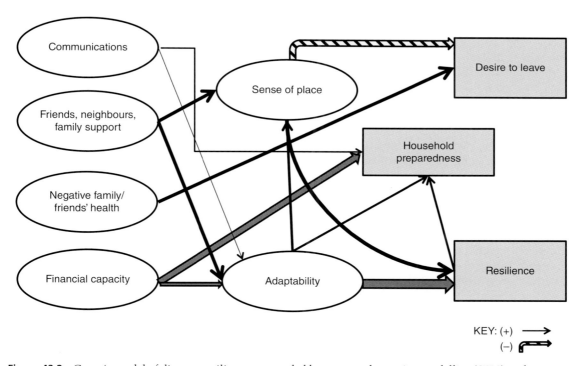

Figure 43.3 Generic model of disaster resilience as revealed by structural equation modelling (SEM) analyses.

social connectedness or a sense of place. Financial assistance received from government and charity organisations (i.e. support from macrosystem entities) also predicted leaving. This highlighted the link between economic vulnerability and mobility. A desire to leave is tempered by the capacity to leave, and both dictate actual mobility. From the perspective of macro-level community resilience, community stability pre- and post-disaster determined by census data was likely maintained by providing economic assistance. Although it is possible that stoicism rather than resilience *per se* might have kept some individuals in these communities post-disaster, the levels of functioning of the communities at population level suggest that the communities were resilient. It is important to note that the focus was on community resilience, and how resilience of individuals contributed to it, rather than individual resilience *per se* which in some cases might be better characterised by leaving a community rather than staying put. These results provide empirical evidence for the factors that predict disaster resilience in individuals and support the contentions of prior researchers (Norris et al. 2008a).

More importantly, results also showed that individuals' disaster resilience across all sites was both a personal trait and a process facilitated by adaptability and community factors, a result providing empirical evidence for the contentions of prior theorists (e.g. Tusaie and Dyer 2004; Masten and Obradovic 2006). By far the strongest direct pathways to resilience arose from a sense of place and adaptability. Indirect influences upon resilience, mediated by adaptability and showing that resilience can be developed, were financial capacity, support of family and friends (a microsystem effect), communications about the weather event and climate change knowledge and trust in climate change communication sources. The sources of support for individual and community resilience were distributed across Bronfenbrenner's ecosystem levels with a varying degree of importance. Across all research sites, generic factors that enhanced disaster resilience were: the support of family, friends and neighbours (microsystem); a

sense of place; financial capacity and climate change knowledge, in terms of self-reported knowledge of climate change science; and trust in climate change communications from scientists but not the government or the media.

Individual safety and wellbeing were revealed by SEM analyses to be strong contributors to community resilience and recovery because they negatively predicted leaving, a finding that was consistent at all sites and for all disaster types, echoing prior contentions (e.g. Norris et al. 2008a). On the other hand, climate change concerns, which varied by location, education and age, had mixed effects on the four different communities. Queenslanders were less convinced than Victorians that climate change was a problem or that it was caused by anthropogenic carbon dioxide emissions; this was also the case across all sites for the less-well-educated, older and longer-standing residents in each town. SEM analyses revealed some differences between the different sites in relation to climate change views and their associations. Climate change knowledge and trust of climate change communications predicted adaptability, leaving and a poor sense of place in the Bendigo and Ingham samples. The Innisfail sample followed the same patterns, although knowledge of climate change and trust in climate change communications did not predict leaving. Climate change views did not play a significant role for the Beechworth sample. Overall, individuals most likely to want to leave were those who rented their property and were newcomers to their community. Climate change views and their links highlighted that for some individuals climate change knowledge, if not coupled with adaptability, could possibly lead to fear and leaving a disaster-impacted community if they were not connected to the community or had a sense of place.

The demographic profiles of each of the four communities comparing pre-disaster community data with post-disaster community data supported the hypothesis that individuals who remained in the community were likely to be resilient, and that these communities were resilient to disaster since their populations were stable despite the impact of disasters. For the 5% of individuals who wanted to leave the community whose resilience was not supported by the other community factors, the financial support from state, federal and charity bodies most likely helped them remain in the community. It could therefore be inferred that macrosystem financial aid increases community disaster resilience by helping a town's population remain stable.

43.5   Implications

The study confirmed prior overseas research and extended knowledge for Australian rural and regional settings (Kiem et al. 2010; Boon et al. 2012). Emergency managers and policymakers who wish to enhance community resilience and adaptation to climate change should remember that individual resilience supports community resilience. While it is a character trait, because it is predicted by adaptability it can be developed through community processes and activities that promote social cohesion. The Ingham community sample was most socially connected and also most resilient.

Policies must be tailored to the needs of each community. These must identify and provide targeted assistance to the most vulnerable: the economically marginalised, those over 55 and the less well educated so that they do not remain isolated and are better included into the community (Field et al. 2012).

Accurate and timely communications in advance of an impending weather event (Cutter et al. 2008) were found to be critical to household preparedness and must be temporally and spatially as precise as possible. Positive role models for disaster preparedness, promoted via media or community activities, can increase individuals' preparedness through powerful social learning.

A consistent finding across the four communities was that household preparedness was primarily predicted by individuals' financial capacity, as defined by their capacity to meet the costs of the impacts of the event and their insurance cover for the event's damage, supporting other research findings (e.g. Davidson and McFarlane 2006). Policies and programs therefore need to provide specific assistance to those whose financial circumstances prevent them from adequately preparing for disasters. This may take the form of subsidised insurance and low-cost subsidised emergency kits to reduce dependence upon assistance from charities and build self-efficacy for disaster resilience.

A most important finding was the effect of social connectedness upon disaster resilience, empirical evidence new to the disaster literature. The strongest predictor of emotional preparedness across all sites was assistance from neighbours, while a sense of place was the key to residents banding together during the response phase and keeping residents in the community after a disaster. Policies and initiatives must recognise the importance of social connectedness in building community resilience, by fostering stronger connections between neighbours and increasing individual's sense of place through local community programs.

Education is needed to promote adaptation to climate change and enhance disaster resilience. Findings showed gaps in awareness and understanding of climate change in the communities, which could prevent appropriate adaptation to climate change risks. These gaps were linked to education levels. Preparedness was also predicted by adaptability and resilience, a finding new to the literature of disaster resilience. This highlights an opportunity for educational interventions to be introduced at community level. Schools could be used as centres for information dissemination about climate change, with up-to-date evidence-based about the risks and responses needed for climate change. There is a corresponding need to ensure that current and future teachers are prepared and aware of climate change science (Boon 2010) by developing appropriate training in tertiary teacher training institutions to correct gaps in their knowledge and understanding.

The policy proposals suggested can be applied across diverse cultural groups and are equally appropriate for developed and developing countries. Finally, as Cutter et al. (2008) have stressed, the resilience of a community is based on macro-level or population factors. If the community shrinks because of out-migration as a result of a disaster, then this is not an indicator of a resilient community. On the other hand, if those remaining in the community do so with increased levels of hardship, displaying stoicism (Kiem et al. 2010) but not adapting or able to cope, this does not render the community resilient either. For resilience, the community must be able to learn from their experience to adapt accordingly (Arbon et al. 2012). For some entities this might mean a completely new way of operating.

References

Adger, W.N. (2000) Social and ecological resilience; are they related? *Progressive Human Geography* 24 (3), 347–364.

Arbon, P., Gebbie, K., Cusack, L., Perera, M.D. and Becon, V.S. (2012) *Developing a Model and Tool to Measure Community Disaster Resilience.* Torrens Resilience Institute, Adelaide.

Boon, H.J. (2010) Climate change? Who knows? A comparison of secondary students and pre-service teachers. *Australian Journal of Teacher Education* 35 (1), 9.

Boon, H.J., Millar, J., Lake, D., Cottrell, A. and King, D. (2012) *Recovery from Disaster: Resilience, Adaptability and Perceptions of Climate Change.* National Climate Change Adaptation Research Facility, Gold Coast.

Bronfenbrenner, U. (1979) *The Ecology of Human Development: Experiments by Nature and Design.* Harvard University Press, Cambridge, MA.

Bureau of Meteorology and CSIRO (2012) *State of the Climate 2012.* Bureau of Meteorology, Melbourne.

Colten, C.E., Kates, R.W. and Laska, S.B. (2008) Three years after Katrina: Lessons for community resilience. *Environment* 50 (5), 36–47.

Cutter, S.L., Barnes, L., Berry, M. et al. (2008) A place-based model for understanding community resilience to natural disasters. *Global Environmental Change* 18, 598–606.

Davidson, J. and McFarlane, A. (2006) The extent and impact of mental health problems after disaster. *Journal of Clinical Psychiatry* 67(2), 9–14.

Field, C.B., Barros, V., Stocker, T.F. et al. (eds) (2012) *Summary for Policymakers, Managing the Risks of Extreme Events and Disasters to Advance Climate Change Adaptation. A Special Report of Working Groups I and II of the Intergovernmental Panel on Climate Change.* Cambridge University Press, Cambridge.

Gissing, A., Keys, C. and Opper, S. (2010) Towards resilience against flood risks. *Australian Journal of Emergency Management* 25 (2), 39–45.

IPCC (2007) *Climate Change 2007: Impacts, Adaptation and Vulnerability. Contribution of Working Group II to the Fourth Assessment Report of the Intergovernmental Panel on Climate Change.* Parry, M.L., Canziani, O.F., Palutikof, J.P., van der Linden, P.J. and Hanson, C.E. (eds) Cambridge University Press, Cambridge.

Kiem, A.S., Verdon-Kidd, D.C., Boulter, S.L. and Palutikof, J.P. (2010) *Learning from Experience: Historical Case Studies and Climate Change Adaptation.* National Climate Change Adaptation Research Facility, Gold Coast.

Kumagai, Y., Carroll, M.S. and Cohn, P. (2004) Coping with interface wildfire as a human event: lessons from the disaster/hazards literature. *Journal of Forestry* 102(6), 28–32.

Luthar, S.S. and Cicchetti, D. (2000) The construct of resilience: implications for interventions and social policies. *Developmental Psychopathology* 12, 857–85.

Masten, A.S. and Obradovic, J. (2006) Competence and resilience in development. *Annals of the New York Academy of Sciences* 1094, 13–27.

Norris, F.H., Sherrieb, K., Galea, S. and Pfefferbaum, B. (2008a) Capacities that promote community resilience: can we assess them? Paper presented at the *2nd Annual Department of Homeland Security University Network Summit, Washington, DC.* Available at https://www.orau.gov/DHSsummit/2008/presentations/Mar20/Norris.pdf (accessed 12 June 2014).

Norris, F.H., Stevens, S.P., Pfefferbaum, B., Wyche, K.F. and Pfefferbaum, R.L. (2008b) Community resilience as a metaphor, theory, set of capacities, and strategy for disaster readiness. *American Journal of Community Psychology* 41, 127–150.

Roggema, R., Martin, J., Remnant, M., Alday, G. and Mansfield, P. (2014) Design charrettes in two days: Sea Lake and Bendigo. In: Roggema, R. (ed.) *The Design Charrette*, Springer, Netherlands, pp. 117–149.

Tusaie, K. and Dyer, J. (2004) Resilience: a historical review of the construct. *Holistic Nursing Practice* 18, 3–8.

Verdon-Kidd, D.C. and Kiem, A.S. (2009) Nature and causes of protracted droughts in Southeast Australia: Comparison between the Federation, WWII and Big Dry droughts. *Geophysical Research Letters* 36, L22707.

44 Sink or swim? Response, recovery and adaptation in communities impacted by the 2010/11 Australian floods

DEANNE K. BIRD[1], DAVID KING[2], KATHARINE HAYNES[1], PAMELA BOX[3] AND TETSUYA OKADA[1]

[1]Risk Frontiers, Macquarie University, Australia
[2]Centre for Disaster Studies, James Cook University, Australia
[3]Department of Environment and Geography, Macquarie University, Australia

44.1 Introduction

The Queensland (northeast Australia) wet season and Victorian (southeast Australia) summer of 2010/11 were record-breaking periods (Fig. 44.1). La Niña drove the weather systems in early 2011, but had been dominant for much of the period from mid-2007 (BoM 2013). La Niña events typically produce wetter conditions across much of Australia. The strong La Niña peak over 2010/11 produced the wettest December on record in Queensland and the wettest January on record in Victoria (BoM 2012). River levels had already risen before the main impact of very heavy rainfall, resulting in widespread flooding in Queensland in December 2010 and both states throughout January 2011.

Based on climate models, the IPCC (2012) argues with medium confidence that anthropogenic influence has contributed to rainfall extremes. However, Kundzewicz et al. (2013) acknowledge that it is difficult to link an increase in magnitude and/or frequency of flooding to climate change. Nevertheless, exposure to flooding is increasing simply by virtue of increasing development in flood-prone areas (Chapter 4 by McAneney et al.). It is therefore imperative that policymakers, planners and emergency managers promote flood adaptation measures within such communities. Moreover, community-appropriate flood risk reduction measures must be developed with an understanding and appreciation of behavioural choices, adaptive capacities and underlying vulnerabilities of at-risk communities (Wisner et al. 2004; Haynes et al. 2009; Apan et al. 2010; Bird et al. 2011a; O'Sullivan et al. 2012; Phillips et al. 2013).

In order to provide an evidence base on community flood risk in Australia, this project aimed to document:
- people's experience of the 2010/11 Queensland and Victorian floods;
- the behavioural choices made to reduce risk; and
- underlying factors that enabled and inhibited adaptation.

This work builds on a flood survey undertaken in 2008 following flooding in central Queensland

Applied Studies in Climate Adaptation, First Edition. Edited by Jean P. Palutikof, Sarah L. Boulter, Jon Barnett and David Rissik.
© 2015 John Wiley & Sons, Ltd. Published 2015 by John Wiley & Sons, Ltd.

Figure 44.1 Case study locations: Brisbane and Emerald, Queensland, and Donald, Victoria. The insert shows the Brisbane suburbs of Chelmer, Graceville, Rocklea and Tennyson. Source: Reproduced with permission of James O'Brien, Risk Frontiers. For colour details please see Plate 31.

(Apan et al. 2010). In the present project, sampling was conducted in three locations: Brisbane and Emerald in Queensland and Donald in Victoria (Fig. 44.1). Brisbane suburbs included Chelmer, Graceville, Tennyson and Rocklea (Table 44.1).

44.2   Research methods

Information was primarily gathered from: (1) households in flood-affected areas; and (2) officials including emergency management staff, planners, engineers and administrators in local councils and state government departments responsible for flood risk reduction and emergency response. The research applied a mixed methods approach whereby different qualitative (face-to-face interviews, open response questions within questionnaires) and quantitative (closed questions within questionnaires) methods contributed to different aspects of the study. Respondents were asked questions on adjustments made prior to and during the flood, their experience of and the impacts of the flood, adaptive behaviour and intention in relation to mitigating the impact of future events. This methodological approach was pre-tested to ensure the results addressed the aims of the study.

Fieldwork was undertaken in August and September 2011. Interviewees were recruited

Table 44.1 Case study location demographics (ABS 2013) and flood statistics (due to limited gauging on the Richardson River, Victoria, accurate flood data for Donald is unavailable).

	Brisbane				Emerald	Donald
	Chelmer	Graceville	Rocklea	Tennyson		
Area (km²)	1.5	1.8	13.3	8.7	610	1004.5
Population	2594	4213	1255	431	13 884	1693
Median age	38	37	36	42	29	48
Median total household income (AU$/year)	123,968	111,072	54,704	96,668	119,340	38,428
Median rent (AU$/week)	330	400	285	280	310	110
Last major flood		1974			2008	1992
Recent flood						
Date		Jan 11			Dec 10	Jan 11
Duration (hrs)		c. 32			c. 52	–
Height (m)		4.46			16.05	–
No. homes inundated		c. 14,000			c. 1000	13

through door knocking in flood-affected communities and via an opportunistic technique, where the initial respondent (official or resident) suggested others who might be willing to participate in the research. Hard-copy questionnaires were delivered to households for self-completion and an internet link was provided to those who wished to complete an online version. Overall, 18 interviews and 62 questionnaires were completed in Brisbane, 16 interviews and 53 questionnaires in Donald and 21 interviews and 95 questionnaires in Emerald (Table 44.2). Questionnaire data were transferred into Microsoft Excel® and SPSS®. Interview and open response data contained within the questionnaires were transcribed into Microsoft Word®. Quantitative data were coded and analysed using frequency and cross-tabulation tables. Qualitative data were coded by tagging sections of text in order to identify themes. For more details on the methods used, see Bird et al. (2013a).

The following section presents a selection of the quantitative results based on closed questions within the questionnaire. These are followed by descriptive text based on qualitative data gleaned from open-response questions

within the questionnaires and from interviews. Specifically, these questions asked respondents the following questions.

• What do you think are, if any, the main factors that prevent you from making changes to reduce the impact of future flood to your home/property?

• What do you think are, if any, the main factors that encourage you to make changes to reduce the impact of future flood to your home/property?

• What are the three main things you think can be done from a council perspective to help reduce your risk from future flood?

• What are the three main things you think can be done from an emergency services perspective to help reduce your risk from future flood?

• Overall, what do you think are the three best measures or strategies to help reduce your risk from future flood?

A full analysis of the results, including a demographic breakdown of responses and copies of the questionnaires, can be found in Bird et al. (2013a) and a comparative analysis of the similarities and differences between each case study location is available in Bird et al. (2013b).

Table 44.2 Respondent demographics based on questionnaire data. Note that not all categories total 100% due to rounding or where respondents have marked, where indicated, all options that apply.

Demographic variables	Proportion of respondents (%)
Gender	
• Male	43
• Female	57
Age group	
• <35 years	18
• 35–44 years	24
• 45–54 years	25
• >55 years	34
English not main language spoken at home	5
Highest achieved education	
• No formal qualifications	11
• Higher school certificate or interstate equivalent	27
• Vocational qualifications	33
• Bachelor degree	21
• Postgraduate qualification	6
• Other	2
Annual household income	
• <AU$ 50,000	31
• AU$ 50,000–100,000	26
• AU$ 100,000–150,000	16
• >AU$ 150,000	15
• Don't know/don't want to answer	13
Working status	
• Full-time (>30 hours/week)	51
• Part-time (9–29 hours/week)	21
• Looking after house/children/other dependants	8
• Retired	17
• Other	8
Composition of household	
• Couple with children or other dependants	45
• One adult with children or other dependants	3
• Couple without children or other dependants	27
• One person household	11
• Shared house with other adults	10
• Other (including aged care facility)	4
Housing status at time of flood	
• Owned/mortgage	78
• Renting	15
• Other	7
Structure of house	
• Single-storey house (not raised on stumps/stilts)	37
• Single-storey house (raised on stumps/stilts)	30
• Multiple storey house or duplex	22
• Unit ground floor or single-storey duplex	5
• Other	5
Length of time lived at this address	
• <1 year	16
• 1–5 years	22
• 5–10 years	25
• 10–20 years	17
• >20 years	20

44.3   Results and discussion

Many respondents (43%) were living in a single-storey house (i.e. not raised on stumps or stilts), a single-storey duplex or a ground floor unit. At the time of the flood, 36% of respondents indicated that they *thought* their insurance covered them

Table 44.3 Adjustments made to help protect family and home prior to and during the flood.

Hazard adjustment	Proportion of respondents (%)
Devised an evacuation plan	22
Prepared an evacuation kit	16
Followed warning advice on radio/ television/internet	42
Sandbagged house	30
Built temporary flood barriers around property	8
Kept drainage clear of debris	15
Raised household items up off floor	51
Moved household items to a safe place	38
Evacuated to a safe house or centre	44

for all types of flood and a further 37% knew that they were not covered. The majority of respondents (61%) had not experienced flooding in any location where they had lived and 49% were not aware that their home was vulnerable to flood. It is therefore not surprising that many did not make adjustments to protect their family and home in response to the imminent threat (Table 44.3), although many were subsequently impacted (Fig. 44.2).

While the floods raised awareness of climate change, both in the media and among the general public, very few people were prompted to carry out adaptive measures to reduce future risk following these local events (Fig. 44.3). This is despite the fact that many respondents (47%) believed that a flood, causing damage to homes and properties, was likely or very likely to impact their area within the next year (Fig. 44.4).

While the qualitative data (presented in the following sections) reveal evidence of community resilience, the factors that inhibit adaptive change emerged more strongly than enabling factors. These findings have obvious policy implications that put greater responsibility on all levels of

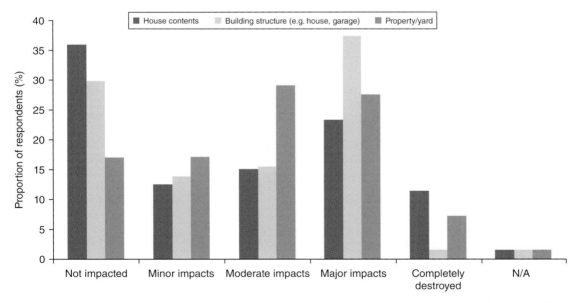

Figure 44.2 Degree of damage to house contents, building structure and property/yard as reported by respondents.

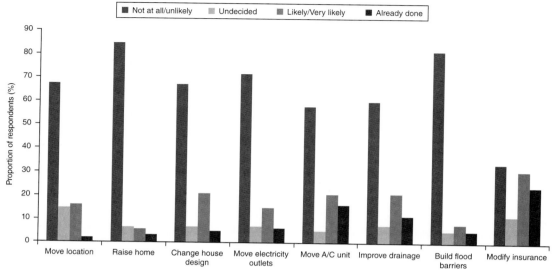

Figure 44.3 The reported likelihood of respondents making specified changes as a result of the 2010–11 Australian floods.

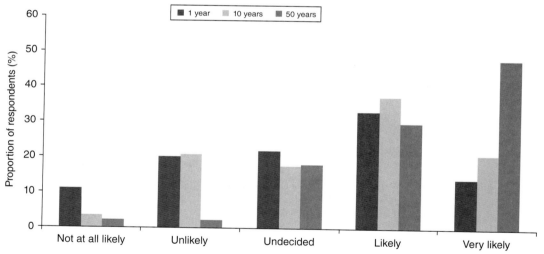

Figure 44.4 Respondents' belief in the likelihood of another damaging flood occurring in their area within 1 year, 10 years and 50 years.

government to educate and better support the community to facilitate change. Each of the enabling and inhibiting factors are discussed in the following, adapted from chapter 9 of Bird et al. (2013a).

44.3.1 Direct experience

The key driver enabling response, recovery and adaptation initiatives was people's direct experience of flood, either in 2010/11 or from a past event. Respondents related negative feelings of stress, heartache and inconvenience contributing to their desire to reduce future risk. However, residents can become complacent in relation to flood risk if they lack direct experience of the hazard. In order to combat this, personal narratives of disaster experience can be used as powerful tools to educate people about the necessity and options for reducing disaster risk (Kurita et al. 2006; Dudley et al. 2009; Apan et al. 2010; Bird et al. 2011b). Anniversaries and times of remembrance are useful moments to focus people's attention on hazard risk, and thereby to educate them on adaptation strategies. It is important for emergency managers to capture personal stories while they are still fresh in people's minds. One example of this occurred in Emerald immediately following the 2010 flood with local residents producing a photographic account of people's experience. The initiative behind the publication, entitled 'With a little help', was to ensure that the impact of the 2010 flood was not forgotten while money raised from sales of the book was used to assist those families impacted by the flood. See www.judigraphics.com/withalittlehelp.htm for details.

44.3.2 Outcome expectancy

Another factor driving adaptation is people's desire to have peace of mind. They are therefore willing to implement strategies or measures to reduce their vulnerability to flood and protect their family, homes and property during future events. For example, sandbagging around the house to protect assets or evacuating to a safe location to protect family. This is a positive outcome expectation. However, there was also a negative outcome expectancy whereby people were fatalistic about risk reduction efforts or protection. In this instance, respondents believed that flood risk reduction was beyond their means so it was not worth the effort.

Building back better is a positive outcome, but the fact that most institutions rebuild structures and infrastructure as they existed before generally sets a bad example. Where a negative fatalism sets in, residents are liable to ignore simple actions such as improving furnishings (e.g. tiling instead of carpeting). Other considerations that are nothing to do with risk reduction, such as fashions or obtaining better value, override decisions and require specific hazard education.

44.3.3 Communications and information

There was a very strong demand from residents for more information and communication from authorities. The request for more information concerned the situation both before and after the flood. People received warnings, but many were unsure of what to do although they were willing and able to take measures to reduce risk. However, the results suggest that people were more likely to react to the immediate threat, rather than to adopt adaptation strategies to reduce flood risk. This is manifested through limited intentions to make significant change to property, even though many people expected that another flood was likely to occur within the next year.

Respondents noted that the warnings were insufficient because they did not contain information on what to do and were often too general. People wanted to know about the flood threat in their own location, that is, their street or neighbourhood. People received flood updates from both SMS and landlines but there was also a need for information after the flood, guiding people in what to do and providing information on insurance issues and suitable flood-proofing measures.

44.3.4 Governance, legislation and protection

People cited issues connected to planning, building codes and information. They also called for more dams, levees and protective barriers. They demanded better management of dams, although at the time of the survey this was a major media controversy and a central issue in the Queensland Flood Inquiry. Some residents had constructed their own walls and drains to reduce the risk of future floods. While this is an example of community resilience there is a danger of protective measures on one property causing additional problems to neighbouring properties. Some council regulations prevented such actions. Otherwise, there were demands for improved drainage, backflow and creek management as well as more sandbags.

Residents also complained about a lack of State Emergency Service (SES) workers and council officers. This reflects a much more general problem as there tends to be an assumption that SES volunteers are an unofficial part of government, and that therefore it is the right of the public to receive their help. More information about the voluntary status of the SES needs to be made available. The SES would like to be able to recruit more active members, which could be a positive outcome of an educational campaign.

44.3.5 Insurance

The desire to repair and rebuild was hampered by the slowness of insurance assessments and payouts. Extensive anger was directed towards insurance companies, their assessments and lack of information. People did not know whether they could commence cleaning up before the insurance assessment had taken place; this delay often increased the damage sustained. The uncertain and protracted insurance assessment and approval process also made it difficult for people to book tradesmen for repair and rebuilding work.

Minimal direction came from insurance companies on building more flood-resistant structures. A small number of examples were cited where insurers had given support to flood risk reduction by reducing premiums or extending coverage for households that had flood-proofed their properties.

The flood events of 2011 were so extensive and costly that flood risk reduction needs a multi-institutional approach that includes insurance companies and places a responsibility on insurers to encourage flood-proofing measures. This will improve options for residents and reduce the extent of future claims. The Insurance Council of Australia produced a wishlist to improve governance and insurance responses to future floods. Their 10-point recommendations suggest: a standard definition of flood; improved disclosure; provision of adequate flood data; removal of insurance

taxes; improved land-use planning; improved building standards; improved community infrastructure; education and financial literacy campaign; measurement of the effectiveness of disaster relief payments; and better advice to consumers (see http://www.floodcommission.qld.gov.au/__data/assets/file/0005/6494/Insurance_Council_of_Australia_2.pdf for details).

44.3.6 Finances and relief payments

Many people were not covered by insurance and were unable either to rebuild or to improve their property. Additionally, many were unable to obtain government assistance such as from the Premier's Flood Appeal in Queensland. The availability of relief assistance was divisive in many communities and can be detrimental to disaster resilience if people come to expect that government relief will always be available.

Future assistance may be constrained however, because:
1. people are unwilling to donate to future extensive hazard events, especially if they become more frequent;
2. governments will be unable to continuously assist disaster-stricken communities; and
3. many residents were taxed with a flood levy and consequently felt that their voluntary donations had been a waste of effort and money.

There was a further problem whereby people waited for insurance claims to be settled and did not claim government relief. Unfortunately, when they were eventually told they were not eligible for an insurance payout the government schemes had expired.

44.3.7 House design issues

People living in rented properties had the least capacity to make any changes to the designs of their homes and people felt that there was little they could do to flood-proof their dwellings. For homeowners, the most commonly cited reason for the inability to raise buildings was the dominant house design of a block-built house constructed on a concrete slab on the ground. Many residents felt it was unwise for such houses to continue to be built in flood zones. Earlier flood studies in Mackay and Charleville (Apan et al. 2010) suggested that builders were no longer experienced in building high-set homes and were not privy to appropriate house designs. This highlights the need to educate the building industry and to strengthen codes and bylaws.

During rebuilding residents experienced a shortage of builders and tradespeople, especially in remote places such as Emerald where housing is already in short supply. Following the 2011 floods, the Queensland Reconstruction Authority was established and has produced information materials such as 'Planning for stronger, more resilient floodplains', in which they state 'The traditional 'Queenslander' style home was designed to allow the cool breezes to circulate through the house in the hot summer and let flood waters flow underneath' (QRA 2011, p. 14). On the other hand, the dominant block structures (slab on ground, brick veneer construction) are the cheapest to build and are not adaptable to the old Queenslander style (high-set timber construction on posts).

44.3.8   Health and wellbeing

Mental and physical health issues hampered community recovery. In Brisbane and Emerald it was the middle- and high-income earners who experienced the strongest negative impacts on their wellbeing. Possibly these were residents who had lost more and may have been underinsured. This was a common problem, with many better-off residents ineligible for government assistance but having inadequate insurance for reconstruction.

Since the floods many people have remained nervous of bad weather, with heavy rain prompting visits to a Brisbane flood recovery centre. There were also many instances of people not accessing public services, either because they did not consider themselves to be in need or through personal pride. Community networks are very important in building up this area of resilience.

44.3.9   Relocation

In both Brisbane and Emerald houses had been abandoned or remained empty. Respondents were asked about their intention to relocate to a flood-safe location with 18% indicating that they were likely to or had already done so. Although this is a minority, any such exodus (which would arise where there is little available housing stock outside of flood-prone land) would impact upon the economic viability of local communities as well as social capital and community resilience. The relocation intent also supports the idea of a government retreat policy whereby hazard-prone areas may need to be re-zoned as non-residential.

People who considered relocation as an option were predominantly younger to middle-aged adults with families earning mid-range incomes with vocational qualifications (what may be regarded as 'middle Australia' or ordinary average residents). Although the numbers in the survey were small, Apan et al. (2010) reported a similar pattern after the 2008 floods in Mackay and Charleville. Overall however most respondents had no intention to leave, especially in Donald, which is what we may expect of resilient communities.

44.3.10   Volunteers and community

The 2011 floods prompted a surge of goodwill and volunteerism and, while respondents were greatly appreciative of their efforts, there was some criticism related to a lack of systematic control over the clean-up of their homes and properties. That is, many people wanted to help and in Brisbane there were many volunteers wandering the streets offering assistance. Often these volunteers were focused on completing the task at hand as rapidly as possible; as a result, residents did not have a chance to sort through their belongings in the hope of saving those things that were not damaged by floodwater. To overcome this issue, it is important to have a more structured approach whereby teams of volunteers are kept active throughout the recovery process and are on standby for when the next disaster strikes. An excellent example of this occurred online through

Facebook, with community groups providing information and assistance where needed during the Queensland and Victorian floods (see Bird et al. 2012 for details). Other examples occurred in response to the Canterbury earthquakes in New Zealand with the establishment of the Farmy Army and the student volunteer army (see http://www.fedfarm.org.nz/services/The-Farmy-Army.asp and http://www.sva.org.nz for details).

While the SES is an appropriate volunteer group, its membership is small and not available in every community (e.g. there is no SES unit in Donald). Instead, local volunteer initiatives that could be mobilised in times of disaster could evolve from community groups that have a non-hazard-related primary purpose such as a sports club. These groups are the core of small communities and country towns. While they can assist response and recovery, community groups are also able to provide community support long after the event.

As many Australian disasters occur during the summer when people are on holiday or absent from the community, a network of contacts would be extremely useful. Friends or neighbours could be provided with access to the vacated homes to raise or rescue valuable items or to sandbag in the event of a flood. This is especially important in places like Emerald where the population is extremely transient. Employers in mining or agricultural industries, for example, may also provide support to workers. Such support networks may also provide assistance to the less-able members of the community. A good example of a community support network called 'The Go List' was established in Victoria as an online resource to provide support for people living in bushfire-prone communities (cross-referenced at http://www.dpc.vic.gov.au/index.php/featured/victorian-emergency-management-reform-white-paper/emerald-house-building-community-resilience-to-disasters).

The overall impression of the surveys was a strong motivation on the part of residents to carry out their own recovery and do their best to reduce the risk of flood for the next event. This was most notable in Donald, where residents created the Donald Flood Recovery Group that secured AU$ 135,000 from the government to carry out a hydrological survey.

44.4 Conclusions

Flood-impacted respondents called overwhelmingly for communication and information from insurance companies, local councils, state and federal governments and non-government organisations (NGOs). People in North Queensland impacted by category four/five Cyclone Yasi a short while later made no such demand. Cyclone warnings, preparation, behaviour and recovery phases are all routines that long-term residents in cyclone-vulnerable areas are used to and which they generally adhere to. There is no such routine of well-publicised warnings and ingrained risk-reducing behaviour for severe floods. In particular, people need information that is relevant to their own localities to help them make correct decisions in response to warnings. They need information on what to do during the flood and its aftermath, and finally what to do during recovery and rebuilding. People called for a range of methods of communication and information, as diverse as the community in which they lived. Communications therefore need to be better targeted to social, demographic and cultural groups.

Respondents identified many barriers to change and adaptation, with the most common related to financial constraints. People also identified house style and construction (especially the block and concrete slab on the ground style of house) that makes flood-proofing difficult and actions such as raising the house impossible. Where people faced building constraints, the alternative of relocating elsewhere became even more realistic. Alongside the idea of relocating away from the flood risk were widely held opinions and criticism of the poor planning decisions that had led to the flooding of their properties or communities.

There is a need for:
• an avoidance of planning new developments in flood-prone areas;
• a process of rezoning land use in existing hazard zones, within the constraints of land and property security
• a process of rezoning existing hazard zones to incorporate improved hazard and risk knowledge, within the constraints of land and property security

• strategies to enhance resiliency of homes located in flood-prone areas to encourage, for example, retrofitting and house raising by providing tax breaks, grants, insurance benefits and better information.

The study shows that there are more constraints that inhibit adaptation to flood risk than there are opportunities that lead to adaptive behaviour and change. Inevitably after a disaster, people are apt to be critical of authorities. The survey timing, eight months after the floods occurred, was at the time of greatest despondency during recovery (when the excitement had worn off but the loss, difficulties and obstacles remained). As this study was primarily concerned with longer-term future adaptation to disastrous floods, it was an appropriate time to engage in reflection of the event and of the future with the residents of Brisbane, Emerald and Donald.

Despite criticism and negativity towards insurance companies and authorities, the flooded communities were strongly resilient with people making progress and driving their own recovery. The survey recorded extensive qualitative comments that demonstrated stoic endurance and acceptance of the disaster. However, this does not mean that people will behave the same or be so accepting in the future. Rather, it is an indicator of strong resilience that may be built upon by government agencies and NGOs to encourage longer-term climate change adaptation behaviours.

References

ABS (2013) *2011 Census Community Profiles: State Suburb (SSC): Basic Community Profile.* Australian Bureau of Statistics. Available at http://www.abs.gov.au/websitedbs/censushome.nsf/home/communityprofiles (accessed 7 June 2014).

Apan, A., Keogh, D.U., King, D. et al. (2010) *The 2008 Floods in Queensland: A Case Study of Vulnerability, Resilience and Adaptive Capacity.* National Climate Change Adaptation Research Facility, Gold Coast.

Bird, D.K., Gísladóttir, G. and Dominey-Howes, D. (2011a) Different communities, different perspectives: Issues affecting residents' response to a volcanic eruption in southern Iceland. *Bulletin of Volcanology* 73, 1209–1227.

Bird, D.K., Chagué-Goff, C. and Gero, A. (2011b) Human response to extreme events: A review of three post-tsunami disaster case studies. *Australian Geographer* 42, 225–239.

Bird, D., Ling, M. and Haynes, K. (2012) Flooding Facebook: the use of social media during the Queensland and Victorian floods. *The Australian Journal of Emergency Management* 27, 27–33.

Bird, D., King, D., Haynes, K., Box, P., Okada, T. and Naim, K. (2013a) *Impact of the 2010/11 Floods and the Factors that Inhibit and Enable Household Adaptation Strategies.* National Climate Change Adaptation Research Facility, Gold Coast.

Bird, D.K., Box, P., Okada, T., Haynes, K. and King, D. (2013b) Response, recovery and adaptation in flood affected communities in Queensland and Victoria. *Australian Journal of Emergency Management* 28, 36–43.

Bureau of Meteorology (2012) *Special Climate Statement 38. Australia's Wettest Two-Year Period on Record: 2010–2011.* National Climate Centre, Bureau of Meteorology, Melbourne.

Bureau of Meteorology (2013) *La Niña: Detailed Australian Analysis.* Bureau of Meteorology, Victoria. Available at http://www.bom.gov.au/climate/enso/lnlist/index.shtml (accessed 7 June 2014).

Dudley, W., Goff, J., Chagué-Goff, C. and Johnston, J. (2009) Capturing the next generation of cultural memories--the process of video interviewing tsunami survivors. *Science of Tsunami Hazards* 28, 154–170.

Haynes, K., Coates, L., Leigh, R. et al. (2009) 'Shelter-in-place' vs. evacuation in flash floods. *Environmental Hazards* 8, 291–303.

IPCC (2012) *Managing The Risks of Extreme Events and Disasters to Advance Climate Change Adaptation. A Special Report of Working Groups I and II of the Intergovernmental Panel on Climate Change.* Field, C.B., Barros, V., Stocker, T.F. et al. (eds) Cambridge University Press, Cambridge, UK, and New York, NY, USA.

Kundzewicz, Z.W., Kanae, S., Seneviratne, S.I. et al. (2013) Flood risk and climate change: global and regional perspectives. *Hydrological Sciences Journal* 59, 1–28.

Kurita, T., Nakamura, A., Kodama, M. and Colombage, S.R.N. (2006) Tsunami public awareness and the disaster management system of Sri Lanka. *Disaster Prevention and Management* 15, 92–110.

O'Sullivan, J.J., Bradford, R.A., Bonaiuto, M. et al. (2012) Enhancing flood resilience through improved risk communications. *Natural Hazards Earth Systems Science* 12, 2271–2282.

Phillips, E., Bird, D., O'Brien, J. et al. (2013) *An Integrated Research Assessment of the Physical and Social Aspects of the March 2011 Flash Flooding in Shellharbour, Kiama and Bega Valley, NSW. Report for the New South Wales Emergency Service* (NSW SES), Risk Frontiers, Sydney.

Queensland Reconstruction Authority (2011) *Planning for a Stronger, More Resilient North Queensland. Part 1 Rebuilding in Storm Tide Prone Areas: Tully Heads and Hull Heads.* Queensland Government, Brisbane.

Wisner, B., Blaikie, P., Cannon, T. and Davis, I. (2004) *At Risk: Natural Hazards, People's Vulnerability, and Disasters,* 2nd edition. Routledge, New York.

45 Disaster risk management and climate change adaptation revisited

MICHAEL HOWES

Griffith School of Environment, Griffith University, Australia

45.1 Introduction

Governments around the world are increasingly under pressure to do more with less. On the one hand, there are demands to address an ever-growing list of issues. On the other hand, they are expected to keep taxes low, get budgets into surplus and reduce the level of public debt. To complicate matters further, some issues (such as responding to floods and bushfires) demand immediate action, while others (such as adapting to the impacts of climate change) require a long-term policy commitment. Making the most effective, efficient and appropriate use of scarce public resources is therefore a high priority. One promising strategy is to integrate related policy responses in order to allow for the pooling of resources. In this chapter opportunities to create such an integrated approach to disaster risk management and climate change adaptation (Howes et al. 2013) are considered. Our research centred on a series of Australian case studies, but the findings are relevant to governments around the world due to the global nature of the problems addressed.

45.2 Background

Policy responses focused on disaster risk management and climate change adaptation are linked in a number of ways. First, both deal with similar biophysical and socio-economic impacts on society and the environment. Second, both require a coordinated response across and between different levels of government that will draw on overlapping sets of agencies and departments. Third, as the climate changes weather-related disasters are likely to increase in terms of frequency, intensity and/or duration. Finally, they share a common goal: to increase the resilience of society to extreme weather events (APSC 2007; COAG 2007; IPCC 2012; Howes et al. 2013).

In 2012 teams from Griffith University and the Royal Melbourne Institute of Technology undertook a three-way comparative case study of the 2009 Victorian 'Black Saturday' bushfires, the 2011 Perth Hills bushfire and the 2011 Brisbane floods (Howes et al. 2013). The project was focused on achieving climate change adaptation though improved disaster risk management policies, plans and management strategies, including

Applied Studies in Climate Adaptation, First Edition. Edited by Jean P. Palutikof, Sarah L. Boulter, Jon Barnett and David Rissik.
© 2015 John Wiley & Sons, Ltd. Published 2015 by John Wiley & Sons, Ltd.

the development of foundations for a nationally consistent approach and a set of appropriate reforms for governing institutions and tools.

The Victorian case involved 316 fires across the state that claimed the lives of 173 people and damaged or destroyed 2133 homes (VBRC 2010b). The Perth Hills Bushfire destroyed 71 homes and damaged a further 39 (GWA 2011). Both of these events were exacerbated by the millennium drought 2001–2010 that affected the whole country. The 2011 Brisbane floods occurred after this drought was broken by a strong La Niña event that brought heavy deluges and flooding across Queensland. The floodwaters resulted in 35 deaths state-wide and affected 20,000 homes in Brisbane alone (QFCI 2012). These case studies were selected for three reasons. First, they are the kinds of events that are likely to increase in frequency, duration and/or intensity under the impacts of climate change (IPCC 2012). Second, they dealt with two different kinds of disasters in three different states that were geographically dispersed around the country, so any findings should be easier to generalise (Howes et al. 2013). Third, they were all major events that placed significant pressure on the emergency services and were subject to formal inquiries, making them a rich source of data regarding what needs to change (VBRC 2010a–c; GWA 2011; QFCI 2011, 2012).

45.3   Research method

The research into these events began by working through the official inquiry reports for each case study in parallel with the relevant research literature and key policies/plans dealing with disaster risk management, climate change adaptation, governance within the federal political system and the process of public policymaking. This produced a synthesis of four themes that identified the need to: (1) improve interagency communication and collaboration; (2) foster institutional improvement and learning; (3) enhance community engagement and communication; and (4) create a renewed common focus on resilience (Howes et al. 2012).

These four themes were then used to design semi-structured interviews that were conducted with 22 key public sector stakeholders. From the analysis of the interviews, a set of proposed reforms and tools was developed. These were then peer-review tested at workshops in Melbourne, Perth and Brisbane. A broad range of stakeholders (26 in total) drawn from both the public sector and community organisations attended these workshops (Howes et al. 2013). The findings are summarised in the following and organised according to the identified themes. While our study focused on disaster risk management and climate change adaptation, the recommendations made here could be usefully applied to many areas of public policy and planning.

45.4   Improving interagency communication and collaboration

The research literature on public policy generally supports a less hierarchical, more collaborative model of governance (Waugh and Streib 2006). This is particularly true with regards to environmental policy (Ross and Dovers 2008), disaster risk management (Mitchell et al. 2010) and climate change adaptation (Head 2008). Problems with interagency communication and collaboration were identified as inhibiting effective responses in all three inquiries (VBRC 2010a, p. 8; GWA 2011, p.133; QFCI 2012, p. 28; Goode, et al. 2012). The interviews supported these findings and, as one respondent in Perth stated, the challenge was about 'working in partnership, recognising the skills of the various agencies and how they can actually complement each other but having a common goal'.

Barriers to this improvement were explored in the interviews. Several respondents identified the problem of the 'silo mentality' whereby staff worked largely in isolation, focusing on their particular job and missing the opportunities and benefits of collaboration. A second problem was where experts use different, specialised language that can make effective communication difficult. Another issue was that of 'turf wars', where agencies jealously guard their jurisdiction and view moves towards collaboration with suspicion. Finally, there was the problem of the lack of commitment from senior decision-makers who do not support change for one reason or another. These findings were in accord with previous research (Liebrecht and Howes 2006; Howes 2008; Rolfe et al. 2009).

Participants at the workshops agreed that there was a need to improve interagency communication and collaboration. They also identified the same barriers that had emerged from the interviews. Three key factors arose during the discussions that were seen as needing to be addressed. First was the need to develop clear roles and shared goals for both the emergency services and other agencies. Second was the need to get political and executive support so that: (1) there was a clear 'message from the top' supporting collaboration; (2) there was the ability to create a network of collaboration champions empowered to work across agencies; and (3) there was a broadening of committee membership to include relevant non-government organisations. Finally, workshop participants advocated a local/regional approach with local government playing a key role in both disaster risk management and climate change adaptation.

45.5 Fostering institutional improvement and learning

Public sector institutions at all levels are increasingly being asked to respond to rapidly changing environmental, economic and social circumstances. The impacts of climate change will add to both the extent and rate of these changes around the world (IPCC 2007). Australia is particularly vulnerable, especially with regards to extreme weather-related events such as bushfires and floods (IPCC 2012). Responding to such changes will require significant institutional improvement and learning across the public sector to assist with the integration of areas such as disaster risk management and climate change adaptation (Waugh and Streib 2006; Birkmann and von

Teichman 2011). To some extent this has already been acknowledged by the sector (APSC 2007; COAG 2007) and was noted in the inquiry reports (VBRC 2010c, pp. 81, 86, 229; QFCI 2011, pp. 24, 62; GWA 2011, p. 188; Goode et al. 2012, p. 16).

The interviews revealed a common acceptance of the need to improve institutions and learning. As one respondent in Brisbane noted, there was a particular need to learn from experience and: 'there's a window of opportunity after any major event in a place to say this is what we have to embed in the corporate knowledge and understanding'. The emergency management practitioners also acknowledged that they need to learn more about the implications of climate change. There was general agreement that institutions, policies and plans need to be flexible as the needs of communities vary. Further, public concerns and local knowledge need to be brought into the decision-making process.

The workshops backed up the interview findings but noted that institutional learning is difficult in high-pressure situations. Participants felt that there was a need to create a space for learning from both mistakes and successes. Two key concepts emerged. First was the idea of learning from each other by fostering both formal and informal interactions between researchers, practitioners, policymakers and the broader community. This would require improved communication along the lines outlined in the previous theme. The second idea was that there needs to be a high level of commitment to learning at all levels, as has happened with the successful implementation of water demand management programs during the millennium drought.

45.6 Enhancing community engagement and communication

As can be seen in the previous sections, the idea of enhancing community engagement and communication was a strong underlying theme. While a well-coordinated public sector is an essential feature of good public policy across many areas of government, on its own it is not sufficient as the sector simply does not have the resources needed to address all of the issues faced (APSC 2007; Productivity Commission 2012). The community needs to be well informed, supported and empowered in order to contribute to a more effective response, particularly when it comes to disaster risk management and climate change adaptation (Dovers 1998; O'Brien et al. 2006; Waugh and Streib 2006; Beck 2011; Gero et al. 2012). There are challenges in making this happen (Burton and Mustelin 2011), but there are also benefits in terms of improving resilience (Handmer et al. 2011). Improving the community's understanding of the risk of disasters such as floods and bushfires is also important (Goode et al. 2012). The inquiry into the 2009 Victorian bushfires made several references throughout its reports on the need for enhanced community communication and engagement (VBRC 2010c, pp. 3, 31, 34, 37, 230, 352). The 2011 Perth Hills bushfires inquiry extended the point to make the case for a shared responsibility for disaster risk management between the community and public sectors (GWA 2011, pp. 13, 46). Further, the inquiry into the 2011 Queensland floods discussed the need to improve community preparedness as well as assist local groups in responding to major events (QFCI 2012, pp. 118, 122).

The interview participants supported the idea of enhanced community communication and engagement, but there were some questions about how this might be achieved. They saw the need to move beyond simple public education campaigns, where information is just distributed by various media. As one Victorian respondent stated: 'it means people being actively involved right across the state, not just in SES headquarters or CFA headquarters or regional offices'. It was acknowledged that while there had been some success in the area of disaster risk management, climate change adaptation poses a more difficult challenge because of the complexity of the underlying science.

The workshop participants supported these findings, but noted that improvements would be resource intensive. Five changes were discussed: (1) the need to tailor programs to the situation of the community, making use of the available assets and addressing unique risks; (2) supporting locally driven planning using volunteers and existing networks; (3) identifying the key groups of people that need to be engaged in the process, to allow for a more carefully targeted response; (4) the need to have a broader focus that looked at the combination of risks together, rather than in isolation; and (5) the importance of a good media strategy before, during and after the event. All of these applied equally well to community disaster risk management and climate change adaptation, and could help to integrate them into an overall strategy.

45.7   Creating a renewed focus on resilience

The fourth theme of creating a renewed focus on resilience was an essential step to pulling the framework together. Emergency services have generally relied upon the prevent–prepare–respond–recover (PPRR) model, but much of the research proposes taking a more proactive approach and one way to achieve this would be to shift the emphasis to building resilience (Cronstedt 2002; Handmer and Dovers 2007; Prosser and Peters 2010; Handmer et al. 2011; Rogers 2011). The *National Strategy for Disaster Resilience* developed by the Council of Australian Governments has taken some steps in this direction (COAG 2011). Such a move facilitates the integration with climate change adaptation as it shares the goal of building resilience. While the inquiry reports related to our three case studies touch on resilience, they demonstrate a lack of consensus about how it should be defined (VBRC 2010c, pp.31, 34, 230; GWA 2011, pp.13, 46; QFCI 2011, pp. 115, 118, 122; Goode et al. 2012). This lack of consensus extends to climate change adaptation policies and plans internationally (Davoudi 2012).

While the interview participants generally supported the move to focus on resilience there was an acknowledgement of the lack of an agreed definition of both this term and the nature of communities. A respondent in Perth suggested that: 'A resilient community is one which pulls together. A not so resilient community would be one that just evaporates and people go their own separate ways'. There was a general feeling that there was an over-reliance on the emergency services, so a community engagement strategy was needed to inform people about the risks and empower them to build up their own resilience. Participants agreed that this means moving beyond the existing PPRR model.

The workshop participants agreed that there had been an emphasis on responding to emergencies and that it would be good to refocus efforts. Three concepts were put forward for building community resilience: (1) recognising and supporting the important role that volunteers have to play; (2) the importance of taking into account demographic shifts, such as new people who lack local knowledge, moving into an area; and (3) the need to foster self-reliance and shared responsibility for building resilience to both disasters and climate change.

45.8 Four reforms to integrate disaster management with climate change adaptation

On the basis of these findings, four reforms were formulated and put through the review process of the three workshops. There was general support for these reforms although there were some reservations about the detail of their implementation. The first three reforms were put forward by the research team while the fourth originated with the workshop participants themselves.

1. Create a new collaborative public sector funding model, which sees part of the funding for each level of the public sector put into a common pool and allocated to fund (cross-sector) solutions to specific disaster risk management and climate change adaptation issues. Agencies, offices and departments at all levels would then be encouraged to form consortiums (that may also include the businesses and/or community organisations) to bid for funding to address the nominated issue. This would provide an ongoing financial incentive to improve collaboration across the public sector, foster institutional change and learning

and enhance community engagement. In terms of practicality, there are already grant programs that provide a useful model, such as the *Landcare* and the *Natural Disaster Resilience Program* schemes. Funds might be made available from Infrastructure Australia and the COAG is currently reviewing its *National Partnerships Agreements*, so the timing is opportune.

2. Create a set of local community resilience grants to be administered by local councils. Part of the existing grant program could be set aside to fund the new scheme, in which the community would be asked to propose simple projects to improve local resilience to disasters and climate change. The council could hold public meetings and get residents to vote on which projects to fund. This would raise awareness, enhance community engagement and renew the focus on resilience. The types of projects that might be funded could include establishing a network of volunteers to assist elderly neighbours during a flood or bushfire. On the practical side, this would simply be redirecting existing local government grant schemes into a new area.

3. Embed climate change researchers within emergency service organisations to assist with the revision of risk assessments. For larger organisations this would mean bringing them into the existing research offices. For smaller organisations this might mean forming working partnerships with researchers at other institutions (e.g. universities). In some situations it may also be appropriate to form research consortiums involving several organisations. This proposal would greatly assist with fostering institutional change and learning and may also assist with improving communication/ collaboration across the public sector. There are already practical examples of this kind of program, such as Australia's National Climate Change Adaptation Facility (NCCARF) and the Australian Research Council *Linkage Grant* scheme.

4. Implement a broad set of organisational changes, starting at the top and using the COAG to get all levels of government aiming to integrate disaster risk management and climate change adaptation. This could be supported by revamped interagency senior officer groups that translate the executive commitment into day-to-day management changes. Finally, there would need to be a network of field operatives across all the relevant organisations that were willing to act as 'champions' for increased communication, collaboration and engagement across the public sector as well as with business and community groups. There would therefore be both top-down and bottom-up changes pushing in the same direction.

45.9   Conclusions

For the foreseeable future, governments around the world will continue to be expected to do more with less while the impacts of both disasters and climate change exacerbate the demand for public goods and services. Given the unlikelihood of a major restructuring of the public sector, or a significant boost in funding, a high priority will be to implement reforms that can appropriately boost effectiveness and efficiency in the use of scarce public resources. Integrating related public policy areas such as disaster risk management and climate change adaptation offers one opportunity for this kind of reform. This project identified the need to: (1) improve interagency communication and collaboration; (2) foster institutional improvement and learning; (3) enhance community engagement and communication; and (4) create a renewed common focus on resilience. This may be achieved by implementing reforms such as: developing a collaborative public sector funding model; providing community resilience grants at the local level; embedding climate change researchers within emergency service organisations; and adopting organisational changes that network collaboration from top to bottom. Some precedents have been mentioned for these reforms, which suggest that they are feasible. While they are not a global panacea, such reforms could be extended to other areas of public policy and planning around the world.

Acknowledgements

The author would like to thank his fellow research team members whose hard work provided a foundation that made this chapter possible: Deanna Grant-Smith (at QUT), Kim Reis, Peter Tangney, Michael Heazle and Paul Burton (at Griffith University) and Darryn McEvoy and Karyn Bosomworth (at RMIT University). This work was carried out with financial support from the Australian Government (Department of Climate Change and Energy Efficiency) and the National Climate Change Adaptation Research Facility (NCCARF). Further support was given by the Urban Research Program (URP) at Griffith University, RMIT University and the Queensland Department of Community Safety.

References

Australian Public Service Commission (2007) *Tackling Wicked Problems: A Public Policy Perspective.* Australian Government, Canberra.

Beck, S. (2011) Moving beyond the linear model of expertise? IPCC and the test of adaptation. *Regional Environmental Change* 11, 297–306.

Birkman, J. and von Teichman, K. (2011) Integrating disaster risk reduction and climate change adaptation: key challenges--scales, knowledge, and norms. *Sustainability Science* 5(2), 171–184.

Burton, P. and Mustelin, J. (2011) Planning for climate adaptation: is public participation the key to success? Paper presented at *State of Australian Cities Conference, 29 November–2 December 2011, Melbourne.* University of New South Wales. Available at http://soac.fbe.unsw.edu.au/2011/papers/SOAC2011_0048_final(3).pdf (accessed 9 June 2014).

Council of Australian Governments (2007) *National Climate Change Adaptation Framework.* Department of Climate Change and Energy Efficiency, Canberra.

Council of Australian Governments (2011) *National Strategy for Disaster Resilience. Building the Resilience of our Nation to Disasters.* Commonwealth of Australia, Canberra.

Crondstedt, M. (2002) Prevention, preparedness, response, recovery: an out dated concept? *Australian Journal of Emergency Management* 17(2), 10–13.

Davoudi, S. (2012) Resilience: a bridging concept or a dead end? *Planning Theory and Practice* 13(2), 299–307.

Dovers, S. (1998) Community involvement in environmental management: thoughts for emergency management. *Australian Journal of Emergency Management* Winter 1998, 6–11.

Gero, A. Méheux, K. and Dominey-Howes, D. (2012) *Disaster Risk Reduction and Climate Change Adaptation in the Pacific: The Challenge of Integration.* ATRC-NHRLMIC Report 4, University of New South Wales and Australian Tsunami Research Centre Natural Hazards Laboratory, Sydney.

Goode, N., Spencer, C., Archer, F., McArdle, D., Salmon, P. and McClure, R. (2012) *Review of Recent Australian Disaster Inquiries.* Monash University, Melbourne.

Government of Western Australia (2011) *A Shared Responsibility: The Report of the Perth Hills Bushfire Review.* Government of Western Australia, Perth.

Handmer, J. and Dovers, S. (2007) *Handbook of Disaster and Emergency Policies and Institutions.* Earthscan, London.

Handmer, J., McLennan, B., Towers, B. et al. (2011) *Emergency Management and Climate Change: National Climate Change Adaptation Research Plan--an updated review of the literature.* National Climate Change Adaptation Research Facility, Gold Coast.

Head, B. (2008) Wicked problems in public policy. *Public Policy* 3(2), 101–118.

Howes, M. (2008) *Rethinking Governance: Lessons in Collaboration from Environmental Policy.* Australasian Political Studies Association Conference, 6-9 July, University of Queensland, Hilton Hotel, Brisbane. Available at http://www.polsis. uq.edu.au/apsa2008/Refereed-papers/Howes.pdf (accessed 9 June 2014).

Howes, M., Grant-Smith, D., Bosomworth, K. et al. (2012) *The Challenge of Integrating Climate Change Adaptation and Disaster Risk Management: Lessons from Bushfire and Flood Inquiries in an Australian Context,* Urban Research Program, Issues Paper 17, Griffith University, Brisbane.

Howes, M., Grant-Smith, D., Reis, K. et al. (2013) *Rethinking Disaster Risk Management and Climate Change Adaptation: Final Report.* National Climate Change Adaptation Facility, Griffith University, Gold Coast.

IPCC (2007) *Climate Change 2007: Impacts, Adaptation and Vulnerability, Contribution of Working Group 2 to the Fourth Assessment Report of the Intergovernmental Panel on Climate Change.* M.L. Parry, Canziani, O.F. Palutikof, J.P. et al. (eds) Cambridge University Press, Cambridge, UK and New York, USA.

IPCC (2012) *Managing the Risks of Extreme Events and Disasters to Advance Climate Change Adaptation: A Special Report of Working Groups I and II of the Intergovernmental Panel on Climate Change.* Field, C.B., V. Barros, T.F. Stocker, D. et al. (eds) Cambridge University Press: Cambridge, UK and New York, USA.

Liebrecht, T. and Howes, M. (2006) Collaboration: A solution to inter-jurisdictional strife? *Governments and Communities in Partnership Conference,* Centre for Public Policy, 25– 27 September 2006, University of Melbourne. Available at http://www.researchgate. net/publication/29462753_Collaboration_a_solution_ to_inter-jurisdictional_strife (accessed 9 June 2014).

Mitchell, T., van Aalst, M. and Villanueva, P.S. (2010) *Assessing Progress on Integrating Disaster Risk Reduction and Climate Change Adaptation in Development Processes.* Institute of Development Studies, University of Sussex, Brighton.

O'Brien, G., O'Keefe, P., Rose, J. et al. (2006) Climate change and disaster management. *Disasters,* 30(1), 64–80.

Productivity Commission (2012) *Barriers to Effective Climate Change Adaptation: Productivity Commission Draft Report.* Productivity Commission, Melbourne.

Prosser, B. and Peters, C. (2010) Directions in disaster resilience policy. *Australian Journal of Disaster Management* 25(3), 8–11.

Queensland Floods Commission of Inquiry (2011) *Queensland Floods Commission of Inquiry: Interim Report.* Queensland Government, Brisbane.

Queensland Floods Commission of Inquiry (2012) *Queensland Floods Commission of Inquiry: Final Report.* Queensland Government, Brisbane.

Rogers, P. (2011) Development of resilient Australia: enhancing the PPRR approach with anticipation, assessment and registration of risks. *Australian Journal of Emergency Management* 26(1), 54–58.

Rolfe, J., Bishop, P., Cheshire, L. et al. (2009) *Engaged Government: A Study of Government-Community Engagement for Regional Outcomes: Final Report.* Central Queensland University, Rockhampton.

Ross, A. and Dovers, S. (2008) Making the Harder Yards: Environmental Policy Integration in Australia. *The Australian Journal of Public Administration* 67(3), 245–260.

Victorian Bushfires Royal Commission (2010a) *Final Report: Summary.* Parliament of Victoria, Melbourne.

Victorian Bushfires Royal Commission (2010b) *Final Report: Volume 1: The fires and the fire-related deaths*. Parliament of Victoria, Melbourne.

Victorian Bushfires Royal Commission (2010c) *Final Report: Volume 11: Fire preparation, response and recovery*. Parliament of Victoria, Melbourne.

Waugh, W. and Streib, G. (2006) Collaboration and leadership for effective emergency management. *Public Administration Review* December 2006, 131–140.

Section 9
Business

46 Adaptation to climate change by business organisations

FRANS BERKHOUT

Department of Geography, King's College, UK

46.1 Introduction

Climate variability and change generate new conditions to which social actors (people, households, businesses and public sector agencies) respond through managing risks or by exploiting new opportunities. These responses – adaptation – are usually seen as being specific to the places and contexts in which actors exist and function. This is partly because of the variability of climate change impacts over time and place, and partly because of the diverse features and capabilities of social actors themselves. This variability of climate vulnerability and adaptive capacity (the capacity to influence exposure to risk; cope with damages as a result of a hazardous event; and take opportunities to profit from climate change; IPCC 2007 p. 727) poses an analytical challenge when comparing between individual organisations.

In this brief commentary I summarise some literature on organisational adaptation, including adaptation to climate variability and change (this paper is a shortened and updated version of Berkhout 2012). This literature draws on social and economic research on the micro-foundations of organisational structure, strategy, behaviour and change. The organisations referred to include private sector businesses, public–private organi-sations such as water boards, public sector organisations such as municipal governments and civil society organisations. Organisations are the primary actors involved in choosing and enacting societal responses to climate change. Understanding how organisations might adapt to the threats or possibilities represented by climate change is therefore fundamental to studies of the economics and governance of climate adaptation.

46.2 Internal and external factors in organisational adaptation

A central theoretical question dealt with in the organisational adaptation literature is: to what extent can organisations be seen as acting auton-omously? Are climate vulnerability and adaptive capacity features of an organisation in itself, or are they explained by factors external to and outside the control of the organisation? Although from some perspectives it may be useful to assume that organisations have choice and act autonomously, much organisational research on adaptation stresses the embeddedness of organi-sations in social, institutional and cultural con-texts (Granoveter 1985). Research on business adaptation reflects a well-established debate in organisational studies about the extent to which organisational change is an outcome of internal adaptation or external selection (Astley and van de Ven 1983). Most contemporary theories in organisational studies agree that strategic choice and environment interact to constitute the adaptive responses of organisations (Hrebiniak and Joyce 1985). Understanding how organisations are embedded in their social and institutional environments and how this shapes their goals, structure and ways of doing things is also a primary concern of research on organisational adaptation to climate variability and change.

Related to the question of embeddedness is the question of whether climate change is a motivator of organisational responses and change. Organisational behaviour and change may be viewed from a number of different perspectives. One starting point is to see organisations as autonomous actors set up to achieve specific organisational goals, whether that is producing food crops or offering transport or retail services. To achieve their objectives, organisations have human and capital resources, capabilities and routines, a culture shaping a way of doing things and a system of governance. Climate variability and change may influence the ability of organisations to achieve their objectives effectively and efficiently. Such influences may be indirect, through changes in the operating environment of the business (such as regulations, insurance premiums, or customer preferences) or direct, for instance through the experience of damage caused by an extreme weather event (Chapter 48 by Kuruppu et al.). Sometimes the limits or opportunities represented by climate change are so pressing that they lead to a fundamental reconsideration of organisational strategies and behaviour (Linnenluecke and Griffiths 2010; Pelling 2011).

Tracing the many ways in which beliefs, structures, business models, strategies and activities of businesses are influenced by climate change is therefore a complex task of disentangling the primary from the secondary and the direct from the indirect. Many factors play a role in shaping the decisions and actions of organisations, of which perceptions about climate change as an influence on resources, capabilities and operating conditions will be one. An analysis of business adaptation needs to start with the complex reality of businesses and their operating environment, rather than with an observed climate signal with a putative influence on organisational behaviour. The analysis needs to be done 'inside-out', rather than 'outside-in'.

46.3   Three perspectives on organisational adaptation

We start with the question: why do businesses adapt? As we have argued, a simple answer is that adaptation relates to the functional goals of an organisation. Adaptations can be viewed as adjustments designed to sustain the organisation's capacity to meet a particular objective. For example, an infrastructure company may build coastal assets to withstand higher design standards in anticipation of higher peak wind speeds during tropical storms (Chapter 49 by West). Continuity in terms of organisational structure, business model and performance would be preserved, although at the cost of greater investment or insurance costs. However, the common definition of climate change adaptation, '... adjustments in natural or human systems in response to actual or expected climate stimuli or their effects, which moderate harm or exploit beneficial opportunities' (IPCC 2001, p. 879) suggests that adaptation may not always aim to sustain the performance levels of an organisation. The additional effort involved in maintaining existing functions and standards of performance may be disproportionate, or conversely there may be opportunities to change taking advantage of climate change (Mills 2009). Adaptation may therefore imply a more radical organisational, strategic or business model transformation.

There are broadly three approaches to analysing organisational adaptation: the utility-maximising, behavioural and institutional approaches (Berkhout 2012), described in the following sections.

46.3.1   *Utility-maximising approach*

Utility-maximising approaches assume that adaptive behaviour is a question of optimal choices between a set of alternatives whose costs and benefits are known and can be discounted over time (this sort of analytical strategy is applied in Chapter 49). These choices are made by organisations pursuing their own self-interest, hence the notion of autonomous adaptation (Fankhauser et al. 1999). This position holds that adaptation can be regarded as efficient if '... the [private] cost of making the effort is less than the resulting [private] benefits ...' (Mendelsohn 2000 p. 585). A utility-maximising position suggests that most organisations will adapt only once they have experienced the effects of climate change, and will then adapt to the extent needed to maximise their overall utility for some period into the future relevant to their investment and operating time horizons. The costs of inaction (the damage costs to the organisation without adaptation) and of adaptive responses will be known, as will the stream of benefits of adaptive action through time. On this basis, utility-maximising organisations will choose inaction when that is economically optimal while making timely investments in adaptation when that is economically justified, given uncertainties. Only private costs and benefits are relevant to this assessment.

The utility-maximising approach has been criticised for making invalid assumptions about the nature of climate impacts (Schneider et al. 2000), for misunderstanding the decision-making by adapting actors (Risbey et al. 1999) and for ignoring the broader social benefits of adaptation by individual organisations.

46.3.2   *Behavioural approach*

Taking a behavioural approach, Berkhout et al. (2006) argue that issues of perception, interpretation, problem-solving and learning are central to understanding organisational adaptation. Drawing on research in behavioural economics and organisational studies (Cyert and March 1963; Levinthal and March 1993), they stress the importance of uncertainty and the bounded rationality of social actors (including organisations). Berkhout et al. propose that actors do not conform to the tenets of expected utility theory, use 'rules of thumb' in responding to new situations and that they exhibit satisficing behaviour; that is, in the face of uncertainty, organisations choose 'good enough' responses that conform to normative ideas of appropriate behaviour, rather than optimising across the full universe of potential options.

For this 'behavioural' strand of adaptation research, the adaptive response of organisations will be determined by the perceptions and capabilities of the organisation. The strategy chosen will depend less on an objective assessment of costs and benefits, and more on a messy process of sense-making, learning-by-doing and organisational adjustment. Organisations frequently exhibit inertia to change, so that modifications in structure, goals and activities can meet resistance. Organisational adaptation and change will therefore be seen as serving the goals of specific groups within organisations that may be distant from the objective problem of responding to climate vulnerabilities.

46.3.3   *Institutional approach*

A third institutionalist approach of research places greater emphasis on the role of the institutional context in which the organisation is embedded. For these researchers, drawing on a tradition of institutional economics and governance studies (cf. Ostrom 1990; North 1991), the adaptive capacity of an organisation rests not only on utility-maximising behaviour or on the perceptions and capabilities of the organisation, but is enabled and constrained by factors (often called 'barriers') imposed by external social, cultural, political and economic structures and processes (see Chapter 47 by Shearer et al.). Although institutionalist perspectives have been applied across many contexts of climate adaptation, they are especially powerful in sectors such as water resource management, coastal protection and flood risk management where national and regional governments play an important role in defining and enforcing the formal and informal 'rules of the game' by which organisations operate (Naess et al. 2005; Moser and Luers 2008; McDonald 2011).

Institutional contexts are overlapping and multi-layered, some being proximate and others distant. A central concern of this research is the flexibility offered to adapting agents by institutionally framed rules to determine and carry out adaptive responses. Contexts may provide the knowledge, resources, incentives and legitimacy needed for (collective) adaptive action, but they may also promote climate vulnerability and constrain adaptive responses. For this strand of the literature, the prevailing external incentives and disincentives and how they influence organisational behaviour will be the critical determinants of adaptive action.

46.3.4 Overlap between the three types of approach

While there are clear differences between the utility-maximising, behavioural and institutional approaches, there are also overlaps. For instance, economists working from a utility-maximising position acknowledge that uncertainties about climate change impacts have an important bearing on organisational strategies towards climate vulnerabilities, something stressed in behaviourist perspectives. Likewise, behaviourist organisational studies accept that organisations operate in the light of a fairly stable set of preferences and goals (such as the need to turn a profit) and that maladapted organisations will fail to survive in the long run. Finally, although some scholars argue that organisational behaviour is primarily conditioned by external environments, most studies find that organisational behaviour and change is a product of endogenous and exogenous factors and processes. Behavioural and institutional perspectives therefore share common ground.

46.4 Conclusions

Organisations are the primary actors in societal responses to the impacts of climate variability and change. Our understanding of how organisations perceive, make sense of and respond to signals

about climatic change is still developing, however. I have briefly discussed a number of theoretical perspectives that are currently applied to the study of organisational adaptation to climate change, characterising them as utility-maximising, behavioural and institutionalist perspectives.

Empirical and theoretical studies of organisational adaptation, including those in this volume, offer findings with some common emerging patterns. First, organisational adaptation needs to be understood from the perspective of the goals and perceptions of the organisation itself. Organisational change and adaptation takes place in response to many stimuli from the market and from competitors, with climate risks and opportunities being one set of stimuli among many that may affect business performance and continuity. Second, a complex set of organisational processes is involved in perceiving, evaluating, enacting and learning about climate impacts and adaptive responses. The climate signal will often be difficult to read for organisations against the noise of other signals from their market and institutional environment. Insofar as there is a climate signal, this will often be perceived indirectly through, for instance, a regulatory requirement (Chapter 48 by Kuruppu et al.). Third, few changes made by organisations will be a response to a climate signal alone. Climate change considerations often play a supplementary role in decisions about technological, organisational or strategic innovation, even in sectors such as water management in which climate change is highly salient. Fourth, organisational adaptation is always strongly influenced by the institutional context in which the organisation is embedded. Adaptations will emerge as an interaction between factors internal to the organisation and a range of technological, market and regulatory factors external to the organisation. Lastly, while many organisations – even those in highly vulnerable settings – may have a wide range of potential adaptive responses, their willingness and capacity to exercise these options will vary greatly. Whether or not the institutional context is enabling or constraining will play a role, but leadership, resources and culture of the organisation will also be important.

References

Astley, W.G. and van de Ven, A.H. (1983) Central perspectives and debates in organizational theory. *Administrative Science Quarterly* 28, 245–273.

Berkhout, F. (2012) Adaptation to climate change by organizations. *WIREs Climate Change* 3, 91–106.

Berkhout, F., Hertin, J. and Gann, D.M. (2006) Learning to adapt: Organisational adaptation to climate change impacts. *Climatic Change* 78,135–156.

Cyert, R.M. and March, J.G. (1963) *A Behavioral Theory of the Firm*. Prentice Hall, Englewood Cliffs, NJ.

Fankhauser, S., Smith, J.B. and Tol, R.S.J. (1999) Weathering climate change: some simple rules to guide adaptation decisions. *Ecological Economics* 30, 67–76.

Granoveter, M. (1985) Economic action and social structure: The problem of embeddedness. *American Journal of Sociology* 91, 481–510.

Hrebiniak, L.G. and Joyce, W.F. (1985) Organizational adaptation: Strategic choice and environmental determinism. *Administrative Science Quarterly* 30, 336–349.

IPCC (2001) *Climate Change 2001: Impacts, Adaptation and Vulnerability*. McCarthy, J.J., Canziani, O.F., Leary, N.A., Dokken, D.J. and White, K.S. (eds) Cambridge University Press, Cambridge.

IPCC (2007) *Climate Change 2007: Impacts, Adaptation and Vulnerability. Contribution of Working Group II to the Fourth Assessment Report of the Intergovernmental Panel on Climate Change*. Parry, M.L., Canziani, O.F., Palutikof, J.P., van der Linden, P.J. and Hanson, C.E. (eds) Cambridge University Press, Cambridge.

Levinthal, D.A. and March, J.G. (1993) The myopia of learning. *Strategic Management Journal* 14, 95–112.

Linnenluecke, M.K. and Griffiths, A. (2010) Beyond adaptation: Resilience for business in light of climate change and weather extremes. *Business and Society* 49, 477–511.

McDonald, J. (2011) The role of law in adapting to climate change. *WIREs Climate Change* 2, 283–295.

Mendelsohn, R. (2000) Efficient adaptation to climate change. *Climatic Change* 45, 583–600.

Mills, E. (2009) A global review of insurance industry responses to climate change. *The Geneva Papers on Risk and Insurance* 34, 323–359.

Moser, S.C. and Luers, A.L. (2008) Managing climate risks in California: the need to engage resource managers for successful adaptation to change. *Climatic Change* 87, 309–322.

Naess, L.O., Bang, G., Eriksen, S. and Vevatne, J. (2005) Institutional adaptation to climate change: Flood responses in the municipal level in Norway. *Global Environmental Change* 15, 125–138.

North, D.C. (1991) Institutions. *Journal of Economic Perspectives* 5, 97–112.

Ostrom, E. (1990) *Governing the Commons: The Evolution of Institutions for Collective Action*. Cambridge University Press, Cambridge.

Pelling, M. (2011) *Adaptation to Climate Change: From Resilience to Transformation*. Routledge, London.

Risbey, J., Kandlikar, M. and Dowlatabati, H. (1999) Scale, context and decision making in agricultural adaptation to climate variability and change. *Mitigation and Adaptation Strategies for Global Change* 4, 137–165.

Schneider, S. H., Easterling, W. E. and Mearns, L. O. (2000) Adaptation: sensitivity to natural variability, agent assumptions and dynamic climatic changes. *Climatic Change* 45, 203–221.

47 Capacities of private developers in urban climate change adaptation

HEATHER SHEARER[1], JAGO DODSON[2],
EDDO COIACETTO[3] AND PAZIT TAYGFELD[1]

[1] Urban Research Program, Griffith University, Australia
[2] Global, Urban and Social Studies, RMIT University, Australia
[3] Griffith School of Environment, Griffith University, Australia

The biggest climate change challenge is the inability of the human system to respond in time to the challenge. (MD1, Developer)

47.1 Introduction

Anthropogenic climate change has the potential to severely impact the natural and human environment (Norman 2010; World Bank 2011). The urban environment is particularly exposed to the impacts of climate change because of a range of geographic and social factors. Humanity is now urban and many cities are situated in areas which are directly vulnerable to sea-level rise, floods and storm surge, such as the coastal plain or floodplains of large rivers, and are therefore indirectly vulnerable to climate impacts in external regions. Moreover, due to the concentration of the population and the density of the built environment, the impacts of extreme climate events are often more severe in urban areas (Bambrick et al. 2011). Recent extreme climate events have highlighted this vulnerability. In October 2012, Superstorm Sandy caused US$ 82 billion worth of damage to the 'megalopolitan' northeastern USA (Tamari 2012) and the 2011 Queensland floods in Australia, which were focused on urbanised South East Queensland, cost US$ 5.3 billion (QRA and World Bank 2011).

Urban regions can respond to climate change in a number of ways including mitigation, adaptation, rehabilitation, partial or total relocation and resettlement (Imura and Shaw 2009; Vasey-Ellis 2009). Many options are unfeasible because of economic, social and other costs: the built environment and urban infrastructure, such as buildings, roads and rail, which comprises the fabric of a city, is spatially fixed and cannot easily or cost-effectively be relocated to less vulnerable locations. While it may be possible to resettle small communities at high risk of climate impacts (such as Grantham in Australia, see Coates 2012), it would be extremely expensive and disruptive to relocate the population of a city such as Dhaka in Bangladesh which has a population of 15 million, 80% of which live in areas at risk of flooding. In Bangkok serious questions have been raised about the possibility of relocating that city to less vulnerable terrain (AFP 2011).

Although mitigation is considered an essential part of responding to the challenge of climate change, many climate researchers consider that a range of climate impacts are unavoidable (Matthews 2011; IPCC 2007). Moreover, recent

Applied Studies in Climate Adaptation, First Edition. Edited by Jean P. Palutikof, Sarah L. Boulter, Jon Barnett and David Rissik.
© 2015 John Wiley & Sons, Ltd. Published 2015 by John Wiley & Sons, Ltd.

climate modelling has indicated a potentially catastrophic mean global temperature rise of 4°C or greater (Stafford Smith et al. 2011; World Bank 2012). It is therefore essential that humanity learns to adapt to the urban impacts of climate change by adjusting or modifying systems such as the urban built environment (IPCC 1996).

The built environment itself has long-term climate impacts through the design, location and building materials used and these in turn influence the degree to which urban residents are exposed to climate change hazards (World Bank 2011). Further, cities are dynamic systems that are continually changing; new buildings and suburbs are being constructed, areas of urban decay are being revitalised and new infrastructure is being laid. The urban areas of the future will be modified or created by the actions of both state and non-state actors (e.g. property developers; Bosher et al. 2007). However, while a large literature exists on the role of government actors in climate adaptation, limited current research exists on the adaptive capacity of non-state actors, such as the private property development sector (Taylor et al. 2012; Bulkeley and Betsill 2013).

Adaptive capacity is the ability to prepare for hazards and opportunities in advance (known as anticipatory or planned adaptation) and to respond and cope with the effects (known as autonomous or reactive adaptation; Smit and Pilifosova 2003; Smit and Wandel 2006). As such, it includes the ability to improve resilience, decrease vulnerability and even exploit the benefits of a changing climate (Smit and Wandel 2006). Climate risks such as floods can drive adaptation but indirect climate risks, such as changes to planning regulation in response to past or future climate events, can also drive adaptation. For example, developers generally respond more to indirect than direct climate risks, especially through changes to the regulatory environment (Taylor et al. 2010), and many do not consider climate change as a significant threat to their business (Bosher et al. 2007). Another study found that developers are reluctant to incorporate sustainability measures into buildings because of a perceived lack of market demand or the likelihood of climate

events occurring within short development timeframes (Häkkinen and Belloni 2011).

The private property development sector has a pivotal role in creating the future built environment, and is a major potential source of improved adaptive activity. An improved understanding of the capacity of this sector to respond to climate change is therefore critical to responsive climate change adaptation. To this end, this study aimed to investigate the institutional capacity of the private urban development sector in South East Queensland (SEQ), Australia to respond to the task of climate change adaptation and, in turn, investigate the role of private financial institutions in funding climate adaptive urban development.

47.2   Method and analysis

The project methodology included a desktop literature review, an online questionnaire survey and a series of semi-structured interviews and focus groups held with members of the SEQ development sector that included developers, consultants, state and local government staff, architects, solicitors, planners and financiers. A total of 62 people responded to the online survey and 21 interviews and 3 focus groups (a total of 9 participants) were held. The majority of interviews and focus groups were recorded (with the permission of the attendees), and the data were analysed using the software program NVivo, using a form of content analysis that identifies common themes in the data.

The questionnaire included items on developer characteristics, type of development practised, knowledge of and attitudes to climate change. Participants were also questioned about whether their firm had incorporated climate adaptive measures and if they were likely to do so in the future. The interviews and focus groups were semi-structured and questions expanded on those in the questionnaire survey. The first section focused on development type and related descriptors and the second focused on climate change and included questions on the potential

Table 47.1 Factors contributing to the adaptive capacity of the South East Queensland property development sector.

Internal factors		External factors
Developer characteristics	**Development characteristics**	**External factors**
Firm size and structure		Access to consultants and other experts
Economic resources		Economic environment
Developer standards	Temporal and spatial scale of development	Financiers and finance
Leadership, legacy and brand	Type of development	Market demand
Attitude		Regulation and governance
Communication and information		Insurance industry

impact of climate change on their business, attitudes and perceptions towards various climate change risks and how these might be managed. Interviewees were also questioned on finance availability, insurance, litigation, market demand for greener products and attitude to the regulatory environment.

47.3 Results and discussion

A diverse range of firms participated in the study including large Australian Stock Exchange listed development companies, a number of small family firms and two government-run development corporations. A diverse range of consultants to the development sector participated, ranging from multinational engineering firms to single-person practices, two of which specialised in climate change adaptation. Two financiers, two architects, a solicitor and state and local government representatives also participated.

This study found that the adaptive capacity of the urban property development industry was dependent on a combination of complex and interconnected factors. These factors could be broadly characterised into internal and external factors (see Table 47.1) and included: (1) economic resources and financing; (2) market conditions; (3) government regulation and industry self-regulation; (4) firm size and structure; (5) the type and spatiotemporal scale of developments; (6) attitude, leadership and

legacy; and (7) the insurance sector. These are briefly discussed below.

First, economic resources and market conditions were highly significant. Subsequent to the Global Financial Crisis (GFC) of 2007–2008, market conditions and the availability of finance largely dictated the type of development produced (Bryant 2012). Post the GFC period, financiers began to impose a range of increasingly onerous conditions on development loans, requiring a high percentage of pre-sales (often greater than 75%), more stringent reporting requirements and a higher proportion of equity (Bryant 2012). In this study, all respondents reported difficulty in obtaining bank finance for developments with some developers resorting to alternative funding sources, such as from high net worth individuals or international private equity. Although larger firms had access to a wider range of funding sources, smaller firms were more dependent on finance from the major banks. The combination of conservative lending by the major banks and problems with property valuation led many developers to abandon innovative or 'green' developments in favour of more conventional building.

Second, increased marked demand drives development. Prior to 2008, SEQ had experienced high population growth due to a combination of factors such as high levels of inter-state migration, relatively affordable property in comparison to New South Wales and Victoria and a buoyant economy resulting from a mining boom

(BITRE 2013). Subsequent to the GFC, the demand for property plummeted and many firms went out of business or moved to other ventures. The purchasers who were still in the market were primarily price-conscious and there was little if any demand for properties with innovative or sustainability features, particularly if these came at higher cost. In general, developers considered that market risk and finance risk were the most important factors.

Third, government regulation was seen more as a barrier than a driver of adaptive capacity. The regulatory environment was viewed as having a plethora of complex, confusing and inflexible legislation at all levels of government. Respondents particularly noted that dealing with local government authority (LGA) rules and regulations were often difficult, complicated and time consuming. The issue of time was striking throughout the research, but particularly in the context of LGA processes. These delays caused major problems for many developers, and could cost them significant sums of money. A policy specifically aimed at climate change adaptation, the now superseded *Queensland Coastal Plan 2012 (the Coastal Plan)* (State of Queensland 2012), was singled out for the majority of criticism.

> The Coastal Plan is a very poor piece of legislation, has not been well drafted, it seems to be a knee jerk reaction to whatever the government thought they were trying to do. But it has created difficulties in the profession because it is such a difficult document to interpret. (CS8, Consultant)

On the other hand, the Building Code of Australia (BCA) (Australian Building Codes Board 2011) was viewed by many as the regulatory avenue with the most potential for compelling climate change adaptation. For example, past climate events, such as Darwin's Cyclone Tracey (1974) had resulted in rapid changes to the BCA, to which all developers have to adhere.

> Climate change is no bigger [risk] factor than anything else. The sustainability of houses has now been mandated in the building code. (SD2, Developer)

Industry self-regulation also appeared to have potential to drive adaptation. Developers regularly used industry-specific certification schemes such as Green Star (GBCA 2014) and EnviroDevelopment (http://www.envirodevelopment.com.au/). Industry certification tools incorporate a range of sustainable practices with respect to waste, energy and materials (UDIA 2011; GBCA 2014). These schemes were all viewed positively and are a potentially valuable driver for increased adaptation to climate change.

> It would be useful to take that general market perception and how to translate into real benefits and the Green Star – because you can say there is a real benefit here, we are delivering extra environmental goodies, and there is an independence to that and people are convinced that it is real. (MD2, Developer)

Fourth, the size and company structure of developers was an indication of adaptive capacity. Larger developers that produced a range of diverse products had more flexibility to introduce adaptive measures as they were less dependent on bank funding, and could also afford to employ or contract specialist consultants. Such firms were also conscious of their enduring legacy, and intended their development products to showcase their brand. Some smaller developers were also driven to produce high-quality products however, and this motivation largely depended on personal attitude.

Fifth, the type and spatiotemporal scale of development were also major determinants of adaptive capacity. In general, the larger the development and the longer the development time frames, the more likely it was that developers were concerned about the potential risks of climate change and would implement adaptive measures. The converse was true for smaller developments with shorter time frames. For these, the developer's main goal was to complete the development as quickly as possible in order to reimburse bank loans. Lending policies exacerbated this, as some major banks were not prepared to loan money to developers for longer than a maximum of two years.

With regards to type of development, commercial developers regularly incorporated sustainability features into buildings in response to tenant demand, reduced operating costs and government subsidies. However, there was little demand for sustainability features from the residential market, unless these incurred no additional cost. For the retirement sector, the desire to reduce costs, particularly energy costs, drove a small demand for sustainability features. Market demand therefore served as both a driver and a barrier to adaptive capacity.

Sixth, attitudes and knowledge were potentially important avenues for adaptation. For example, a number of respondents expressed confusion and scepticism regarding climate change and its potential impacts. They associated climate change primarily with sea-level rise and saw this happening, if at all, in the distant future and thus posing minimal risk. Despite this, many commented on the need to incorporate environmentally sustainable features into developments, particularly if there was a market demand for them. To some extent, the development industry appears willing to adapt to climate change, but felt the risks had to be better communicated.

> At the moment sea level rise is not really high enough...the rate of sea level rise around Australia is such that it is hard to get confidence that the predicted sea-level rises will occur. We are potentially spending a significant amount of money for a phenomenon that is predicted to occur not for another 40 years. (CS8, Consultant)

Finally, the insurance industry has the potential to drive adaptive capacity. In North Queensland, extreme weather events such as Cyclone Yasi (2011) resulted in large premium rises, particularly for strata titled units (Australian Government Actuary 2012). Developers reported great difficulty in selling these properties, as the Body Corporate fees were extremely high due to the insurance costs. Unless insurers link premiums to the introduction of adaptive measures however price rises are likely to be maladaptive, resulting in properties with inadequate or no insurance.

47.4   Conclusion

In conclusion the research we undertook identified some major barriers and drivers of adaptive capacity for the urban property development industry. Larger developers tended to have more capacity to include climate adaptive features in their developments. This was due to their having a wider range of financing instruments, corporate policies requiring sustainability features, a strong focus on brand and legacy and the ability to land bank. Some smaller developers were also strongly motivated to produce more sustainable developments, but had been hampered by a combination of post-GFC conservative bank lending practices, lack of market demand and government 'red tape'. As Häkkinen and Belloni (2011) also found, our study revealed that property developers were primarily motivated by client demand. As they reported little or no demand for sustainable or adaptive buildings, particularly if these came at a higher cost, they were unwilling to provide these. Similarly, many considered that the relatively short times that they were exposed to a development would protect them from the risk of climate change (Häkkinen and Belloni 2011).

The majority of government regulation dedicated to climate adaptation was seen more as a barrier to than a driver of improved adaptive responses. The development sector viewed most legislation and policy, as well as government processes, as hindering their ability to produce climate-adaptive and sustainable developments. Legislation aimed specifically at climate change, such as the now superseded Queensland Coastal Plan, bore the brunt of criticism as being too complex and unfeasible to implement. In contrast to the findings of Taylor et al. (2012), government regulation such as the Coastal Plan was not seen as a motivator for action, rather for increased lobbying against adaptation legislation. Local governments were also seen as hindering adaptive development, with many councils perceived as reactive and unwilling to approve innovative developments. On the other hand, the use of building codes such as the BCA was seen as a potential avenue for rapidly and flexibly incorporating adaptive measures into developments.

Industry self-regulation schemes, on the other hand, were identified as a potential driver of adaptation. These were widely accepted within the industry, and developers often competed among themselves to achieve higher sustainability ratings. The clients, especially commercial clients, often demanded these before renting or purchasing in a development. For the residential market, encouraging demand for sustainable features in the residential market may be achieved by quantifying long-term savings or subsidising developers (Häkkinen and Belloni 2011).

Finally, neither financiers nor the insurance industry are currently driving adaptation; indeed, their risk-averse nature was seen as hindering rather than helping developers incorporate adaptive measures. However, both have significant potential to drive major changes. For example, financiers could tie loan conditions and/or the insurance industry could link premiums to the provision of adaptive measures.

More research is needed on many of these, particularly on the potential for the financial and insurance sectors to drive adaptation. At present, both these industries are serving more as agents of inertia and at worst maladaptation, with financiers putting time and cost pressure on developers and insurers raising premiums without requiring adaptive changes. The urban development sector delivers most of Australia's urban built environment. Greater attention from policymakers and regulators to the constraints and capacities of this sector to contribute to responsive adaptation will be necessary if national climate adaptation goals are to be achieved.

References

Agence France Presse (2011) *Thai Lawmakers Submit Motion on Moving Capital*, AFP newswire, 15 November 2011, via Dow Jones News Database.

Australian Building Codes Board (2011) *The Building Code of Australia, National Construction Code Series*. ABCB, Canberra.

Australian Government Actuary (2012) *Report on Investigation into Strata Title Insurance Price Rises in North Queensland*. Australia Government, Canberra.

Bambrick, H.J., Capon, A.G., Barnett, G.B., Beaty, R.M. and Burton, A.J. (2011) Climate change and health in the urban environment: adaptation opportunities in Australian cities. *Asia-Pacific Journal of Public Health* 23, 67S.

Bosher, L., Dainty, A., Carrillo, P., Glass, J. and Price, A. (2007) Integrating disaster risk management into construction: a UK perspective. *Building Research and Information* 35(2), 163–177.

Bryant, L. (2012) An assessment of development funding for new housing post GFC in Queensland, Australia. *International Journal of Housing Markets and Analysis* 5(2), 118–133

Bulkeley, H. and Betsill, M.M. (2013) Revisiting the urban politics of climate change. *Environmental Politics* 22(1), 136–154.

Bureau of Infrastructure, Transport and Regional Economics (2013) *Population Growth, Jobs Growth and Commuting Flows in South East Queensland, Report 134*. Commonwealth of Australia, Canberra.

Coates, L. (2012) Moving Grantham? Relocating flood-prone towns is nothing new. *The Conversation* 10 January 2012. Available at https://theconversation.com/moving-grantham-relocating-flood-prone-towns-is-nothing-new-4878 (accessed 9 June 2014).

Green Building Council of Australia (2014) *Green Star*. Available at http://www.gbca.org.au/green-star/ (accessed 9 June 2014).

Häkkinen, T. and Belloni, K. (2011) Barriers and drivers for sustainable building. *Building Research and Information* 39(3), 239–255.

Imura, M. and Shaw, R. (2009) Challenges and potentials of post-disaster relocation. *Asian Journal of Environment and Disaster Management* 1(2), 197–221.

IPCC (1996) *Climate Change 1995 Impacts, Adaptations and Mitigation of Climate Change: Scientific-Technical Analyses*. Cambridge University Press, Cambridge.

IPCC (2007) *Climate Change 2007: Synthesis Report. Contribution of Working Groups I, II and III to the Fourth Assessment Report of the Intergovernmental Panel on Climate Change*. Pachauri, R.K. and Reisinger, A. (eds) Cambridge University Press, Cambridge.

Matthews, T. (2011) *Climate Change Adaptation in Urban Systems: Strategies for Planning Regimes*. Urban Research Program, Griffith University, Brisbane.

Norman, B. (2010) *A Low Carbon and Resilient Urban Future; A Discussion Paper on an Integrated Approach* to *Planning for Climate Change.* Commonwealth of Australia, Canberra.

Queensland Reconstruction Authority and World Bank (2011) *Queensland Recovery and Reconstruction in the Aftermath of the 2010/2011 Flood Events and Cyclone Yasi.* A report prepared by the World Bank in collaboration with the Queensland Reconstruction Authority. World Bank, Washington; Queensland Reconstruction Authority, Brisbane.

Smit, B. and Pilifosova, O. (2003) Adaptation to climate change in the context of sustainable development and equity. In: Smit, J.B. Klein, R.J.T. and Hug, S. (eds) *Climate Change Adaptive Capacity and Development.* Imperial College Press, London.

Smit, B., and Wandel, J. (2006) Adaptation, adaptive capacity and vulnerability. *Global Environmental Change* 16, 282–292.

Stafford Smith, M., Horrocks, L., Harvey, A. and Hamilton, C. (2011) Rethinking adaptation for a 4°C world. *Philosophical Transactions A Math Physics Engineering and Science* 369(1934), 196–216.

State of Queensland (2012) *Queensland Coastal Plan.* Department of Environment and Heritage Protection, Brisbane.

Tamari, J. (2012) Obama requests $60.4 billion to help area recover from superstorm Sandy. McClatchy–Tribune Business News, 2012 Dec 08.

Taylor, B., Harman, B., Heyenga, S. and McAllister, R. (2010) *Property Developers' Perspective on Urban Adaptation to Climate Change.* Industry Summary Report. CSIRO, Brisbane.

Taylor, B.M., Harman, B.P., Heyenga, S. and McAllister, R.J. (2012) Property developers and urban adaptation: conceptual and empirical perspectives on governance. *Urban Policy and Research* 30(1), 5–24.

Urban Development Institute of Australia (2011) About EnviroDevelopment. Available at http://www.envirodevelopment.com.au/01_cms/details.asp?ID=1 (accessed 9 June 2014).

Vasey-Ellis, N. (2009) Planning for climate change in coastal Victoria. *Urban Policy and Research* 27(2), 157–169.

World Bank (2011) *Guide to Climate Change Adaptation in Cities.* World Bank, Washington DC.

World Bank (2012) *Turn Down the Heat: Why a 4°C Warmer World Must be Avoided.* A report for the World Bank by the Potsdam Institute for Climate Impact Research and Climate Analytics. International Bank for Reconstruction and Development/World Bank, Washington.

48   Ensuring small business continuity under a changing climate: the role of adaptive capacity

NATASHA KURUPPU, PIERRE MUKHEIBIR
AND JANINA MURTA

Institute for Sustainable Futures, University of Technology Sydney, Australia

48.1   Introduction

Research examining adaptation and adaptive capacity of the private sector in general has been underexplored and sparse, but is vital for framing strategies and policies supporting adaptation (Ingirige and Jones 2008; Vogel 2009; Busch 2011). Small-to-medium enterprises (SMEs) and the private sector may need to take a key role in mobilising particular adaptation interventions and innovations recommended for communities and government, for example, where SMEs are able to provide the technologies required to help communities to adapt. SMEs are critically important to the global economy and form an important sector of society; they provide employment, goods and services and tax revenue for communities (Howe 2011). This chapter adopts the Australian Bureau of Statistics definition of SMEs (ABS 2011), which includes those enterprises employing less than 200 people. This can be broken down further to medium business (20–199 employees), small business (5–19 employees) and micro business (1–4 employees).

SMEs comprise 96% of all private businesses in Australia, and represent the economy's largest employer and contributor to GDP (Gross Domestic Product; DISR 2010). The location of SMEs in the broader socio-economic system, with its associated spatial interconnections and dependencies, means that specific risks or stresses (e.g. global financial crisis or political upheaval) experienced at a particular scale are likely to shape the economic viability of SMEs. Within this context, SMEs are confronted with the new challenge of the expected impacts of climate change and increased variability. These impacts are likely to affect the supply and production chains of SMEs through business interruptions, increased insurance costs, property damage and decline in financial measures such as value, return and growth (Reynolds 2013).

SMEs are likely to face greater short-term losses after natural disasters and have lower adaptive capacity than larger enterprises for various reasons (UNISDR 2013). They tend to not have diverse portfolios of products or operations to mitigate their risks, nor comprehensive

Applied Studies in Climate Adaptation, First Edition. Edited by Jean P. Palutikof, Sarah L. Boulter, Jon Barnett and David Rissik.
© 2015 John Wiley & Sons, Ltd. Published 2015 by John Wiley & Sons, Ltd.

contingency plans to offset losses. They also tend to have smaller cash reserves, and are less able to distribute risk through methods such as insurance against property damage and business interruptions (Runyan 2006).

This chapter argues that scholarship on climate adaptation within the private sector, particularly the SME sector, requires further consideration. The overall aim of the study on which this chapter is based was to identify the underlying processes and factors shaping adaptive capacity of SMEs in Australia to climate change (Kuruppu et al. 2013). Specifically, the study sought to determine: (1) how SMEs have considered and integrated adaptation into business planning; and (2) the key underlying processes that constrain and influence the adaptive capacity of SMEs.

48.2   Examining adaptive capacity of SMEs

Adaptive capacity is often defined as:

> ...the set of resources (natural, financial, institutional or human, and including access to ecosystems, information, expertise, and social networks) available for adaptation, as well as the ability or capacity of that system to use these resources effectively in the pursuit of adaptation. (Brooks and Adger 2004, p. 168)

Figure 48.1 demonstrates that an understanding of adaptive capacity is central to reducing vulnerability and exposure of SMEs to climate change and other stresses. Vulnerability to climatic change 'arises through particular levels of exposure to underlying socio-economic changes, as well as biophysical changes' (Lorenzoni et al. 2000, p. 149). Adaptation within this context is recognised as a process that transforms the system by re-orienting the deep social structures that drive vulnerability (Watts 1983; Pelling 2011).

The authors also acknowledge that adaptation actions that enhance adaptive capacity will simultaneously support resilience building of SMEs and the larger system it operates. Resilience

is 'the potential of a particular system to maintain its function [which may not be by the most efficient means] in the face of disturbance and the ability of the system to re-organise following disturbance-driven change' (Holling and Walker 2003, p. 1) and involves learning, self-organising and adapting.

In identifying particular determinants that underpin the concept of adaptive capacity, this research draws on the literature on climate adaptation, social resilience and organisational change (Folke et al. 2002; Yohe and Tol 2002). Literature on organisational change confirms the significance of understanding the functioning of a business as a complex open system in which its internal characteristics are constantly interacting with the broader external context in which it is embedded (Burke 2008). It also suggests determinants that characterise the internal (specific determinants) and external context of the SME (generic determinants), which are likely to shape the adaptive capacity of SMEs (Fig. 48.1).

To understand the causal processes that mediate adaptive capacity such as social structure, power, cultural norms and human agency, this chapter expands the conceptual framework to include theories from political ecology, which pays attention to spatial and temporal scales, human agency and power. Agency is defined as 'the capacity of human actors to project alternative future possibilities, and then to actualise those possibilities within the context of current contingencies' (McLaughlin and Dietz 2008, p. 105). The agency aspect also encourages a focus on external relations and the broader social context in which SMEs operate; these are seen as conversion factors or means of further expansion of agency and opportunities (Alkire 2007).

Additionally, the study draws on the capability approach to conceptualise and evaluate the adaptive capacity of SMEs. A key aspect of the approach pertinent to this study is that capabilities reflect the range of opportunities/ choices available to a small business to generate outcomes that they value, factoring in relevant personal and external factors (Sen 1999). Enhancing adaptive capacity can therefore be

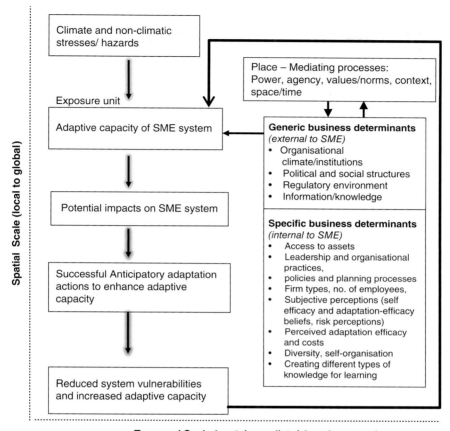

Figure 48.1 A framework for conceptualising how adaptive capacity of SMEs reduces overall vulnerability to climate and other stresses. Source: Kuruppu 2009.

seen as a process by which the opportunity or choice sets that SMEs value are expanded.

48.3 Research approach and activities

In addition to a background literature review of the contextual constraints experienced by SMEs in Australia and the extent of adaptation within the sector both nationally and internationally, a number of participatory methods were adopted in conducting the research over a period of one year, including the following.

- An online survey with a mix of multiple choice and open-ended questions targeting SMEs in various sectors in Australia. A total of 45 businesses responded. All industry categories from the Australian Bureau of Statistics classification system for SMEs were represented in the survey apart from the electricity, gas water and waste services, mining and wholesale trade.
- Semi-structured interviews with 35 representatives from a wide range of government and non-government organisations who provide direct and/or indirect support to the SME sector.

• Five case studies illustrating the experiences and capacity constraints of SMEs in dealing with extreme climatic events. For each of the case studies, 4–6 SMEs were interviewed using semi-structured interviews.
• A half-day workshop held with 20 participants towards the end of the data collection period with the purpose of bringing the research participants together to share their experiences and discussing the preliminary results.

It must be noted that the sample size in this study is only a fraction of total SMEs in Australia and therefore has statistical uncertainty associated with such a sample. The study adopted extreme weather events as a proxy for climate change impacts and drew on rich qualitative data gathering techniques mentioned above to understand adaptive capacity.

48.4   SMEs coping with and adapting to climate change

By focusing on the temporal scale of the conceptual diagram presented in Figure 48.1, the research initially examined how SMEs had coped with past climate extremes and variability and whether this had provided an impetus for adaptation planning. The research distinguishes between adaptation and coping. The latter term refers to short-term strategies to deal with the immediate impacts of a climate-related stress, whilst the former is associated with long-term strategies.

48.4.1   Coping with past climate extremes and variability

The types of climate-related stresses experienced by the SMEs in this study varied according to geographical contexts. A majority of SMEs seemed to recall past experience with rapid-onset extreme events in the form of bushfires and cyclones, rather than the slower-onset events such as drought or sea-level rise. The results suggest that extreme climatic events are not a new stress that SMEs are dealing with; rather, they are ingrained in their social memory alongside other business risks they may experience (e.g. changes to regulation).

The impacts from the climatic events mentioned above produced both direct and indirect impacts on SMEs. The most commonly reported direct impacts included: infrastructure damage; production processes and service disruptions; effects on staff working conditions; supply chains disruptions; disruption of transport for staff and customers; and increased insurance costs. This was compounded by the restrained spending patterns of communities in the affected areas. Across the four case studies, the indirect or 'flow-on' impacts from the extreme climatic events were significant, exposing the interdependencies between multiple stakeholders within the landscape in which SMEs operated. For example, many of the SMEs in Victoria purchased fresh produce from local suppliers in neighbouring towns and when the fires affected these SMEs, it had significant economic impacts on those neighbouring suppliers. Turner and Slatter (2012) confirm that most SMEs go back to being start-ups after a natural disaster, taking approximately five years before breaking even.

48.4.2   Integrating climate risks into business planning

Most surveyed respondents (75%) believed that extreme weather events would become more frequent and intense with 20% believing that they will be less frequent and intense. A majority of respondents (67%) believed that climate change was a problem for Australia, with 30% thinking that it posed no real issue.

The data from the survey suggest that SMEs past experience with climate extremes may shape attitudes and perceptions to future climate change. Such findings have been reported in other sectors and regions (Grothmann and Patt 2005; Jones and Boyd 2011; Kuruppu and Liverman 2011). SMEs which had not been greatly affected by extreme events felt that climate change does not pose a real threat to their business and therefore had no intention adapt. On the other hand,

SMEs who had directly or indirectly been affected by extreme climatic events had incorporated climate risks into business planning initiatives. However, these SMEs did not necessarily associate these initiatives as a method of addressing climate change risks specifically.

Of the case study SMEs, 90% had considered and started to integrate climate risks into their business plans through adopting various adaptation strategies. Many of these were directly related to their core business operations and ensuring business continuity. Most SME respondents had a planning horizon of 2–5 years. The motivation to consider climate risks seemed to be largely driven by SMEs' past experience with climate extremes rather than identifying climate change as a new threat to business continuity. In contrast, those SMEs unaffected by extreme climatic events had not proactively considered climate risk planning.

In addition to the short-term planning horizons, other key challenges faced by SMEs when planning for future extreme events included: lack of up-to-date climate information; lack of finance; competing priorities; and lack of knowledge. Only one of the case study SMEs had developed a formal plan that outlined how the business would continue to operate under various stresses (i.e. a business continuity plan). This may suggest that they are either unaware of the process or do not see the value in such a plan.

48.5 Processes shaping adaptive capacity of small businesses

In drawing out the key processes constraining and influencing adaptive capacity of SMEs, the study undertook a deeper analysis of the context in which organisations supporting SMEs operate as well examining the day-to-day realities confronting SMEs in general.

The results revealed that in their day-to-day operations, multiple challenges confronted SMEs. These included limited access to financial and human resources, lack of education and training in operating a small business, under-insurance

and distinct challenges associated with operating in a rural versus urban setting. Together, these challenges undermined the capacity of SMEs to implement adaptive strategies. For example, without adequate insurance respondents confirmed that many SMEs had limited capacity to cover the costs of impacts from extreme events on their businesses; in some cases, this led to SMEs having to close down their businesses. Further, four SMEs mentioned that there was also limited support for innovation and business planning to help businesses develop, which was largely attributed to the reduction in government funding to support innovation.

Various challenges or constraints were highlighted by organisations involved in supporting and delivering programs for SMEs. These included processes that were internal to the support organisations as well as those present in the wider landscape in which they operated.

48.5.1 *Limitations in funding and human resources*

Whether representing government, private or non-government organisations, many of the stakeholders mentioned that decreased government funding and availability of human resource capacity often hampered the extent of services that could be offered to SMEs. For example, four non-government stakeholders (e.g. Business Growth Centres, Business Associations, etc.) mentioned that they were dependent on government grants to offer particular support programmes such as business advice for SMEs or to employ staff to keep their organisations running.

48.5.2 *Struggles over power*

A familiar constraint highlighted by support organisations was the fragmented co-ordination of support activities for SMEs between the various government, non-government and private sector stakeholders. Support organisations agreed that disconnections between different levels of government had often led to poor engagement between each other, limited information sharing and missed opportunities for social learning. In some cases, when information was shared it was done so cautiously so that it would not provide others with a competitive advantage in the overcrowded landscape in which support was delivered to SMEs. For SMEs this created confusion as to which support mechanisms were associated with particular organisations and, at times, this stood to deter SMEs from accessing available support.

48.5.3 *Climate change information needs*

There seemed to be limited urgency to deal with climate risk reduction initiatives for SMEs among the majority of support organisations. This may be related to support organisations' perceptions of climate change. Over half of the respondents perceived that climate change as an issue that was not a priority for SMEs within the short-term (i.e. next five years) but rather a long-term issue that would have to be addressed eventually, as part of their support programmes for SMEs. The few support organisations that were working on climate risk reduction programmes were largely focused on disaster recovery rather than long-term climate adaptation. The majority of the respondents from the support organisations also believed that because small businesses tend not to undertake formal business planning or adopt business continuity plans, it was difficult for them to talk about climate adaptation.

48.6 Conclusion

A central finding of this study is that underlying contextual processes are critical to enhancing the adaptive capacity of the SME system. These processes include: social relations between SMEs and support organisations as well as within support organisations; the agency of SMEs to convert resources to build resilience into business continuity; SMEs' perceptions of climate risks; and power struggles between support organisations. A combination of these processes has the potential to limit the adaptive choices available

to SMEs in responding to climate change and other related stresses on business continuity.

Two main lessons are drawn from examining the extent to which SMEs had considered and integrated adaptation into business planning. Firstly, due to the distinct meanings they attach to the terms 'climate change' and 'climate extremes', SMEs are likely to adopt adaptation strategies which aim to address climate extremes rather than climate change. Moreover, because of the short-term planning horizons of SMEs (2–5 years), they are likely to plan for climate extremes rather than long-term 'climate change' which is perceived as being outside these traditional planning horizons. Unlike other sectors such as water and agriculture, the results of this study indicate that it is perhaps viable and practical for SMEs, particularly those with less than 30 employees, to plan and adapt to climate change impacts within a horizon of 5–10 years rather than long-term impacts (beyond 20 years). However, SMEs who possess critical long-lived infrastructure or other assets and larger SMEs may need to plan more than 10 years ahead. Secondly, the past experiences of SMEs with climate extremes act as motivators for introducing measures to adapt to future climate change. Many of the SMEs in this study had experienced extreme events such as bushfires, drought and cyclones, and the direct and indirect impacts of these events had changed their operating environment, leaving them vulnerable to future impacts.

Two further lessons are drawn from examining the underlying processes that constrain and influence the adaptive capacity of SMEs. First, the adaptive capacity of SMEs is to a large extent shaped by the adaptive capacity of the organisations that support them. This limits the agency of SMEs in securing business continuity. For example, the lack of a sense of urgency among support organisations about the need to assist SMEs to respond to climate change is likely to hinder the opportunities available for SMEs to use their agency and implement proactive adaptation measures. Second, in examining constraints of the adaptive capacity of SMEs, it is perhaps vital to understand the extent to which those constraints limit the opportunity or choice sets of SMEs, as well as their ability to exercise agency and transform their assets to supporting measures that will promote business continuity under uncertainty.

This study suggests that the success of efforts to build the adaptive capacity of SMEs to future climate and related stresses will depend on how they address these underlying processes which affect their agency in pursuing adaptive choices they value.

Previous studies on disaster management indicate low to moderate levels of preparedness among SMEs. For example, in his study of SMEs located on the coastline of Florida, Howe (2011) found just over half the respondents had developed business emergency management plans. This finding resonates with the results in our study in which only one SME had prepared a business continuity plan. In their study of direct and indirect impacts of climate change on business organisations in England, Berkhout et al. (2004) conclude that businesses rarely take proactive adaptation actions autonomously; their behaviour is largely influenced by policy and market conditions and tends to draw upon resources external to the business. This finding also resonated with this study in which the adaptive capacity of an SME was largely dependent on the adaptive capacity of the support organisations. In their study of SME managers in England, Williams and Schaefer (2013) found personal values and beliefs acted as key drivers in dealing with climate change impacts on business continuity. Our study also revealed the importance of cognitive factors such as self-efficacy beliefs in coping with impacts of extremes events, particularly in the disaster recovery phase. Vogel (2009) explored the extent of adaptation within the business community in South Africa, stressing the overwhelming need for climate scientists and those stakeholders involved in formal adaptation efforts to interact with small businesses so that best-practice cases can be promoted and constraints and opportunities can be documented and shared. Similar insights were

also found in our study, particularly the need to create spaces for cross-fertilisation of knowledge between support organisations and SMEs.

References

Alkire, S. (2007) *Choosing Dimensions: The Capability Approach and Multidimensional Poverty*. Chronic Poverty Research Centre Working Paper No. 88. Oxford Poverty and Human Development Initiative, University of Oxford, England.

Australian Bureau of Statistics (2011) *Counts of Australian Businesses, including Entries and Exits, June 2007 to June 2009*. Australia Bureau of Statistics, Canberra.

Berkhout, F., Hertin, J. and Arnell, N. (2004) *Business and Climate Change: Measuring and Enhancing Adaptive Capacity. Tyndal Centre for Climate Change Research, Technical Report 11*. Tyndal Centre for Climate Change Research, East Anglia.

Brooks, N. and Adger, N. (2004) Assessing and enhancing adaptive capacity. In: Lim, B. and Spanger-Siegfried, E. (eds) *Adaptation Policy Framework for Climate Change*. Cambridge University Press, Cambridge, UK, pp. 165–182.

Burke, W. (2008) *Organisation Change: Theory and Practice*. Sage Publications, London.

Busch, T. (2011) Organizational adaptation to disruptions in the natural environment: The case of climate change. *Scandinavian Journal of Management* 27(4), 389–404.

Department of Innovation Science and Research (2010) *Small to Medium Enterprises and Productivity*. Department of Innovation Science and Research, Canberra.

Folke, C., Colding, J. and Berkes, F. (2002) Building resilience for adaptive capacity in social-ecological systems. In: Berkes, F., Colding, J. and Folke, C. (eds) *Navigating Social-Ecological Systems: Building Resilience for Complexity and Change*. Cambridge University Press, Cambridge, pp. 352–387.

Grothmann, T. and Patt, A. (2005) Adaptive capacity and human cognition: The process of individual adaptation to climate change. *Global Environmental Change Part A* 15, 199–213.

Holling, C. and Walker, B. (2003) *Resilience Defined. Entry Prepared for the Internet Encyclopedia of Ecological Economics*. International Soeciety for Ecological Economics. Available at http://isecoeco.org/pdf/resilience.pdf (accessed 10 June 2014).

Howe, P. (2011) Hurricane preparedness as anticipatory adaptation: A case study of community businesses. *Global Environmental Change* 21(2), 711–720.

Ingirige, B. and Jones, K. (2008) Investigating SME resilience and their adaptive capacities to extreme weather events: A literature review and synthesis. In: *CIB International Conference on Building Education and Research*. 11–15 February 2008. Salford, UK: University of Salford, pp. 582–593.

Jones, L. and Boyd, E. (2011) Exploring social barriers to adaptation: Insights from Western Nepal. *Global Environmental Change* 21(4), 1262–1274.

Kuruppu, N. (2009) *Confronting Climate Change and Variability: Enhancing the Adaptive Capacity of Water Resource Management in Kiribati*. PhD Thesis, Environmental Change Institute, University of Oxford.

Kuruppu, N. and Liverman, D. (2011) Mental preparation for climate adaptation: The role of cognition and culture in enhancing adaptive capacity of water management in Kiribati. *Global Environmental Change* 21(2), 657–669.

Kuruppu, N., Murta, J., Mukheibir, P., Chong, J. and Brennan, T. (2013) *Understanding the Adaptive Capacity of Australian Small-to-Medium Enterprises to Climate Change and Variability*. National Climate Change Adaptation Research Facility, Gold Coast.

Lorenzoni, I., Jordan, A., O'Riordan, T., Turner, R.K. and Hulme, M. (2000) A co-evolutionary approach to climate change impact assessment. Part II: A scenario-based case study in East Anglia (UK). *Global Environmental Change* 10(2), 145–155.

McLaughlin, T. and Dietz, T. (2008) Structure, agency and environmnet: Towards an integrated perspective on vulnerbaility. *Global Environmental Change* 18(1), 99–111.

Pelling, M. (2011) *Adaptation to Climate Change: From Resilience to Transformation*. Routledge, London.

Reynolds, L. (2013) *Climate Change Preparedness and the Small Business Sector*. M.J Bradeley and Associates, LLC. American Business Council, Concord, USA.

Runyan, R. (2006) Small business in the face of crisis: Identifying barriers to recovery from a natural disaster. *Journal of Contingencies and Crisis Management* 14(1), 12–26.

Sen, A. (1999) *Development as Freedom*. Oxford University Press, Oxford.

Turner, A. and Slatter, S. (2012) *Are U Ready? Surviving Small Business Disaster*. Slater/Turner Publishing, Melbourne.

United Nations Office for Disaster Risk Reduction (2013) *Global Assessment Report on Disaster Risk Reduction*. United Nations Office for Disaster Risk Reduction, Geneva.

Vogel, C. (2009) Busines and climate change: Initial explorations in South Africa. *Climate and Development* 1, 82–97.

Watts, M. (1983) *Silent Violence: Food, Famine and Peasantry in Northern Nigeria*. University of California, Berkley.

Williams, S. and Schaefer, A. (2013) Small and medium-sized enterprises and sustainability: managers' values and engagement with environmental and climate change issues. *Business Strategy and the Environment* 22(3), 173–186.

Yohe, G. and Tol, R. (2002) Indicators for social and economic coping capacity: moving toward a working definition of adaptive capacity. *Global Environmental Change* 12(1), 25–40.

49 Investing in adaptive capacity: opportunities, risks and firm behaviour

Department of Accounting Finance and Economics, Griffith University, Australia

49.1 Introduction

Companies and other entities face a range of both threats and opportunities from climate change; this is not a matter of conjecture. What is however undecided is the manner in which companies should respond and how they communicate this response with their investors and other stakeholders. There are differing opinions on the extent to which industry sectors may be affected by climate change and what measures companies could or should take now and in the future.

The impacts of climate change will vary by company, industry and location. The extent to which assets and operations will need to adapt, and the form that this adaptation may take, will vary with the nature and level of risk. In fact, many companies already address the risks of climate change without an explicit need to 'ring-fence' adaptation activities. Witness the recent activities of developers constructing long-life infrastructure projects and city residential buildings that are engineered against potential losses from rising sea levels and higher wind speeds. Many firms already make substantial attempts to address these risks, but almost all tend to encounter difficulty accounting for and managing potential risks over their investment horizon. A key concern is the need to convince investors that such adaptation activities are worthwhile in the long term.

Through an industry consultation process we studied the corporate response to climate change. The research revealed that the optimal approach used to manage corporate climate change exposures and maintain best-practice corporate governance is to use a feedback assessment, management and reporting loop. After introducing a proposed 'adaptation framework' for use in the corporate governance of the adaptation response, we introduce a case study of a firm looking to reduce its business risk from climate change through adaptation. The case study highlights the notion of adaptive capacity used in asset construction, the risk–return decision criteria firms use to justify their behaviour and the ways in which such decisions are risk managed, governed and disclosed to investors.

Applied Studies in Climate Adaptation, First Edition. Edited by Jean P. Palutikof, Sarah L. Boulter, Jon Barnett and David Rissik.
© 2015 John Wiley & Sons, Ltd. Published 2015 by John Wiley & Sons, Ltd.

This analysis is based on a series of workshops conducted with multinationals in the energy, resources, construction and transport sectors facing significant climate change risks to their assets and operations (West and Brereton 2013).

49.2   Adaptive capacity

Adaptive capacity is a component of economic capital. Economic capital is the smallest amount that can be invested to insure the value of a firm's net assets against a loss in value relative to the risk-free investment of those net assets (Merton and Perold 1993). Economic capital is therefore the amount of equity capital that a company can allocate to fund operations, to fund a portfolio of assets or indeed fund the entire firm itself. Economic capital provides a buffer against potential losses. It allows, at an acceptable degree of confidence, the firm to continue its operating activities and capital investment program. The cost of economic capital is the credit spread the firm pays in the form of insurance to cover its bankruptcy risk. Like any form of capital, the cost of economic capital depends on adverse selection, moral hazard and agency costs.

Whether adaptive capacity is an asset such as a seawall or levee, or a process such as changed irrigation practices and water rights management in agricultural activities, adaptive capacity is a vital component of economic capital that is rarely explicitly valued. A crucial first step in any climate change adaptation initiative is for companies to assess the implications of climate change on their systems and processes (e.g. productivity, infrastructure damage), workplace environment (e.g. worker health) and external effects (e.g. operational restrictions) to determine whether, and the extent to which, climate change will have an impact, pose a risk or offer beneficial opportunities. A systematic way to address this is through a climate change adaptation framework.

A broad framework is illustrated in Figure 49.1 that can be applied to assist companies through the recognition and formulation process. The framework identifies four activities where busi-

nesses should apply targeted and integrated adaptation efforts:

- risk assessment;
- vulnerability assessment;
- adaptation measures; and
- disclosure and reporting.

These are represented as quadrants of equal size. Each segment is equally necessary to address a firm's adaptation response preparedness. The process is cyclic with the central arrow representing the co-dependence of each activity and continual monitoring and assessment of the process. Supporting functional decision-making in each of the activities is a number of processes and assessment tools, outlined adjacent to each of the quadrants.

49.3   Decision tools for adaptation assessment

For conducting a risk and vulnerability assessment using the above framework, firms generally use one or more of the following straightforward approaches that are proven, simple and effective decision support tools (West and Brereton 2013):

- expected loss assessment (ELA);
- cost benefit analysis (CBA);
- cost effectiveness analysis (CEA); or
- multi-criteria analysis (MCA)

The approach depends on the number of adaptation objectives required by the company's business units and the measurability of the impacts. In this analysis we consider a case study using the expected loss assessment (ELA) approach, which can accommodate the analysis of multiple adaptation objectives assuming a high degree of measurability of likely costs or benefits as well as reasonably well-defined likelihood and severity statistics of certain climate change effects. In the absence of such estimates, other techniques must be used.

49.3.1   Expected loss assessment

Firms already recognise that certain adaptation options are a great deal less expensive than standard insurance options. For instance, the capital

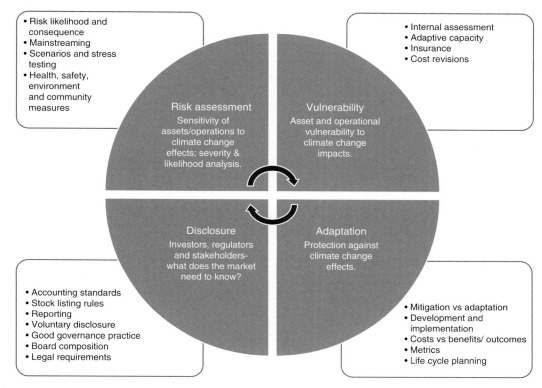

Figure 49.1 Example of a climate change adaptation framework for corporate governance.

expenditures associated with constructing a sea-wall or retrofitting a stronger roof on a building critical for the firm's operations may be significantly lower than the cost of an insurance policy, especially one that is based on the expected loss of not having such protection measures installed. Other adaptation activities are however much more expensive than insurance. For instance, elevating the height of an asset above sea level (e.g. an airport on a coastline) or even relocating a firm's operations will constitute a significant expense that would generally exceed the cost of insuring the asset, even on an annually renewed basis.

In terms of either expected loss or actuarial valuation principles for assets, the costs of adaptation can be defined as the present value of expected losses or, if the asset undergoes adaptive capacity building, the partial or whole future

insurance costs avoided as a result of new adaptive capacity. This approach is most useful for assessing impacts on identifiable assets, supply chains and measurable operational activities where costs and benefits can be quantified to some degree.

The costs of adaptation can be compared against the total amount of insurance avoided assuming a degree of adaptive capacity:

$$\text{PV (Insurance avoided)} = \text{Adaptation costs}$$
$$= \sum_{t=1}^{n} \frac{A_t}{\left(1+r_t\right)^t} \approx \frac{A_0\left(1+g\right)}{r-g}$$

where A_t is the additional annual cost of insuring an asset without inbuilt adaptive capacity (i.e. its baseline value) in year t, r_t is the discount rate during year t (generally defined as the cost of capital), g is an assumed fixed growth rate (representing a constant growth in insurable costs from year to year as the asset's value increases or the pool of insurable asset values increase) and r is an assumed fixed discount rate. This relation naturally assumes that the asset is insurable.

The above relation relies heavily on the ability of the company to calculate expected losses and unexpected losses at a given significance level, or to actuarially compute an annual insurance cost and obtain accurate forecasts of the future cost of capital. If the actual costs of adaptation exceed the estimated present value of insurance costs, then we may interpret that to mean that the asset is being 'over-engineered' above the required design standard.

49.3.2   Design standards

Regulations enforced via building and design codes act to incorporate some degree of loading which then defines the minimum building standards. Building codes are typically based on historically observed usage, weather and earthquake data. Climate-related hazards such as cyclones, extreme winds, intense rain, bushfire and to some extent floods can be quantified. Building codes impose loads to buildings determined mainly by historical records from which design events with annual probabilities of excess for each hazard are specified. The component of adaptive capacity that should be measured is therefore not made in reference to some absolute minimum standard, but with reference to the existing minimum standard that already contains a certain design loading.

An illustration of the difference in the resilience between new buildings and existing buildings (constructed before the adherence to standards when subjected to current and future climate-related hazard events) under a certain 'low' emissions scenario is provided in Figure 49.2. The excess in asset capacity is referred to here as 'adaptive capacity'. The degree of adaptive capacity needed to defend against climate risks can quantified using any of the methods described above, although ELA will yield the most concise loss estimate.

49.4   Case study: wind damages to coastal assets

In 2010 the annual global damage from tropical cyclones was US$ 26 billion. This is equivalent to 0.04% of world GDP and is expected to grow to US$ 56 billion by 2100 (Nordhaus 2010; Mendelsohn et al. 2012). Forecasts project that both income and capital will rise considerably over this period, and that almost all regions will experience at least some increase in damages. Projected damages from tropical cyclones are expected to be less than US$ 1 billion per annum in Europe and South America (due to relatively fewer storms) and are also expected to be subdued in Africa (due to less capital at risk in coastal regions). Damage is also expected to increase modestly in Southeast Asia and Central America, mainly due to forecasts of higher expected economic growth being tempered by a medium degree of coastal development. However East Asia and North America are projected to account for over 88% of global damage forecasts, due to a combination of a greater frequency of high-intensity storms and a high rate of growth in expensive coastal developments (Mendelsohn et al. 2012). Island nations, particularly in the Pacific, are also expected to be greatly impacted.

Given the growth in exposure to tropical storms across almost all global regions, the following case study serves as a simple illustration of one way to evaluate extreme events and adaptation options in damage-prone areas.

Consider a firm concerned with the potential impacts and adaptation alternatives to changes in the distribution of tropical storms that strike the firm's coastal assets as a consequence of climate change. As part of their risk assessment process, they wish to understand the 'adaptation measures' task within the proposed adaptation framework (Fig. 49.1). The firm can obtain a model that measures its exposure to large storms over a given time horizon. This modelling relies upon concepts such as the 'return period' of storms (a statistical measurement based on historical data which estimates the average recurrence interval of an event over a period of time, commonly used

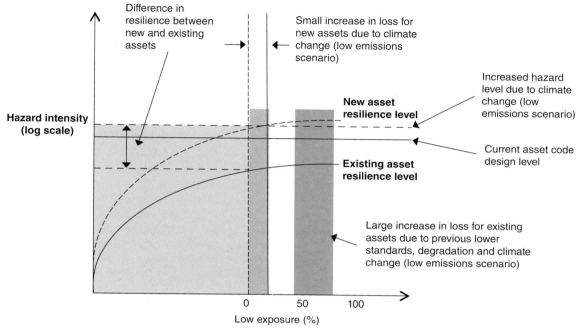

Figure 49.2 A conceptual view of asset resilience and loss (IPCC low emissions scenario). Source: Australian Building Codes Board 2010. Image modified with permission from the Australian Building Codes Board (ABCB) www.abcb.gov.au.

in the design, engineering and construction of roads, rail, buildings and other physical assets) and the associated impacts they cause.

The focus is on the probability distribution of tropical storms, characterised by their peak wind speed sustained over land for a period of at least 10 minutes. We refer to this as peak wind speed, which is closely correlated with structural asset damage caused by high winds, storm surges, intense rainfall and flooding.

Assets (buildings and other infrastructure) that can withstand storms with higher wind speeds typically cost more to construct but the annualised value of damages are expected to be lower than the incremental adaptation cost over its economic life. This additional cost may not be warranted in locations where assets are built with a shorter investment horizon. The firm wishes to scale the design standard of its asset from a 10-year return period up to a 50-year

return period to avoid higher insurance costs and maintain business continuity during severe storms. (Design standards in OECD countries generally require firms to use return periods of 50–100 years for infrastructure construction.)

A simple test can be used to consider whether the current design standards for the asset, based on a 10-year period, are appropriate. Based on historical data, this corresponds to a peak wind speed of around 67 mph. The firm compares the reduction in the expected annual losses from storm damage if a higher design standard were adopted, with the annualised value of the additional investment costs required to construct the asset to the higher standard. The firm also wants to determine the expected annual value of losses from the new design standards based on protecting buildings and other infrastructure from storms with return periods of 50 years. This corresponds to a peak wind speed of 91 mph using historical estimates. If the investment horizon of the asset extends to 50 years, the firm needs to conduct a cost–benefit analysis that scales the design standard from a 10-year return period to a 50-year return period. The model needs to estimate the expected annual benefit of adopting higher design standards as a percentage of the annualised cost of the capital stock (the asset).

49.4.1   The model

In any year, an asset is subject to a series of more or less severe storms. The distribution of the maximum peak wind speed in each year can be estimated through observations collected over many years. In this example we apply a variant of the generalised extreme value (GEV) distribution to describe the distribution of extreme wind speeds caused by severe storms. This distribution assumption applies to other natural events such as floods, storm surges and earthquakes. Specifically, a Gumbel distribution (a two-parameter version of the GEV distribution) is used because it offers a more flexible distribution in the presence of limited data.

An estimate of the financial damage caused by a storm can be represented by a power function of the positive difference between peak wind speed and the wind speed that the asset is designed to resist without damage S_D (the design standard). We choose parameters that reflect historical empirical observations of damage caused by storms that have affected similar assets.

Financial damage estimates caused by a storm with peak wind speed S is represented here by the power function

$$D = \gamma \max\left(S - S_D, 0\right)^{\lambda} A$$

where A is the asset value and the parameters γ and λ are chosen to represent historical empirical observations of damage caused by storms that have affected similar assets. The wind speed that the asset is designed to resist without damage is S_D. The design standard set for this asset allows it to resist storms with a 10-year return period $(S_D \approx 81$ mph). We select $\gamma = 1.5$ and $\lambda = 0.004$

to estimate the damage parameters for the power function above (typical OECD Country Engineering Standards). The impact of a shift in the probability distribution of storms on the expected value of storm damage can be largely offset by changing the design standards that are applied when building new assets.

49.4.2   Results

Figure 49.3 shows that the two forecast curves predict a fall in the return period if it is assumed that the distribution parameters (known as the location and scale parameters) for storms increase by either 10% (climate forecast: low) or 25% (climate forecast: high) due to climate change. The low and high forecasts are roughly in line with the low- and high-impact scenarios published by the IPCC (Adger et al. 2007). For the firm's asset, the graph shows that under a low climate change scenario the return period for a storm with a peak wind speed of 81 mph reduces from 10 years to around 5 years for the low scenario, and to 2 years for the high scenario. In the absence of climate change, to scale the asset up to resist peak wind speeds with a 50-year return period, the design standard must be able to withstand speeds of around 105 mph. This design standard would need to increase the resistance level to 117 mph under a low-climate scenario, and to 126 mph for the high-climate scenario. If there was no allowance for climate change in the design of this asset after being retrofitted to a 50-year return period standard, the low-climate scenario would reduce the return period to 23 years and the high-climate scenario would reduce it to 11 years. This represents a significant risk of financial damage to this asset if climate projections are ignored.

If there were no change in design standards, then the expected value of annual storm damage would increase from around 4.8% of asset value to 11.6% of asset value using a base design standard of a one-in-ten-year storm. This was inferred from the probability of loss as described above and represents a material loss exposure. If the construction cost of building adaptive capacity into this asset was less than 6.8% then it is

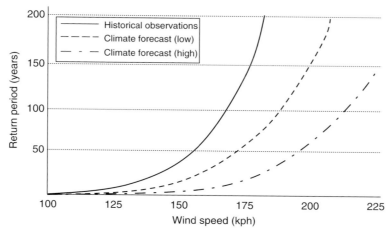

Figure 49.3 Severe storm peak wind speed return periods using a Gumbel distribution.

economically rational to adapt rather than insure, since annual insurance costs will rise in line with the hazard rate associated with the asset.

This case study only examines a single asset. Larger firms are likely to have a range of assets that are exposed to climate change, which marginally complicates the analysis. Different distributions can be used to model different loss exposures. Nevertheless, the basic principles remain and damage estimates can be obtained and compared with insurance costs and other forms of loss mitigation to yield a cost–benefit outcome.

49.5   Disclosure and governance

The disclosure and reporting task under the adaptation framework (Fig. 49.1) sits squarely within the firm's corporate governance duties. Building adaptive capacity directly impacts asset values, depreciation, future insurance costs, expected losses and financing costs. In the initial stages, adaptive capacity built into an infrastructure asset is likely to increase the capital cost of the asset. Subsequent accounting treatment is critical in recognising both the current and future value of the asset. There are fundamental differences between capacity in assets that have been

designed to cater for climate change impacts and assets that are truly impaired. These outcomes may not necessarily attract the same accounting treatment.

The first problem arises when accounting for the possible impairment of long-lived assets based on the fair value of the asset being less than its carrying value. An asset with adaptive capacity may be subject to very significant impairment provisions. This would be of great concern for companies who buy or build large long-lived assets with contingencies to cope with expected climate change impacts. The recoverable amount of the asset is heavily dependent on the revaluation assessment used. At each reporting date, companies must assess whether there is any indication that an asset may be impaired; in addition to the 'base value' of an asset, any additional adaptive capacity must be quantified.

Market prices are not available for many long-lived assets so the base value estimate must be judged using the best information available, including prices for similar assets. While firms may use other valuation techniques, present value is likely to be the most popular method for estimating the fair value for assets that have been adapted.

Another key concern for firms is the fact that while building adaptive capacity is an alternative to insurance, insurance costs can be expensed but adaptive capacity costs cannot. This puts firms who adopt adaptive capacity activities (that is, firms who self-insure) at a disadvantaged relative to those who simply obtain external insurance coverage.

In the above case study we assumed that the construction cost for upgrading this asset would be less than 6.8% on an annualised basis to improve the protection standard from a 10-year to a 50-year return period. If the firm were to go ahead with this investment, and if annual insurance premiums avoided decline by more than 6.8%, it could be argued that the new design standard does not represent excess capacity and thus the asset fair value can be represented by its book value (cost of construction). If no partial relief is obtained from a commensurate insurance discount (because the insurer's pool of assets are dominated by older design standards), the balance sheet may need to annually provision the asset by at least 6.8% and the firm would then need to provide sufficient reasoning and conduct regular value assessments over the asset's life. The immediate effect will be a reduction in balance sheet equity, which may have implications for firm or project debt covenants and other performance metrics or ratios used to control credit exposure.

49.6 Conclusion

Using the adaptation framework of West and Brereton (2013) as a guideline, firms can better consider the strengths and weaknesses of the various approaches for assessing adaptation options. In some situations a number of approaches could be applied in a complementary fashion (Fankhauser et al. 1999). Regardless of which assessment approach a firm chooses, each should be practical, relevant, robust, proportional and comprehensive. Above all, the selected approach should be motivated by the need for a decision rather than aiming to make the perfect decision.

Adapting to the effects of climate change will significantly affect companies both operationally

and financially. Boards and company executives largely understand the need to proactively take steps to address the projected impacts of climate change and respond to regulations that address risk and financial performance. Leadership in addressing and acting on climate change may, in various ways, create competitive advantage for a company, as long as such decisions are done rationally. A systematic framework for adaptation planning coupled with economically rational investment decisions are key to preserving firm value over the long term.

References

Adger, W.N., Agrawala, S., Mirza, M.M.Q. et al. (2007) Assessment of adaptation practices, options, constraints and capacity. In: Parry, M.L., Canziani, O.F., Palutikof, J.P., van der Linden, P.J. and Hanson, C.E. (eds) *Climate Change 2007: Impacts, adaptation and vulnerability. Contribution of Working Group II to the Fourth Assessment Report of the Intergovernmental Panel on Climate Change.* Cambridge University Press, Cambridge, UK, pp. 717–743.

Australian Building Codes Board (2010) *An Investigation of Possible Building Code of Australia (BCA) Adaptation Measures for Climate Change.* Australian Government, Canberra, Australia.

Fankhauser, S., Smith, J. and Tol, R.S.J. (1999) Weathering climate change: some simple rules to guide adaptation decisions. *Ecological Economics* 30(1), 67–78.

Mendelsohn, R., Emanuel, K., Chonabayashi, S. and Bakkensen, L. (2012) The impact of climate change on global tropical cyclone damage. *Nature Climate Change* 2(3), 205–209.

Merton, R.C. and Perold, A.F. (1993) Theory of risk capital in financial firms. *Journal of Applied Corporate Finance* 6, 16–32.

Nordhaus, W. (2010) The economics of hurricanes and implications of global warming. *Climate Change Economics* 1, 1–24.

West, J.M. and Brereton, D. (2013) *Climate Change Adaptation in Industry and Business: A Framework for Best Practice in Financial Risk Assessment, Governance and Disclosure.* National Climate Change Adaptation Research Facility, Gold Coast, Australia.

Index

References to tables are given in bold type. References to figures are given in italic type.

Applied Studies in Climate Adaptation, First Edition. Edited by Jean P. Palutikof, Sarah L. Boulter, Jon Barnett and David Rissik.
© 2015 John Wiley & Sons, Ltd. Published 2015 by John Wiley & Sons, Ltd.